T0319454

Computational Geomechanics

Computational Geomechanics

Theory and Applications

Second Edition

Andrew H. C. Chan
University of Tasmania

Manuel Pastor
ETS de Ingenieros de Caminos
Universidad Politécnica de Madrid, Spain
formerly at Centro de Estudios y Experimentación de Obras Públicas, Madrid, Spain

Bernhard A. Schrefler
University of Padua

Tadahiko Shiomi
Mind Inc., 3D Laboratory, Japan

O. C. Zienkiewicz
CINME, UNESCO former Professor of Numerical Methods in Engineering at Technical University of Catalonia (UPC), Spain

This edition first published 2022

Registered Offices
John Wiley & Sons, Inc., 111 River Street, Hoboken, NJ 07030, USA
John Wiley & Sons Ltd., The Atrium, Southern Gate, Chichester, West Sussex, PO19 8SQ, UK

Editorial Office
The Atrium, Southern Gate, Chichester, West Sussex, PO19 8SQ, UK

For details of our global editorial offices, customer services, and more information about Wiley products visit us at www.wiley.com.

Wiley also publishes its books in a variety of electronic formats and by print-on-demand. Some content that appears in standard print versions of this book may not be available in other formats.

Library of Congress Cataloging-in-Publication Data

Names: Chan, Andrew H. C., author. | Pastor, Manuel, author. | Schrefler, B. A., author. | Shiomi, Tadahiko, author. | Zienkiewicz, O. C., author.
Title: Computational geomechanics : theory and applications / Andrew H. C Chan, University of Tasmania, Manuel Pastor, Bernard Schrefler, University of Padua, Tadahiko Shiomi, Olgierd C. Zienkiewicz, CINME, UNESCO Professor of Numerical Methods in Engineering at Technical University of Catalonia (UPC), Spain.
Description: Second edition. | Hoboken, NJ : John Wiley & Sons, Inc., 2022. | Revised edition of: Computational geomechanics with special reference to earthquake engineering / O.C. Zienkiewicz ... [et al.]. 1999.
Identifiers: LCCN 2021049320 (print) | LCCN 2021049321 (ebook) | ISBN 9781118350478 (cloth) | ISBN 9781118535318 (adobe pdf) | ISBN 9781118535301 (epub)
Subjects: LCSH: Geotechnical engineering–Mathematics. | Earthquake engineering–Mathematics.
Classification: LCC TA705 .C46 2022 (print) | LCC TA705 (ebook) | DDC 624.1/51–dc23/eng/20211208
LC record available at https://lccn.loc.gov/2021049320
LC ebook record available at https://lccn.loc.gov/2021049321

Cover Design: Wiley
Cover Image: Courtesy of Sendai City, Fire Bureau

Set in 9.5/12.5pt STIXTwoText by Straive, Pondicherry, India

Printed and bound by CPI Group (UK) Ltd, Croydon CR0 4YY

C9781118350478_140322

Contents

Preface

Our first text on this subject *Computational Geomechanics with Special Reference to Earthquake Engineering* was published 23 years ago and has been out of print for much of the past decade. It was the first book of its kind having as the main topic *Computational Dynamic Aspects of Geomechanics* which obviously comprise statics also. In the intervening period, there was a rapid expansion in the research and practical applications of these types of problems, which has prompted us to write this new and thoroughly updated version.

It contains not only the results of research carried out at our four institutions but also reports on the work done elsewhere. The chapters from the previous edition have been extensively updated and new chapters have been added to give a much broader coverage of recent research interests. The Preface to the first edition was written by the Late Professor Oleg Cecil Zienkiewicz. Its validity is still fully conserved today. So, we reprinted large parts of it.

Although the concept of effective stress in soils is accepted by all soil mechanicians, practical predictions and engineering calculations are traditionally based on total stress approaches. When the senior author began, in the early seventies, the application of numerical approaches to the field of soil mechanics in general and to soil dynamics in particular, it became clear to him that a realistic prediction of the behavior of soil masses could only be achieved if the total stress approaches were abandoned. The essential model should consider the coupled interaction of the soil skeleton and of the pore fluid. Indeed, the phenomena of weakening and of "liquefaction" in soil, when subjected to repeated loading such as that which occurs in earthquakes, can only be explained by considering this "two-phase" action and the quantitative analysis and prediction of real behaviour can only be achieved by sophisticated computation. The simple limit methods often applied in statics are no longer useful. It, therefore, seems necessary at the present time to present, in a single volume, the basis of such computational approaches because a wider audience of practitioners and engineering students will require the knowledge which hitherto has only been available through scientific publications scattered throughout many journals and conferences. The present book is an attempt to provide a rapid answer to this need. Since 1975, a large number of research workers, both students and colleagues, have participated both at Swansea and elsewhere in laying the foundations of numerical predictions which were based largely on concepts introduced in the early forties by Biot. However, the total stress calculation continues to be used by some engineers for earthquake response analysis, often introduced with linear approximations. Such simplifications are generally not useful and can lead to erroneous predictions. In recent years, centrifuge experiments have permitted the study of some soil problems involving both statics and dynamics. These provide a useful set of benchmark predictions. Here a validation of the two-phase approach was available and a close agreement between computation and experiment was found. A very important landmark was a workshop held at the University of California, Davis, in 1993, which reported results of the VELACS project (Verification of Liquefaction Analysis by Centrifuge Studies) sponsored by the National Science Foundation of the USA.

At this workshop, a full vindication of the effective stress, two-phase approaches was clearly available and it is evident that these will be the bases of future engineering computations and prediction of behavior for important soil problems. The book shows some examples of this validation and also indicates examples of the practical application of the procedures described. During numerical studies, it became clear that the geomaterial – soil would often be present in a state of incomplete saturation when part of the void was filled with air. Such partial saturation is responsible for the presence of negative pressures which allow some "apparent" cohesion to be developed in noncohesive soils. This phenomenon may be present at the outset of loading or may indeed develop during the dynamic process. We have therefore incorporated its presence in the treatment presented in this book and thus achieved wider applicability for the methods described.

Despite a large number of authors, we have endeavored to present a unified approach and have used the same notation, style, and spirit throughout. The first three chapters present the theory of porous media in the saturated and unsaturated states and thus establish general backbone to the problem of soil mechanics.

Even though the fundamental nature of the basic theory remains unchanged as shown in Chapters 2 and 3, many of the other chapters have been substantially updated. The following part of the book has been extensively restructured, reworked, and updated, and new chapters have been added such as to cover essentially all the important aspects of computational soil mechanics.

Chapter 4, essential before numerical approximation, deals with the very important matter of the quantitative description of soil behavior which is necessary for realistic computations. This chapter has been substantially rewritten such as to introduce new developments. It is necessarily long and devotes a large part to generalized plasticity and critical-state soil mechanics and also includes a simple plasticity model. The generalized plasticity model is then extended to partially saturated soil mechanics. Presentation of alternative advanced models such as bounding surface models and hypoplasticity concludes the chapter.

Chapter 5 addresses some special aspects of analysis and formulation such as far-field solutions in quasi-static problems, input for earthquake analysis and radiation damping, adaptive finite element requirements, the capture of localized phenomena, regularization aspects and stabilization for nearly incompressible soil behavior both in dynamics and consolidation permitting to use equal order interpolation for displacements and pressures.

Chapter 6 presents applications to static problems, seepage, soil consolidation, hydraulic fracturing, and also examples of dynamic fracturing in saturated porous media. Validation of the predictions by dynamic experiments in a centrifuge is dealt with in Chapter 7.

Chapter 8 is entirely devoted to application in unsaturated soils, including the dynamic analysis with a full two-phase fluid flow solution, analysis of land subsidence related to exploitation of gas reservoirs, and initiation of landslides.

Chapter 9 addresses practical prediction, application, and back analysis of earthquake engineering examples. Finally, Chapter 10 pushes the limits of the analysis beyond failure showing the modeling of fluidized geomaterials with application to fast catastrophic landslides.

We are indebted to many of our coworkers and colleagues and, in particular, we thank the following people who over the years have contributed to the work (in alphabetical order of their surnames):

T. Blanc,
G. Bugno,
T.D. Cao,
P. Cuéllar,
S. Cuomo (MP),
P. Dutto,
E. González,
B. Haddad,
M.I. Herreros,
Maosong Huang,
E. Kakogiannou,
M. Lazari,
Chuan Lin,
Hongen Li,
Li Tongchun,
Liu Xiaoqing,
D. Manzanal,
M. Martín Stickle
A. Menin,
J.A. Fernández Merodo,
E. Milanese,
P. Mira,
M. Molinos,
S. Moussavi,
R. Ngaradoumbe Nanhorngué,
P. Navas,
T. Ni,
Jianhua Ou,
M. Passarotto,
M.J. Pastor,
C. Peruzzo,
F. Pisanò,
M. Quecedo,
V. Salomoni,
L. Sanavia,
M. Sánchez-Morles,
R. Santagiuliana,
R. Scotta,

S. Secchi,
Y. Shigeno,
L. Simoni,
C. Song,
A. Yagüe,
Jianhong Ye,
M. Yoshizawa,
H.W. Zhang.

Finally, we would like to dedicate this edition to the memory of the Late Oleg Cecil Zienkiewicz. Without his inspiration and enthusiasm, we would not have undertaken the research work reported here. We would also like to thank our beloved Late Helen Zienkiewicz, wife of Professor Zienkiewicz, who kindly allowed us to celebrate Oleg's decades of pioneering and research field defining achievements in computational geomechanics.

Andrew H. C. Chan
Manuel Pastor
Bernhard A. Schrefler
Tadahiko Shiomi
Hobart, Madrid, Padua, Tokyo, January 2022

1

Introduction and the Concept of Effective Stress

1.1 Preliminary Remarks

The engineer designing such soil structures as embankments, dams, or building foundations should be able to predict the safety of these against collapse or excessive deformation under various loading conditions which are deemed possible. On occasion, he may have to apply his predictive knowledge to events in natural soil or rock outcrops, subject perhaps to new, man-made conditions. Typical of this is the disastrous collapse of the mountain (Mount Toc) bounding the Vajont reservoir which occurred on 9 October 1963 in Italy (Müller 1965). Figure 1.1 shows both a sketch indicating the extent of the failure and a diagram indicating the cross section of the encountered ground movement.

In the above collapse, the evident cause and the "straw that broke the camel's back" was the filling and the subsequent drawdown of the reservoir. The phenomenon proceeded essentially in a static (or quasi-static) manner until the last moment when the moving mass of soil acquired the speed of "an express train" at which point, it tumbled into the reservoir, displacing the water dynamically and causing an unprecedented death toll of some 4000 people from the neighboring town of Longarone.

Such *static* failures which occur, fortunately at a much smaller scale, in many embankments and cuttings are subjects of typical concern to practicing engineers. However, dynamic effects such as those frequently caused by earthquakes are more spectacular and much more difficult to predict.

We illustrate the dynamic problem by the near-collapse of the Lower San Fernando dam near Los Angeles during the 1971 earthquake (Figure 1.2) (Seed, 1979; Seed et al. 1975). This failure, fortunately, did not involve any loss of life as the level to which the dam "slumped" still contained the reservoir. Had this been but a few feet lower, the overtopping of the dam would indeed have caused a major catastrophe with the flood hitting a densely populated area of Los Angeles.

It is evident that the two examples quoted so far involved the interaction of pore water pressure and the soil skeleton. Perhaps the particular feature of this interaction, however, escapes immediate attention. This is due to the "weakening" of the soil–fluid composite during the periodic motion such as that which is involved in an earthquake. However, it is this

Computational Geomechanics: Theory and Applications, Second Edition. Andrew H. C. Chan, Manuel Pastor, Bernhard A. Schrefler, Tadahiko Shiomi and O. C. Zienkiewicz.

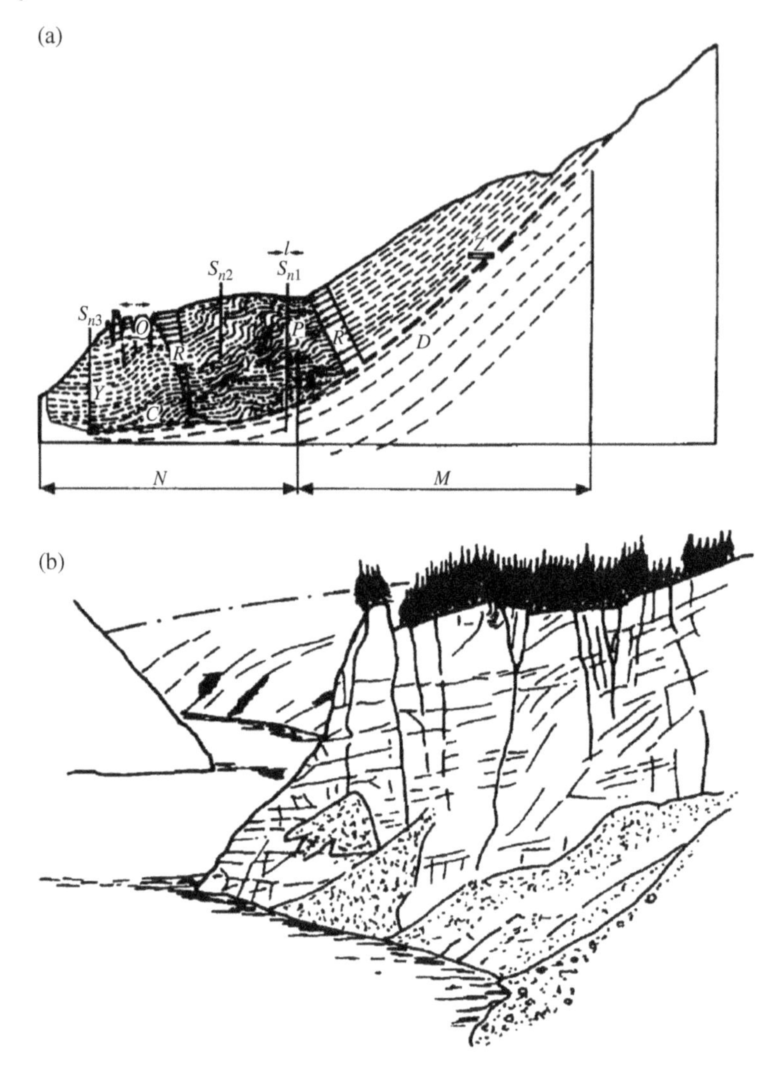

Figure 1.1 The Vajont reservoir, failure of Mant Toc in 1963 (9 October): (a) hypothetical slip plane; (b) downhill end of the slide (Müller, 1965). Plate 1 shows a photo of the slides (front page).

rather than the overall acceleration forces which caused the collapse of the Lower San Fernando dam. What appears to have happened here is that during the motion, the interstitial pore pressure increased, thus reducing the interparticle forces in the solid phase of the soil and its strength.[1]

This phenomenon is well documented and, in some instances, the strength can drop to near-zero values with the soil then behaving almost like a fluid. This behavior is known as *soil liquefaction* and Plate 2 shows a photograph of some buildings in Niigata, Japan taken after the 1964 earthquake. It is clear here that the buildings behaved as if they were floating during the active part of the motion.

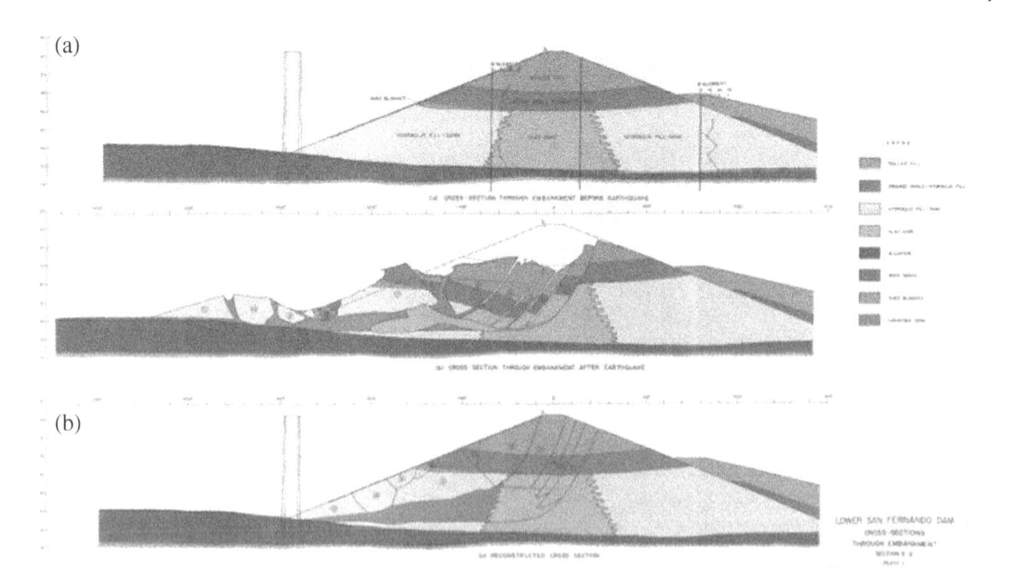

Figure 1.2 Failure and reconstruction of original conditions of Lower San Fernando dam after 1971 earthquake, according to Seed (1979): (a) cross section through embankment after the earthquake; (b) reconstructed cross section. *Source:* Based on Seed (1979).

In this book, we shall discuss the nature and detailed behavior of the various static, quasi-static and dynamic phenomena which occur in soils and will indicate how a computer-based, finite element, analysis can be effective in predicting all these aspects quantitatively.

1.2 The Nature of Soils and Other Porous Media: Why a Full Deformation Analysis Is the Only Viable Approach for Prediction

For single-phase media such as those encountered in structural mechanics, it is possible to predict the ultimate (failure) load of a structure by relatively simple calculations, at least for static problems. Similarly, for soil mechanics problems, such simple, limit-load calculations are frequently used under static conditions, but even here, full justification of such procedures is not generally valid. However, for problems of soil dynamics, the use of such simplified procedures is almost never admissible.

The reason for this lies in the fact that the behavior of soil or such a rock-like material as concrete, in which the pores of the solid phase are filled with one fluid, cannot be described by behavior of a single-phase material. Indeed, to some, it may be an open question whether such porous materials as shown in Figure 1.3 can be treated at all by the methods of continuum mechanics. Here we illustrate two apparently very different materials. The first has a granular structure of loose, generally uncemented, particles in contact with each other. The second is a solid matrix with pores that are interconnected by narrow passages.

From this figure, the answer to the query concerning the possibility of continuum treatment is self-evident. Provided that the dimension of interest and the so-called

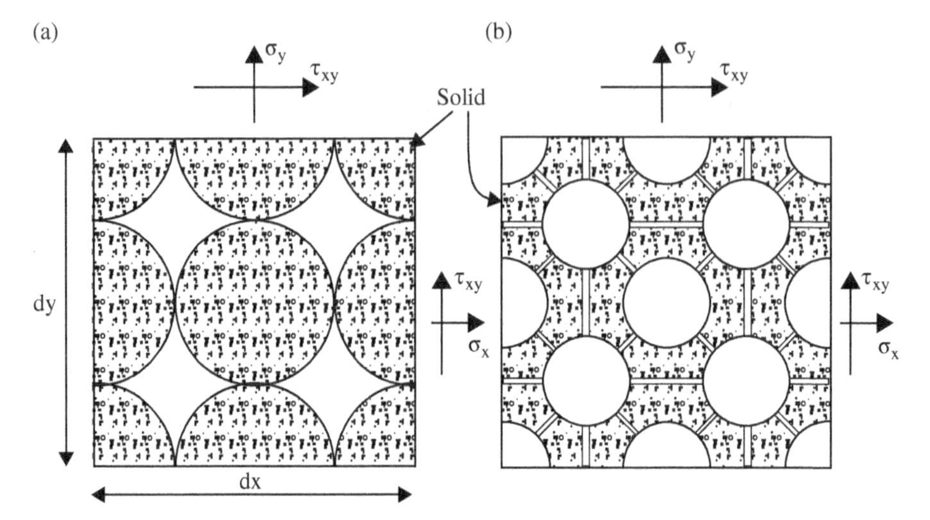

Figure 1.3 Various idealized structures of fluid-saturated porous solids: (a) a granular material; (b) a perforated solid with interconnecting voids.

"infinitesimals" d*x*, d*y*, etc., are large enough when compared to the size of the grains and the pores, it is evident that the approximation of a continuum behavior holds. However, it is equally clear that the intergranular forces will be much affected by the pressures of the fluid–p in single phase (or p_1, p_2, etc., if two or more fluids are present). The strength of the solid, porous material on which both deformations and failure depend can thus only be determined once such pressures are known.

Using the concept of *effective stress*, which we shall discuss in detail in the next section, it is possible to reduce the soil mechanics problem to that of the behavior of a single phase once all the pore pressures are known. Then we can again use the simple, single-phase analysis approaches. Indeed, on occasion, the limit load procedures are again possible. One such case is that occurring under long-term load conditions in the material of appreciable permeability when a steady-state drainage pattern has been established and the pore pressures are independent of the material deformation and can be determined by uncoupled calculations.

Such *drained behavior*, however, seldom occurs even in problems that we may be tempted to consider as static due to the slow movement of the pore fluid and, theoretically, the infinite time required to reach this asymptotic behavior. In very finely grained materials such as silts or clays, this situation may never be established even as an approximation.

Thus, in a general situation, the complete solution of the problem of solid material deformation coupled to a transient fluid flow needs to be solved generally. Here no shortcuts are possible and full *coupled analyses* of equations which we shall introduce in Chapter 2 become necessary.

We have not mentioned so far the notion of the so-called *undrained behavior*, which is frequently assumed for rapidly loaded soil. Indeed, if all fluid motion is prevented, by zero permeability being implied or by extreme speed of the loading phenomena, the pressures developed in the fluid will be linked in a unique manner to deformation of the solid material and a single-phase behavior can again be specified. While the artifice of simple undrained behavior is occasionally useful in static studies, it is not applicable to dynamic phenomena such as those which occur in earthquakes as the pressures developed will, in general, be linked again to the straining (or loading) history and this must always be taken into account. Although in early attempts to deal with earthquake analyses and to predict the damage and response, such undrained analyses were invariably used, adding generally a linearization of the total behavior and a heuristic assumption linking the pressure development with cycles of loading and the behavior predictions were poor. Indeed, comparisons with centrifuge experiments confirmed the inability of such methods to predict either the pressure development or deformations (VELACS – Arulanandan and Scott 1993). For this reason, we believe that the only realistic type of analysis is of the type indicated in this book. This was demonstrated in the same VELACS tests to which we shall frequently refer in Chapter 7.

At this point, perhaps it is useful to interject an observation about the possible experimental approaches. The question which could be addressed is whether a scale model study can be made relatively inexpensively in place of elaborate computation. A typical civil engineer may well consider here the analogy with hydraulic models used to solve such problems as spillway flow patterns where the cost of a small-scale model is frequently small compared to equivalent calculations.

Unfortunately, many factors conspire to deny in geomechanics a readily accessible model study. Scale models placed on shaking tables cannot adequately model the main force acting on the soil structure, i.e. that of gravity, though, of course, the dynamic forces are reproducible and scalable.

To remedy this defect, centrifuge models have been introduced and, here, though, at considerable cost, gravity effects can be well modeled. With suitable fluids substituting water, it is indeed also possible to reproduce the seepage timescale and the centrifuge undoubtedly provides a powerful tool for modeling earthquake and consolidation problems in fully saturated materials. Unfortunately, even here a barrier is reached which appears to be insurmountable. As we shall see later under conditions when two fluids, such as air and water, for instance, fill the pores, capillary effects occur and these are extremely important. So far, no significant success has been achieved in modeling these and, hence, studies of structures with free (phreatic) water surface are excluded. This, of course, eliminates the possible practical applications of the centrifuge for dams and embankments in what otherwise is a useful experimental procedure.

1.3 Concepts of Effective Stress in Saturated or Partially Saturated Media

1.3.1 A Single Fluid Present in the Pores – Historical Note

The essential concepts defining the stresses which control the strength and constitutive behavior of a porous material with internal pore pressure of fluid appear to have been defined, at least qualitatively two centuries ago. The work of Lyell (1871), Boussinesq (1876), and Reynolds (1886) was here of considerable note for problems of soils. Later, similar concepts were used to define the behavior of concrete in dams (Levy 1895 and

Fillunger 1913a, 1913b, 1915) and indeed for other soil or rock structures. In all of these approaches, the concept of division of the total stress between the part carried by the solid skeleton and the fluid pressure is introduced and the assumption made that the strength and deformation of the skeleton is its intrinsic property and not dependent on the fluid pressure.

If we thus define the *total stress* $\boldsymbol{\sigma}$ by its components σ_{ij} using indicial notation, these are determined by summing the appropriate forces in the i-direction on the projection, or cuts, dx_j (or dx, dy, and dz in conventional notation). The surfaces of cuts are shown for two kinds of porous material structure in Figure 1.3 and include the total area of the porous skeleton.

In the context of the finite element computation, we shall frequently use a vectorial notation for stresses, writing

$$\boldsymbol{\sigma} \equiv \left[\sigma_{11}, \sigma_{22}, \sigma_{33}, \sigma_{12}, \sigma_{23}, \sigma_{31}\right]^{\mathrm{T}} \tag{1.1a}$$

or

$$\boldsymbol{\sigma} \equiv \left[\sigma_x, \sigma_y, \sigma_z, \tau_{xy}, \tau_{yz}, \tau_{zx}\right]^{\mathrm{T}} \tag{1.1b}$$

This notation reduces the components to six rather than nine and has some computational merit.

Now if the stress in the solid skeleton is defined as the *effective stress* σ' again over the whole cross sectional area, then the hydrostatic stress due to the pore pressure, p acting, only on the pore area should be

$$-\delta_{ij} np \tag{1.2}$$

where n is the porosity and δ_{ij} is the Kronecker delta. The negative sign is introduced as it is a general convention to take tensile components of stress as positive.

The above, plausible, argument leads to the following relation between total and effective stress with total stress

$$\sigma_{ij} = \sigma'_{ij} - \delta_{ij} np \tag{1.3}$$

or if the vectorial notation is used, we have

$$\boldsymbol{\sigma} = \boldsymbol{\sigma}' - \mathbf{m} np \tag{1.4}$$

where $\mathbf{m}$ is a vector written as

$$\mathbf{m} = [1 \;\; 1 \;\; 1 \;\; 0 \;\; 0 \;\; 0]^{\mathrm{T}} \tag{1.5}$$

The above arguments do not stand the test of experiment as it would appear that, with values of porosity n with a magnitude of 0.1–0.2, it would be possible to damage a specimen of a porous material (such as concrete, for instance) by subjecting it to external and internal pressures simultaneously. Further, it would appear from Equation (1.3) that the strength of the material would be always influenced by the pressure p.

Fillunger introduced the concepts implicit in (1.3) in 1913 but despite conducting experiments in 1915 on the tensile strength of concrete subject to water pressure in the pores, which gave the correct answers, he was not willing to depart from the simple statements made above.

It was the work of Terzaghi and Rendulic (1934) and by Terzaghi (1936) which finally modified the definition of effective stress to

$$\boldsymbol{\sigma} = \boldsymbol{\sigma}' - \mathbf{m} n_w p \tag{1.6}$$

where n_w is now called the *effective area coefficient* and is such that

$$n_w \approx 1 \tag{1.7}$$

Much further experimentation on such porous solids as the concrete had to be performed before the above statement was generally accepted. Here the work of Leliavsky (1947), McHenry (1948), and Serafim (1954, 1964) made important contributions by experiments and arguments showing that it is more rational to take sections for determining the pore water effect through arbitrary surfaces with minimum contact points.

Bishop (1959) and Skempton (1960) analyzed the historical perspective and, more recently, de Boer (1996) and de Boer et al. (1996) addressed the same problem showing how an acrimonious debate between Fillunger and Terzaghi terminated in the tragic suicide of the former in 1937.

Zienkiewicz (1947, 1963) found that interpretation of the various experiments was not always convincing. However, the work of Biot (1941, 1955, 1956a, 1956b, 1962) and Biot and Willis (1957) clarified many concepts in the interpretation of effective stress and indeed of the coupled fluid and solid interaction. In the following section, we shall present a somewhat different argument leading to Equations (1.6) and (1.7).

If the quantity $\boldsymbol{\sigma}'$ of (1.3) and (1.4) is interpreted as the volume-averaged solid stress $(1-n)\,\mathbf{t}_s$ used in the mixture theory (partial stress), see Gray et al. (2009), then we recover the stress split introduced in Biot (1955). There the fluid pressure, as opposed to the effective stress concept, is weighted by the porosity. Biot (1955) declares that "the remaining components of the stress tensor are the forces applied to that portion of the cube faces occupied by the solid." In this book, we use the much more common concept of effective stress.

1.3.2 An Alternative Approach to Effective Stress

Let us now consider the effect of the simultaneous application of a total external hydrostatic stress and a pore pressure change, both equal to Δp, to any porous material. The above requirement can be written in tensorial notation as requiring that the total stress increment is defined as

$$\Delta\sigma_{ij} = -\delta_{ij}\Delta p \tag{1.8a}$$

or, using the vector notation

$$\Delta\boldsymbol{\sigma} = -\mathbf{m}\Delta p \tag{1.8b}$$

In the above, the negative sign is introduced since "pressures" are generally defined as being positive in compression, while it is convenient to define stress components as positive in tension.

It is evident that for the loading mentioned, only a very uniform and small volumetric strain will occur in the skeleton and the material will not suffer any damage provided that the grains of the solid are all made of identical material. This is simply because all parts of the porous medium solid component will be subjected to identical compressive stress.

However, if the microstructure of the porous medium is composed of different materials, it appears possible that nonuniform, localized stresses, can occur and that local grain damage may be suffered. Experiments performed on many soils and rocks and rock-like materials show, however, that such effects are insignificant. Thus, in general, the grains and, hence, the total material will be in a state of pure volumetric strain

$$\Delta\varepsilon_v \approx \Delta\varepsilon_{ii} = \Delta\varepsilon_{11} + \Delta\varepsilon_{22} + \Delta\varepsilon_{33} = -\frac{1}{K_s}\Delta p \tag{1.9}$$

where K_s is the average material bulk modulus of the solid components of the skeleton. Alternatively, adopting a vectorial notation for strain in a manner involved in (1.1)

$$\Delta\varepsilon_v = \mathbf{m}^{\mathrm{T}}\Delta\boldsymbol{\varepsilon} = -\frac{1}{K_s}\Delta p \tag{1.10a}$$

where $\boldsymbol{\varepsilon}$ is the vector defining the strains in the manner corresponding to that of stress increment definition. Again, assuming that the material is isotropic, we shall have

$$\Delta\boldsymbol{\varepsilon} = -\mathbf{m}\frac{1}{3K_s}\Delta p \tag{1.10b}$$

Those not familiar with soil mechanics may find the following hypothetical experiment illustrative. A block of porous, sponge-like rubber is immersed in a fluid to which an increase in pressure of Δp is applied as shown in Figure 1.4. If the pores are connected to the fluid, the volumetric strain will be negligible as the solid components of the sponge rubber are virtually incompressible.

If, on the other hand, the block is first encased in a membrane and the interior is allowed to drain freely, then again a purely volumetric strain will be realized but now of a much larger magnitude.

The facts mentioned above were established by the very early experiments of Fillunger (1915) and it is surprising that so much discussion of "area coefficients" has since been necessary.

From the preceding discussion, it is clear that if the material is subject to a simultaneous change of total stress $\Delta\sigma$ and of the total pore pressure Δp, the resulting strain can always be written incrementally in tensorial notation as

$$\Delta\varepsilon_{kl} = C_{klij}\left(\Delta\sigma_{ij} + \delta_{ij}\Delta p\right) - \delta_{kl}\frac{1}{3K_s}\Delta p + \Delta\varepsilon_{kl}^{0} \tag{1.11a}$$

or in vectoral notation

$$\Delta\boldsymbol{\varepsilon} = \mathbf{D}^{-1}(\Delta\boldsymbol{\sigma} + \mathbf{m}\Delta p) - \mathbf{m}\frac{1}{3K_s}\Delta p + \Delta\boldsymbol{\varepsilon}^{0} \tag{1.11b}$$

with

$$C_{ijkl}D_{mnop} = \delta_{im}\delta_{jn}\delta_{ko}\delta_{lp} \tag{11.11c}$$

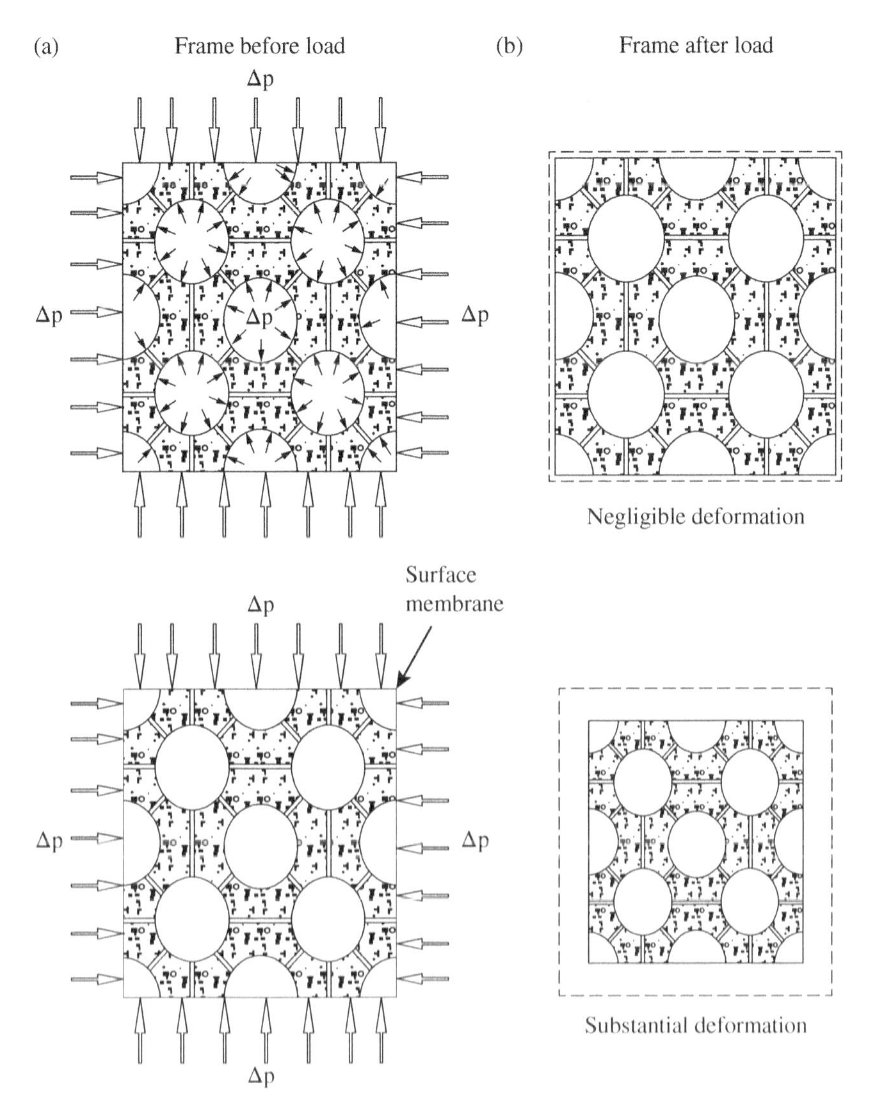

Figure 1.4 A porous material subject to external hydrostatic pressure increases Δ_p, and (a) internal pressure increment Δ_p; (b) internal pressure increment of zero.

The last term in (1.11a) and (1.11b), $\Delta\varepsilon^0$, is simply the increment of an initial strain such as may be caused by temperature changes, etc., while the penultimate term is the strain due to the grain compression already mentioned, viz. Equation (1.10). **D** is a tangent matrix of the solid skeleton implied by the constitutive relation with corresponding compliance coefficient matrix $\mathbf{D}^{-1} = \mathbf{C}$. These, of course, could be matrices of constants, if linear elastic behavior is assumed, but generally will be defined by an appropriate nonlinear relationship of the type which we shall discuss in Chapter 4 and this behavior can be established by fully drained ($p = 0$) tests.

Although the effects of skeleton deformation due to the effective stress defined by (1.6) with $n_w = 1$ have been simply added to the uniform volumetric compression, the principle of superposition requiring linear behavior is not invoked and in this book, we shall almost exclusively be concerned with nonlinear, irreversible, elastoplastic and elastoviscoplastic responses of the skeleton which, however, we assume incremental properties.

For assessment of the strength of the saturated material, the effective stress previously defined with $n_w = 1$ is sufficient. However, we note that the *deformation* relation of (1.11) can always be rewritten incorporating the small compressive deformation of the particles as (1.12).

It is more logical at this step to replace the finite increment by an infinitesimal one and to invert the relations in (1.11) writing these as

$$\mathrm{d}\sigma''_{ij} = \mathrm{d}\sigma_{ij} + \alpha\delta_{ij}\mathrm{d}p = D_{ijkl}\left(\mathrm{d}\varepsilon_{kl} - \mathrm{d}\varepsilon^{0}_{kl}\right) \tag{1.12a}$$

or

$$\mathrm{d}\sigma'' = \mathrm{d}\sigma + \alpha\mathbf{m}\mathrm{d}p = \mathbf{D}(\mathrm{d}\boldsymbol{\varepsilon} - \mathrm{d}\boldsymbol{\varepsilon}_0) \tag{1.12b}$$

where a new "effective" stress, σ'', is defined. In the above

$$\alpha\delta_{ij} = \delta_{ij} - D_{ijkl}\delta_{kl}\frac{1}{3K_s} \tag{1.13a}$$

or

$$\alpha\mathbf{m} = \mathbf{m} - \mathbf{D}\mathbf{m}\frac{1}{3K_s} \tag{1.13b}$$

and the new form eliminates the need for separate determination of the volumetric strain component. Noting that, in three dimensions,

$$\delta_{ij}\delta_{ij} = 3$$

or

$$\mathbf{m}^{\mathrm{T}}\mathbf{m} = 3$$

we can write

$$\alpha\mathbf{m}^{\mathrm{T}}\mathbf{m} = \mathbf{m}^{\mathrm{T}}\mathbf{m} - \mathbf{m}^{\mathrm{T}}\mathbf{D}\mathbf{m}\frac{1}{3K_s} \tag{1.14a}$$

or simply

$$\alpha = 1 - \frac{\mathbf{m}^{T}\mathbf{D}\mathbf{m}}{K_s}$$

Alternatively, in tensorial form, the same result is obtained as

$$\alpha\delta_{ij}\delta_{ij} = \delta_{ij}\delta_{ij} - \delta_{ij}D_{ijkl}\delta_{kl}\frac{1}{3K_s} \tag{1.14b}$$

and

$$\alpha = 1 - \frac{\delta_{ij} D_{ijkl} \delta_{kl}}{9K_s}$$

For isotropic materials, we note that,

$$\frac{\mathbf{m}^{\mathrm{T}}\mathbf{D}\mathbf{m}}{9} = \frac{\delta_{ij} D_{ijkl} \delta_{kl}}{9} = \frac{\delta_{ij}\left(\lambda \delta_{ij} \delta_{kl} + \mu\left(\delta_{ik}\delta_{jl} + \delta_{il}\delta_{jk}\right)\right)\delta_{kl}}{9} = \frac{9\lambda + 6\mu}{9} = K_T \tag{1.15a}$$

which is the tangential bulk modulus of an isotropic elastic material with λ and μ being the Lamé's constants. Thus we can write

$$\alpha = 1 - \frac{K_T}{K_s} \tag{1.15b}$$

The reader should note that in (1.12), we have written the definition of the effective stress increment which can, of course, be used in a non-incremental state as

$$\sigma''_{ij} = \sigma_{ij} + \alpha \delta_{ij} p \tag{1.16a}$$

or

$$\boldsymbol{\sigma}'' = \boldsymbol{\sigma} + \alpha \mathbf{m}\, p \tag{1.16b}$$

assuming that all the stresses and pore pressure started from a zero initial state (for example, material exposed to air is taken as under zero pressure). The above definition corresponds to that of the effective stress used by Biot (1941) but is somewhat more simply derived. In the above, α is a factor that becomes close to unity when the bulk modulus K_s of the grains is much larger than that of the whole material. In such a case, we can write, of course

$$\sigma''_{ij} = \sigma'_{ij} \equiv \sigma_{ij} + \delta_{ij} p \tag{1.17a}$$

or

$$\boldsymbol{\sigma}'' = \boldsymbol{\sigma}' \equiv \boldsymbol{\sigma} + \mathbf{m} p \tag{1.17b}$$

recovering the common definition used by many in soil mechanics and introduced by Terzaghi (1936). Now, however, the meaning of α is no longer associated with an effective area.

It should have been noted that in some materials such as rocks or concrete, it is possible for the ratio K_T/K_s to be as large as 1/3 with $\alpha = 2/3$ being a fairly common value for determination of deformation.

We note that in the preceding discussion, the only assumption made, which can be questioned, is that of neglecting the local damage due to differing materials in the soil matrix. We have also implicitly assumed that the fluid flow is such that it does not separate the contacts of the soil grains. This assumption is not totally correct in soil liquefaction or flow in the soil-shearing layer during localization; therefore, it is not clear if Terzaghi's definition of effective stress still applies when the soil is liquefied.

1.3.3 Effective Stress in the Presence of Two (or More) Pore Fluids – Partially Saturated Media

The interstitial space, or the pores, may, in a practical situation, be filled with two or more fluids. We shall, in this section, consider only two fluids with the degree of saturation by each fluid being defined by the proportion of the total pore volume n (porosity) occupied by each fluid. In the context of soil behavior discussed in this book, the fluids will invariably be *water* and *air*, respectively. Thus, we shall refer to only two saturation degrees, S_w that for water and S_a that for air, but the discussion will be valid for any two fluids.

It is clear that if both fluids fill the pores completely, we shall always have

$$S_w + S_a = 1 \tag{1.18}$$

Clearly, this relation will be valid for any other pair of fluids, e.g. oil and water and indeed the treatment described here is valid for any fluid conditions.

The two fluids may well present different areas of contact with the solid grains of the material in the manner illustrated in Figure 1.5a and b. The average *pressure* reducing the interstitial contact and relevant to the definition of effective stress found in the previous section (Equations (1.16) and (1.17)) can thus be taken as

$$p = \chi_w p_w + \chi_a p_a \tag{1.19}$$

where the coefficients χ_w and χ_a refer to water and air, respectively, and are such that

$$\chi_w + \chi_a = 1 \tag{1.20}$$

The individual pressures p_w and p_a are again referring to water and air and their difference, i.e.

$$p_c = p_a - p_w \tag{1.21}$$

is dependent on the magnitude of surface tension or capillarity and on the degree of saturation (p_c is often referred to, therefore, as capillary pressure).

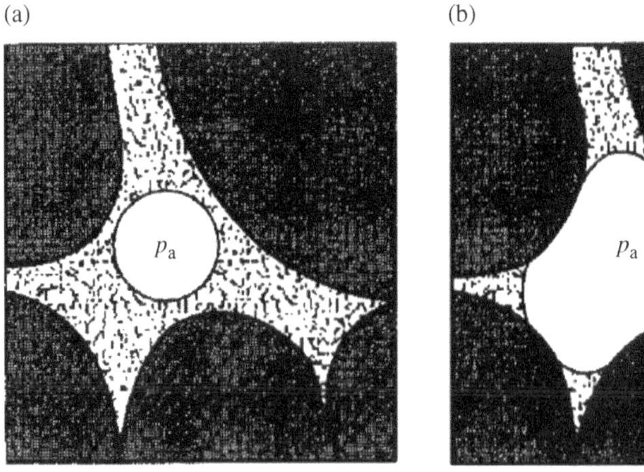

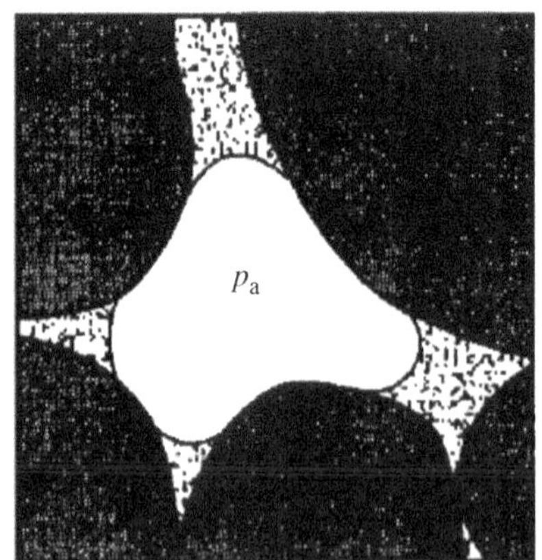

Figure 1.5 Two fluids in pores of a granular solid (water and air). (a) Air bubble not wetting solid surface (effective pressure $p = p_w$); (b) Both fluids in contact with solid surfaces (effective pressure $p = \chi_w p_w + \chi_a p_a$).

Depending on the nature of the material surface, the contact surface may take on the shapes shown in Figure 1.5 with

$$\chi_w = \chi_w(S_w) \tag{1.22a}$$

and

$$\chi_a = \chi_a(S_a) \tag{1.22b}$$

Occasionally, the contact of one of the phases and the solid may disappear entirely as shown in Figure 1.5a giving isolated air bubbles and making in this limit

$$\chi_a = 0 \ \chi_w = 1 \tag{1.23}$$

In many situations, in soil mechanics, it is sufficient to take χ equal to the respective degrees of saturation (Lewis and Schrefler 1982; Nuth and Laloui 2008).

Whatever the nature of the contact, we shall find, neglecting the hysteresis during the wetting and drying cycles, that a unique relationship between p_c and the saturation S_w can be written, i.e.

$$p_c = p_c(S_w) \tag{1.24}$$

Indeed, the degree of saturation will similarly affect flow parameters such as the permeability k to which we shall make reference in the next chapter, giving

$$\begin{aligned} k_w &= k_w(S_w) \\ k_a &= k_a(S_a) \end{aligned} \tag{1.25}$$

Many studies of such relationship are reported in the literature (Liakopoulos 1965; Neuman 2017; Van Genuchten et al. 1977; Narasimhan and Witherspoon 1978; Safai and Pinder 1979; Lloret and Alonso 1980; Bear et al. 1984; Alonso et al. 1987). Figure 1.6 shows a typical relationship.

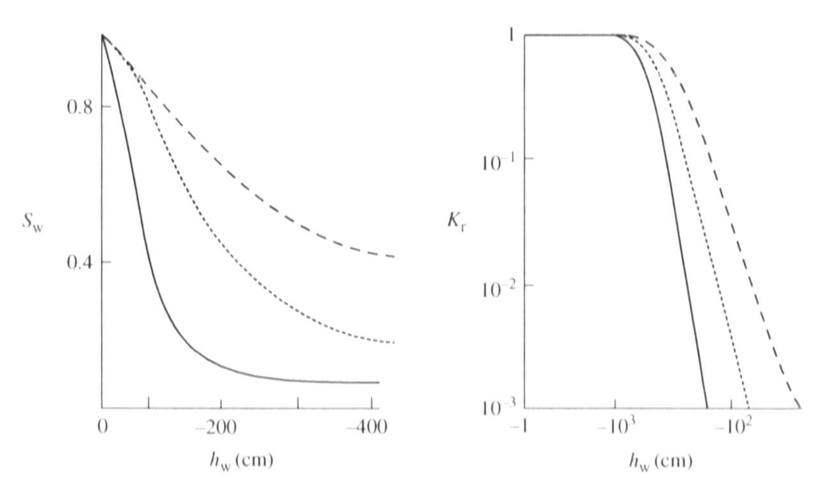

Figure 1.6 Typical relations between pore pressure head, $h_w = p_w/\chi_w$, saturation, S_w, and relative permeability, $k_r = k_w(S_w)/k_w(1)$ (Safai and Pinder 1979). Note that relative permeability decreases very rapidly as saturation decreases. *Source:* From Safai and Pinder (1979).

The concepts of dealing with the effects of multiple pore pressure by introducing an *average pressure* and using the standard definition of *effective stress* (1.19, 1.16, and 1.17) were first introduced by Bishop (1959). Certainly, the arguments for thus extending the original concepts are less clear than is the case when only a single fluid is present. However, the results obtained by this extension are quite accurate. We shall, therefore, use such a definition in the study of partially saturated media.

In many cases occurring in practice, the air pressure is close to zero (atmospheric datum) as the pores are interconnected. Alternatively, negative air pressure occurs as cavitation starts and here the datum is the vapor pressure of water. In either case, the effect of p_a can be easily neglected as the water pressure simply becomes negative from Equation (1.24). Such negative pressures are responsible for the development of certain cohesion by the soil and are essential in the study of free surface conditions occurring in embankments, as we shall see later.

Note

1 Such strength reduction phenomena are mainly evident in essentially non-cohesive materials such as sand and silt. Clays in which negative capillary pressure provide an apparent cohesion are less liable to such strength reduction.

References

Alonso, E. E., Gens, A., and Hight, D. W. (1987). Special problem soils. General report. *Proc. 9th European Conf. Soil Mechanics and Foundation Eng.*, **3**, 1087–1146.

Arulanandan, K. and Scott, R. F. (Eds) (1993). *Proceedings of the VELACS Symposium, 1, A.* A. Balkema, Rotterdam.

Bear, J., Corapcioglu, M. Y. and Balakrishna, J., (1984). Modeling of centrifugal filtration in unsaturated deformable porous media, *Adv. Water Resour.*, **7**, 150–167.

Biot, M. A. (1941). General theory of three-dimensional consolidation, *J. Appl. Phys.*, **12**, 155–164.

Biot, M. A. (1955). Theory of elasticity and consolidation for a porous anisotropic solid, *J. Appl. Phys.*, **26**, 182–185.

Biot, M. A. (1956a). Theory of propagation of elastic waves in a fluid-saturated porous solid, part I-low-frequency range, *J. Acoust. Soc. Am.*, **28**, 2, 168–178.

Biot, M. A. (1956b). Theory of propagation of elastic waves in a fluid-saturated porous solid, part-II-higher frequency range, *J. Acoust. Soc. Am.*, **28**, 2, 179–191.

Biot, M. A. (1962). Mechanics of deformation and acoustic propagation in porous media, *J. Appl. Phys.*, **33**, 4, 1482–1498.

Biot, M. A. and Willis, P. G. (1957). The elastic coefficients of the theory consolidation, *J. Appl. Mech.*, **24**, 594–601.

Bishop, A. W. (1959). The principle of effective stress, *Teknisk Ukeblad*, **39**, 859–863.

Boussinesq, J. (1876). Essai théorique sur l'equilibre d'elasticité des massif pulvérulents, Mem. savants étrangers, *Acad. Belgique*, **40**, 1–180.

De Boer, R. (1996). Highlights in the historical development of the porous media theory, *Appl. Mech. Rev.*, **49**, 201–262.

De Boer, R., Schiffman, R. L. and Gibson, R. E. (1996). The origins of the theory of consolidation: the Terzaghi-Fillunger dispute, *Géotechnique*, **46**, 2, 175–186.

Fillunger, P. (1913a). *Der Auftrieb in Talsperren*. Österr. Wochenschrift öffentlichen Baudienst, 532–556.

Fillunger, P. (1913b). *Der Auftrieb in Talsperren*. Österr. Wochenschrift öffentlichen Baudienst, 567–570.

Fillunger, P. (1915). Versuch über die Zugfestigkeit bei allseitigem Wasserdruck, *Österr. Wochenschrift offentl. Baudienst*, **H29**, 443–448.

Gray, W. G., Schrefler, B. A., and Pesavento, F. (2009). The solid stress tensor in porous media mechanics and the Hill-Mandel condition, *J. Mech. Phys. Solids.*, **57**, 539–544.

Leliavsky, S. (1947). Experiments on effective area in gravity dams, *Trans. Am. Soc. Civ. Eng.*, **112**, 444.

Levy, M. M. (1895). Quelques Considerations sur la construction des grandes barrages. *Comptes Rendus De L'Academie Des Sciences* Serie I-Mathematique, 288.

Lewis, R. W. and Schrefler, B. A. (1982). A finite element simulation of the subsidence of gas reservoirs undergoing a waterdrive, in *Finite Element in Fluids*, **4** R. H. Gallagher, D. H. Norrie, J. T. Oden and O. C. Zienkiewicz(Eds), Wiley, 179–199.

Liakopoulos, A. C. (1965). Transient flow through unsaturated porous media. D. Eng. dissertation, University of California, Berkeley.

Lloret, A. and Alonso, E. E. (1980). Consolidation of unsaturated soils including swelling and collapse behaviour, *Géotechnique*, **30**, 449–447.

Lyell, C. (1871). Student's elements of geology, London.

McHenry, D. (1948). The effect of uplift pressure on the shearing strength of concrete—R.48. *International Congress Large Dams*, 3rd, Stockholm, Vol. I.

Müller, L. (1965). The Rock slide in the Vajont Valley, *Fels Mechanik*, **2**, 148–212.

Narasimhan, T. N. and Witherspoon, P. A. (1978). Numerical model for saturated-unsaturated flow in deformable porous media 3. Applications. *Water Resour. Res.*, **14**, 1017–1034.

Neuman, S. P. (2017). Galerkin approach to saturated-unsaturated flow in porous media. *International Symposium on the Finite Element Methods in Flow Problems, Finite Elem in Fluids* – Swansea, Wales (1 January 2017), 201–217.

Nuth, M., Laloui, L. (2008). Effective stress concept in unsaturated soils: clarification and validation of a unified framework. *Int. J. Numer. Anal. Methods Geomech.*, **32**, 771–801.

Reynolds, O. (1886). Experiments showing dilatancy, a property of granular material, *Proc. R. Inst.*, **11**, 354–363.

Safai, N. M. and Pinder, G. F. (1979). Vertical and horizontal land deformation in a desaturating porous medium, *Adv. Water Resour.* **2**, 19–25.

Seed, H. B. (1979). Consideration in the earthquake resistant design of earth and rockfill dams, *Géotechnique*, **29**, 3, 215–263.

Seed, H. B., Idriss, I. M., Lee, K. L. and Makdisi, F. I. (1975). Dynamic analysis of the slide in the Lower San Fernando dam during the earthquake of February 9, 1971, *J. Geotech. Eng. Div.*, ASCE, **101**, 9, 889–911.

Serafim, J. L. (1954). A subpressëo nos barreyens—Publ. 55, Laboratorio Nacional de Engenheria Civil, Lisbon.

Serafim, J. L. (1964). The 'uplift area' in plain concrete in the elastic range—C. 17. *International Congress on Large Dams, 8th Congress*, Edinburgh, Vol. 5, Comm. 17, pp. 599–622; also: "Deformations du b4ton dues aux press ions dans les pores," Bull. RILEM, No. 27, 1965, pp. 73–76.

Skempton, A. W. (1960). Effective stress in soils, concretes, and rocks. *Proceedings Conference Pore Pressures and Suction in Soils*, 4–16. London: Butterworths.

Terzaghi K. von (1936). The shearing resistance of saturated soils, *Proc. 1st ICSMFE*, **1**, 54–56.

Terzaghi, K. von and Rendulic, L. (1934). Die wirksame Flächenporosität des Betons, *Z. Öst. Ing.-u. ArchitVer.*, **86**, 1/2, 1–9.

Van Genuchten, M. T., Pinder, G. F., and Saukin, W. P. (1977). Modeling of leachate and soil interactions in an aquifer—EPA-600/9-77-026. *Management of Gas and Leachate in Landfills: Proceedings of the Third Annual Municipal Solid Waste Research Symposium*, St Louis, Missouri (14, 15 and 16 March 1977), 95–103.

Zienkiewicz, O. C. (1947). The stress distribution in gravity dams, *J. Inst. Civ. Eng.*, **27**, 244–271.

Zienkiewicz, O. C. (1963). Stress analysis of hydraulic structures including pore pressures effects, *Water Power*, **15**, 104–108.

2

Equations Governing the Dynamic, Soil–Pore Fluid, Interaction

2.1 General Remarks on the Presentation

In this chapter, we shall introduce the reader to the equations which govern both the static and dynamic phenomena in soils containing pore fluids. We shall divide the presentation into three sections: Section 2.2 will deal with soil, or indeed any other porous medium, saturated with a single fluid. This most common problem contains all the essential features of soil behavior and the equations embrace and explain the vast majority of problems encountered in practice.

We shall show here how the dynamic equations, which are essential for the study of earthquakes, reduce to those governing the quasi-static situations of consolidating soils and indeed to purely static problems without modification. This feature will be used when discretization is introduced and computer codes are derived since a single code will be capable of dealing with most phenomena encountered in soil and rock mechanics.

The limitations of the approximating simplification are discussed in Section 2.2 by using a simple linearized example and deriving conclusions on the basis of an available analytical solution. The same discussion will show the domain of the validity of the assumptions of *undrained* and *fully drained* behavior.

In the same section, we shall introduce a simplification which is valid for the treatment of most low-frequency phenomena – and this simplified form will be used in the subsequent Section 2.3 dealing with partially saturated soil in which the air pressure is assumed constant and also, finally, in Section 2.4 dealing with simultaneous water and airflow in the pores.

The notation used throughout this chapter will generally be of standard, tensorial form. Thus:

u_i will be the displacement of the solid matrix with $i = 1, 2$ in two dimensions or

$i = 1, 3$ in three dimensions

Alternatively, the form

$$\mathbf{u} = [u_1, u_2, u_3]^{\mathrm{T}}$$

Computational Geomechanics: Theory and Applications, Second Edition. Andrew H. C. Chan, Manuel Pastor, Bernhard A. Schrefler, Tadahiko Shiomi and O. C. Zienkiewicz.

will also be used for the same quantity in vectorial notation.

Similarly, we shall use w_i and v_i or $\mathbf{w}$ and $\mathbf{v}$ to denote the velocities of water and air relative to the solid components. These velocities are calculated on the basis of dividing the appropriate flow by the total cross sectional area of the solid–fluid composite.

As mentioned in Chapter 1, σ_{ij} and σ''_{ij} refer to the appropriate total and effective stresses, with $\boldsymbol{\sigma}$ and $\boldsymbol{\sigma}''$ being the vectorial alternatives.

Similarly, ε_{ij} or $\boldsymbol{\varepsilon}$ refers to the strain components. Further, p_a, p_w, and $p = \chi_w p_w + \chi_a p_a$ will stand for air and water pressure and the "effective" pressure defined in the effective stress concept in Equation (1.11) when two fluids are present.

S_a and S_w are the relative degrees of saturation and k_a and k_w are the permeabilities for air and water flow.

Other symbols will be added and defined in the text as the need arises.

The derivation of the equations in this chapter follows a physical approach which establishes clearly the interactions involved in the manner presented by Zienkiewicz and Shiomi (1984), Zienkiewicz (1982), Zienkiewicz et al. (1990a, 1990b), etc. This is a slightly different approach from that used in the earlier presentations of Biot (1941, 1955, 1956a, 1956b, 1962) and Biot and Willis (1957) but we believe it is slightly easier to follow as it explores the physical meaning of each term.

Later, it became fashionable to derive the equations in the form of the so-called mixture theories (see Green and Adkin 1960; Green 1969; Bowen 1976). The equations derived were subsequently recast in varying forms. Here an important step forward was introduced by Morland (1972) who used extensively the concept of volume fractions. Derski (1978) introduced a different derivation of coupled equations and Kowalski (1979) compared various parameters occurring in Derski's equations with those of Biot's equations. A full discussion of the development of the theory is given in the paper by de Boer (1996).

For completeness, we shall include such mixture derivations of the equations in Section 2.5. If correctly applied, the mixture theory establishes, of course, identical equations with appropriately chosen parameters and rheological relations.

It seems that despite much sophistication of various sets of coupled equations, most authors limited their works to conventional, linear elastic, behavior of the solid. Indeed, de Boer and Kowalski (1983) found it necessary to develop a special plasticity theory for porous, saturated solids. In the equations of Zienkiewicz (1982) and Zienkiewicz et al. (1990a), any nonlinear behavior can be specified for the skeleton and, therefore, realistic models can be incorporated. Indeed we shall find that such models are essential if practical conclusions are to be drawn from this work.

2.2 Fully Saturated Behavior with a Single Pore Fluid (Water)

2.2.1 Equilibrium and Mass Balance Relationship (*u*, *w*, and *p*)

We recall first the effective stress and constitutive relationships as defined in Equation (1.16) of Chapter 1 which we repeat below.

$$\sigma''_{ij} = \sigma_{ij} + \alpha\delta_{ij}p \tag{2.1a}$$

or

$$\boldsymbol{\sigma}'' = \boldsymbol{\sigma} + \alpha\mathbf{m}p \tag{2.1b}$$

This effective stress is conveniently used as it can be directly established from the total strains developed.

However, it should be remembered here that this stress definition was derived in Chapter 1 as a corollary of using the effective stress defined as below:

$$\sigma'_{ij} = \sigma_{ij} + \delta_{ij}p \tag{2.2a}$$

or

$$\boldsymbol{\sigma}' = \boldsymbol{\sigma} + \mathbf{m}p \tag{2.2b}$$

which is responsible for the major part of the deformation and certainly for failure.

In soils, the difference between the two effective stresses is negligible as $\alpha \approx 1$. However, for such materials as concrete or rock, the value of α in the first expression can be as low as 0.5 but experiments on tensile strength show that the second definition of effective stress is there much more closely applicable as shown by Leliavsky (1947), Serafim (1954), etc.

For soil mechanics problems, to which we will devote most of the examples, $\alpha = 1$ will be assumed. Constitutive relationships will still, however, be written in the general form using an incremental definition

$$\mathrm{d}\sigma''_{ij} = D_{ijkl}\left(\mathrm{d}\varepsilon_{kl} - \mathrm{d}\varepsilon^0_{kl}\right) \tag{2.3a}$$

$$\mathrm{d}\boldsymbol{\sigma}'' = \mathbf{D}\left(\mathrm{d}\boldsymbol{\varepsilon} - \mathrm{d}\boldsymbol{\varepsilon}^0\right) \tag{2.3b}$$

The vectorial notation used here follows that corresponding to stress components given in (1.1). We thus define the strains as

$$\mathrm{d}\boldsymbol{\varepsilon}^{\mathrm{T}} = \left(\mathrm{d}\varepsilon_x, \mathrm{d}\varepsilon_y, \mathrm{d}\varepsilon_z, \mathrm{d}\gamma_{xy}, \mathrm{d}\gamma_{yz}, \mathrm{d}\gamma_{zx}\right)^{\mathrm{T}} \tag{2.4}$$

In the above, $\mathbf{D}$ is the "tangent matrix" and $\mathrm{d}\boldsymbol{\varepsilon}^0$ is the increment of the thermal or similar autogenous strain and of the grain compression $\mathbf{m}\dot{p}/3K_s$. The latter is generally neglected in soil problems.

If large strains are encountered, this definition needs to be modified and we must write

$$\mathrm{d}\sigma''_{ij} = D_{ijkl}\left(\mathrm{d}\varepsilon_{kl} - \mathrm{d}\varepsilon^0_{kl}\right) + \sigma''_{ik}\mathrm{d}\omega_{kj} + \sigma''_{jk}\mathrm{d}\omega_{ki} \tag{2.5}$$

where the last two terms account for simple rotation (via the definition in 2.6b) of the existing stress components and are known as the Zaremba (1903a, 1903b)–Jaumann (1905) stress changes. We omit here the corresponding vectorial notation as this is not easy to implement.

The large strain rotation components are small for small displacement computation and can be frequently neglected. Thus, in the derivations that follow, we shall do so – though their inclusion presents no additional computational difficulties.

The strain and rotation increments of the soil matrix can be determined in terms of displacement increments du_i as

$$d\varepsilon_{ij} = \frac{1}{2}\left[du_{i,j} + du_{j,i}\right] \tag{2.6a}$$

and

$$d\omega_{ij} = \frac{1}{2}\left[du_{i,j} - du_{j,i}\right] \tag{2.6b}$$

The comma in the suffix denotes differentiation with respect to the appropriate coordinate specified. Thus

$$u_{i,j} \equiv \frac{\partial u_i}{\partial x_j} \text{ etc.}$$

If the vectorial notation is used, as is often the case in the finite element analysis, the so-called engineering strains are used in which (with the repeated index of $\partial u_{i,i}$ not summed)

$$d\varepsilon_i = du_{i,i} \tag{2.7a}$$

or

$$d\varepsilon_x = \frac{d}{\partial x} du_x \text{ etc.}$$

However, the shear strain increments will be written as

$$d\gamma_{ij} = 2d\varepsilon_{ij} = du_{i,j} + du_{j,i} \tag{2.7b}$$

or

$$d\gamma_{xy} = \frac{\partial du_y}{\partial x} + \frac{\partial du_x}{\partial y}$$

We shall usually write the process of strain computation using matrix notation as

$$d\boldsymbol{\varepsilon} = \mathbf{S}d\mathbf{u} \tag{2.8}$$

where

$$\mathbf{u} = \left[u_x\, u_y\, u_z\right]^{\mathrm{T}} \tag{2.9}$$

And for two dimensions, the strain matrix is defined as:

$$\mathbf{S} = \begin{bmatrix} \frac{\partial}{\partial x} & 0 \\ 0 & \frac{\partial}{\partial y} \\ \frac{\partial}{\partial y} & \frac{\partial}{\partial x} \end{bmatrix} \tag{2.10}$$

with corresponding changes for three dimensions (as shown in Zienkiewicz et al. 2005).

We can now write the overall equilibrium or momentum balance relation for the soil–fluid “mixture” as

$$\sigma_{ij,j} - \rho\ddot{u}_i - \underline{\rho_f\left[\dot{w}_i + w_j w_{i,j}\right]} + \rho b_i = 0 \tag{2.11a}$$

or

$$\begin{gathered}\mathbf{S}^T\boldsymbol{\sigma} - \rho\ddot{\mathbf{u}} - \underline{\rho_f\left[\dot{\mathbf{w}} + \mathbf{w}\nabla^T\mathbf{w}\right]} + \boldsymbol{\rho}\mathbf{b} = 0\\ \text{where } \dot{w}_i \equiv \frac{\mathrm{d}w_i}{\mathrm{d}t} \text{ etc.}\end{gathered} \tag{2.11b}$$

In the above, w_i (or $\mathbf{w}$) is the average (Darcy) velocity of the percolating water.

The underlined terms in the above equation represent the fluid acceleration relative to the solid and the convective terms of this acceleration. This acceleration is generally *small* and we shall frequently omit it. In derivations of the above equation, we consider the solid skeleton and the fluid embraced by the usual control volume: $\mathrm{d}x \cdot \mathrm{d}y \cdot \mathrm{d}z$.

Further, ρ_f is the density of the fluid, $\mathbf{b}$ is the body force per unit mass (generally gravity) vector, and ρ is the density of the total composite, i.e.

$$\rho = n\rho_f + (1-n)\rho_S \tag{2.12}$$

where ρ_S is the density of the solid particles and n is the porosity (i.e. the volume of pores in a unit volume of the soil).

The second equilibrium equation ensures the momentum balance of the fluid. If again we consider the same unit control volume as that assumed in deriving (2.11) (and we further assume that this moves with the solid phase), we can write

$$-p_{,i} - R_i - \rho_f\ddot{u}_i - \underline{\rho_f\left[\dot{w}_i + w_j w_{i,j}\right]/n} + \rho_f b_i = 0 \tag{2.13a}$$

or

$$-\nabla p - \mathbf{R} - \rho_f\ddot{\mathbf{u}} - \underline{\rho_f\left[\dot{\mathbf{w}} + \mathbf{w}\nabla^T\mathbf{w}\right]/n} + \rho_f\mathbf{b} = 0 \tag{2.13b}$$

In the above, we consider only the balance of the fluid momentum and $\mathbf{R}$ represents the viscous drag forces which, assuming the Darcy seepage law, can be written as

$$k_{ij}R_j = w_i \tag{2.14a}$$

$$\mathbf{kR} = \mathbf{w} \tag{2.14b}$$

Note that the underlined terms in (2.13) represent again the convective fluid acceleration and are generally small. Also note that, throughout this book, the permeability $\mathbf{k}$ is used with dimensions of [length]^3·[time]/[mass] which is different from the usual soil mechanics convention k' which has the dimension of velocity, i.e. [length]/[time]. Their values are related by $k = k'/\rho'_f g'$ where ρ'_f and g' are the fluid density and gravitational acceleration at which the permeability is measured.

The final equation is one accounting for the mass balance of the flow. Here we balance the flow divergence $w_{i,i}$ by the augmented storage in the pores of a unit volume of soil occurring in time dt. This storage is composed of several components given below in order of importance:

i) the increased volume due to a change in strain, i.e.: $\delta_{ij}\mathrm{d}\varepsilon_{ij} = \mathrm{d}\varepsilon_{ii} = \mathbf{m}^T\mathrm{d}\boldsymbol{\varepsilon}$
ii) the additional volume stored by compression of void fluid due to fluid pressure increase: $n\mathrm{d}p/K_f$
iii) the additional volume stored by the compression of grains by the fluid pressure increase: $(1-n)\mathrm{d}p/K_S$

and

iv) the change in volume of the solid phase due to a change in the intergranular effective contact stress $\left(\sigma'_{ij} = \sigma_{ij} + \delta_{ij}p\right)$: $-\frac{1}{3}\delta_{ij}\mathrm{d}\sigma'_{ij}/K_S = -\frac{K_T}{K_S}(\mathrm{d}\varepsilon_{ii} + \mathrm{d}p/K_S)$.

Here K_T is the average bulk modulus of the solid skeleton and ε_{ii} the total volumetric strain.

Adding all the above contributions together with a source term and a second-order term due to the change in fluid density in the process, we can finally write the flow conservation equation

$$w_{i,i} + \dot{\varepsilon}_{ii} + \frac{n\dot{p}}{K_f} + \frac{(1-n)\dot{p}}{K_S} - \frac{K_T}{K_S}\left(\dot{\varepsilon}_{ii} + \frac{\dot{p}}{K_S}\right) + n\frac{\dot{\rho}_f}{\rho_f} + \dot{s}_0 = 0 \tag{2.15}$$

This can be rewritten using the definition of α given in Equation (1.15b) as

$$w_{i,i} + \alpha\dot{\varepsilon}_{ii} + \frac{\dot{p}}{Q} + \underline{n\frac{\dot{\rho}_f}{\rho_f} + \dot{s}_0} = 0 \tag{2.16a}$$

or in vectorial form

$$\nabla^{\mathrm{T}}\mathbf{w} + \alpha\mathbf{m}^T\dot{\boldsymbol{\varepsilon}} + \frac{\dot{p}}{Q} + \underline{n\frac{\dot{\rho}_f}{\rho_f} + \dot{s}_0} = 0 \tag{2.16b}$$

where

$$\frac{1}{Q} \equiv \frac{n}{K_f} + \frac{\alpha - n}{K_s} \cong \frac{n}{K_f} + \frac{1-n}{K_s} \tag{2.17}$$

In (2.16), the last two (underlined) terms are those corresponding to a change of density and rate of volume expansion of the solid in the case of thermal changes and are negligible in general. We shall omit them from further consideration here.

Equations (2.11), (2.13), and (2.16) together with appropriate constitutive relations specified in the manner of (2.3) define the behavior of the solid together with its pore pressure in both static and dynamic conditions. The unknown variables in this system are:

The pressure of fluid (water), $p \equiv p_w$
The velocities of fluid flow w_i or $\mathbf{w}$
The displacements of the solid matrix u_i or $\mathbf{u}$.

The boundary condition imposed on these variables will complete the problem. These boundary conditions are:

1) For the total momentum balance on the part of the boundary Γ_t, we specify the total traction $t_i(\mathbf{t})$ (or in terms of the total stress $\sigma_{ij}n_j$ ($\boldsymbol{\sigma}\cdot\mathbf{G}$) with n_i being the ith component of the normal at the boundary and $\mathbf{G}$ is the appropriate vectorial equivalence) while for Γ_u, the displacement $u_i(\mathbf{u})$, is given.
2) For the fluid phase, again the boundary is divided into two parts Γ_p on which the values of p are specified and Γ_w where the normal outflow w_n is prescribed (for instance, a zero value for the normal outward velocity on an impermeable boundary).

Summarizing, for the overall assembly, we can thus write

$$\begin{aligned}&\Gamma = \Gamma_t \cup \Gamma_u\\ &\mathbf{t} = \boldsymbol{\sigma}\cdot\mathbf{G} = \bar{\mathbf{t}} \quad \text{on} \quad \Gamma = \Gamma_t\end{aligned} \tag{2.18}$$

and

$$\mathbf{u} = \bar{\mathbf{u}} \quad \text{on} \quad \Gamma = \Gamma_u$$

Further

$$\begin{aligned}&\Gamma = \Gamma_p \cup \Gamma_w\\ &p = \bar{p} \quad \text{on} \quad \Gamma = \Gamma_p\end{aligned} \tag{2.19a}$$

and

$$\mathbf{n}^{\mathrm{T}}\mathbf{w} = w_n \quad \text{on} \quad \Gamma = \Gamma_w \tag{2.19b}$$

It is of interest to note, as shown by Zienkiewicz (1982), that some typical soil constants are implied in the formulation. For instance, we note from (2.16) that for *undrained behavior*, when $w_{i,i} = 0$, i.e. with no net outflow, we have (neglecting the last two terms which are of the second order).

$$\mathrm{d}p = -Q\alpha \mathrm{d}\varepsilon_{ii}$$

or

$$\mathrm{d}p = -Q\alpha\mathbf{m}^{\mathrm{T}}\mathrm{d}\boldsymbol{\varepsilon}$$

and

$$\begin{aligned}\mathrm{d}\sigma_{ij} = \mathrm{d}\sigma''_{ij} - \alpha\delta_{ij}\mathrm{d}p &= D_{ijkl}\mathrm{d}\varepsilon_{kl} + \alpha^2 Q\mathrm{d}\varepsilon_{ij}\\ &= \left(D_{ijkl} + \alpha^2 Q\delta_{ik}\delta_{jl}\right)\mathrm{d}\varepsilon_{kl}\end{aligned}$$

or

$$\mathrm{d}\boldsymbol{\sigma} = \left(\mathbf{D} + \alpha^2\mathbf{m}Q\mathbf{m}^{\mathrm{T}}\right)\mathrm{d}\boldsymbol{\varepsilon}$$

If the pressure change $\mathrm{d}p$ is considered as a fraction of the mean total stress change $\mathbf{m}^{\mathrm{T}}$ $\mathrm{d}\sigma/3$ or $\mathrm{d}\sigma_{ii}/3$, we obtain the so-called B soil parameter (Skempton 1954) as

$$B \equiv \frac{-3\mathrm{d}p}{\mathrm{d}\sigma_{ii}} = \frac{3\alpha Q\mathrm{d}\varepsilon_{kk}}{D_{iikl}\mathrm{d}\varepsilon_{kl} + 3\alpha^2 Q\mathrm{d}\varepsilon_{kk}}$$

$$B \equiv \frac{-3\mathrm{d}p}{\mathbf{m}^{\mathrm{T}}\mathrm{d}\boldsymbol{\sigma}} = \frac{3\alpha Q\mathbf{m}^{\mathrm{T}}\mathrm{d}\boldsymbol{\varepsilon}}{\mathbf{m}^{\mathrm{T}}(\mathbf{D} + \alpha^2\mathbf{m}Q\mathbf{m}^{\mathrm{T}})\mathrm{d}\boldsymbol{\varepsilon}}$$

Using the assumption that the material is isotropic so that

$$D_{ijkl} = \left(K_T - \frac{2}{3}\mu\right)\delta_{ij}\delta_{kl} + \mu\left(\delta_{ik}\delta_{jl} + \delta_{il}\delta_{jk}\right)$$

$$B = \frac{3\alpha Q \mathrm{d}\varepsilon_{kk}}{((3K_T - 2\mu)\delta_{kl} + 2\mu\delta_{kl})\mathrm{d}\varepsilon_{kl} + 3\alpha^2 Q \mathrm{d}\varepsilon_{kk}} = \frac{\alpha Q}{K_T + \alpha^2 Q} = \frac{\alpha}{(K_T/Q) + \alpha^2}$$

$$B = \frac{3\alpha Q \mathbf{m}^T\mathbf{m}}{\mathbf{m}^T(\mathbf{D} + \alpha^2\mathbf{m}Q\mathbf{m}^T)\mathbf{m}} = \frac{9\alpha Q}{\mathbf{m}^T\mathbf{D}\mathbf{m} + 9\alpha^2 Q} = \frac{9\alpha Q}{9K_T + 9\alpha^2 Q} = \frac{\alpha Q}{K_T + \alpha^2 Q} = \frac{\alpha}{(K_T/Q) + \alpha^2}$$

where K_T is (as defined in equation (1.10), the bulk modulus of the solid phase and μ is once again Lamé's constant. B has, of course, a value approaching unity for soil but can be considerably lower for concrete or rock. Further, for unsaturated soils, the value will be much lower (Terzaghi 1925; Lambe and Whitman 1969; Craig 1992).

2.2.2 Simplified Equation Sets (*u–p* Form)

The governing equation set (2.11), (2.13), and (2.16) together with the auxiliary definition system can, of course, be used directly in numerical solution as shown by Zienkiewicz and Shiomi (1984). This system is suitable for explicit time-stepping computation as shown by Sandhu and Wilson (1969) and Ghaboussi and Wilson (1972) and later by Chan et al. (1991). However, in implicit computation, where large algebraic equation systems arise, it is convenient to reduce the number of variables by neglecting the apparently small (underlined) terms of equations (2.11) and (2.13). These contain the variable $w_i(\mathbf{w})$ which now can be eliminated from the system.

The first equation of the reduced system becomes (from (2.11))

$$\sigma_{ij,\ j} - \rho\ddot{u}_i + \rho b_i = 0 \tag{2.20a}$$

or

$$\mathbf{S}^{\mathrm{T}}\boldsymbol{\sigma} - \rho\ddot{\mathbf{u}} + \rho\mathbf{b} = 0 \tag{2.20b}$$

The second equation is obtained by coupling (2.13) and (2.16) using the definition (2.14) and thus eliminating the variable $w_i(\mathbf{w})$. We now have, omitting density changes

$$\left(k_{ij}\left(-p_{,j} - \rho_f\ddot{u}_j + \rho_f b_j\right)\right)_{,i} + \alpha\dot{\varepsilon}_{ii} + \frac{\dot{p}}{Q} + \dot{s}_0 = 0 \tag{2.21a}$$

or

$$\nabla^{\mathrm{T}}\mathbf{k}\left(-\nabla p - \rho_f\ddot{\mathbf{u}} + \rho_f\mathbf{b}\right) + \alpha\mathbf{m}^{\mathrm{T}}\dot{\boldsymbol{\varepsilon}} + \frac{\dot{p}}{Q} + \dot{s}_0 = 0 \tag{2.21b}$$

This reduced equation system is precisely the same as that used conventionally in the study of consolidation if the dynamic terms are omitted or even of static problems if the steady state is reached and all the time derivatives are reduced to zero. Thus, the formulation conveniently merges with procedures used for such analyses. However, some loss of accuracy will be evident for problems in which high-frequency oscillations are important. As we shall show in the next section, these are of little importance for earthquake analyses.

In eliminating the variable $w_i(\mathbf{w})$, we have neglected several terms but have achieved an elimination of two or three variable sets depending on whether the two- or three-dimensional problem is considered. However, another possibility exists for obtaining a reduced equation set without neglecting any terms provided that the fluid (i.e. water in this case) is compressible.

With such compressibility assumed, Equation (2.16) can be integrated in time, provided that we introduce the water displacement $U_i^{\mathrm{R}} = \left(\mathbf{U}^{\mathrm{R}}\right)$ in place of the velocity $w_i(\mathbf{w})$. We define

$$\dot{U}_i^R = w_i/n \tag{2.22a}$$

or

$$\dot{\mathbf{U}}^R = \mathbf{w}/n \tag{2.22b}$$

where the division by the porosity n is introduced to approximate the true rather than the averaged fluid displacement. We now can rewrite (2.16) after integration with respect to time as

$$p = -Q\left(\alpha\varepsilon_{ii} + nU_{i,i}^R\right) \tag{2.23a}$$

or

$$p = -Q\left(\alpha\mathbf{m}^{\mathrm{T}}\boldsymbol{\varepsilon} + n\nabla^{\mathrm{T}}\mathbf{U}^R\right) \tag{2.23b}$$

and thus we can eliminate p from (2.11) and (2.13).

The resulting system which is fully discussed in Zienkiewicz and Shiomi (1984) is not written down here as we shall derive this alternative form in Chapter 3 using the total displacement of water $\mathbf{U} = \mathbf{U}^R + \mathbf{u}$ as the variable. It presents a very convenient basis for using a fully explicit temporal scheme of integration (see Chan et al. 1991) but it is not applicable for long-term studies leading to steady-state conditions, as the water displacement $\mathbf{U}$ then increases indefinitely.

It is fortunate that the inaccuracies of the $\mathbf{u}$–p version are pronounced only in high-frequency, short-duration, phenomena, since, for such problems, we can conveniently use explicit temporal integration. Here a very small time increment can be used for the short time period considered (see Chapter 3).

Table 2.1 summarizes various forms of governing equations used.

2.2.3 Limits of Validity of the Various Approximations

It is, of course, important to know the degree of approximation involved in various differential equation systems. Thus, it is of interest to know under what circumstances *undrained conditions* can be assumed, to define the behavior of the material and when the simplified equation system discussed in the previous section is applicable, without introducing serious

Table 2.1 Comparative sets of coupled equations governing deformation and flow.

A) $\mathbf{u}-\mathbf{w}-p$ equations (exact) [(2.11), (2.13), and (2.16)]

$$\mathbf{S}^T\sigma-\rho\ddot{\mathbf{u}}-\underline{\rho_f\left[\dot{\mathbf{w}}+\mathbf{w}\nabla^{\mathrm{T}}\mathbf{w}\right]}+\rho\mathbf{b}=0$$

$$-\nabla p-\mathbf{R}-\rho_f\ddot{\mathbf{u}}-\underline{\rho_f\left[\dot{\mathbf{w}}+\mathbf{w}\nabla^{T}\mathbf{w}\right]/n}+\rho_f\mathbf{b}=0$$

$$\nabla^T\mathbf{w}+\alpha\mathbf{m}^T\dot{\boldsymbol{\varepsilon}}+\frac{\dot{p}}{Q}+n\frac{\dot{\rho}_f}{\rho_f}+\dot{s}_0=0$$

B) $\mathbf{u}-p$ approximation for dynamics of lower frequencies. Exact for consolidation [(2.20), (2.21)]

$$\mathbf{S}^T\boldsymbol{\sigma}-\rho\ddot{\mathbf{u}}+\rho\mathbf{b}=0$$

$$\nabla^T\mathbf{k}\left(-\nabla p-\rho_f\ddot{\mathbf{u}}+\rho_f\mathbf{b}\right)+\alpha\mathbf{m}^T\dot{\boldsymbol{\varepsilon}}+\frac{\dot{p}}{Q}+\dot{s}_0=0$$

C) $\mathbf{u}-\mathbf{U}$, only convective terms neglected (3.72)

$$\mathbf{S}^T\boldsymbol{\sigma}+\alpha Q(\alpha-n)\nabla(\nabla\mathbf{u})+\alpha Qn\nabla\left(\nabla^{T}\mathbf{U}\right)-(1-n)\rho\ddot{\mathbf{u}}-\rho_f n\ddot{\mathbf{U}}+\rho\mathbf{b}=0$$

$$(\alpha-n)Q\nabla\left(\nabla^T\mathbf{u}\right)+nQ\nabla\left(\nabla^T\mathbf{U}\right)-\mathbf{k}^{-1}(n\mathbf{U}-n\mathbf{u})-\rho_f\mathbf{U}+\rho_f\mathbf{b}=0$$

In all the above

$\boldsymbol{\sigma}''=\boldsymbol{\sigma}+\alpha\mathbf{m}p$ and $d\boldsymbol{\sigma}''=\mathbf{D}d\boldsymbol{\varepsilon}=\mathbf{DS}d\mathbf{u}$

error. An attempt to answer these problems was made by Zienkiewicz et al. (1980). The basis was the consideration of a one-dimensional set of linearized equations of the full systems (2.11), (2.13), and (2.16) and of the approximations (2.20) and (2.21). The limiting case in which $w_{i,i}=0$ (representing undrained conditions) was also considered.

For all these conditions, the exact solution of the equation is possible. We consider thus that the only physical variation is in the vertical, x_1, direction ($x_1=x$) and then we have

$$\sigma_x''=\sigma_x'=\sigma_x+p$$
$$\sigma_x'=D\varepsilon_x\quad\left(\varepsilon_y=\varepsilon_z=0\right)$$
$$D=\frac{E(1-\nu)}{(1+\nu)(1-2\nu)}$$

where D is called the one-dimensional constrained modulus, E is Young's modulus and ν is Poisson's ratio of the linear elastic soil matrix, also

$$\varepsilon^0=0,\quad S_0=0,\quad b=0,\quad u_1=u,\quad u_2=u_3=0$$

The differential equations are, in place of (2.11):

$$\frac{\partial\sigma}{\partial x}-\rho\frac{\partial^2u}{\partial t^2}-\rho_f\frac{\partial w}{\partial t}=0$$

In place of (2.13):

$$-\frac{\partial p}{\partial x}-\frac{w}{k}-\rho_f\frac{\partial^2u}{\partial t^2}-\frac{\rho_f}{n}\frac{\partial w}{\partial t}=0$$

and in place of (2.16):

$$\frac{\partial w}{\partial x} + \frac{\partial \varepsilon}{\partial t} + \frac{1}{Q}\frac{\partial p}{\partial t} = 0$$

with

$$\sigma = D\varepsilon - p$$

Taking $K_s \to \infty$

$$\frac{1}{Q} = \frac{n}{K_f}$$

For a periodic applied surface load

$$q = \overline{q}e^{iwt}$$

a periodic solution arises after the dissipation of the initial transient in the form

$$u = \bar{u}e^{iwt}$$

$$p = \overline{p}e^{iwt} \text{ etc.}$$

and a system of ordinary linear differential equations is obtained in the frequency domain which can be readily solved by standard procedure.

The boundary conditions imposed are as shown in Figure 2.1. Thus, at $x = L$, $u = 0$, $w = 0$, and at $x = 0$, $\sigma_x = q$, $p = 0$.

In Figure 2.1 (taken from Zienkiewicz et al. (1980)), we show a comparison of various numerical results obtained by various approximations:

i) exact solution (Biot's, labelled B)
ii) the u–p equation approximation (labelled Z)
iii) the undrained assumption ($w = 0$) and
iv) the consolidation equation obtained by omitting all acceleration terms (labelled C).

The reader will note that the results are plotted against two nondimensional coefficients:

$$\Pi_1 = \frac{k' V_c^2}{g\beta\omega L^2} = \frac{2}{\beta\pi}\frac{k'}{g}\frac{T}{\hat{T}^2} = \frac{2k\rho}{\pi}\frac{T}{\hat{T}^2}$$

where k' and k are the two definitions of permeability discussed earlier.

In the above

$$\hat{T} = \frac{2L}{V_c}$$

where L is a typical length such as the length of the one-dimensional soil column under consideration, g is the gravitational acceleration,

$$V_c = \sqrt{\frac{D + (K_f/n)}{\rho}}$$

is the speed of sound, $\hat{T}$ is the natural vibration period and T is the period of excitation.

The second nondimensional parameter is defined as

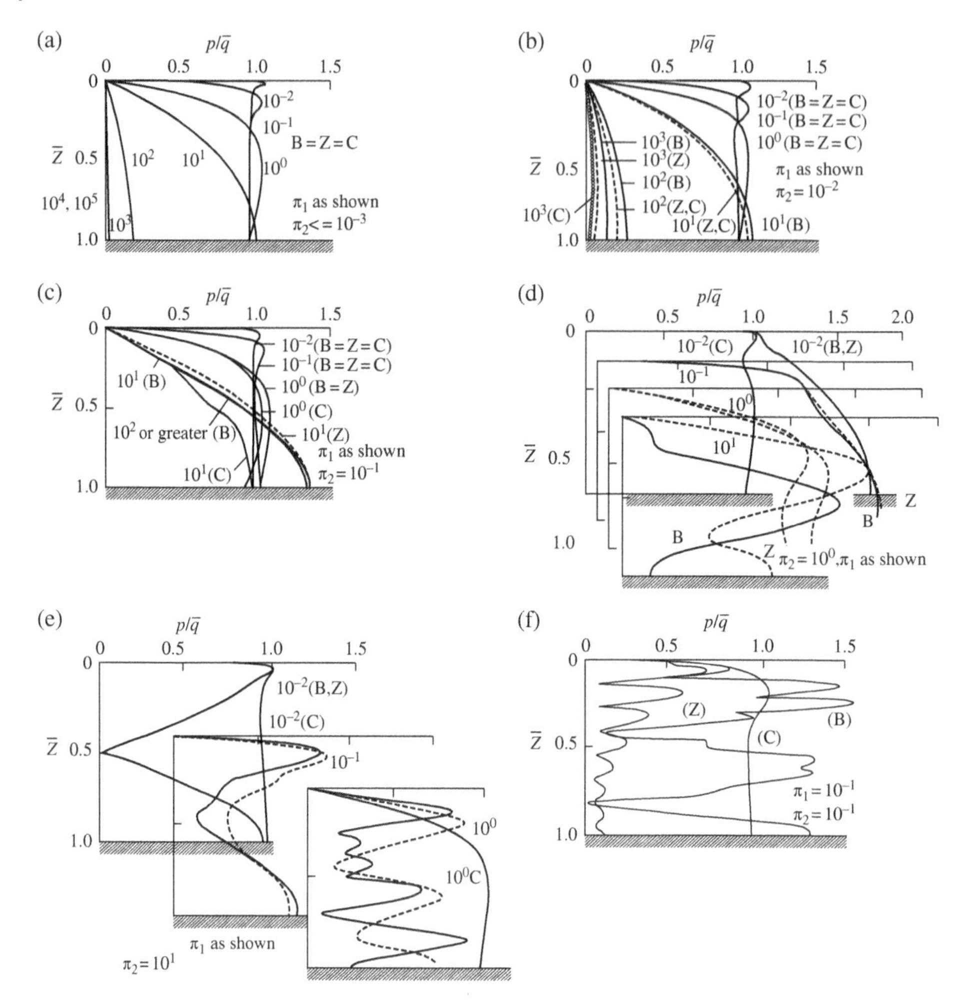

Figure 2.1 The soil column – variation of pore pressure with depth for various values of π_1 and π_2 ---- B (Biot theory) ---- Z (u–p approximation theory) -- C (Consolidation theory) (Solution (C) is independent of π_2). *Source:* Reproduced from Zienkiewicz et al. (1980) by permission of the Institution of Civil Engineers. (a) $\pi_2 \leq 10^{-3}$. (b) $\pi_2 = 10^{-2}$. (c) $\pi_2 = 10^{-1}$. (d) $\pi_2 = 10^{0}$. (e) $\pi_2 = 10^{1}$. (f) $\pi_1 = 10^{-1}$ $\pi_2 = 10^{2}$. Reproduced from Zienkiewicz (1980) by permission of the Institution of Civil Engineers.

$$\Pi_2 = \pi^2\left(\frac{\hat{T}}{T}\right)^2$$

In the study, the following values were assumed:

$$\beta \equiv \rho_f/\rho = 0.333, \quad n\,(\text{porosity}) = 0.333,$$

and

$$\kappa \equiv \frac{K_f/n}{D + \left(K_f/n\right)} = 0.973$$

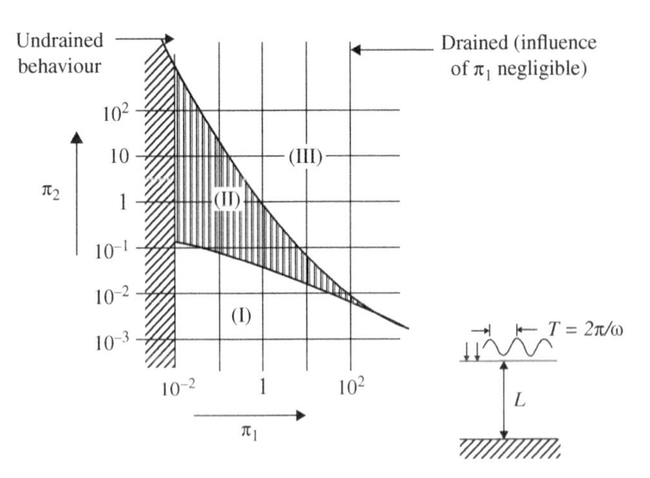

Figure 2.2 Zones of sufficient accuracy for various approximations: Zone 1, $B = Z = C$, slow phenomena ($\ddot{w}$ and $\ddot{u}$ can be neglected); Zone 2, $B = Z \neq C$, moderate speed ($\ddot{w}$ can be neglected); Zone 3, $B \neq Z \neq C$, fast phenomena ($\ddot{w}$ cannot be neglected, only full Biot equation valid)

Figure 2.2 summarizes the conclusions by indicating three zones in which various approximations are sufficiently accurate.

We note that, for instance, fully undrained behavior is applicable when $\Pi_1 < 10^{-2}$ and when $\Pi_1 > 10^2$, the drainage is so free that fully drained condition can be safely assumed.

To apply this table in practical cases, some numerical values are necessary. Consider, for instance, the problem of the earthquake response of a dam in which the typical length is characterized by the height $L = 50$ m, subject to an earthquake in which the important frequencies lie in the range

$$0.05 \text{ seconds} < T < 5 \text{seconds}$$

Thus, with the wave speed taken as

$$V_s = 1000\,\text{m/s}$$

we have

$$\hat{T} = \frac{2L}{V_s} = 0.1 \text{ seconds}$$

the parameter Π_2 is, therefore, in the range $3.9 \times 10^{-3} < \Pi_2 < 39$

$$\Pi_2 = \pi^2 \left(\frac{\hat{T}}{T}\right) = \pi^2 \left(\frac{0.1}{0.05 \sim 5}\right)^2 \approx 39 \sim 0.0039$$

and Π_1 is dependent on the permeability k with the range defined by

$$0.97k' < \Pi_1 < 97k'$$

$$\Pi_1 = \frac{2}{\beta\pi}\frac{k'}{g}\frac{T}{\hat{T}^2} = \frac{2k'\pi}{\beta g T}\frac{1}{\pi^2}\left(\frac{T}{\hat{T}}\right)^2 = \frac{2\pi}{\beta g}\frac{1}{\Pi_2 T}k' \approx \frac{2\pi}{\beta g}\left(\frac{1}{39\times 0.05 \sim 0.0039\times 5}\right)k' \approx (0.97 \sim 97)k'$$

According to Figure 2.2, we can, with reasonable confidence:

i) assume fully undrained behavior when $\Pi_1 = 97k' < 10^{-2}$ or the permeability $k' < 10^{-4}$ m/s.
ii) We can assume **u**–p approximation as being valid when $k' < 10^{-3}$ m/s to reproduce the complete frequency range. However, when $k' < 10^{-1}$ m/s, periods of less than 0.5 seconds are still well modeled.

We shall, therefore, typically use the **u**–p formulation appropriately in what follows reserving the full form for explicit transients where shocks and very high frequency are involved.

2.3 Partially Saturated Behavior with Air Pressure Neglected ($p_a = 0$)

2.3.1 Why Is Inclusion of Partial Saturation Required in Practical Analysis?

In the previous, fully saturated, analysis, we have considered both the water pore pressure and the solid displacement as problem variables. In the general case of nonlinear nature, which is characteristic of the problems of soil mechanics, both the effective stresses and pressures will have to be determined incrementally as the solution process (or computation) progresses step by step. In many soils, we shall encounter a process of "densification" implied in the constitutive soil behavior. This means that the history of straining (associated generally with shear strain) induces the solid matrix to contract (or the material to densify). Such densification usually will cause the pore pressure to increase, leading finally to a decrease of contact stresses in the soil particles to near-zero values when complete liquefaction occurs. Indeed, generally, failure will occur prior to the liquefaction limit. However, the reverse may occur where the soil "dilation" during the deformation history is imposed. This will imply the development of *negative pressures* which may reach substantial magnitudes. Such negative pressures cannot exist in reality without the presence of separation surfaces in the fluid which is contained in the pores, and consequent capillary effects. Voids will therefore open up during the process in the fluid which is essentially incapable of sustaining tension. This opening of voids will probably occur when zero pressure (or corresponding vapor pressure of water) is reached. Alternatively, *air will come out of the solution* – or indeed ingress from the free water surface if this is open to the atmosphere. The pressure will not then be vapor pressure but simply atmospheric.

We have shown in Chapter 1 that once voids open, a unique relationship exists between the degree of saturation S_w and the pore pressures p_w (see Figure 1.6). Using this relation, which can be expressed by formula or simply a graph, we can modify the equation used in

Section 2.2 to deal with the problem of partial saturation *without introducing any additional variables* assuming that the air throughout is at constant (atmospheric) pressure. Note that both phenomena of densification and dilation will be familiar to anybody taking a walk on a sandy beach after the tide has receded leaving the sand semi-saturated. First, one can note how when the foot is placed on the damp sand, the material appears to *dry* in the vicinity of the applied pressure. This obviously is the *dilation* effect. However, if the pressure is not removed but reapplied several times, the sand "*densifies*" and becomes quickly almost fluid. Clearly, liquefaction has occurred. It is surprising how much one can learn by keeping one's eyes open!

The presence of negative water pressures will, of course, increase the strength of the soil and thus have a beneficial effect. This is particularly true above the free water surface or the so-called *phreatic line*. Usually, one is tempted to assume simply a zero pressure throughout that zone but for non-cohesive materials, this means almost instantaneous failure under any dynamic load. The presence of negative pressure in the pores assures some cohesion (of the same kind which allows castles to be built on the beach provided that the sand is damp). This cohesion is essential to assure the structural integrity of many embankments and dams.

2.3.2 The Modification of Equations Necessary for Partially Saturated Conditions

The necessary modification of Equations (2.20) and (2.21) will be derived below, noting that generally we shall consider partial saturation only in the slower phenomena for which **u**–p approximation is permissible.

Before proceeding, we must note that the effective stress definition is modified and the effective pressure now becomes (viz Section 1.3.3)

$$p = \chi_w p_w + (1-\chi_w)p_a \approx \chi_w p_w \tag{2.24}$$

with the effective stress still defined by (2.1).

Equation (2.20) remains unaltered in form whether or not the material is saturated but the overall density ρ is slightly different now. Thus in place of (2.12), we can write

$$\rho = nS_w\rho_w + (1-n)\rho_s \tag{2.25}$$

neglecting the weight of air. The correction is obviously small and its effect insignificant.

However, (2.21) will now appear in a modified form which we shall derive here.

First, the water momentum equilibrium, Equation (2.13), will be considered. We note that its form remains unchanged but with the variable p being replaced by p_w. We thus have

$$-p_{w,i} - R_i - \rho_f\ddot{u}_i + \rho_f b_i = 0 \tag{2.26a}$$

$$-\nabla p_w - \boldsymbol{R} - \rho_f\ddot{\mathbf{u}} + \rho_f\mathbf{b} = 0 \tag{2.26b}$$

As before, we have neglected the relative acceleration of the fluid to the solid.

Equation (2.14), defining the permeabilities, remains unchanged as

$$k_{ij}R_j = w_i \tag{2.27a}$$

$$\mathbf{kR} = \mathbf{w} \tag{2.27b}$$

However, in general, only scalar, i.e. isotropic, permeability will be used here

$$k_{ij} = k\delta_{ij} \tag{2.28a}$$

$$\mathbf{k} = k\mathbf{I} \tag{2.28b}$$

where $\mathbf{I}$ is the identity matrix. The value of k is, however, dependent strongly on S_w and we note that:

$$k = k(S_w) \tag{2.29}$$

Such typical dependence is again shown in Figure 1.6.

Finally, the conservation Equation (2.16) has to be restructured, though the reader will recognize similarities.

The mass balance will once again consider the divergence of fluid flow $w_{i,i}$ to be augmented by terms previously derived (and some additional ones). These are

i) Increased pore volume due to change of strain assuming no change of saturation: $\delta_{ij}\mathrm{d}\varepsilon_{ij} = \mathrm{d}\varepsilon_{ii}$
ii) An additional volume stored by compression of the fluid due to fluid pressure increase: $nS_w\mathrm{d}p_w/K_f$
iii) Change of volume of the solid phase due to fluid pressure increase: $(1-n)\chi_w\mathrm{d}p_w/K_s$
iv) Change of volume of solid phase due to change of intergranular contact stress: $-K_T/K_s(\mathrm{d}\varepsilon_{ii} + \chi_w\mathrm{d}p_w/K_s)$
v) And a new term taking into account the change of saturation: $n\mathrm{d}S_w$

Adding to the above, as in Section 2.2, the terms involving density changes, on thermal expansion, the conservation equation now becomes:

$$\begin{aligned} & w_{i,i} + \alpha\dot{\varepsilon}_{ii} + S_w\frac{n}{K_f}\dot{p}_w + \frac{\alpha - n}{K_s}\chi_w\dot{p}_w + n\dot{S}_w + nS_w\frac{\dot{\rho}_w}{\rho_w} + \dot{s}_0 \\ & \equiv w_{i,i} + \alpha\dot{\varepsilon}_{ii} + \frac{\dot{p}_w}{Q^*} + nS_w\frac{\dot{\rho}_w}{\rho_w} + \dot{s}_0 = 0 \end{aligned} \tag{2.30a}$$

or

$$\nabla^T\mathbf{w} + \alpha\mathbf{m}^T\dot{\boldsymbol{\varepsilon}} + \frac{\dot{p}_w}{Q^*} + nS_w\frac{\dot{\rho}_w}{\rho_w} + \dot{s}_0 = 0 \tag{2.30b}$$

Now, however, Q^* is different from that given in Equation 2.17 and we have in its place

$$\frac{1}{Q^*} \equiv C_s + \frac{nS_w}{K_f} + \frac{(\alpha - n)\chi_w}{K_s} \tag{2.30c}$$

which, of course, must be identical with (2.17) when $S_w = 1$ and $\chi_w = 1$, i.e. when we have full saturation. The above modification is mainly due to an additional term to those defining the increased storage in (2.17). This term is due to the changes in the degree of saturation and is simply:

$$n\frac{\mathrm{d}S_w}{\mathrm{d}t} \tag{2.31}$$

but here we introduce a new parameter C_S defined as

$$n\frac{\mathrm{d}S_w(p_w)}{dt} = n\frac{\mathrm{d}S_w(p_w)}{\mathrm{d}p_w}\frac{\mathrm{d}p_w}{\mathrm{d}t} = C_s\dot{p}_w \tag{2.32}$$

The final elimination of **w** in a manner identical to that used when deriving (2.21) gives (neglecting density variation):

$$\left(k_{ij}\left(-p_{w,\ j} - S_w\rho_f\ddot{u}_j + S_w\rho_f b_j\right)\right)_{,i} + \alpha\dot{\varepsilon}_{ii} + \frac{\dot{p}}{Q^*} + \dot{s}_0 = 0 \tag{2.33a}$$

or

$$\nabla^{\mathrm{T}}\mathbf{k}\left(-\nabla p_w - S_w\rho_f\ddot{\mathbf{u}} + S_w\rho_f\mathbf{b}\right) + \alpha\mathbf{m}^T\dot{\boldsymbol{\varepsilon}} + \frac{\dot{p}}{Q^*} + \dot{s}_0 = 0 \tag{2.33b}$$

The small changes required here in the solution process are such that we found it useful to construct our computer program for the partially saturated form, with the fully saturated form being a special case.

In the time-stepping computation, we still always assume that the parameters S_w, k_w, and C_s change slowly and hence we will compute these at the start of the time interval keeping them subsequently constant.

Previously, we mentioned several typical cases where pressure can become negative and hence saturation drops below unity. One frequently encountered example is that of the flow occurring in the capillary zone during *steady-state seepage*. The solution to the problem can, of course, be obtained from the general equations simply by neglecting all acceleration and fixing the solid displacements at zero (or constant) values.

If we consider a typical dam or a water-retaining embankment shown in Figure 2.3, we note that, on all the surfaces exposed to air, we have apparently incompatible boundary conditions. These are:

$$p_w = 0 \text{ and } w_n = 0 \text{ (i.e.net zero inflow)}$$

Clearly, both conditions cannot be simultaneously satisfied and it is readily concluded that only the second is true above the area where the flow emerges. Of course, when the flow leaves the free surface, the reverse is true.

Computation will easily show that negative pressures develop near the surface and that, therefore, a partially saturated zone with very low permeability must exist. The result of such a computation is shown in Figure 2.3 and indeed it will be found that very little flow occurs above the zero-pressure contour. This contour is, in fact, the well-known Phreatic line and the partially saturated material procedure has indeed been used frequently purely as a numerical device for its determination (see Desai 1977a, 1977b; Desai and Li 1983 etc.). Another example is given in Figure 2.4. Here a numerical solution of Zienkiewicz et al. (1990b) is given for a problem for which experimental data are available from Liakopoulos (1965).

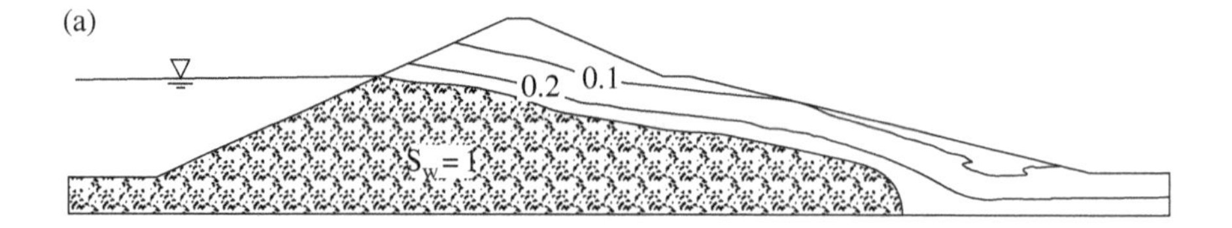

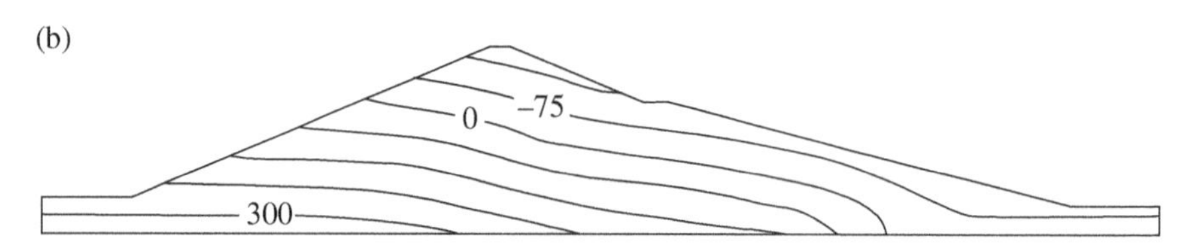

Figure 2.3 A partially saturated dam. Initial steady-state solution. Only saturation (a) and pressure contours (b) are shown. Contour interval in (b) is 75 kPa. The Phreatic line is the boundary of the fully saturated zone in (a)

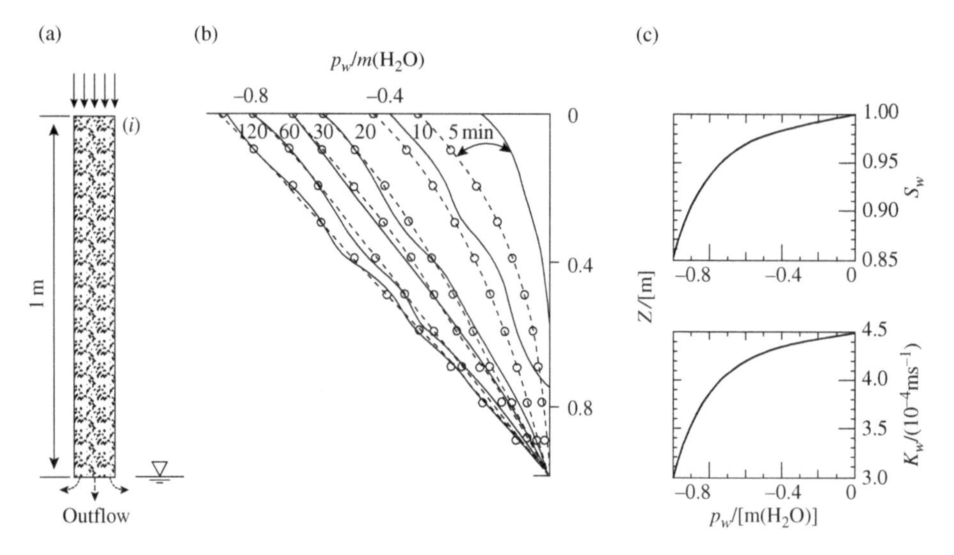

Figure 2.4 Test example of partially saturated flow experiment by Liakopoulos (1965). (a) Configuration of test (uniform inflow interrupted at t = 0); (b) pressures with – – –, computed; ———, recorded; (c) data (linear elastic analysis, E = 3000 kPa). *Source:* From Liakopoulos (1965)

In the practical code used for earthquake analysis, we shall use this partially saturated flow to calculate a wide range of soil mechanics phenomena. However, for completeness in Section 2.4, we shall show how the effects of air movement can be incorporated into the analysis.

2.4 Partially Saturated Behavior with Air Flow Considered ($p_a \geq 0$)

2.4.1 The Governing Equations Including Air Flow

This part of the chapter is introduced for completeness – though the effects of the air pressure are insignificant in most problems. However, in some cases of consolidation and confined materials, the air pressures play an important role and it is useful to have means for their prediction. Further, the procedures introduced are readily applicable to other pore–fluid mixtures. For instance, the simultaneous presence of water and oil is important in some areas of geomechanics and coupled problems are of importance in the treatment of hydrocarbon reservoirs. The procedures used in the analysis follow precisely the same lines as introduced here.

In particular, the treatment following the physical approach used in this chapter has been introduced by Simoni and Schrefler (1991), Li et al. (1990) and Schrefler and Zhan (1993) for the flow of water with air.

The alternative approach of using the mixture theory in these problems was outlined by Li and Zienkiewicz (1992) and Schrefler (1995).

Some simple considerations will allow the basic equations for the dynamics of the soil containing two pore fluids to be derived. They have been solved by Schrefler and Scotta (2001) and an example will be shown in Section 8.5.

2.4.2 The Governing Equation

The dynamics of the total mixture can, just as in Section 2.3, be written in precisely the same form as that for a single fluid phase (see (2.11)). For completeness, we repeat that equation here (now, however, *a priori* omitting the small convective terms)

$$\sigma_{ij,\ j} - \rho\ddot{u}_i + \rho b_i = 0 \tag{2.34a}$$

$$\mathbf{S}^T\boldsymbol{\sigma} - \rho\ddot{\mathbf{u}} + \rho\mathbf{b} = 0 \tag{2.34b}$$

However, just as in Equation (2.25), we have to write

$$\rho = nS_w\rho_w + n(1 - S_w)\rho_a + (1 - n)\rho_s \tag{2.35}$$

noting that

$$S_a = 1 - S_w$$

For definition of effective stress, we again use (2.24) now, however, without equating the air pressure to zero, i.e. writing

$$p = \chi_w p_w + (1 - \chi_w)p_a \tag{2.36}$$

For the flow of water and air, we can write the Darcy equations separately, noting that

$$k_w R_i^w = w_i \tag{2.37a}$$

$$k_w \mathbf{R}^w = \mathbf{w} \tag{2.37b}$$

for water as in (2.27) and for the flow of air:

$$k_a R_i^a = v_i \tag{2.38a}$$

$$k_a \mathbf{R}^a = \mathbf{v} \tag{2.38b}$$

Here we introduced appropriate terms for coefficients of permeability for water and air, while assuming isotropy. A new variable **v** now defines the air velocity.

The approximate momentum conservation Equation (see 2.13) can be rewritten in a similar manner using isotropy but omitting acceleration terms for simplicity. We therefore have for water

$$w_i = k_w\left(-p_{w,i} + \rho_w b_i\right) \tag{2.39a}$$

$$\mathbf{w} = k_w(-\nabla p_w + \rho_w \mathbf{b}) \tag{2.39b}$$

and for air

$$v_i = k_a\left(-p_{a,i} + \rho_a b_i\right) \tag{2.40a}$$

$$\mathbf{v} = k_a(-\nabla p_a + \rho_a \mathbf{b}) \tag{2.40b}$$

Finally, the mass balance equations for both water and air have to be written. These are derived in a manner identical to that used for Equation (2.30). Thus, for water, we have

$$w_{i,i} + \alpha S_w \dot{\varepsilon}_{ii} + S_w \frac{n}{K_w} \dot{p}_w + \frac{\alpha - n}{K_s} \chi_w \dot{p}_w + n\dot{S}_w + nS_w \frac{\dot{\rho}_w}{\rho_w} + \dot{s}_0 = 0 \tag{2.41a}$$

$$\nabla^T \mathbf{w} + \alpha S_w m^T \dot{\boldsymbol{\varepsilon}} + S_w \frac{n}{K_w} \dot{p}_w + \frac{\alpha - n}{K_s} \chi_w \dot{p}_w + n\dot{S}_w + nS_w \frac{\dot{\rho}_w}{\rho_w} + \dot{s}_0 = 0 \tag{2.41b}$$

and for air

$$v_{i,i} + \alpha S_a \dot{\varepsilon}_{ii} + S_a \frac{n}{K_w} \dot{p}_a + \frac{\alpha - n}{K_s} \chi_a \dot{p}_a + n\dot{S}_a + nS_a \frac{\dot{\rho}_a}{\rho_a} + \dot{s}_0 = 0 \tag{2.42a}$$

or

$$\nabla^T \mathbf{v} + \alpha S_a \mathbf{m}^T \dot{\boldsymbol{\varepsilon}} + S_a \frac{n}{K_a} \dot{p}_a + \frac{\alpha - n}{K_s} \chi_a \dot{p}_a + n\dot{S}_a + nS_a \frac{\dot{\rho}_a}{\rho_a} + \dot{s}_0 = 0 \tag{2.42b}$$

Now, in addition to the solid phase displacement $u_i(\mathbf{u})$, we have to consider the water pressure p_w and the air pressure p_a as independent variables.

However, we note that now (see (1.21))

$$p_a - p_w = p_c \tag{2.43}$$

and that the relation between p_c and S_W is unique and of the type shown in Figure 1.6. p_c now defines S_W and from the fact that

$$S_w + S_a = 1 \quad \text{and} \quad \chi_w + \chi_a = 1 \tag{2.44}$$

air saturation can also be found.

We have now the complete equation system necessary for dealing with the flow of air and water (or any other two fluids) coupled with the solid phase deformation.

2.5 Alternative Derivation of the Governing Equation (of Sections 2.2–2.4) Based on the Hybrid Mixture Theory

It has already been indicated in Section 2.1 that the governing equations can also be derived using mixture theories. The classical mixture theories (viz. Green 1969; Morland 1972; Bowen 1980, 1982) start from the macro-mechanical level, i.e. the level of interest for our computations, while the so-called hybrid mixture theories (viz. Whitaker 1977; Hassanizadeh and Gray 1979a, 1979b, 1980) start from micro-mechanical level. The equations at the macro-mechanical level are then obtained by spatial averaging procedures. Further, there exists a macroscopic thermodynamical approach to Biot's theory proposed by Coussy (1995). All these theories lead to a similar form of the balance equations. This was in particular shown by de Boer et al. (1991) for the mixture theory, the hybrid mixture theories and the classical Biot's theory.

The theories differ in the constitutive equations, usually obtained from the entropy inequality. This is here shown in particular for the effective stress principle because this was extensively discussed in Chapter 1. Let's consider first the fully saturated case. Runesson (1978) shows, for instance, that the principle of effective stress follows from the mixture theory under the assumption of incompressible grains. This means that α in Equation (1.16a) or (2.1a)

$$\sigma''_{ij} = \sigma_{ij} + \alpha\delta_{ij}p \tag{2.45a}$$

has to be equal to one, which results in

$$\sigma'_{ij} = \sigma_{ij} + \delta_{ij}p \tag{2.45b}$$

Only in this case, the two formulations given by Biot's theory and the mixture theory coincide.

In the hybrid mixture theories, the concept of the solid pressure p_s, i.e. the pressure acting on the solid grains is introduced (viz. Hassanizadeh and Gray 1990). From the application of the second principle of thermodynamics, it appears that under conditions close to thermodynamic equilibrium

$$\frac{p_s}{\theta_s} - \frac{p_w}{\theta_w} = -\bar{k}\dot{\varepsilon}_{ii} \tag{2.46}$$

where $\bar{k}$ is a coefficient and θ_s and θ_w the absolute temperatures of solid and water, respectively. Under the assumption of local thermodynamic equilibrium, where temperatures are equal and rate terms vanish, it follows that

$$p_s = p_w = p \tag{2.47}$$

and the effective stress principle in the form of (2.45b) can be derived following Hassanizadeh and Gray (1990). Equation (2.47) holds also under nonequilibrium conditions; it has however to be assumed that the solid grains remain incompressible, the same assumption as with the mixture theory.

Let us consider now the partially saturated case. There are again several conflicting expressions in literature. The first expression for partially saturated soils was developed by Bishop (1959) and may be written as follows, see Equations (2.36), (2.24), and (1.19)

$$\sigma'_{ij} = \sigma_{ij} + \delta_{ij}p_s = \sigma_{ij} + \delta_{ij}(p_a - \chi_w(p_a - p_w)) \tag{2.48}$$

where χ_w is the Bishop parameter, usually a function of the degree of saturation, see (1.22a) and (1.22b). The same expression, but with Bishop's parameter equal to the water degree of saturation was derived by Lewis and Schrefler (1982, 1987) using volume averaging. Hassanizadeh and Gray (1990) find for the partially saturated case under the assumption of the following form of the Helmholtz free energy for the solid $A^s = A^s(\rho^s, \varepsilon_{ij}, \theta^s, S_w)$ that

$$\frac{p_s}{\theta_s} - \left(\frac{S_a p_a}{\theta_a} + \frac{S_w p_w}{\theta_w}\right) = \bar{k}\dot{\varepsilon}_{ii} \tag{2.49}$$

which considering thermodynamic equilibrium conditions or nonequilibrium conditions but incompressible solid grains reduces to

$$p_s = S_a p_a + S_w p_w \tag{2.50}$$

Taking into account that $S_a + S_w = 1$, the solid pressure (2.50) coincides with that of Equation (2.48) if $\chi_w = S_w$, which is often the case in soil mechanics as shown by Nuth and Laloui (2008). The effective stress can then simply be written as

$$\sigma'_{ij} = \sigma_{ij} + \delta_{ij}(p_a - S_w p_c) \tag{2.51}$$

Coussy (1995) obtains under the assumption of a simpler form of the functional dependence of the Helmholtz free energy of the solid phase $A^s = A^s(\varepsilon_{ij}, \theta^s, S_w)$ as above:

$$\mathrm{d}\sigma'_{ij} = \mathrm{d}\sigma_{ij} + \delta_{ij}(\mathrm{d}p_a - S_w \mathrm{d}p_c) \tag{2.52}$$

where $\mathrm{d}p_c = \mathrm{d}p_a - \mathrm{d}p_w$ is the capillary pressure increment. This equation has an incremental form and differs substantially from the previous ones, i.e. it is not an exact differential and its use in soil dynamics is not straightforward because the solid pressure-like term has to be integrated in each time step even with a linear elastic effective stress-strain relationship, being the capillary pressure-saturation relationship in general nonlinear, see Figure 1.6. The practical implication of these different formulations for slow phenomena has been investigated in detail by Schrefler and Gawin (1996). It was concluded that in many soil mechanics situations, the resulting differences are small and appear usually after long-lasting variations of the moisture content. Only several cycles of drying and wetting would produce significant differences. The new stress tensor of Coussy (2004) in finite form coincides with (2.52) when passing to the differential form.

The thermodynamic consistency of the different formulations has been investigated in Gray and Schrefler (2001, 2007) and Borja (2004): (2.45b), (2.51), and (2.52) are thermodynamically consistent while the conditions under which (2.45a) and (2.49) are consistent are given in Gray and Schrefler (2007).

For the sake of completeness, we recall two other formulations of the stress tensor in partially saturated soils which are currently used.

A previously commonly used form of a stress tensor in partially saturated soil mechanics is the net stress, introduced by Fredlund and Morgenstern (1977). The net stress is defined as the difference between the total stress and the air pressure (no assumption is needed for the grain compressibility)

$$\boldsymbol{\sigma}' = \boldsymbol{\sigma} + \mathbf{m}p^a \tag{2.53}$$

Its success is due to a great part to experimental reasons: in many problems, the air pressure may be considered constant and equal to the atmospheric pressure and in laboratory experiments, it is preferable to vary water pressure. However, it has to be made clear that the choice of the stress variables in constitutive modeling is a different problem from the choice of controlling variables in the experimental investigation (Jommi 2000). The transformation of the experimentally measured quantities is straightforward from the net stress to the form (2.51), see (Bolzon and Schrefler 1995).

Its drawback lies in the fact that in the presence of saturated and unsaturated states, the stress tensor changes between the two states; when the pore air is absent, the constitutive equations for saturated states cannot be recovered from those for unsaturated states without additional control (Sheng et al. 2004). This precludes practically its application in soil dynamics; capturing liquefaction becomes a problem.

The compressibility of the solid grains was also considered by Khalili et al. (2000) in their stress tensor

$$\boldsymbol{\sigma} = \boldsymbol{\sigma}' - a_1\mathbf{m}p^g - a_2\mathbf{m}p^w \tag{2.54}$$

where a_1, a_2 are the effective stress parameters defined as $a_1 = \frac{c_m}{c} - \frac{c_s}{c}$, $a_2 = 1 - \frac{c_m}{c}$, c_s is the grain compressibility, c the drained compressibility of the soil structure, and c_m the tangent compressibility of the soil structure with respect to a change in capillary pressure.

In the following, we make use of (2.51) with the assumption that solid grains are incompressible, i.e. $\alpha = 1$. The governing equations are derived again, using the hybrid mixture theory, as has been done by Schrefler (1995) and Lewis and Schrefler (1998). Isothermal conditions are assumed to hold, as throughout this book. For the full non-isothermal case, the interested reader is referred to Lewis and Schrefler (1998) and Schrefler (2002).

We first recall briefly the kinematics of the system.

2.5.1 Kinematic Equations

As indicated in Chapter 1, a multiphase medium can be described as the superposition of all π phases, $\pi = 1, 2, \ldots, \kappa$, i.e. in the current configuration, each spatial point $\mathbf{x}$ is simultaneously occupied by material points $\mathbf{X}^\pi$ of all phases. The state of motion of each phase is however described independently.

In a Lagrangian or material description of motion, the position of each material point $\mathbf{x}^\pi$ at time t is a function of its placement in a chosen reference configuration, $\mathbf{X}^\pi$ and of the current time t

$$\mathrm{x}_i^\pi = \mathrm{x}_i^\pi\left(\mathrm{X}_1^\pi, \mathrm{X}_2^\pi, \mathrm{X}_3^\pi, \mathrm{t}\right) \tag{2.55}$$

To keep this mapping continuous and bijective at all times, the determinant of the Jacobian of this transformation must not equal zero and must be strictly positive, since it is equal to the determinant of the deformation gradient tensor $\mathbf{F}^{\pi}$

$$F^{\pi}_{ij} = x^{\pi}_{i,j} = R^{\pi}_{ik}\, U^{\pi}_{kj} = V^{\pi}_{ik}\, R^{\pi}_{kj} \tag{2.56}$$

where $\mathbf{U}^{\pi}$ is the right stretch tensor, $\mathbf{V}^{\pi}$ the left stretch tensor, and the skew-symmetric tensor $\mathbf{R}^{\pi}$ gives the rigid body rotation. Differentiation with respect to the appropriate coordinates of the reference or actual configuration is respectively denoted by comma or slash, i.e.

$$\frac{\partial f}{\partial X_i} = f_{,i} \quad \text{or} \quad \frac{\partial f}{\partial x_i} = f_{/i} \tag{2.57}$$

Because of the non-singularity of the Lagrangian relationship (2.55), its inverse can be written and the Eulerian or spatial description of motion follows

$$X^{\pi}_{i} = X^{\pi}_{i}\left(X^{\pi}_{1}, X^{\pi}_{2}, X^{\pi}_{3}, t\right) \tag{2.58}$$

The material time derivative of any differentiable function $f^{\pi}(\mathbf{X}, t)$ given in its spatial description and referred to a moving particle of the π phase is

$$\frac{\overset{\pi}{D} f^{\pi}}{Dt} = \frac{\partial f^{\pi}}{\partial t} + f^{\pi}_{/i} v^{\pi}_{i} \tag{2.59}$$

If superscript α is used for $\frac{\overset{\alpha}{D}}{Dt}$, the time derivative is taken moving with the α phase.

2.5.2 Microscopic Balance Equations

In the hybrid mixture theories, the microscopic situation of any π phase is first described by the classical equations of continuum mechanics. At the interfaces to other constituents, the material properties and thermodynamic quantities may present step discontinuities. As throughout the book, the effects of the interfaces are here not taken into account explicitly. These are introduced, e.g. in Schrefler (2002) and Gray and Schrefler (2001, 2007).

For a thermodynamic property ψ, the conservation equation within the π phase may be written as

$$\frac{\partial(\rho\ \psi)}{\partial t} + \rho\ \psi \dot{r}_{i/i} - i_{i/i} - \rho\ g_i = \rho\ G_i \tag{2.60}$$

where $\dot{\mathbf{r}}$ is the local value of the velocity field of the π phase in a fixed point in space, $\mathbf{i}$ is the flux vector associated with Ψ, $\mathbf{g}$ the external supply of Ψ, and $\mathbf{G}$ is the net production of Ψ. The relevant thermodynamic properties Ψ are mass, momentum, energy, and entropy. The values assumed by $\mathbf{i}$, $\mathbf{g}$, and $\mathbf{G}$ are given in Table 2.2 (Hassanizadeh and Gray 1980, 1990; Schrefler 1995). The constituents are assumed to be microscopically nonpolar; hence, the angular momentum balance equation has been omitted. This equation shows, however, that the stress tensor is symmetric.

Table 2.2 Thermodynamic properties for the microscopic mass balance equations.

Quantity	ψ	i	g	G
Mass	1	0	0	0
Momentum	$\dot{\mathbf{r}}$	$\mathbf{t}_m$	$\mathbf{g}$	0
Energy	$E + 0.5\,\dot{\mathbf{r}}\cdot\dot{\mathbf{r}}$	$\mathbf{t}_m\,\dot{\mathbf{r}} - \mathbf{q}$	$\mathbf{g}\cdot\dot{\mathbf{r}} + \mathbf{h}$	0
Entropy	Λ	Φ	S	φ

Sources: Adapted from Hassanizadeh and Gray (1980, 1990), and Schrefler (1995).

2.5.3 Macroscopic Balance Equations

For isothermal conditions, as here assumed, the macroscopic balance equations for mass, linear momentum, and angular momentum are then obtained by systematically applying the averaging procedures to the microscopic balance Equation (2.60) as outlined in Hassanizadeh and Gray (1979a, 1979b, 1980). The balance equations have here been specialized for a deforming porous material, where the flow of water and of moist air (mixture of dryair and vapor) is taking place (see Schrefler 1995).

The local thermodynamic equilibrium hypothesis is assumed to hold because the time scale of the modeled phenomena is substantially larger than the relaxation time required reaching equilibrium locally. The temperatures of each constituent in a generic point are hence equal. Further, the constituents are assumed to be immiscible and chemically non-reacting. All fluids are assumed to be in contact with the solid phase. As throughout this book, stress is defined as tension positive for the solid phase, while pore pressure is defined as compressively positive for the fluids.

In the averaging procedure, the volume fractions η^{π} appear which are identified as follows: for solid phase $\eta^{s} = 1 - n$, for water $\eta^{w} = nS_{w}$, and for air $\eta^{a} = nS_{a}$.

The averaged macroscopic mass balance equations are given next. For the solid phase, this equation reads

$$\frac{\overset{s}{D}\,[(1-n)\rho^{s}]}{Dt} + \rho^{s}(1-n)\dot{u}_{i/i} = 0 \tag{2.61}$$

where $\dot{\boldsymbol{u}}$ is the mass averaged solid phase velocity and ρ^{π} is the intrinsic phase averaged density. The intrinsic phase averaged density ρ^{π} is the density of the π phase averaged over the part of the control volume (Representative Elementary Volume, REV) occupied by the π phase. The phase averaged density ρ_{π}, on the contrary, is the density of the π phase averaged over the total control volume. The relationship between the two densities is given by

$$\rho_{\pi} = \eta_{\pi}\rho^{\pi} \tag{2.62}$$

For water, the averaged macroscopic mass balance equation reads

$$\frac{\overset{w}{D}\,(nS_{w}\rho^{w})}{Dt} + nS_{w}\rho^{w}\, v^{w}_{i/i} = nS_{w}\rho^{w}e^{w}(\rho) \tag{2.63}$$

where $nS_w\rho^w e^w(\rho) = -\dot{m}$ is the quantity of water per unit time and volume, lost through evaporation and $\mathbf{v}^w$ the mass averaged water velocity.

For air, this equation reads

$$\frac{\overset{a}{D}(nS_a\rho^a)}{Dt} + nS_a\rho^a v^a_{i/i} = \dot{m} \tag{2.64}$$

where $\mathbf{v}^a$ is the mass averaged air velocity.

The linear momentum balance equation for the fluid phases is

$$t^\pi_{ji/j} + \rho_\pi(b^\pi_i - a^\pi_i) + \rho_\pi[e^\pi(\rho\dot{r}^\pi_i) + \hat{t}^\pi_i] = 0 \tag{2.65}$$

where t^π is the partial stress tensor, $\rho_\pi\mathbf{b}^\pi$ the external momentum supply due to gravity, $\rho_\pi\mathbf{a}^\pi$ the volume density of the inertia force, $\rho_\pi e^\pi(\rho\,\dot{\mathbf{r}})$ the sum of the momentum supply due to averaged mass supply, and the intrinsic momentum supply due to a change of density and referred to the deviation $\dot{\mathbf{r}}^\pi$ of the velocity of constituent π from its mass averaged velocity, and $\hat{\mathbf{t}}^\pi$ accounts for exchange of momentum due to mechanical interaction with other phases. $\rho_\pi e^\pi(\rho\,\dot{\mathbf{r}})$ is assumed to be different from zero only for fluid phases. For the solid phase, the linear momentum balance equation is hence

$$t^s_{ji/j} + \rho^s(b^s_i - a^s_i) + \rho^s\hat{t}^s_i = 0 \tag{2.66}$$

The average angular momentum balance equation shows that for nonpolar media, the partial stress tensor is symmetric $t^\pi_{ji} = t^\pi_{tj}$ at the macroscopic level also and the sum of the coupling vectors of angular momentum between the phases vanishes.

2.5.4 Constitutive Equations

Constitutive models are selected here which are based on quantities currently measurable in laboratory or field experiments and which have been extensively validated. Most of them have been obtained from entropy inequality; see Hassanizadeh and Gray (1980, 1990).

It can be shown that the stress tensor in the fluid is

$$t^\pi_{ij} = -\eta^\pi p_\pi\delta_{ij} \tag{2.67}$$

where p^π is the fluid pressure, and in the solid phase is

$$t^s_{ij} = \sigma'_{ij} - (1-n)p_s\delta_{ij} \tag{2.68}$$

with $p_s = p_wS_w + p_aS_a$ in the case of thermodynamic equilibrium or for incompressible solid grains (2.50).

The sum of (2.67), written for air and water and of (2.68), gives the total stress σ, acting on a unit area of the volume fraction mixture

$$\sigma_{ij} = \sigma'_{ij} - (S_wp_w + S_ap_a)\delta_{ij} \tag{2.69}$$

This is the form of the effective stress (2.51), also called generalized Bishop stress (Nuth and Laloui 2008), employed in the following, as already explained.

Moist air in the pore system is assumed to be a perfect mixture of two ideal gases, dry air and water vapour, with $\pi = ga$ and $\pi = gw$, respectively. The equation of a perfect gas is hence valid

$$p_{ga} = \rho^{ga}\theta\frac{R}{M_a} \quad p_{gw} = \rho^{gw}\theta\frac{R}{M_w} \tag{2.70}$$

where M_π is the molar mass of constituent π, R the universal gas constant, and θ the common absolute temperature. Further, Dalton's law applies and yields the molar mass of moisture

$$\rho^a = \rho^{ga} + \rho^{gw}, \quad p_a = p_{ga} + p_{gw}, \quad M_g = \left(\frac{\rho^{gw}}{\rho^a}\frac{1}{M_w} + \frac{\rho^{ga}}{\rho^a}\frac{1}{M_a}\right)^{-1} \tag{2.71}$$

Water is usually present in the pores as a condensed liquid, separated from its vapor by a concave meniscus because of surface tension. The capillary pressure is defined as $p_c = p_g - p_w$, see Equation (2.43).

The momentum exchange term of the linear momentum balance equation for fluids has the form

$$\rho_\pi \hat{t}_i^\pi = -R_{ij}^\pi \eta^\pi v_j^{\pi s} + p^\pi \eta^\pi_{/i} \tag{2.72}$$

where $\mathbf{v}^{\pi s}$ is the velocity of the π phase relative to the solid.

It is assumed that $\mathbf{R}^\pi$ is invertible, its inverse being $(\mathbf{R}^\pi)^{-1} = \frac{\mathbf{k}^\pi}{\eta^\pi}$ and $\mathbf{k}^\pi$ is defined by the following relation:

$$k_{ij}^\pi = \frac{K_{ij}}{\mu^\pi}(\rho^\pi, \eta^\pi, T) \tag{2.73}$$

where μ_π is the dynamic viscosity with dimensions [mass] $[\text{length}]^{-1}$ $[\text{time}]^{-1}$, K the intrinsic permeability $[\text{length}]^2$ and T the temperature above some datum. The permeability of (2.73) is the dynamic permeability $[\text{length}]^3$ $[\text{mass}]^{-1}$ [time]; to obtain the soil mechanics permeability k' [length]/[time], Equation (2.73) has to be multiplied by the specific weight of water γ_w of dimensions [mass] $[\text{length}]^{-2}$ $[\text{time}]^{-2}$. In the case of more than one fluid flowing, the intrinsic permeability is modified as

$$K_{ij}^\pi = K^{r\pi} K_{ij} \tag{2.74}$$

where $K^{r\pi}$ is the relative permeability, a function of the degree of saturation. For the water density, the following holds:

$$\frac{1}{\rho^w}\frac{D^w\rho^w}{Dt} = \frac{1}{K_w}\frac{D^w p^w}{Dt} \tag{2.75}$$

where K_w is the bulk modulus of water.

2.5.5 General Field Equations

The macroscopic balance laws are now transformed and the constitutive equations introduced, to obtain the general field equations.

The linear momentum balance equation for the fluid phases is obtained first. In Equation (2.65), the fluid acceleration is expressed, taking into account Equation (2.59), and introducing the relative fluid acceleration $\mathbf{a}^{\pi s}$. Further, Equations (2.67) and (2.68) are introduced. The terms dependent on the gradient of the fluid velocity and the effects of phase change are neglected and a vector identity for the divergence of the stress tensor is used. Finally, Equations (2.73) and (2.74) are included, yielding

$$\eta^{\pi} v_i^{\pi s} = \frac{K_{ij}K^{r\pi}}{\mu^{\pi}}\left[-p_{\pi/j} + \rho^{\pi}\left(g_j - a_j^s - a_j^{\pi s}\right)\right] \tag{2.76}$$

The linear momentum balance equation for the solid phase is obtained in a similar way, taking into account Equations (2.68) instead of (2.67).

By summing this momentum balance equation with Equation (2.76) written for water and air and by taking into account the definition of total stress (2.69), assuming continuity of stress at the fluid-solid interfaces and by introducing the averaged density of the multiphase medium

$$\rho = (1-n)\rho^s + nS_w\rho^w + nS_a\rho^a \tag{2.77}$$

we obtain the linear momentum balance equation for the whole multiphase medium

$$-\rho\ \ddot{u}_i - nS_w\rho^w\left(a_i^{ws} + v_k^{ws}v_{k/i}^w\right) - nS_a\rho^a\left(a_i^{as} + v_k^{as}v_{k/i}^a\right) + \sigma_{ji/j} + \rho b_i = 0 \tag{2.78}$$

The mass balance equations are derived next.

The macroscopic mass balance equation for the solid phase (2.61), after differentiation and dividing by ρ^s is obtained as

$$\frac{1-n}{\rho^s}\frac{\overset{s}{D}\rho^s}{Dt} - \frac{\overset{s}{D}n}{Dt} + (1-n)\dot{u}_{i/i} = 0 \tag{2.79}$$

This equation is used in the subsequent mass balance equations to eliminate the material time derivative of the porosity. For incompressible grains, as assumed here, $\frac{\overset{s}{D}\rho^s}{Dt} = 0$. For compressible grains, see Equation (2.89) and related remarks.

The mass balance equation for water (2.63) is transformed as follows. First in Equation (2.75), the material time derivative of the water density with respect to the moving solid phase and the relative velocity $\mathbf{v}^{ws}$ are introduced. Then the derivatives are carried out, the quantity of water lost through evaporation is neglected and the material time derivative of the porosity is expressed through Equation (2.79), yielding

$$n\rho^w\frac{\overset{s}{D}S_w}{Dt} + \rho^w S_w\dot{u}_{i/i} + \frac{n\rho^w S_w}{K_w}\frac{\overset{s}{D}p_w}{Dt} + \left(n\rho^w S_w v_i^{ws}\right)_{/i} = 0 \tag{2.80}$$

The mass balance equation for air is derived in a similar way

$$n\frac{\overset{s}{D}S_a}{Dt} + S_a\dot{u}_{i/i} + \frac{S_a n}{\rho^a}\frac{\overset{s}{D}}{Dt}\left(\frac{M_g}{\theta R}p_a\right) + \frac{1}{\rho^a}\left(nS_a\rho^a v_i^{as}\right)_{/i} = 0 \tag{2.81}$$

To obtain the equations of Section 2.4.2, further simplifications are needed, which are introduced next.

An updated Lagrangian framework is used where the reference configuration is the last converged configuration of the solid phase. Further, the strain increments within each time step are small. Because of this, we can neglect the convective terms in all the balance equations. Neglecting in the linear momentum balance Equation (2.78) further the relative accelerations of the fluid phases with respect to the solid phase yields the equilibrium Equation (2.34a)

$$\sigma_{ij,j} - \rho \ddot{u}_i + \rho b_i = 0 \tag{2.82}$$

The linear momentum balance equation for fluids (2.76) by omitting all acceleration terms, as in Section 2.2.2, can be written for water

$$w_i = k_{wij}\left(- p_{w_{,j}} + \rho^w b_j \right) \tag{2.83}$$

where

$$\eta^w v_i^{ws} = w_i$$

and

$$k_{wij} = \frac{K_{ij} K^{rw}}{\mu^w} \tag{2.84}$$

and for air

$$v_i = k_{aij}\left(- p_{a_{,j}} + \rho^a b_j \right) \tag{2.85}$$

where

$$\eta^a v_i^{as} = v_i$$

and

$$k_{aij} = \frac{K_{ij} K^{ra}}{\mu^a} \tag{2.86}$$

The phase densities appearing in Sections 2.2–2.4 are intrinsic phase averaged densities as indicated above.

The mass balance equation for water is obtained from Equation (2.80), taking into account the reference system chosen, dividing by ρ^w, developing the divergence term of the relative velocity and neglecting the gradient of water density. This yields

$$n\dot{S}_w + S_w \dot{u}_{i,i} + S_w \frac{n}{K^w} \dot{p}_w + w_{i,i} = 0 \tag{2.87}$$

where the first of Equation (2.84) has been taken into account. This coincides with Equation (2.41a) for incompressible grains ($\alpha = 1$) except for the source term and the second-order term due to the change in fluid density. This last one could be introduced in the constitutive relationship (2.75).

Similarly, the mass balance equation for air becomes

$$n\dot{S}_a + S_a \dot{u}_{i,i} + \frac{S_a n}{\rho^a} \frac{\partial}{\partial t} \left(\frac{M_g}{\theta R} p_a \right) + v_{i,i} = 0 \tag{2.88}$$

where again the first of Equation (2.86) has been taken into account and the gradient of water density has been neglected. Similar remarks as for the water mass balance equation apply. In particular, the constitutive relationships for moist air, Equations (2.70) and (2.71), have been used.

Finally, if, for the solid phase, the following constitutive relationship is used (viz. Lewis and Schrefler 1998)

$$\frac{1}{\rho^s}\frac{\overset{s}{D}\rho^s}{Dt} = \frac{1}{1-n}\left[(\alpha-n)\frac{1}{K_s}\frac{\overset{s}{D}p_s}{Dt} - (1-\alpha)\dot{u}_{i,i}\right] \tag{2.89}$$

where K_s is the bulk modulus of the grain material, then the mass balance equations are obtained in the same form as in Section 2.4 (with $\chi_w = S_w$), though this is not in agreement with what was assumed here for the effective stress.

2.5.6 Nomenclature for Section 2.5

As this section does not follow the notations use of the book, we summarize below for purposes of nomenclature:

$\mathbf{a}^\pi$ mass averaged acceleration of π phase
$\mathbf{a}^{\pi s}$ acceleration relative to the solid
$\mathbf{b}$ external momentum supply
$\frac{D}{Dt}$ material time derivative
E specific intrinsic energy
F^π deformation gradient tensor
f^π differentiable function
$f_{/i}$ $\partial f/\partial x_i$
$f_{,i}$ $\partial f/\partial X_i$
$\mathbf{g}$ external momentum supply related to gravitational effects
$\mathbf{G}$ net production of thermodynamic property Ψ
h intrinsic heat source
i flux vector associated with thermodynamic property Ψ
$\overline{k}$ constitutive coefficient
$\mathbf{K}$ intrinsic permeability tensor
$\mathbf{K}^{r\pi}$ relative permeability of π phase
$\mathbf{k}^\pi$ dynamic permeability tensor
K_w water bulk modulus
$\dot{m}$ mass rate of water evaporation
M_π molar mass of constituent π
n porosity
p_c capillary pressure
p_{ga} dry air pressure

p_{gw} vapor pressure
p_a air pressure
p_s solid pressure
p_w pressure of liquid water
p_π macroscopic pressure of the π phase
$\dot{\mathbf{r}}$ local value of the velocity field
R universal gas constant
$\mathbf{R}$ constitutive tensor
$\mathbf{R}^\pi$ rigid body rotation tensor
S intrinsic entropy source
S_w degree of water saturation
S_a degree of air saturation
t current time
$\mathbf{t}_m$ microscopic stress tensor
$\mathbf{t}^\pi$ partial stress tensor
$\hat{\mathbf{t}}^\pi$ exchange of momentum due to mechanical interaction of the π phase with other phases
$\dot{\mathbf{u}}$ mass averaged solid phase velocity
$\mathbf{U}^\pi$ right stretch tensor
$\mathbf{v}^{\pi s}$ velocity of the π phase with respect to the solid phase
$\mathbf{v}^w$ mass averaged water velocity
$\mathbf{v}^a$ mass averaged air velocity
$\mathbf{v}$ volume averaged water relative velocity
$\mathbf{w}$ volume averaged air relative velocity
$\mathbf{V}^\pi$ left stretch tensor
$\mathbf{x}^\pi$ material point
$\mathbf{X}^\pi$ reference configuration
$\boldsymbol{\varepsilon}$ linear strain tensor
$\boldsymbol{\Phi}$ entropy flux
φ increase of entropy
η^π volume fraction of the π phase
χ_w Bishop parameter
λ specific entropy
μ^π dynamic viscosity
θ_π absolute temperature of constituent π
ρ averaged density of the multi-phase medium
ρ^π intrinsic phase averaged density of the π phase
ρ_π phase averaged density of the π phase
ρ microscopic density
$\boldsymbol{\sigma}'$ effective stress tensor
Ψ generic thermodynamic property or variable

Superscripts or subscripts

ga =	dry air
gw =	vapor
a =	air
w =	water
s =	solid

2.6 Conclusion

The equations derived in this chapter together with appropriately defined constitutive laws allow (almost) all geomechanical phenomena to be studied. In Chapter 3, we shall discuss in some detail approximation by the finite element method leading to their solution.

References

Biot, M. A. (1941). General theory of three-dimensional consolidation, *J. Appl. Phys.*, **12**, 155–164.

Biot, M. A. (1955). Theory of elasticity and consolidation for a porous anisotropic solid, *J. Appl. Phys.*, **26**, 182–185.

Biot, M. A. (1956a). Theory of propagation of elastic waves in a fluid-saturated porous solid, Part I: Low-frequency range, *J. Acoust. Soc. Am.*, **28**, 2, 168–178.

Biot, M. A. (1956b). Theory of propagation of elastic waves in a fluid-saturated porous solid, Part II: Low-frequency range, *J. Acoust. Soc. Am.*, **28**, 2, 179–191.

Biot, M. A. (1962). Mechanics of deformation and acoustic propagation in porous media, *J. Appl. Phys.*, **33**, 4, 1482–1498.

Biot, M. A. and Willis, P. G. (1957). The elastic coefficients of the theory consolidation, *J. Appl. Mech.*, **24**, 594–601.

Bishop, A. W. (1959). The principle of effective stress, *Teknisk Ukeblad*, **39**, 859–863.

Bolzon, G., Schrefler, B. A. (1995). State surfaces of partially saturated soils: an effective pressure approach. *Appl. Mech. Rev.*, **48** (10), 643–649.

Borja, R. I. (2004). Cam-Clay plasticity. Part V: A mathematical framework for three-phase deformation and strain localization analyses of partially saturated porous media. *Comp. Methods Appl. Mech. Eng.*, **193**, 5301–5338.

Bowen, R. M. (1976). *Theory of Mixtures in Continuum Physics*, Academic Press, New York.

Bowen, R. M. (1980). Incompressible porous media models by use of the theory of mixtures, *Int. J. Eng. Sci.* **18**, 1129–1148.

Bowen, R. M. (1982). Compressible porous media models by use of theories of mixtures, *Int. J. Eng. Sci.* **20**, 697–735.

Chan, A. H. C., Famiyesin, O. O., and Muir Wood, D. (1991). A fully explicit u-w schemes for dynamic soil and pore fluid interaction. Asian Pacific Conference on Computational Mechanics, Hong Kong, Vol. 1, Balkema, Rotterdam, 881–887 (11–13 December 1991).

Coussy, O. (1995). *Mechanics of Porous Media*, John Wiley & Sons, Chichester.

Coussy, O. (2004). *Poromechanics*, John Wiley & Sons, Chichester.

Craig, R. F. (1992). *Soil Mechanics* (5), Chapman & Hall, London.

De Boer, R. (1996). Highlights in the historical development of the porous media theory, *Appl. Mech. Rev.*, **49**, 201–262.

De Boer, R. and Kowalski, S. J. (1983). A plasticity theory for fluid saturated porous solids, *Int. J. Eng. Sci.*, **21**, 1343–1357.

De Boer, R., Ehlers, W., Kowalski, S. and Plischka, J. (1991). *Porous Media, a Survey of Different Approaches*, Forschungsbericht aus dem Fachbereich Bauwesen, 54, Universitaet-Gesamthochschule Essen.

Derski, W. (1978). Equations of motion for a fluid saturated porous solid, *Bull. Acad. Polish Sci. Tech.*, **26**, 11–16.

Desai, C. S. (1977a). Discussion—Finite element, residual schemes for unconfined flow, *Int. J. Numer. Methods Eng.*, **11**, 80–81.

Desai, C. S. (1977b). Finite element, residual schemes for unconfined flow, *Int. J. Numer. Methods Eng.* **10**, 1415–1418.

Desai, C. S. and Li, G. C. (1983). A residual flow procedure and application for free surface flow in porous media, *Adv. Water Resour.*, **6**, 27–35.

Fredlund, D. G., Morgenstern, N. R. (1977). Stress state variables for unsaturated soils, *J. Geotech. Eng. Div. ASCE*, **103**, 447–466.

Ghaboussi, J. and Wilson, E. L. (1972). Variational formulation of dynamics of fluid saturated porous elastic solids, *ASCE EM*, **98**, EM4, 947–963.

Gray, W. G. and Schrefler, B. A. (2001). Thermodynamic approach to effective stress in partially saturated porous media, *Eur. J. Mech. A/Solids*, **20**, 521–538.

Gray, W.G. and Schrefler, B. A. (2007). Analysis of the solid phase stress tensor in multiphase porous media, *Int. J. Numer. Anal. Meth. Geomech.*, **31**, 541–581.

Green, A. E. (1969). On basic equations for mixtures, *Quart. J. Mech. Appl. Math.*, **22**, 428–438.

Green, A. E. and Adkin, J. E. (1960). *Large Elastic Deformations and Nonlinear Continuum Mechanics*, Oxford University Press, London.

Hassanizadeh, M. and Gray, W. G., (1979a). General conservation equations for multiphase systems: 1 Averaging procedure, *Adv. Water Resour.*, **2**, 131–144.

Hassanizadeh, M. and Gray, W. G. (1979b). General conservation equations for multiphase systems: 2 Mass, momenta, energy and entropy equations, *Adv. Water Resour.*, **2**, 191–203.

Hassanizadeh, M. and Gray, W. G. (1980). General conservation equations for multiphase systems: 3 Constitutive theory for porous media flow, *Adv. Water Resour.*, **3**, 25–40.

Hassanizadeh, M. and Gray, W. G. (1990). Mechanics and thermodynamics of multiphase flow in porous media including interphase transport, *Adv. Water Resour.*, **13**, 169–186.

Jaumann, G. (1905). Die Grundlagen der Bewegungslehre von einem modernen Standpunkte aus, Leipzig.

Jommi, C. (2000). Remarks on the constitutive modeling of unsaturated soils, from experimental evidence and theoretical approaches in unsaturated soils. *Proceedings of International Workshop on Unsaturated Soils*, Trento. Italy (10–12 April 2000). Roterdam: Balkema, 139–153.

Khalili, N, Khabbaz, M. H., Valliappan, S. (2000). An effective stress based numerical model for flow and deformation in unsaturated soils. *Comput. Mech.*, **26** (2), 174–184.

Kowalski, S. J. (1979). Comparison of Biot's equation of motion for a fluid saturated porous solid with those of Derski, *Bull. Acad. Polish Sci. Tech.*, **27**, 455–461.

Lambe, T. W. and Whitman, R. V. (1969). *Soil Mechanics*, (SI Version), John Wiley & Sons, New York.

Leliavsky, S. (1947). Experiments on effective area in gravity dams, *Trans. Am. Soc. Civ. Eng.*, **112**, 444.

Lewis, R. W. and Schrefler, B. A. (1982). A finite element simulation of the subsidence of gas reservoirs undergoing a waterdrive in *Finite Element in Fluids* Vol. **4** R.H. Gallagher, D. H. Norrie, J. T. Oden, O. C. Zienkiewicz (Eds), John Wiley, 179–199.

Lewis, R. W. and Schrefler, B. A. (1987). *The Finite Element Method in the Deformation and Consolidation of Porous Media*, John Wiley & Sons, Chichester.

Lewis, R. W. and Schrefler, B. A. (1998). *The Finite Element Method in the Static and Dynamic Deformation and Consolidation of Porous Media*, John Wiley & Sons, Chichester.

Li, X. K. and Zienkiewicz, O. C. (1992). Multiphase flow in deforming porous-media and finite-element solutions, *Comp. Struct.*, **45**, 2, 211–227.

Li, X. K., Zienkiewicz, O. C. and Xie, Y. M. (1990). A numerical model for immiscible 2-phase fluid-flow in a porous medium and its time domain solution, *Int. J. Numer. Methods Eng.*, **30**, 6, 1195–1212.

Liakopoulos, A. C. (1965). Transient flow through unsaturated porous media. D.Eng. dissertation. University of California, Berkeley, USA.

Morland, L. W. (1972). A simple constitutive theory for fluid saturated porous solids, *J. Geophys. Res.*, **77**, 890–900.

Nuth, M., Laloui, L. (2008). Effective stress concept in unsaturated soils: clarification and validation of a unified framework. *Int. J. Numer. Anal. Methods Geomech.*, **32**, 771–801.

Runesson, K. (1978). On non-linear consolidation of soft clay. Ph.D Thesis. Chalmers University of Technology, Goeteborg.

Sandhu, R. S. and Wilson, E. L. (1969). Finite element analysis of flow in saturated porous elastic media, *ASCE EM*, **95**, 641–652.

Schrefler, B. A. (1995). Finite elements in environmental engineering: coupled thermo-hydro-mechanical process in porous media involving pollutant transport, *Arch. Comput. Methods Eng.*, **2**, 1–54.

Schrefler, B. A., (2002). Mechanics and thermodynamics of saturated-unsaturated porous materials and quantitative solutions, *Appl. Mech. Rev.*, **55**(4), 351–388.

Schrefler, B. A. and Gawin D. (1996). The effective stress principle: incremental or finite form?, *Int. J. Numer. Anal. Methods Geomech.*, **20**, 785–814.

Schrefler, B. A. and Scotta, R. (2001). A fully coupled dynamic model for two phase fluid flow in deformable porous media, *Comput. Methods Appl. Mech. Eng.*, **190**, 3223–3246.

Schrefler, B. A. and Zhan, X. (1993). A fully coupled model for waterflow and airflow in deformable porous media, *Water Resour. Res.*, **29**, 1, 155–167.

Serafim, J. L. (1954). A subpressëo nos barreyens – Publ. 55, Laboratorio Nacional de Engenheria Civil, Lisbon.

Sheng, D., Sloan, S. W., Gens, A. (2004). A constitutive model for unsaturated soils: thermomechanical and computational aspects. *Comput. Mech.*, **33** (6), 453–465.

Simoni, L. and Schrefler, B. A. (1991). A staggered finite element solution for water and gas flow in deforming porous media, *Commun. Appl. Num. Meth.*, **7**, 213–223.

Skempton, A. W. (1954). The pore pressure coefficients A and B, *Géotechnique*, **4**, 143–147.
Terzaghi, K. von (1925). *Erdbaumechanik auf bodenphysikalischer Grundlage*, Deuticke, Vienna.
Whitaker, S. (1977). Simultaneous heat mass and momentum transfer in porous media: a theory of drying, in *Advances in Heat Transfer*, J. P. Hartnett and T. F. Irvine (Eds), **13**, 119–203, Academic Press, New York.
Zaremba, S. (1903a). Le principe des mouvements relatifs et les equations de la mécanique physique. Reponse a M. Natanson. *Bull. Int. Acad. Sci. Cracovie.*, 614–621.
Zaremba, S. (1903b). Sur une generalisation de la theorie classique de la viscosite. *Bull. Int. Acad. Sci. Cracovie.*, 380–403.
Zienkiewicz, O. C. (1982). Field equations for porous media under dynamic loads, numerical methods in geomechanics. Proceedings of the NATO Advanced Study Institute, University of Minho, Braga, Portugal, held at Vimeiro (24 August–4 September 1981). Boston: D. Reidel.
Zienkiewicz, O. C. and Shiomi, T. (1984). Dynamic behaviour of saturated porous media: the generalized Biot formulation and its numerical solution, *Int. J. Num. Anal. Geomech.*, **8**, 71–96.
Zienkiewicz, O. C., Chang, C. T. and Bettess, P. (1980). Drained, undrained, consolidating and dynamic behaviour assumptions in soils, *Géotechnique*, **30**, 4, 385–395.
Zienkiewicz, O. C., Chan, A. H. C., Pastor M., Paul, D. K. and Shiomi, T. (1990a). Static and dynamic behaviour of geomaterials – a rational approach to quantitative solutions, Part I: Fully saturated problems, *Proc. Roy. Soc. London*, **A429**, 285–309.
Zienkiewicz, O. C., Xie, Y. M., Schrefler, B. A., Ledesma, A. and Bicanic, N. (1990b). Static and dynamic behaviour of soils: a rational approach to quantitative solutions, Part II: Semi-saturated problems, *Proc. Roy. Soc. London*, **A429**, 310–323.
Zienkiewicz, O. C., Taylor, R. L., and Zhu, J. Z. (2005). *The Finite Element Method Set* (Sixth Edition) Its Basis and Fundamentals, Elsevier.

3

Finite Element Discretization and Solution of the Governing Equations

3.1 The Procedure of Discretization by the Finite Element Method

The general procedures of the Finite Element discretization of equations are described in detail in various texts. Here we shall use throughout the notation and methodology introduced by Zienkiewicz et al. (2013) which is the seventh edition of the first text for the finite element method published in 1967.

In the application to the problems governed by the equations of the previous chapter, we shall typically be solving partial differential equations which can be written as

$$\mathbf{A}\ddot{\mathbf{\Phi}} + \mathbf{B}\dot{\mathbf{\Phi}} + \mathbf{L}(\mathbf{\Phi}) = 0 \tag{3.1}$$

where $\mathbf{A}$ and $\mathbf{B}$ are matrices of constants and $\mathbf{L}$ is an operator involving spatial differentials such as $\partial/\partial x$, $\partial/\partial y$, etc., which can be, and frequently are, nonlinear.

The dot notation implies time differentiation so that

$$\frac{\partial \mathbf{\Phi}}{\partial t} \equiv \dot{\mathbf{\Phi}} \qquad \frac{\partial^2 \mathbf{\Phi}}{\partial t^2} \equiv \ddot{\mathbf{\Phi}} \text{ etc.} \tag{3.2}$$

In all of the above, $\mathbf{\Phi}$ is a vector of dependent variables (say representing the displacements $\mathbf{u}$ and the pressure p).

The finite element solution of the problem will always proceed in the following pattern:

i) The unknown functions $\mathbf{\Phi}$ are "discretized" or approximated by a finite set of parameters $\overline{\mathbf{\Phi}}_{\kappa}$ and shape function N_k which are specified in spatial dimensions. Thus we shall write

$$\mathbf{\Phi} \cong \mathbf{\Phi}^h = \sum_{k=1}^{n} N_k \overline{\mathbf{\Phi}}_k \tag{3.3}$$

where the shape functions are specified in terms of the spatial coordinates, i.e.

$$N_k = N_k(x, y, z) \tag{3.4a}$$

or

$$N_k = N_k(\mathbf{x})$$

$$\text{and } \overline{\mathbf{\Phi}}_i \equiv \overline{\mathbf{\Phi}}_i(t) \tag{3.4b}$$

Computational Geomechanics: Theory and Applications, Second Edition. Andrew H. C. Chan, Manuel Pastor, Bernhard A. Schrefler, Tadahiko Shiomi and O. C. Zienkiewicz.

are usually the values of the unknown function at some discrete spatial points called nodes which remain as variables in time.

ii) Inserting the value of the approximating function $\hat{\mathbf{\Phi}}$ into the differential equations, we obtain a *residual* which is not identically equal to zero but for which we can write a set of weighted residual equations in the form

$$\int_{\Omega} \mathbf{W}_j^{\mathrm{T}} \left(\mathbf{A}\ddot{\mathbf{\Phi}}^h + \mathbf{B}\dot{\mathbf{\Phi}}^h + \mathbf{L}(\mathbf{\Phi}^h) \right) \mathrm{d}\Omega = 0 \tag{3.5}$$

which on integration will always reduce to a form

$$\mathbf{M}\ddot{\overline{\mathbf{\Phi}}} + \mathbf{C}\dot{\overline{\mathbf{\Phi}}} + \mathbf{P}(\overline{\mathbf{\Phi}}) = 0 \tag{3.6}$$

where **M**, **C**, and **P** are matrices or vectors corresponding in size to the full set of numerical parameters $\overline{\Phi}_K$. A very suitable choice for the weighting function W_j is to take them being the same as the shape function $\mathbf{N}_j$:

$$\mathbf{W}_j = \mathbf{N}_j \tag{3.7}$$

Indeed, this choice is optimal for accuracy in the so-called self-adjoint equations as shown in the basic texts and is known as the Galerkin process.

If time variation occurs, i.e. if the parameters $\overline{\Phi}_i$ are time-dependent, Equation (3.6), which is now an *ordinary differential equation*, requires solution in the time domain. This can be, once again, achieved by discretization in time and use of finite elements there, although many alternative approximations (such as the use of finite differences or other integration schemes) are possible.

Usually, the parameters $\overline{\mathbf{\Phi}}$ represent simply the values of $\mathbf{\Phi}^h$ at specified points – called *nodes* and the shape functions are derived on a polynomial basis of interpolating between the nodal values for elements into which the space is assumed divided.

Typical finite elements involving linear and quadratic interpolation are shown in Figure 3.1.

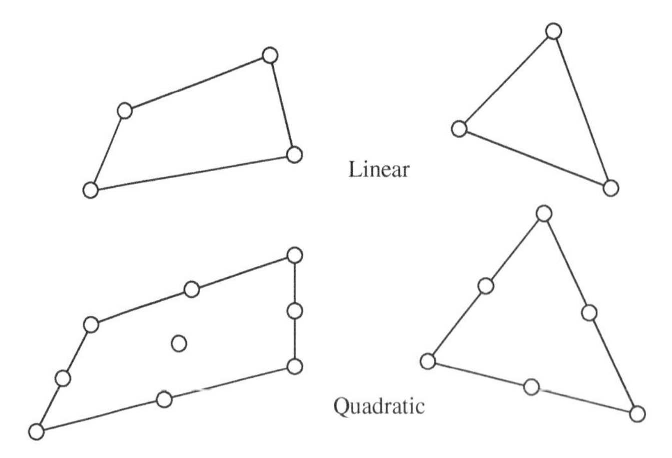

Figure 3.1 Some typical two-dimensional elements for linear and quadratic interpolations from nodal values.

In Section 3.2, we shall consider solution of the approximation based on the **u**–p form in which the dependent variables are the displacement of the soil matrix and the pore pressure characterized by a single fluid, i.e. water. However, we shall allow incomplete saturation to exist assuming that the air pressure is zero.

The formulation thus embraces all the features of the **u**–p approximation of Sections 2.2.2, 2.3.1, and 2.3.2 and is the basis of a code capable of solving all low-frequency dynamic problems, consolidation problems, and static drained or undrained problems of soil mechanics. Only two dimensions will be considered here and in the examples which follow, but extension to three dimensions is obvious.

The code based on the formulation contained in this part of the chapter is named SWAN-DYNE (indicating its Swansea origin) and its outline was presented in literature by Zienkiewicz et al. (1990a, b). The implicit form used allows long periods to be studied. Indeed, with suitable accuracy control, such codes can be used both for earthquake phenomena limited to hundreds of seconds or consolidation problem with a duration of hundreds of days.

3.2 u-p Discretization for a General Geomechanics' Finite Element Code

3.2.1 Summary of the General Governing Equations

We will report here the basic governing equation derived in the previous chapter. However, we shall limit ourselves to the use of the condensed, vectorial form of these which is convenient for finite element discretization. The tensorial form of the equations can be found in Section 3.3.

The overall equilibrium or momentum balance equation is given by (2.11) and is copied here for completeness as

$$\mathbf{S}^T\boldsymbol{\sigma} - \rho\ddot{\mathbf{u}} + \rho\mathbf{b} = 0 \tag{3.8}$$

In the above and in all the following equations, the relative fluid acceleration terms are omitted as only the **u**–p form is being considered.

The strain matrix **S** is defined in two dimensions as (see (2.10))

$$\mathrm{d}\boldsymbol{\varepsilon} \equiv \begin{Bmatrix} \mathrm{d}\varepsilon_x \\ \mathrm{d}\varepsilon_y \\ \mathrm{d}\gamma_{xy} \end{Bmatrix} = \begin{bmatrix} \frac{\partial}{\partial x} & 0 \\ 0 & \frac{\partial}{\partial y} \\ \frac{\partial}{\partial y} & \frac{\partial}{\partial x} \end{bmatrix} \begin{Bmatrix} \mathrm{d}u_x \\ \mathrm{d}u_y \end{Bmatrix} \equiv \mathbf{S}\mathrm{d}\mathbf{u} \tag{3.9}$$

Here **u** is the displacement vector and ρ the total density of the mixture (see (2.19))

$$\rho = nS_\mathrm{w}\rho_\mathrm{w} + (1-n)\rho_\mathrm{s} \tag{3.10}$$

generally taken as constant and $\boldsymbol{\sigma}$ is the total stress with components

$$\boldsymbol{\sigma} = \begin{Bmatrix} \sigma_x \\ \sigma_y \\ \sigma_{xy} \end{Bmatrix} \tag{3.11}$$

The effective stress is defined as in (2.1)

$$\boldsymbol{\sigma}'' = \boldsymbol{\sigma} + \alpha \mathbf{m} p \tag{3.12}$$

where α again is a constant usually taken for soils as

$$\alpha = 1 \tag{3.13}$$

and p the effective pressure defined by (2.24) with $p_{\mathrm{a}} = 0$.

$$p = \chi_{\mathrm{w}} p_{\mathrm{w}} \tag{3.14}$$

The effective stress $\boldsymbol{\sigma}''$ is computed from an appropriate constitutive law generally defined as "increments" by (2.2)

$$\mathrm{d}\boldsymbol{\sigma}'' = \mathbf{D}\left(\mathrm{d}\boldsymbol{\varepsilon} - \mathrm{d}\boldsymbol{\varepsilon}^0\right) \tag{3.15}$$

where $\mathbf{D}$ is the tangent matrix dependent on the state variables and history and $\boldsymbol{\varepsilon}^0$ corresponds to thermal and creep strains.

The main variables of the problem are thus $\mathbf{u}$ and p_{w}. The effective stresses are determined at any stage by a sum of all previous increments and the value of p_{w} determines the parameters S_w (saturation) and χ_{w} (effective area). On occasion, the approximation

$$\chi_{\mathrm{w}} = S_{\mathrm{w}} \tag{3.16}$$

can be used.

An additional equation is supplied by the mass conservation coupled with fluid momentum balance. This is conveniently given by (2.33b) which can be written as

$$\nabla^{\mathrm{T}}\left(\mathbf{k}\left(-\nabla p_w + \mathbf{S}_w \rho_f \mathbf{b}\right)\right) + \alpha \mathbf{m}^{\mathrm{T}} \mathbf{S} \dot{\mathbf{u}} + \frac{\dot{p}_w}{Q^*} + \dot{s}_0 = 0 \tag{3.17}$$

with $\mathbf{k} = \mathbf{k}\ (\mathbf{S}_w)$.

The contribution of the solid acceleration is neglected in this equation. Its inclusion in the equation will render the final equation system nonsymmetric (see Leung 1984) and the effect of this omission has been investigated in Chan (1988) who found it to be insignificant. However, it has been included in the force term of the computer code SWANDYNE-II (Chan 1995) although it is neglected in the left-hand side of the final algebraic equation when the symmetric solution procedure is used.

The above set defines the complete equation system for solution of the problem defined providing the necessary boundary conditions have been specified as in (2.18) and (2.19), i.e.

$$\begin{aligned} \mathbf{t} = \boldsymbol{\sigma}.\mathbf{G} = \tilde{\mathbf{t}} \ \text{ on } \ \Gamma = \Gamma_t \\ \mathbf{u} = \tilde{\mathbf{u}} \ \text{ on } \ \Gamma = \Gamma_u \end{aligned} \tag{3.18}$$

and

$$p_w = \overline{p_w} \text{ on } \Gamma = \Gamma_p$$
$$\mathbf{n}^{\mathrm{T}} w = \mathbf{n}^{\mathrm{T}} \mathbf{k}\left(-\nabla p_w + S_w \rho_f \mathbf{b}\right) = \widetilde{w}_n \text{ on } \Gamma = \Gamma_w$$

Assuming isotropic permeability, the above equation becomes

$$\mathbf{k}\left(-\frac{\partial p_w}{\partial \mathbf{n}} + S_w \rho_f \mathbf{n}^{\mathrm{T}} \mathbf{b}\right) = \widetilde{w}_n = -\overline{q} \text{ on } \Gamma = \Gamma_w$$

where $\overline{q}_n$ is the influx, i.e. having an opposite sign to the outflow $\overline{w}_n$.

The total boundary Γ is the union of its components, i.e.

$$\Gamma = \Gamma_t \cup \Gamma_u = \Gamma_p \cup \Gamma_w$$

3.2.2 Discretization of the Governing Equation in Space

The spatial discretization involving the variables $\mathbf{u}$ and p_w is achieved by suitable shape (or basis) functions, writing

$$\begin{aligned} \mathbf{u} \cong \mathbf{u}^h &= \sum_{k=1}^{n} \mathrm{N}_k^u \overline{\mathbf{u}}_k = \mathbf{N}^u \overline{\mathbf{u}} \\ p_{\mathrm{w}} \cong p_{\mathrm{w}}^h &= \sum_{k=1}^{m} \mathrm{N}_k^p \overline{\mathbf{p}}_k^{\mathrm{w}} = \mathrm{N}^p \overline{\mathbf{p}}^{\mathrm{w}} \end{aligned} \tag{3.19}$$

Note that the nodal values of the pore pressures are indicated with a superscript.

We assume here that the expansion is such that the strong boundary conditions (3.18) are satisfied on Γ_u and Γ_p automatically by a suitable prescription of the (nodal) parameters. As in most other finite element formulations, the natural boundary condition will be obtained by integrating the weighted equation by parts.

To obtain the first equation discretized in space, we premultiply (3.8) by $(\mathbf{N}^u)^{\mathrm{T}}$ and integrate the first term by parts (see for details Zienkiewicz et al (2013) or other texts) giving:

$$\int_\Omega \mathbf{B}^{\mathrm{T}} \boldsymbol{\sigma} \mathrm{d}\Omega + \left[\int_\Omega (\mathbf{N}^{\mathbf{u}})^{\mathrm{T}} \rho \mathbf{N}^{\mathbf{u}} \mathrm{d}\Omega\right] \ddot{\overline{\mathbf{u}}} = \mathbf{f}^{(1)} \tag{3.20}$$

where the matrix $\mathbf{B}$ is given as

$$\mathbf{B} \equiv \mathbf{S}\mathbf{N}^u \tag{3.21}$$

and the "load vector" $\mathbf{f}^{(1)}$, equal in number of components to that of vector $\overline{\mathbf{u}}$ contains all the effects of body forces, and prescribed boundary tractions, i.e.

$$\mathbf{f}^{(1)} = \int_\Omega (\mathbf{N}^u)^{\mathrm{T}} \rho \mathbf{b} \mathrm{d}\Omega + \int_{\Gamma_t} (\mathbf{N}^{\mathrm{u}})^{\mathrm{T}} \overline{\mathbf{t}} \mathrm{d}\Gamma \tag{3.22}$$

At this stage, it is convenient to introduce the effective stress see (3.12) now defined to allow for effects of incomplete saturation as

$$\boldsymbol{\sigma} = \boldsymbol{\sigma}'' - \alpha \chi_{\mathrm{w}} \mathbf{m} p_w \tag{3.23}$$

The discrete, ordinary differential equation now becomes

$$\boxed{\mathbf{M}\ddot{\bar{\mathbf{u}}} + \int_{\Omega} \mathbf{B}^{\mathrm{T}}\boldsymbol{\sigma}'' \mathrm{d}\Omega - \mathbf{Q}\bar{\mathbf{p}}^{w} - \mathbf{f}^{(1)} = 0} \tag{3.24}$$

where

$$\mathbf{M} = \int_{\Omega} (\mathbf{N}^{u})^{\mathrm{T}} \rho \mathbf{N}^{u} \mathrm{d}\Omega \tag{3.25}$$

is the *MASS MATRIX* of the system and

$$\mathbf{Q} = \int_{\Omega} \mathbf{B}^{\mathrm{T}} \alpha \chi_{w} \mathbf{m} \mathbf{N}^{p} \mathrm{d}\Omega \tag{3.26}$$

is the coupling matrix-linking equation (3.23) and those describing the fluid conservation, and

$$\mathbf{f}^{(1)} = \int_{\Omega} (\mathbf{N}^{u})^{\mathrm{T}} \rho \mathbf{b} \mathrm{d}\Omega + \int_{\Gamma_t} (\mathbf{N}^{u})^{\mathrm{T}} \bar{\mathbf{t}} \mathrm{d}\Gamma \tag{3.22 bis}$$

The computation of the effective stress proceeds incrementally as already indicated in the usual way and now (3.15) can be written in discrete form:

$$\boxed{\Delta\boldsymbol{\sigma}'' = \mathbf{D}\left(\mathbf{B}\Delta\bar{\mathbf{u}} - \Delta\boldsymbol{\varepsilon}_{l}^{0}\right)} \tag{3.27}$$

where, of course, $\mathbf{D}$ is evaluated from appropriate state and history parameters.

Finally, we discretize Equation (3.17) by pre-multiplying by $(\mathbf{N}^{p})^{\mathrm{T}}$ and integrating by parts as necessary. This gives the ordinary differential equation

$$\boxed{\tilde{\mathbf{Q}}\dot{\bar{\mathbf{u}}} + \mathbf{H}\bar{\mathbf{p}}^{w} + \tilde{\mathbf{S}}\dot{\bar{\mathbf{p}}}^{w} - \mathbf{f}^{(2)} = 0} \tag{3.28}$$

where the various matrices are as defined below

$$\tilde{\mathbf{Q}} = \int_{\Omega} \boldsymbol{B}^{T} \alpha \mathbf{m} \mathbf{N}^{p} \mathrm{d}\Omega \tag{3.29}$$

$$\mathbf{H} = \int_{\Omega} (\nabla \mathbf{N}^{p})^{\mathrm{T}} \mathbf{k} \nabla \mathbf{N}^{p} \mathrm{d}\Omega \tag{3.30}$$

$$\tilde{\mathbf{S}} = \int_{\Omega} (\mathbf{N}^{p})^{\mathrm{T}} \frac{1}{Q^{*}} \mathbf{N}^{p} \mathrm{d}\Omega \tag{3.31}$$

$$\mathbf{f}^{(2)} = -\int_{\Omega} (\mathbf{N}^{p})^{\mathrm{T}} \nabla^{\mathrm{T}} (\mathbf{k} S_{\mathrm{w}} \rho_{\mathrm{f}} \mathbf{b}) \mathrm{d}\Omega + \int_{\Gamma_{\mathrm{w}}} (\mathbf{N}^{p})^{\mathrm{T}} \bar{\mathbf{q}} \mathrm{d}\Gamma \tag{3.32}$$

where $\mathbf{Q}^{*}$ is defined as in (2.30c), i.e.

$$\frac{1}{Q^{*}} \equiv \mathrm{C_s} + \frac{\mathrm{n}S_{\mathrm{w}}}{\mathrm{K_f}} + \frac{(\alpha - n)S_{\mathrm{w}}}{\mathrm{K_s}} \tag{3.33}$$

and C_{S}, S_{w}, C_{w} and $\mathbf{k}$ depend on p_{w}.

3.2.3 Discretization in Time

To complete the numerical solution, it is necessary to integrate the ordinary differential Equations (3.23), (3.27), and (3.28) in time by one of the many available schemes. Although there are many multistep methods available (see, e.g., Wood 1990), they are inconvenient as most of them are not self-starting and it is more difficult to incorporate restart facilities which are required frequently in practical analyses. On the other hand, the single-step methods handle each step separately and there is no particular change in the algorithm for such restart requirements.

Two similar, but distinct, families of single-step methods evolved separately. One is based on the finite element and weighted residual concept in the time domain and the other based on a generalization of the Newmark or finite difference approach. The former is known as the SSpj – Single Step p^{th} order scheme for j^{th} order differential equation ($p \geq j$). This was introduced by Zienkiewicz et al. (1980b, 1984) and extensively investigated by Wood (1984a, 1984b, 1985a, 1985b). The SSpj scheme has been used successfully in SWANDYNE-I (Chan, 1988). The later method, which was adopted in SWANDYNE-II (Chan 1995) was an extension to the original work of Newmark (1959) and is called Beta-m method by Katona (1985) and renamed the Generalized Newmark (GNpj) method by Katona and Zienkiewicz (1985). Both methods have similar or identical stability characteristics. For the SSpj, no initial condition, e.g. acceleration in dynamical problems, or higher time derivatives are required. On the other hand, however, all quantities in the GNpj method are defined at a discrete time station, thus making transfer of such quantities between the two equations easier to handle. Here we shall use the later (GNpj) method, exclusively, due to its simplicity.

In all time-stepping schemes, we shall write a recurrence relation linking a known value $\boldsymbol{\Phi}_n$ (which can either be the displacement or the pore water pressure), and its derivatives $\dot{\boldsymbol{\Phi}}_n$, $\ddot{\boldsymbol{\Phi}}_n$,... at time station t_n with the values of $\boldsymbol{\Phi}_{n+1}$, $\dot{\boldsymbol{\Phi}}_{n+1}$, $\ddot{\boldsymbol{\Phi}}_{n+1}$,..., which are valid at time $t_n + \Delta t$ and are the unknowns. Before treating the ordinary differential equation system (3.23), (3.27), and (3.28), we shall illustrate the time-stepping scheme on the simple example of (3.6) by adding a forcing term:

$$\mathbf{M}\ddot{\boldsymbol{\Phi}} + \mathbf{C}\dot{\boldsymbol{\Phi}} + \mathbf{P}(\boldsymbol{\Phi}) = \mathbf{f}(t) \tag{3.34}$$

From the initial conditions, we have the known values of $\boldsymbol{\Phi}_n$, $\dot{\boldsymbol{\Phi}}_n$. We assume that the above equation has to be satisfied at each discrete time, i.e. t_n and t_{n+1}. We can thus write:

$$\mathbf{M}\ddot{\Phi}_n + \mathbf{C}\dot{\Phi}_n + \mathbf{P}(\Phi_n) = \mathbf{f}(t_n) \tag{3.35}$$

and

$$\mathbf{M}\ddot{\boldsymbol{\Phi}}_{n+1} + \mathbf{C}\dot{\boldsymbol{\Phi}}_{n+1} + \mathbf{P}(\boldsymbol{\Phi}_{n+1}) = \mathbf{f}(t_{n+1}) \tag{3.36}$$

From the first equation, the value of the acceleration at time t_n can be found and this solution is required if the initial conditions are different from zero.

The link between the successive values is provided by a truncated series expansion taken in the simplest case as GN22 as Equation (3.34) is a second-order differential equation j and the minimum order of the scheme required is then two: as ($p \geq j$)

$$\begin{aligned}
\ddot{\mathbf{\Phi}}_{n+1} &= \ddot{\mathbf{\Phi}}_n + \Delta\ddot{\mathbf{\Phi}}_n \\
\dot{\mathbf{\Phi}}_{n+1} &= \dot{\mathbf{\Phi}}_n + \ddot{\mathbf{\Phi}}_n\Delta t + \beta_1\Delta\ddot{\mathbf{\Phi}}_n\Delta t \\
\mathbf{\Phi}_{n+1} &= \mathbf{\Phi}_n + \dot{\mathbf{\Phi}}_n\Delta t + \frac{1}{2}\ddot{\mathbf{\Phi}}_n\Delta t^2 + \frac{1}{2}\beta_2\Delta\ddot{\mathbf{\Phi}}_n\Delta t^2
\end{aligned} \tag{3.37}$$

Alternatively, a higher order scheme can be chosen such as GN32 and we shall have:

$$\begin{aligned}
\dddot{\mathbf{\Phi}}_{n+1} &= \dddot{\mathbf{\Phi}}_n + \Delta\dddot{\mathbf{\Phi}}_n \\
\ddot{\mathbf{\Phi}}_{n+1} &= \ddot{\mathbf{\Phi}}_n + \dddot{\mathbf{\Phi}}_n\Delta t + \beta_1\Delta\dddot{\mathbf{\Phi}}_n\Delta t \\
\dot{\mathbf{\Phi}}_{n+1} &= \dot{\mathbf{\Phi}}_n + \ddot{\mathbf{\Phi}}_n\Delta t + \frac{1}{2}\dddot{\mathbf{\Phi}}_n\Delta t^2 + \frac{1}{2}\beta_2\Delta\dddot{\mathbf{\Phi}}_n\Delta t^2 \\
\mathbf{\Phi}_{n+1} &= \mathbf{\Phi}_n + \dot{\mathbf{\Phi}}_n\Delta t + \frac{1}{2}\ddot{\mathbf{\Phi}}_n\Delta t^2 + \frac{1}{3\times 2}\dddot{\mathbf{\Phi}}_n\Delta t^3 + \frac{1}{3\times 2}\beta_3\Delta\dddot{\mathbf{\Phi}}_n\Delta t^2
\end{aligned} \tag{3.38}$$

In this case, an extra set of equations is required to obtain the value of the highest time derivatives. This is provided by differentiating (3.35) and (3.36).

$$\mathbf{M}\dddot{\mathbf{\Phi}}_n + \mathbf{C}\ddot{\mathbf{\Phi}}_n + \left.\frac{d\mathbf{P}(\mathbf{\Phi})}{d(\mathbf{\Phi})}\right|_{t=t_n}\dot{\mathbf{\Phi}}_n = \dot{\mathbf{f}}(t_{n+1}) \tag{3.39}$$

and

$$\mathbf{M}\dddot{\mathbf{\Phi}}_{n+1} + \mathbf{C}\ddot{\mathbf{\Phi}}_{n+1} + \left.\frac{d\mathbf{P}(\mathbf{\Phi})}{d(\mathbf{\Phi})}\right|_{t=t_{n+1}}\dot{\mathbf{\Phi}}_{n+1} = \dot{\mathbf{f}}(t_{n+1}) \tag{3.40}$$

In the above equations, the only unknown is the incremental value of the highest derivative and this can be readily solved for.

Returning to the set of ordinary differential equations we are considering here, i.e. (3.23), (3.27), and (3.28) and writing (3.23) and (3.28) at the time station t_{n+1}, we have:

$$\mathbf{M}\ddot{\bar{\mathbf{u}}}_{n+1} + \left(\int_\Omega \mathbf{B}^{\mathrm{T}}\boldsymbol{\sigma}''d\Omega\right)_{n+1} - \mathbf{Q}\bar{\mathbf{p}}^w_{n+1} - \mathbf{f}^{(1)}_{n+1} = 0 \tag{3.41}$$

$$\tilde{\mathbf{Q}}\dot{\bar{\mathbf{u}}}_{n+1} + \mathbf{H}\bar{\mathbf{p}}^w_{n+1} + \mathbf{S}\dot{\bar{p}}^w_{n+1} - \mathbf{f}^{(2)}_{n+1} = 0 \tag{3.42}$$

assuming that (3.27) is satisfied.

Using GN22 for the displacement parameters $\bar{\mathbf{u}}$ and GN11 for the pore pressure parameter $\bar{\mathbf{p}}^w$, we write:

$$\begin{aligned}
\ddot{\bar{\mathbf{u}}}_{n+1} &= \ddot{\bar{\mathbf{u}}}_n + \Delta\ddot{\bar{\mathbf{u}}}_n \\
\dot{\bar{\mathbf{u}}}_{n+1} &= \dot{\bar{\mathbf{u}}}_n + \ddot{\bar{\mathbf{u}}}_n\Delta t + \beta_1\Delta\ddot{\bar{\mathbf{u}}}_n\Delta t \\
\bar{\mathbf{u}}_{n+1} &= \bar{\mathbf{u}}_n + \dot{\bar{\mathbf{u}}}_n\Delta t + \frac{1}{2}\ddot{\bar{\mathbf{u}}}_n\Delta t^2 + \frac{1}{2}\beta_2\Delta\ddot{\bar{\mathbf{u}}}_n\Delta t^2
\end{aligned} \tag{3.43a}$$

and

$$\begin{aligned}
\dot{\overline{\mathbf{p}^w}}_{n+1} &= \dot{\overline{\mathbf{p}^w}}_n + \Delta\dot{\overline{\mathbf{p}^w}}_n \\
\overline{\mathbf{p}^w}_{n+1} &= \overline{\mathbf{p}^w}_n + \dot{\overline{\mathbf{p}^w}}_n\Delta t + \bar{\beta}_1\Delta\dot{\overline{\mathbf{p}^w}}_n\Delta t
\end{aligned} \tag{3.43b}$$

where $\Delta\ddot{\bar{\mathbf{u}}}_n$ and $\Delta\dot{\overline{\mathbf{p}^w}}_n$ are as yet undetermined quantities. The parameters β_1, β_2, and $\bar{\beta}_1$ are usually chosen in the range of 0 to 1. For $\beta_2 = 0$ and $\bar{\beta}_1 = 0$, we shall have an explicit form if both the mass and damping matrices are diagonal. If the damping matrix is non-diagonal, an explicit scheme can still be achieved with $\beta_1 = 0$, thus eliminating the contribution of the damping matrix. The well-known central difference scheme is recovered from (3.41) if $\beta_1 = 1/2, \beta_2 = 0$ and this form with an explicit $\bar{\mathbf{u}}$ and implicit $\overline{\mathbf{p}^w}$ scheme has been considered in detail by Zienkiewicz et al. (1982) and Leung (1984). However, such schemes are only conditionally stable and for unconditional stability of the recurrence scheme, we require

$$\beta_2 \geq \beta_1 \geq \frac{1}{2} \quad \text{and} \quad \overline{\beta}_1 \geq \frac{1}{2}$$

The optimal choice of these values is a matter of computational convenience, the discussion of which can be found in literature. In practice, if the higher order accurate "trapezoidal" scheme is chosen with $\beta_2 = \beta_1 = 1/2$ and $\overline{\beta}_1 = 1/2$, numerical oscillation may occur if no physical damping is present. Usually, some algorithmic (numerical) damping is introduced by using such values as

$$\beta_2 = 0.605 \quad \beta_1 = 0.6 \quad \text{and} \quad \overline{\beta}_1 = 0.6$$

or

$$\beta_2 = 0.515 \quad \beta_1 = 0.51 \quad \text{and} \quad \overline{\beta}_1 = 0.51.$$

Dewoolkar (1996), using the computer program SWANDYNE II in the modelling of a free-standing retaining wall, reported that the first set of parameters led to excessive algorithmic damping as compared to the physical centrifuge results. Therefore, the second set was used and gave very good comparisons. However, in cases involving soil, the physical damping (viscous or hysteretic) is much more significant than the algorithmic damping introduced by the time-stepping parameters and the use of either sets of parameters leads to similar results.

Inserting the relationships (3.43) into Equations (3.41) and (3.42) yields a general nonlinear equation set in which only $\Delta\ddot{\bar{\mathbf{u}}}_n$ and $\Delta\dot{\overline{\mathbf{p}^w}}_n$ remain as unknowns.

This set can be written as

$$\mathbf{\Psi}^{(1)}_{n+1} = \mathbf{M}_{n+1}\Delta\ddot{\bar{\mathbf{u}}}_n + \mathbf{P}(\bar{\mathbf{u}}_{n+1}) - \mathbf{Q}_{n+1}\overline{\beta}_1\Delta t\Delta\dot{\bar{\mathbf{p}}}^w_n - \mathbf{F}^{(1)}_{n+1} = 0 \tag{3.44a}$$

$$\mathbf{\Psi}^{(2)}_{n+1} = \tilde{\mathbf{Q}}^{\mathrm{T}}_{n+1}\beta_1\Delta t\Delta\ddot{\bar{\mathbf{u}}}_n + \mathbf{H}_{n+1}\overline{\beta}_1\Delta t\Delta\dot{\bar{\mathbf{p}}}^w_n + \mathbf{S}_{n+1}\Delta\dot{\bar{\mathbf{p}}}^w_n - \mathbf{F}^{(2)}_{n+1} = 0 \tag{3.44b}$$

where $\mathbf{F}^{(1)}_{n+1}$ and $\mathbf{F}^{(2)}_{n+1}$ can be evaluated explicitly from the information available at time t_n and

$$\overline{\mathbf{P}}(\bar{\mathbf{u}}_{n+1}) = \int_\Omega \mathbf{B}^{\mathrm{T}}\boldsymbol{\sigma}''_{n+1}\mathrm{d}\Omega = \int_\Omega \mathbf{B}^{\mathrm{T}}\Delta\boldsymbol{\sigma}''_n\,\mathrm{d}\Omega + \mathbf{P}(\bar{\mathbf{u}}_n) \tag{3.45}$$

In this, $\Delta\boldsymbol{\sigma}''_n$ must be evaluated by integrating (3.27) as the solution proceeds. The values of $\bar{\mathbf{u}}_{n+1}$ and $\overline{\mathbf{p}^w}_{n+1}$ at the time t_{n+1} are evaluated by Equation (3.43).

The equation will generally need to be solved by a convergent, iterative process using some form of Newton–Raphson procedure typically written as

$$\mathbf{J}\begin{Bmatrix} \delta\Delta\ddot{\bar{\mathbf{u}}}_n \\ \delta\Delta\dot{\overline{\mathbf{p}^w}}_n \end{Bmatrix}^{l+1} = -\begin{Bmatrix} \mathbf{\Psi}^{(1)}_{n+1} \\ \mathbf{\Psi}^{(2)}_{n+1} \end{Bmatrix}^{l} \tag{3.46a}$$

where 1 is the iteration number and

$$\begin{Bmatrix} \Delta\ddot{\mathbf{u}}_n \\ \Delta\dot{\overline{\mathbf{p}}^w}_n \end{Bmatrix}^{l+1} = \begin{Bmatrix} \Delta\ddot{\mathbf{u}}_n \\ \Delta\dot{\overline{\mathbf{p}}^w}_n \end{Bmatrix}^{l} + \begin{Bmatrix} \delta\Delta\ddot{\mathbf{u}}_n \\ \delta\Delta\dot{\overline{\mathbf{p}}^w}_n \end{Bmatrix}^{l+1}$$

The Jacobian matrix can be written as

$$\mathbf{J} = \begin{bmatrix} \dfrac{\partial\mathbf{\Psi}^{(1)}_{n+1}}{\partial\Delta\ddot{\mathbf{u}}_n} & \dfrac{\partial\mathbf{\Psi}^{(1)}_{n+1}}{\partial\Delta\dot{\overline{\mathbf{p}}}^w_n} \\ \dfrac{\partial\mathbf{\Psi}^{(2)}_{n+1}}{\partial\Delta\ddot{\mathbf{u}}_n} & \dfrac{\partial\mathbf{\Psi}^{(2)}_{n+1}}{\partial\Delta\dot{\overline{\mathbf{p}}}^w_n} \end{bmatrix} = \begin{bmatrix} \mathbf{M}_{n+1} + \dfrac{1}{2}\mathbf{K}_{Tn+1}\beta_2\Delta t^2 - \mathbf{Q}_{n+1}\overline{\beta}_1\Delta t \\ \widetilde{\mathbf{Q}}^T_{n+1}\beta_1\Delta t\mathbf{S}_{n+1} + \mathbf{H}_{n+1}\beta_1\Delta t \end{bmatrix} \tag{3.47}$$

where

$$\mathbf{K}_{\mathrm{T}} = \int_\Omega \mathbf{B}^{\mathrm{T}}\mathbf{D}_{\mathrm{T}}\mathbf{B}d\Omega + \underline{\int_\Omega (\mathbf{AB})^{\mathrm{T}}\boldsymbol{\sigma}''\mathbf{AB}d\Omega}$$

which are the well-known expressions for tangent stiffness matrix. The underlined term corresponds to the "initial stress" matrix evaluated in the current configuration as a result of stress rotation defined in (2.5).

Two points should be made here:

a) that in the linear case, a single "iteration" solves the problem exactly
b) that the matrix can be made symmetric by a simple scalar multiplication of the second row (provided $\mathbf{K}_{\mathrm{T}}$ is itself symmetric).

In practice, it is found that the use of various approximations of the matrix $\boldsymbol{J}$ is advantageous such as, for instance, the use of "secant" updates (see, for instance, Crisfield (1979), Matthies, and Strang (1979) and Zienkiewicz et al (2013).

A particularly economical computation form is given by choosing $\beta_2 = 0$ and representing matrix $\boldsymbol{M}$ in a diagonal form. This explicit procedure was first used by Leung (1984) and Zienkiewicz et al. (1980a). It is, however, only conditionally stable and is efficient only for phenomena of short duration.

The process of the time-domain solution of (3.44) can be amended to that of successive separate solutions of the time equations for variables $\Delta\ddot{\mathbf{u}}_n$ and $\Delta\dot{\overline{\mathbf{p}}^w}_n$, respectively, using an approximation for the remaining variable. Such staggered procedures, if stable, can be extremely economical as shown by Park and Felippa (1983) but the particular system of equations presented here needs stabilization. This was first achieved by Park (1983) and, later, a more effective form was introduced by Zienkiewicz et al. (1988).

Special cases of solution are incorporated in the general solution scheme presented here without any modification and indeed without loss of computational efficiency.

Thus, for static or quasi-static, problems, it is merely necessary to put $\mathbf{M} = 0$ and immediately the transient consolidation equation is available. Here time is still real and we have omitted only the inertia effects (although with implicit schemes, this *a priori* assumption is not necessary and inertia effects will simply appear as negligible without any substantial increase of computation). In pure statics, the time variable is still retained but is then purely an artificial variable allowing load incrementation.

In static or dynamic undrained analysis, the permeability (and compressibility) matrices are set to zero, i.e. $\mathbf{H} \cdot \mathbf{f}^{(2)} = 0$, and usually $\mathbf{S} = 0$ resulting in a zero-matrix diagonal term in the Jacobian matrix of Equation (3.47).

The matrix to be solved in such a limiting case is identical to that used frequently in the solution of problems of incompressible elasticity or fluid mechanics and, in such studies, places limitations on the approximating functions $\mathbf{N}^u$ and $\mathbf{N}^p$ used in (3.19) if the Babuska–Brezzi (Babuska 1971, 1973, Brezzi 1974) convergence conditions or their equivalent (Zienkiewicz et al. 1986b) are to be satisfied. Until now, we have not referred to any particular element form, and, indeed, a wide choice is available to the user if the limiting (undrained) condition is never imposed. Due to the presence of first derivatives in space in all the equations, it is necessary to use "C_0-continuous" interpolation functions and Figure 3.2 shows some elements incorporated in the formulation. The form of most of the elements used satisfies the necessary convergence criteria of the undrained limit

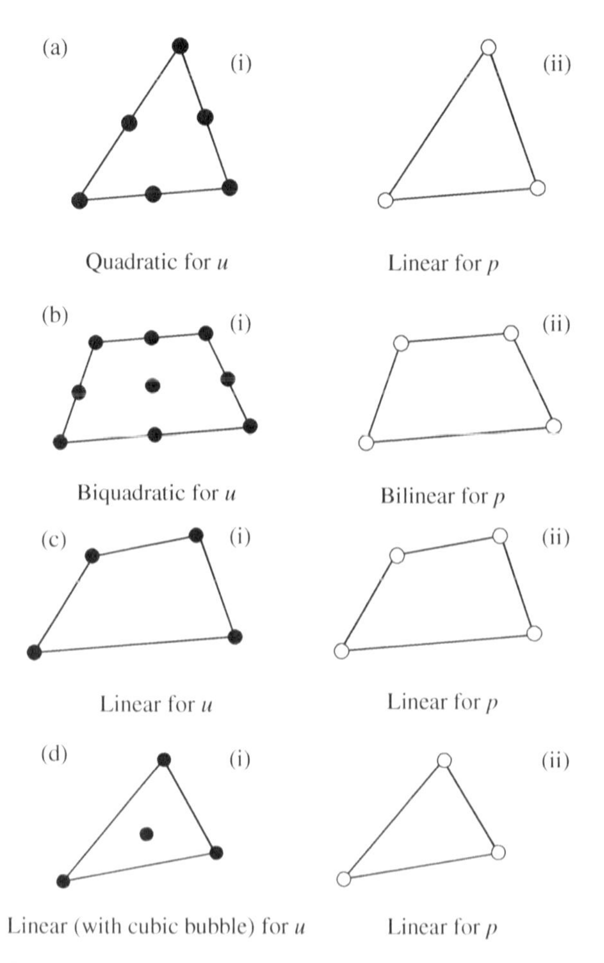

Figure 3.2 Elements used for coupled analysis, displacement (u) and pressure (p) formulation: (a) (i) quadratic for u; (ii) linear for p; (b) (i) biquadratic for u; (ii) bilinear for p: (c) (i) linear for u; (ii) linear for p: (d) (i) linear (with cubic bubble) for u; (ii) linear for Element (c) is not fully acceptable at incompressible–undrained limits.

(Zienkiewicz 1984). Though the bi-linear u and p quadrilateral does not, it is, however, useful when the permeability is sufficiently large.

We shall return to this problem in Chapter 5 where a modification is introduced allowing the same interpolations to be used for both u and p.

In that chapter, we shall discuss a possible amendment to the code permitting the use of identical $\mathbf{u}$-p interpolation even in incompressible cases.

We note that all computations start from known values of $\overline{\mathbf{u}}$ and $\overline{\mathbf{p}}^w$ possibly obtained as the result of static computations by the same program in a manner which will be explained in the next Section 3.2.4. The incremental computation allows the various parameters, dependent on the solution history, to be updated.

Thus, for known $\Delta\overline{\mathbf{u}}$ increment, the $\Delta\boldsymbol{\sigma}''$ are evaluated by using an appropriate tangent matrix $\mathbf{D}$ and an appropriate stress integration scheme.

Further, we note that if $p_w \geq 0$ (full saturation, described in Section 2.2 of the previous chapter), then we have

$$S_w = \chi_w = 1$$

and the permeability remains at its saturated value

$$k = k_S$$

However, when negative pressures are reached, i.e. when $p_w < 0$, the values of S_w, χ_w, and k have to be determined from appropriate formulae or graphs.

3.2.4 General Applicability of Transient Solution (Consolidation, Static Solution, Drained Uncoupled, and Undrained)

3.2.4.1 Time Step Length

As explained in the previous Section 3.2.3, the computation always proceeds in an incremental manner and in the $\mathbf{u}$-p form in general, the explicit time stepping is not used as its limitation is very serious. Invariably, the algorithm is applied here to the unconditionally stable, implicit form and the equation system given by the Jacobian of (3.46) with variables $\Delta\overline{\mathbf{u}}$ and $\Delta\overline{\mathbf{p}}$ needs to be solved at each time step.

With unconditional stability of the implicit scheme, the only limitation on the length of the time step is the accuracy achievable. Clearly, in the dynamic earthquake problem, short time steps will generally be used to follow the time characteristic of the input motion. In the examples that we shall give later, we shall frequently use simply the time interval $\Delta t = 0.02$ s which is the interval used usually in earthquake records.

However, once the input motion has ceased and its record no longer has to be followed, a much longer length of time step could be adopted. Indeed, after the passage of the earthquake, the remaining motion is caused by something resembling a *consolidation process* which has a slower response allowing longer time steps to be used.

The length of the time step based on accuracy considerations was first discussed in Zienkiewicz et al. (1984), Zienkiewicz and Shiomi (1984), and, later, by Zienkiewicz and Xie (1991), and Bergan and Mollener (1985).

The simplest process is that which considers the expansion for such a variable as $\mathbf{u}$ given by (3.43) and its comparison with a Taylor series expansion.

Clearly, for a scalar variable u, the error term is given by the first omitted terms of the Taylor expansion, i.e., in scalar values

$$e = \frac{1}{6}\Delta t^3 \dddot{u} \tag{3.48}$$

Using an approximation of this third derivative shown below

$$\dddot{u} = \frac{\ddot{u}_{n+1} - \ddot{u}_n}{\Delta t} = \frac{\Delta \ddot{u}_n}{\Delta t} \tag{3.49}$$

we have

$$e = \frac{1}{6}\Delta t^2 \Delta \ddot{u}_n \tag{3.50}$$

For a vector variable $\mathbf{u}$, we must consider its L_2 norm, i.e.

$$\|\mathbf{u}\|_2 = \sqrt{\mathbf{u}^T\mathbf{u}} \text{ etc.} \tag{3.51}$$

and we can limit the error to

$$\|e\|_2 = \frac{1}{6}\Delta t^2 \|\Delta \ddot{u}_n\|_2 \tag{3.52}$$

This limit was reestablished later by Zienkiewicz and Xie (1991) who replaced the leading coefficient of (3.52) as a result of a more detailed analysis by

$$\frac{1}{6} \Rightarrow \frac{3\beta_2 - 1}{6} \tag{3.53}$$

Whatever the form of error estimator adopted, the essence of the procedure is identical and this is given by establishing *a priori* some limits or *tolerance* which must not be exceeded, and modifying the time steps accordingly.

In the above, we have considered only the error in one of the variables, i.e. u but, in general, this suffices for quite a reasonable error control.

The tolerance is conveniently chosen as some percentage η of the maximum value of norm $\|\mathbf{u}\|_2$ recorded. Thus, we write

$$\|e\|_2 < \eta\|\mathbf{u}\|_2 \tag{3.54}$$

with some minimum specified.

The time step can always be adjusted during the process of computation noting, however, not to change the length of the time step by more than a factor of 2 or ½, otherwise unacceptable oscillations may arise.

3.2.4.2 Splitting or Partitioned Solution Procedures

The most common way of solving the coupled system of equations is the monolithic scheme where both the coupled equations are solved simultaneously at each time step as discussed above. However, also splitting solutions are possible where the two (or three) equations are solved separately at each time step. To conserve the coupled nature of the problem, iterations within each time step are necessary even in the linear case. The reason for applying partitioned solutions may be the fact that solvers for the solid and fluid fields are already at hand so that only the interaction has to be taken into account. This type of solution procedure is extensively investigated for the dynamic case by Markert et al. (2010) and, as already mentioned, by Park and Felippa (1983), Park (1983), and Zienkiewicz et al. (1988). A splitting procedure will be used in Section 5.5 both for the dynamic case and consolidation allowing for same interpolation for both **u** and p. Partitioned solutions are quite common in consolidation and have been extensively treated in Lewis and Schrefler (1998). From the investigation of the iteration convergence within a time step, Turska and Schrefler (1993) found the existence of a lower limit for $\Delta t/h^2$ which means that it is not always possible to decrease Δt without also decreasing the mesh size h. Such a limit was also found by Murthy et al. (1989) for Poisson-type equations and by Rank et al. (1983) for transient finite element analyses by invoking the discrete maximum principle.

3.2.4.3 The Consolidation Equation

In the standard treatment of consolidation equation (see, for instance, Lewis and Schrefler 1998), the acceleration terms are generally omitted *a priori*. However, as explained above, there is no disadvantage in writing the full dynamic formulation for solving such a problem. The procedure simply reduces the multiplier of the mass matrix **M** to a negligible value without influencing in any way the numerical stability, provided, of course, that an implicit integration scheme is used.

3.2.4.4 Static Problems – Undrained and Fully Drained Behavior

Steady state (static) conditions will only be reached under the extremes of *undrained* or *fully drained* behavior. This can be deduced by rewriting the two, discrete, governing Equations (3.23) and (3.28) omitting terms involving time derivatives. The equations now become:

$$\int_{\Omega} \mathbf{B}^{\mathrm{T}} \boldsymbol{\sigma}'' \mathrm{d}\Omega - \mathbf{Q}\overline{\mathbf{p}}^{\mathrm{w}} - \mathbf{f}_n^{(1)} = 0 \tag{3.55}$$

and

$$\mathbf{H}\overline{\mathbf{p}}^{\mathrm{w}} - \mathbf{f}_n^{(2)} = 0 \tag{3.56}$$

with the effective stresses given by (3.27) and are defined incrementally as

$$\mathrm{d}\boldsymbol{\sigma}'' = \mathbf{D}\left(\mathrm{d}\boldsymbol{\varepsilon} - \mathrm{d}\boldsymbol{\varepsilon}^0\right) = \mathbf{D}\left(\mathbf{B}\mathrm{d}\overline{\mathbf{u}} - \mathrm{d}\boldsymbol{\varepsilon}^0\right) \tag{3.57}$$

First, we observe that the equations are uncoupled and that the second of these, i.e. (3.56) can be solved independently of the first for the water pressures. Indeed, in this solution, the negative pressure zones and, hence, the partially saturated regions can be readily determined following the procedures outlined in the previous chapter.

With $\overline{\mathbf{p}}^w$ determined as

$$\overline{\mathbf{p}}^w \equiv \mathbf{H}^{-1}\mathbf{f}^{(2)} \tag{3.58}$$

the first Equation (3.55) coupled with the appropriate constitutive law (3.57) can be solved once the history of the load applied has been specified.

The solution so obtained is, of course, the well-known, *drained*, behavior.

The case of *undrained* behavior is somewhat more complex. We note that with $k = 0$, i.e. with totally impermeable behavior

$$\mathbf{H} = \mathbf{0} \text{ and } \mathbf{f}^{(2)} = \mathbf{0} \tag{3.59}$$

But on re-examining Equation (3.28), we find that it becomes

$$\widetilde{\mathbf{Q}}^{\mathrm{T}}\dot{\overline{\mathbf{u}}} + \mathbf{S}\dot{\overline{\mathbf{p}}}^{w} = 0 \tag{3.60}$$

which, on integration, establishes a unique relationship between $\overline{\mathbf{u}}$ and $\overline{\mathbf{p}}_w$ which is not time-dependent

$$\widetilde{\mathbf{Q}}^{\mathrm{T}}\overline{\mathbf{u}} + \mathbf{S}\overline{\mathbf{p}}^{w} = 0 \tag{3.61}$$

assuming that the initial condition of $\overline{\mathbf{u}} = \mathbf{0}$ and $\overline{\mathbf{p}}_w = \mathbf{0}$ coincides.

Equation (3.61) now has to be solved together with (3.55). If $\mathbf{S} = 0$, i.e. no compressibility is admitted, then we have the problem already discussed in the previous Section 3.2.3 in which only certain $\overline{\mathbf{u}} - \overline{\mathbf{p}^{w}}$ interpolations are permissible (as shown in Figure 3.2). However, if $\mathbf{S} \neq \mathbf{0}$ $\overline{\mathbf{p}^{w}}$ can be eliminated directly and the solution concerns only the variable $\overline{\mathbf{u}}$.

Solving (3.61) for $\overline{\mathbf{p}^{w}}$ which can only be done provided that some fluid compressibility is available giving $\mathbf{S} \neq 0$, then (3.55) and the constitutive law are sufficient to obtain the unique undrained condition.

The existence of the two steady states is well known and what we have indicated here is a process by which various matrices given in the original computer program can be used to obtain either of the steady state solutions. However, this does require an alternative to the original computer program. Though, it is possible to obtain such steady states by the code, using the previous time-stepping procedure. Two types of undrained conditions exist: (a) when $k = 0$ throughout; (b) $k \neq 0$ but the complete boundary is impermeable. Both cases can be computed with no difficulties.

Provided that the boundary conditions are consistent with the existence of drained and undrained steady state conditions, the time-stepping process will, in due course, converge with

$$\begin{aligned}
&\Delta\ddot{\overline{\mathbf{u}}}_{n+1} \to 0\\
&\Delta\dot{\overline{\mathbf{u}}}_{n+1} \to 0\\
&\Delta\overline{\mathbf{u}}_{n+1} \to 0\\
&\Delta\dot{\overline{\mathbf{p}}}^{w}_{n+1} \to 0\\
&\Delta\overline{\mathbf{p}}^{w}_{n+1} \to 0
\end{aligned}$$

However, this process may be time-consuming even if large time steps, Δt are used. A simpler procedure is to use the GN00 scheme with

$$\overline{\mathbf{u}}_{n+1} = \overline{\mathbf{u}}_n + \Delta\overline{\mathbf{u}}_n$$

$$\overline{\mathbf{p}}^{w}_{n+1} = \overline{\mathbf{p}}^{w}_n + \Delta\overline{\mathbf{p}}^{w}_n$$

Equations (3.41) and (3.61) now become, for the undrained problem,

$$\int_\Omega \mathbf{B}^{\mathrm{T}}(\boldsymbol{\sigma}''_{n+1} - \boldsymbol{\sigma}''_n)\mathrm{d}\Omega - \mathbf{Q}\overline{\mathbf{p}}^w_{n+1} = \mathrm{f}^{(1)}_{n+1} - \mathbf{Q}\overline{\mathbf{p}}^w_n - \mathbf{f}^{(1)}_n$$

$$\widetilde{\mathbf{Q}}^{\mathrm{T}}\overline{\mathbf{u}}_{n+1} + \mathbf{S}\overline{\mathbf{p}}^w_{n+1} = \widetilde{\mathbf{Q}}^{\mathrm{T}}\overline{\mathbf{u}}_n + \mathbf{S}\overline{\mathbf{p}}^w_n$$

If the material behavior is linearly elastic, then the equation can be solved directly yielding the two unknowns $\overline{\mathbf{u}}_{n+1}$ and $\overline{\mathbf{p}}^w_{n+1}$ and if the material is nonlinear, an iteration scheme such as the Newton–Raphson, Quasi-Newton, Tangential Matrix or the Initial Matrix method can be adopted. With a systematic change of the external loading, problems such as the load–displacement curve of a nonlinear soil and pore–fluid system can be traced.

3.2.5 The Structure of the Numerical Equations Illustrated by their Linear Equivalent

If complete saturation is assumed together with a linear form of the constitutive law, we can write the effective stress simply as

$$\boldsymbol{\sigma}'' = \mathbf{D}\mathbf{B}\overline{\mathbf{u}} \tag{3.62}$$

We can now reduce the governing **u**–p Equations (3.23) and (3.28) to the form given below

$$\mathbf{M}\ddot{\overline{\mathbf{u}}} + \mathbf{K}\overline{\mathbf{u}} - \widetilde{\mathbf{Q}}\overline{\mathbf{p}^w} - \mathbf{f}^{(1)} = 0 \tag{3.63}$$

and

$$\widetilde{\mathbf{Q}}^{\mathrm{T}}\dot{\overline{\mathbf{u}}} + H\overline{\mathbf{p}^w} + \mathrm{S}\dot{\overline{\mathrm{p}}} - \mathbf{f}^{(2)} = 0 \tag{3.64}$$

where $\overline{\mathbf{p}^w} = \overline{\mathbf{p}^w}$

$$\text{and}\quad \mathbf{K} = \int_\Omega \mathbf{B}^{\mathrm{T}}\mathbf{D}\mathbf{B}\mathrm{d}\Omega \tag{3.65}$$

is the well-known elastic stiffness matrix which is always symmetric in form. ***S*** and ***H*** are again symmetric matrices defined in (3.31) and (3.30) and $\widetilde{\boldsymbol{Q}}$ is as defined in (3.29).

The overall system can be written in the terms of the variable set $[\overline{\mathbf{u}}, \overline{\mathbf{p}^w}]^{\mathrm{T}}$ as

$$\begin{bmatrix} \mathbf{M} & 0 \\ 0 & 0 \end{bmatrix}\begin{Bmatrix} \ddot{\overline{\mathbf{u}}} \\ \ddot{\overline{\mathbf{p}^w}} \end{Bmatrix} + \begin{bmatrix} 0 & 0 \\ \widetilde{\mathbf{Q}}^{\mathrm{T}} & S \end{bmatrix}\begin{Bmatrix} \dot{\overline{\mathbf{u}}} \\ \dot{\overline{\mathbf{p}^w}} \end{Bmatrix} + \begin{bmatrix} \mathbf{K} & -\widetilde{\mathbf{Q}} \\ 0 & \mathbf{H} \end{bmatrix}\begin{Bmatrix} \overline{\mathbf{u}} \\ \overline{\mathbf{p}^w} \end{Bmatrix} = \begin{Bmatrix} \mathbf{f}^{(1)} \\ \mathbf{f}^{(2)} \end{Bmatrix} = \begin{Bmatrix} 0 \\ 0 \end{Bmatrix} \tag{3.66}$$

Once again the uncoupled nature of the problem under *drained condition* is evident (by dropping the time derivatives) giving

$$\begin{bmatrix} \mathbf{K} & -\widetilde{\mathbf{Q}} \\ 0 & \mathbf{H} \end{bmatrix}\begin{Bmatrix} \overline{\mathbf{u}} \\ \overline{\mathbf{p}^w} \end{Bmatrix} = \begin{Bmatrix} \mathbf{f}^{(1)} \\ \mathbf{f}^{(2)} \end{Bmatrix} \tag{3.67}$$

in which $\overline{\mathbf{p}^w}$ can be separately determined by solving the second equation. For *undrained behavior*, we can integrate the second equation when $\mathbf{H} = 0$ and obtain an antisymmetric system which can be made symmetric by multiplying the second equation by minus unity (Zienkiewicz and Taylor 1985)

$$\begin{bmatrix} \mathbf{M} & 0 \\ 0 & 0 \end{bmatrix} \begin{Bmatrix} \ddot{\bar{\mathbf{u}}} \\ \ddot{\bar{\mathbf{p}}}^w \end{Bmatrix} + \begin{bmatrix} 0 & 0 \\ 0 & 0 \end{bmatrix} \begin{Bmatrix} \dot{\bar{\mathbf{u}}} \\ \dot{\bar{\mathbf{p}}}^w \end{Bmatrix} + \begin{bmatrix} \mathbf{K} & -\tilde{\mathbf{Q}} \\ \tilde{\mathbf{Q}}^{\mathrm{T}} & \mathbf{H} \end{bmatrix} \begin{Bmatrix} \bar{\mathbf{u}} \\ \bar{\mathbf{p}}^w \end{Bmatrix} = \begin{Bmatrix} \mathbf{f}^{(1)} \\ 0 \end{Bmatrix} \tag{3.68}$$

It is interesting to observe that in the steady state, we have a matrix which, in the absence of fluid compressibility, results in

$$\begin{bmatrix} \mathbf{K} & -\tilde{\mathbf{Q}} \\ \tilde{\mathbf{Q}}^{\mathrm{T}} & 0 \end{bmatrix} \begin{Bmatrix} \bar{\mathbf{u}} \\ \bar{\mathbf{p}}^w \end{Bmatrix} = \begin{Bmatrix} \mathbf{f}^{(1)} \\ 0 \end{Bmatrix} \tag{3.69}$$

which only can have a unique solution when the number of $\bar{\mathbf{u}}$ variables $\mathbf{n}_\mathrm{u}$ is greater than the number of $\overline{\mathbf{p}^w}$ variables n_p. This is one of the requirements of the patch test of Zienkiewicz et al. (1986a, 1986b) and of the Babuska-Brezzi (Babuska 1973 and Brezzi 1974) condition.

3.2.6 Damping Matrices

In general, when dynamic problems are encountered in soils (or other geomaterials), the damping introduced by the plastic behavior of the material and the viscous effects of the fluid flow are sufficient to damp out any nonphysical or numerical oscillation. However, if the solutions of the problems are in the low-strain range when the plastic hysteresis is small or when, to simplify the procedures, purely elastic behavior is assumed, it may be necessary to add system damping matrices of the form $\mathbf{C}\dot{\bar{\mathbf{u}}}$ to the dynamic equations of the solid phase, i.e. changing (3.23) to

$$\mathbf{M}\ddot{\bar{\mathbf{u}}} + \mathbf{C}\dot{\bar{\mathbf{u}}} + \int_\Omega \mathbf{B}^{\mathrm{T}}\boldsymbol{\sigma}''\mathrm{d}\Omega - \mathbf{Q}\bar{\mathbf{p}}^w - \mathbf{f}^{(1)} = 0 \tag{3.70}$$

Indeed, such damping matrices have a physical significance and are always introduced in earthquake analyses or similar problems of structural dynamics. With the lack of any special information about the nature of damping, it is usual to assume the so-called “Rayleigh damping” in which

$$\mathbf{C} = \alpha\mathbf{M} + \beta\mathbf{K} \tag{3.71}$$

where α and β are coefficients determined by experience (see, for instance, Clough and Penzien (1975) or (1993)). In the above, $\mathbf{M}$ is the same mass matrix as given in (3.24) and $\mathbf{K}$ is some representative stiffness matrix of the form given in (3.47).

3.3 Theory: Tensorial Form of the Equations

The equation numbers given here correspond to the ones given earlier in the text.

$$\sigma_{ij,j} - \rho\ddot{u}_i + \rho b_i = 0 \tag{3.8b}$$

$$d\varepsilon_{ij} = \frac{1}{2}(du_{i,\,j} + du_{\,j,i}) \tag{3.9b}$$

Noting that the engineering shear strain $d\gamma_{xy}$ is defined as:

$$d\gamma_{xy} = 2d\varepsilon_{xy}$$

Equation (3.10) is scalar

$$\sigma \equiv \sigma_{ij} \tag{3.11b}$$

$$\sigma''_{ij} = \sigma_{ij} + \alpha\delta_{ij}p \tag{3.12b}$$

Equation (3.13) is scalar.
Equation (3.14) is scalar.

$$\mathrm{d}\sigma''_{ij} = D_{ijkl}\left(\mathrm{d}\varepsilon_{kl} - \mathrm{d}\varepsilon^0_{kl}\right) \tag{3.15b}$$

Equation (3.16) is scalar

$$\begin{aligned} &\left(k_{ij}\left(-p_{w,\ j} + S_w\rho_f b_j\right)\right)_{,i} + \alpha\dot{u}_{i,i} + \frac{\dot{p}_w}{Q^*} + \dot{s}_0 = 0 \\ &t_i = \sigma_{ij}n_j = \bar{t}_i \quad \text{on} \quad \Gamma = \Gamma_t \\ &u_i = \bar{u}_i \quad \text{on} \quad \Gamma = \Gamma_u \end{aligned} \tag{3.17b}$$

and

$$n_i w_i = n_i k_{ij}\left(-p_{w,\ j} + S_w\rho_f b_j\right) = \overline{w}_n \quad \text{on} \quad \Gamma = \Gamma_w \tag{3.18b}$$

and assuming isotropic permeability

$$\begin{aligned} &n_i w_i = k\left(-p_{w,n} + S_w\rho_f n_i b_i\right) = \overline{w}_n = -\overline{q} \quad \text{on} \quad \Gamma = \Gamma_w \\ &u_i \cong \hat{u}_i = N^u_k \overline{u}_{Ki} \\ &p_w \cong \hat{p}_w = N^p_L \overline{p}^w_L \end{aligned} \tag{3.19b}$$

The summation range for the upper-case indices will depend on the number of nodes with solid displacement and pore water pressure degrees-of-freedom (dofs), respectively.

$$\int_\Omega N^u_K\left(\sigma_{ij,j} - \rho\ddot{u}_i + \rho b_i\right)\mathrm{d}\Omega = 0$$

$$\int_\Omega N^u_K\left(\sigma_{ij,\ j} - \rho N^u_L \ddot{\overline{u}}_{Li} + \rho b_i\right)\mathrm{d}\Omega = 0$$

Applying Green's identity to the internal force term (first term on the left-hand side)

$$\int_\Omega\left(-N^u_{K,\ j}\sigma_{ij} - N^u_K\rho N^u_L\ddot{\overline{u}}_{Li} + N^u_K\rho b_i\right)\mathrm{d}\Omega + \int_{\Gamma_t} N^u_K n_j\sigma_{ij}\mathrm{d}\Gamma = 0$$

$$\int_\Omega\left(-N^u_{K,j}\sigma_{ij} - N^u_K\rho N^u_L\ddot{\overline{u}}_{Li} + N^u_K\rho b_i\right)\mathrm{d}\Omega + \int_{\Gamma_t} N^u_K\overline{t}_i\mathrm{d}\Gamma = 0$$

Rearranging

$$\int_\Omega N^u_{K,j}\sigma_{ij}\mathrm{d}\Omega + \left[\int_\Omega N^u_K\rho N^u_L\mathrm{d}\Omega\right]\ddot{\overline{u}}_{Li} = \int_\Omega N^u_K\rho b_i\mathrm{d}\Omega + \int_{\Gamma_t} N^u_K\overline{t}_i\mathrm{d}\Gamma \tag{3.20}$$

The definition of the **B** matrix in Equation (3.21) is not needed in tensorial form.

$$\sigma_{ij} = \sigma''_{ij} - \alpha\chi\delta_{ij}p$$

$$M_{KL}\ddot{\bar{u}}_{Li} + \int_\Omega N^u_{K,j}\sigma''_{ij}\mathrm{d}\Omega - Q_{KiL}\bar{p}^W_L - f^{(1)}_{Ki} = 0 \tag{3.23b}$$

$$M_{KL} = \int_\Omega N^u_K \rho N^u_L \mathrm{d}\Omega \tag{3.24b}$$

$$Q_{KiL} = \int_\Omega N^u_{K,i}\alpha\chi N^p_L \mathrm{d}\Omega \tag{3.25b}$$

$$f^{(1)}_{Ki} = \int_\Omega N^u_K \rho \mathrm{b}_i \mathrm{d}\Omega + \int_{\Gamma_t} N^u_K \bar{t}_i \mathrm{d}\Gamma \tag{3.26b}$$

$$\mathrm{d}\sigma''_{ij} = D_{ijkl}\left(\frac{1}{2}\left(N^u_{K,k}\mathrm{d}\bar{u}_{Kl} + N^u_{K,l}\mathrm{d}\bar{u}_{Kk}\right) - \mathrm{d}\varepsilon^0_{kl}\right) \tag{3.27b}$$

$$\int_\Omega N^p_K\left(\left(k_{ij}\left(-p_{w,j} + S_w\rho_f b_j\right)\right)_{,i} + \alpha\dot{u}_{i,i} + \frac{\dot{p}_w}{Q^*} + \dot{s}_0\right)\mathrm{d}\Omega = 0$$

Neglecting source term and integrating by part the first part of the first term

$$-\int_{\Gamma_w} N^p_K n_i k_{ij} p_{w,j}\mathrm{d}\Gamma + \int_\Omega\left[N^p_{K,i}k_{ij}p_{w,j} + N^p_K\left(k_{ij}S_w\rho_f b_j\right)_{,i} + N^p_K\alpha\dot{u}_{i,i} + N^p_K\frac{\dot{p}_w}{Q^*}\right]\mathrm{d}\Omega = 0$$

Inserting the shape functions

$$-\int_{\Gamma_w} N^p_K n_i q_i \mathrm{d}\Gamma + \int_\Omega\left[N^p_{K,i}k_{ij}N^p_{L,j}\bar{p}^w_L + N^p_K\left(k_{ij}S_w\rho_f b_j\right)_{,i} + N^p_K\alpha N^u_{L,i}\dot{\bar{u}}_L + N^p_K\frac{1}{Q^*}N^p_L\dot{\bar{p}}^w_L\right]\mathrm{d}\Omega = 0$$

$$-\int_{\Gamma_w} N^p_K n_i q_i \mathrm{d}\Gamma + \int_\Omega N^p_K\left(k_{ij}S_w\rho_f b_j\right)_{,i}\mathrm{d}\Omega + \int_\Omega N^p_{K,i}k_{ij}N^p_{L,j}\mathrm{d}\Omega\bar{p}$$

$$+\int_\Omega N^p_K\alpha N^u_{L,i}\mathrm{d}\Omega\dot{\bar{u}}_L + \int_\Omega N^p_K\frac{1}{Q^*}N^p_L\mathrm{d}\Omega\dot{\bar{p}}^w_L = 0$$

$$\widetilde{Q}_{LiK}\dot{\bar{u}}_{Li} + H_{KL}\bar{p}^w_L + S_{KL}\dot{\bar{p}}^W_L - f^{(2)}_K = 0 \tag{3.28b}$$

$$\widetilde{Q}_{KiL} = \int_\Omega N^u_{K,i}\alpha N^p_L \mathrm{d}\Omega \tag{3.29b}$$

$$H_{KL} = \int_\Omega N^p_{K,i}k_{ij}N^p_{L,j}\mathrm{d}\Omega \tag{3.30b}$$

$$S_{KL} = \int_\Omega N^p_K\frac{1}{Q^*}N^p_L\mathrm{d}\Omega \tag{3.31b}$$

$$f^{(2)}_K = -\int_\Omega N^p_{K,i}k_{ij}S_w\rho_f b_j\mathrm{d}\Omega + \int_{\Gamma_w} N^p_K\bar{q}\mathrm{d}\Gamma \tag{3.32b}$$

Equation (3.33) is scalar.

3.4 Conclusions

In this chapter, the governing equations introduced in Chapter 2 are discretized in space and time using various implicit and explicit algorithms. They are now ready for implementation into computer codes. In Chapter 5, we shall address some special modeling aspects and in Chapters 6–8, we shall show some applications for static, quasi-static, and dynamic examples to illustrate the practical applications of the method and to validate and verify the schemes and constitutive models used.

References

Babuska, I. (1971). Error bounds for finite element methods, *Num. Math.*, **16**, 322–333.

Babuska, I. (1973). The finite element method with Lagrange Multipliers, *Num. Math.*, **20**, 179–192.

Bergan, P. G. and Mollener, E. (1985). An automatic time-stepping algorithm for dynamic problems, *Comp. Meth. Appl. Mech. Eng.* **49**, 299–318.

Brezzi, F. (1974). On the existence, uniqueness and approximation of saddle point problems arising from Lagrange multipliers, *R.A.I.R.O. Anal. Numér.*, **8**, R-2, 129–151.

Chan, A. H. C. (1988). *A unified Finite Element Solution to Static and Dynamic Geomechanics problems*. Ph.D. Dissertation, University College of Swansea, Wales.

Chan, A. H. C. (1995). User manual for DIANA SWANDYNE-II, School of Civil Engineering, University of Birmingham, December, Birmingham.

Clough, R. W. and Penzien, J. (1975). *Dynamics of Structures*, McGraw-Hill, New York.

Clough, R. W. and Penzien, J. (1993). *Dynamics of Structures* (2nd edn), McGraw-Hill, Inc., New York.

Crisfield, M. A. (1979). A faster modified Newton-Raphson iteration, *Comp. Meth. Appl. Mech. Eng.*, **20**, 267–278.

Dewoolkar, M. M. (1996). *A study of seismic effects on centiliver-retaining walls with saturated backfill*. Ph.D Thesis. Dept of Civil Engineering, University of Colorado. Boulder, USA.

Katona, M. G. (1985). A general family of single-step methods for numerical time integration of structural dynamic equations, *NUMETA* **85**, 1, 213–225.

Katona, M. G. and Zienkiewicz, O. C. (1985). A unified set of single step algorithms Part 3: The beta-m method, a generalisation of the Newmark scheme, *Int. J. Num. Meth. Eng.*, **21**, 1345–1359.

Leung, K. H. (1984). *Earthquake response of saturated soils and liquefaction*. Ph.D. Dissertation, University College of Swansea, Wales.

Lewis, R. W. and Schrefler, B. A. (1998). *The Finite Element Method in the Static and Dynamic Deformation and Consolidation of Porous Media*, John Wiley & Sons, Chichester.

Markert, B, Heider, Y. and Ehlers, W. (2010) Comparison of monolithic and splitting solution schemes for dynamic porous media problems, *Int. J. Num. Meth. Eng.*, **82**, 1341–1383.

Matthies, H. and Strang, G. (1979) The solution of nonlinear finite element equations, *Int. J. Num. Meth. Eng.*, **14**, 1613–1626.

Murthy, V., Valliappan, S. and Khalili-Naghadeh, N. (1989). Time step constraints in finite element analysis of poisson type equation, *Comput. Struct.*, **31**, 269–271.

Newmark, N. M. (1959). A method of computation for structural dynamics, *Proc. ASCE*, **8**, 67–94.

Park, K. C. (1983). Stabilization of partitioned solution procedure for pore fluid-soil interaction analysis, *Int. J. Num. Meth. Eng.*, **19**, 1669–1673.

Park, K. C. and Felippa, C. A. (1983). 'Partitioned analysis of coupled systems', Chapter 3, in *Computational Methods for Transient Analysis*, T. Belytschko and Thomas J R Hughes (Eds), Elsevier Science Publishers B. V.

Rank, E., Katz, C. and Werner, H. (1983). On the importance of the discrete maximum principle in transient analysis using finite element method, *Int. J. Num. Meth. Eng.*, **19**, 1771–1782.

Turska, E., Schrefler, B. A. (1993). On convergence conditions of partitioned solution procedures for consolidation problems, *Comp. Meth. Appl. Mech. Eng.*, **106**, 51–64.

Wood, W. L. (1984a). A further look at Newmark, Houbolt, etc., time-stepping formulae, *Int. J. Num. Meth. Eng.*, **20**, 1009–1017.

Wood, W. L. (1984b). A unified set of single step algorithms Part 2: Theory, *Int. J. Num. Meth. Eng.*, **20**, 2303–2309.

Wood, W. L. (1985a). Addendum to 'a unified set of single step algorithms, Part 2: Theory', *Int. J. Num. Meth. Eng.*, **21**, 1165.

Wood, W. L. (1985b). A unified set of single-step algorithms Part 4: Backward error analysis applied to the solution of the dynamic vibration equation—Numerical Analysis. Report 6/85, Department of Mathematics, University of Reading, Reading.

Wood, W. L. (1990). *Practical Time-Stepping Schemes*, Clarendon Press, Oxford.

Zienkiewicz, O. C. (1984). Coupled problems and their numerical solution, Chapter 1. in *Numerical Methods in Coupled Systems*, R. L. Lewis, P. Bettess and E. Hinton (Eds), John Wiley and Sons Ltd., Chichester.

Zienkiewicz, O. C. (1985). The coupled problems of soil-pore fluid-external fluid interaction: Basis for a general geomechanics code, *ICONMIG* **5**, 1731–1740.

Zienkiewicz, O. C. and Shiomi, T. (1984). Dynamic behaviour of saturated porous media: the generalized Biot formulation and its numerical solution, *Int. J. Num. Anal. Geomech.*, **8**, 71–96.

Zienkiewicz, O. C. and Taylor, R. L. (1985). Coupled problems – a simple time-stepping procedure, *Comm. Appl. Num. Meth.*, **1**, 233–239.

Zienkiewicz, O. C. and Xie, Y. M. (1991). A Simple error estimator for adaptative time stepping procedure in dynamic analysis, *Int. J. Earth. Struct. Dyn.*, **20**, 871–887.

Zienkiewicz, O. C., Hinton, E., Leung, K. H., and Taylor, R. L. (1980a). Innovative numerical analysis for the applied engineering sciences. *Proceedings of the Second International Symposium on Innovative Numerical Analysis in Applied Engineering Sciences.* https://www.amazon.com/Innovative-Numerical-Analysis-Engineering-Sciences/dp/0813908671

Zienkiewicz, O. C., Wood, W. L. and Taylor, R. L. (1980b). An alternative single-step algorithm for dynamic problems, *Earthq. Eng. Struct. Dyn.*, **8**, 31–40.

Zienkiewicz, O. C., Leung, K. H., Hinton, E. and Chang, C. T. (1982). Liquefaction and permanent deformation under dynamic conditions – numerical solution and constitutive relations, Chapter 5, in *Soil Mechanics – Transient and Cyclic loads*, G. N. Pande and O. C. Ziekiewicz (Eds), John Wiley, Chichester.

Zienkiewicz, O. C., Wood, W. L., Hine, N. W. and Taylor, R. L. (1984). A unified set of single step algorithms Part 1: General formulation and applications, *Int. J. Num. Meth. Eng.*, **20**, 1529–1552.

Zienkiewicz, O. C., Taylor, R. L., Simo, J. C. and Chan, A. H. C. (1986a). The patch test – a condition for assessing FEM. convergence, *Int. J. Num. Meth. Eng.*, **22**, 39–62.

Zienkiewicz, O. C., Qu, S., Taylor, R. L. and Nakazawa, S. (1986b). The patch test for mixed formulation, *Int. J. Num. Meth. Eng.*, **23**, 1873–1883.

Zienkiewicz, O. C., Paul, D. K. and Chan, A. H. C. (1988). Unconditionally stable staggered solution procedure for soil pore fluid interaction problems, *Int. J. Num. Meth. Eng.*, **26**, 5, 1039–1055.

Zienkiewicz, O. C., Chan, A. H. C., Pastor, M., Paul, D. K. and Shiomi, T. (1990a). Static and dynamic behaviour of geomaterials: a rational approach to quantitative solutions, Part I: Fully saturated problems, *Proc. Roy. Soc. Lond.*, **A429**, 285–309.

Zienkiewicz, O. C., Xie, Y. M., Schrefler, B. A., Ledesma, A. and Bicanic, N. (1990b). Static and dynamic behaviour of soils: a rational approach to quantitative solutions, Part II: Semi-saturated problems, *Proc. Roy. Soc. Lond.*, **A429**, 310–323.

Zienkiewicz, O. C., Taylor, R. L., Zhu, J. Z. (2013). *The Finite Element Method: Its Basis and Fundamentals* 7. Butterworth-Heinemann.

4

Constitutive Relations

Plasticity

4.1 Introduction

The purpose of constitutive models is to capture some of the main features of the mechanical behavior of solids under the given conditions of temperature, velocity of load application, level of strain, nature of stress conditions, etc.

Roughly speaking, models can be classified into two main groups:

1) Micromechanical or physical models, based on the behavior of grains or particles, and
2) Macromechanical or phenomenological models.

Most of the models used in computer codes are of the second class.

All materials present a response that depends on time to a greater or lesser degree. For instance, a specimen of soft clay subjected to constant vertical stress shows a vertical deformation that increases monotonically with time.

However, most of the geomaterials under normal engineering conditions present a mechanical behavior that depends more on the level of stress, pore pressure, past history, direction of load increment, and material structure than on time. In fact, the major part of the time dependence observed is generally connected with the pore water flow.

For these, plasticity-based theories provide a consistent framework in which the behavior can be accurately understood and predicted.

It can be said that the history of plasticity began in 1773 with the work of Coulomb in soils, applied later by Poncelet and Rankine to practical soil mechanics problems.

It was not until almost a century later, in 1864, when Tresca, based on experimental results on punching and extrusion tests, proposed a yield criterion dependent on the maximum shear stress. Later, St. Venant, in 1870, introduced the concept of isotropic flow rule, which was generalized by Lévy (1871) to three-dimensional conditions. The principal axes of stress and the increment of strain were assumed to be the same.

The next significant step had to wait until the beginning of the new century when von Mises and Huber, Von Mises (1913) and Huber (1804) independently, proposed a new yield criterion which, with the flow equations derived by Lévy, became known as the Lévy–Von Mises equations. There no distinction between total, elastic, and plastic parts of the strain increment was implied.

Computational Geomechanics: Theory and Applications, Second Edition. Andrew H. C. Chan, Manuel Pastor, Bernhard A. Schrefler, Tadahiko Shiomi and O. C. Zienkiewicz.

The decomposition between elastic and plastic parts was introduced for plane stress conditions by Prandtl in 1924 and, later, in 1930, by Reuss for general conditions of stress. Reuss proposed a flow rule for the plastic component. The idea of plastic potential was suggested by von Mises in a work presented in 1928, where the normal to the yield surface was used to provide the direction of plastic flow.

Hardening plasticity was studied by Melan, who, in 1938, generalized previously established concepts of plasticity to account for this effect.

However, the framework of what is known today as Classical Plasticity was established in 1949 by Drucker, who introduced many of the concepts of modern plasticity such as loading surface, loading and unloading, neutral loading, consistency, and uniqueness.

Since then, much development has taken place, motivated by the development of both faster computers and numerical methods for boundary value problems.

There exist today a great variety of models able to deal with most of the observed features of the mechanical behavior of materials. Describing in detail all the existing approaches would exceed by far the length of a chapter, and would require a full book. Therefore, we have selected generalized plasticity as a suitable framework within which many features of soil behavior can be modeled. We have chosen the approach of starting with simple models and then extending them to deal with more complex materials and loadings. This is why we start presenting classical plasticity as a particular case of generalized plasticity, describing basic concepts such as yield, plastic flow, and failure. These concepts have been applied to many engineering materials. In the case of soils, most models are cast within the framework of Critical State Soil Mechanics theory. We have devoted a section to describe critical state-based models, starting with normally consolidated clays, and explaining how the theory was extended to granular soils and the concept of state parameter. The core of this chapter is Section 4.4 deals with generalized plasticity models, from the very first simple model for clays and sands under monotonic and cyclic loading proposed by the Swansea group led by the late Prof. O.C. Zienkiewicz, Pastor et al (1990) to recent developments for unsaturated soils, and materials with crushable grains. Then, a short description of other advanced modeling frameworks such as bounding surface and hypoplasticity are succinctly presented. The chapter closes with a reflection on how discrete elements are able to properly model complex loading paths.

4.2 The General Framework of Plasticity

4.2.1 Phenomenological Aspects

The uniaxial behavior of materials shows that irreversible strain develops in a way that depends on the type of material. In the case of metals such as mild steel, the observed behavior in tension is schematized in Figure 4.1, where it can be seen that the response is elastic and linear until a point A is reached, from which plastic or irreversible strain upon unloading appears. If the specimen is subjected to an increasing strain, the stress does not change until point E. Along the plateau ABDE, the material behavior is known as perfectly plastic. If the specimen is unloaded, both loading and unloading follow the same path, without

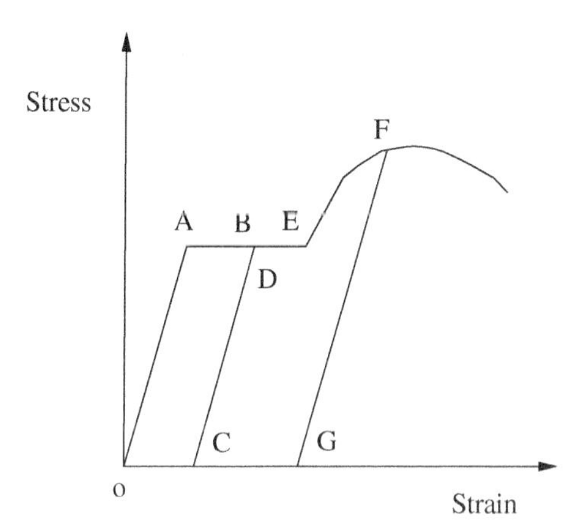

Figure 4.1 Behavior of mild steel

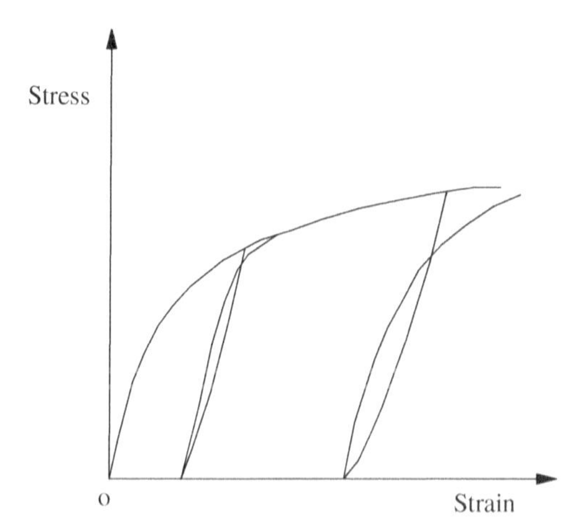

Figure 4.2 Behavior of soft clay

irreversible deformation. The level of stress at which plastic strains appear does not change and the material does not harden.

Once a certain level of strain has been reached (E), the stress again increases. If we unload at some point F and then reload again, the material is able to resist a higher load until new plastic strains develop (hardening). Finally, a maximum load is reached from which the stress decreases until the material fails.

In the case of soft soils such as saturated clays, the stress–strain curve is different, as plastic strains are present from the very early stages of the test (Figure 4.2).

Finally, some geomaterials such as concrete, present degradation due to damage caused by the loading process to the structure of the material (Figure 4.3).

Loading and unloading clearly show how the apparent elastic modulus of the material degrades as the test progresses. A full understanding of this behavior needs to take into account this process of degradation and theories, such as Damage Mechanics, provide a suitable framework. However, plastic models can be developed to reproduce the observed behavior with an acceptable degree of accuracy.

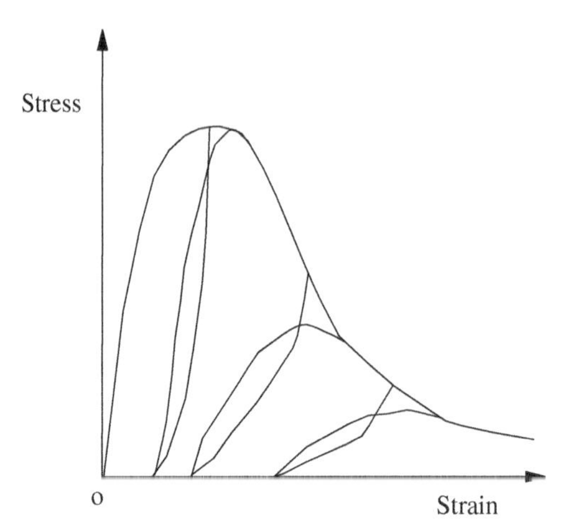

Figure 4.3 Behavior of materials with damage

4.2.2 Generalized Plasticity

In the following, boldface characters will be used for tensors, uppercase (such as $\mathbf{D}$) denoting fourth-order tensors $\mathbf{D}_{\mathrm{ijkl}}$, and lower case (such as $\boldsymbol{\sigma}$) for second-order tensors σ_{ij}.

It is convenient to use a vector–matrix representation of tensorial magnitudes in numerical computations; fourth-order tensors corresponding to matrices and second-order tensors to vectors. In Chapters 2–3, we have introduced this notation. We shall indicate here the small alteration necessary to return to the matrix notation used in the previous chapters where $\boldsymbol{\sigma}$ and $\mathbf{D}$ are vectors and matrices, respectively.

The convention for products and its matrix equivalence is:

a) Double dot denotes contracted product in last two indexes

$$\begin{aligned} \mathbf{A}:\mathbf{B} &\equiv A_{ijkl}B_{klmn} \equiv\gg \mathrm{AB} \equiv A_{ij}B_{jk} \\ \mathbf{A}:\mathbf{b} &\equiv A_{ijkl}b_{kl} \equiv\gg \mathrm{Ab} \equiv A_{ij}b_{j} \\ \mathbf{a}:\mathbf{b} &\equiv a_{ij}b_{ji} \equiv\gg \mathrm{a}^{t}\mathrm{b} \equiv a_{i}b_{i} \end{aligned} \tag{4.1}$$

b) Tensor products are expressed by

$$\mathbf{a}\otimes\mathbf{b} \equiv a_{ij}b_{kl} \equiv\gg \mathbf{ab}^{T} = a_{i}b_{j} \tag{4.2}$$

The behavior of geomaterials depends on effective stresses as shown in Chapter 1, which are denoted by a dash. However, in the first part of this chapter, devoted to the Introduction to Elastoplastic Constitutive Equations, we will not use the dash when referring to stress for the sake of simplicity.

4.2.2.1 Basic Theory

If the response of the material does not depend on the velocity at which the stress varies, the relationship between the increments of stress and strain can be written as

$$\mathrm{d}\varepsilon = \boldsymbol{\Phi}(\mathrm{d}\boldsymbol{\sigma}) \tag{4.3}$$

where $\mathbf{\Phi}$ is a function of the increment of the stress tensor $\mathrm{d}\boldsymbol{\sigma}$ and variables describing the "state" (or history) of the material. This is a general relation embracing most nonlinear, rate-independent constitutive laws.

An inverse form is

$$\mathrm{d}\boldsymbol{\sigma} = \mathbf{\Psi}(\mathrm{d}\boldsymbol{\varepsilon}) \tag{4.4}$$

As the material response does not depend on time,

$$\lambda \mathrm{d}\boldsymbol{\varepsilon} = \mathbf{\Phi}(\lambda \mathrm{d}\boldsymbol{\sigma})$$

where $\lambda \in \Re_+$ is a positive scalar (Darve 1990).

Consequently, $\mathbf{\Phi}$ is a homogeneous function of degree 1, which can be written as

$$\mathbf{\Phi} = \frac{\partial \mathbf{\Phi}}{\partial(\mathrm{d}\boldsymbol{\sigma})} : \mathrm{d}\boldsymbol{\sigma} \tag{4.5}$$

from which the increments of stress and strain are related by

$$\begin{aligned} \mathrm{d}\boldsymbol{\varepsilon} &= \mathbf{C} : \mathrm{d}\boldsymbol{\sigma} \\ \mathrm{d}\boldsymbol{\sigma} &= \mathbf{D} : \mathrm{d}\boldsymbol{\varepsilon} \end{aligned} \tag{4.6}$$

where

$$\mathbf{C} = \frac{\partial \mathbf{\Phi}}{\partial(\mathrm{d}\boldsymbol{\sigma})} \tag{4.7}$$

is a fourth-order tensor, homogeneous, of degree zero in $\mathrm{d}\boldsymbol{\sigma}$. Before continuing, some basic properties of $\mathbf{C}$ will be described.

We will consider a uniaxial loading-unloading-reloading test schematized in Figure 4.4 where the constitutive tensor $\mathbf{C}$ is a scalar, the inverse of the slope at the point considered.

As can be seen, the slope depends on the stress level, being smaller at higher stresses. However, if we compare the slopes at points A_1, A_2, and A_3, they are not the same, and $\mathbf{C}$ depends on past history (stresses, strains, modification of material microstructure, etc.).

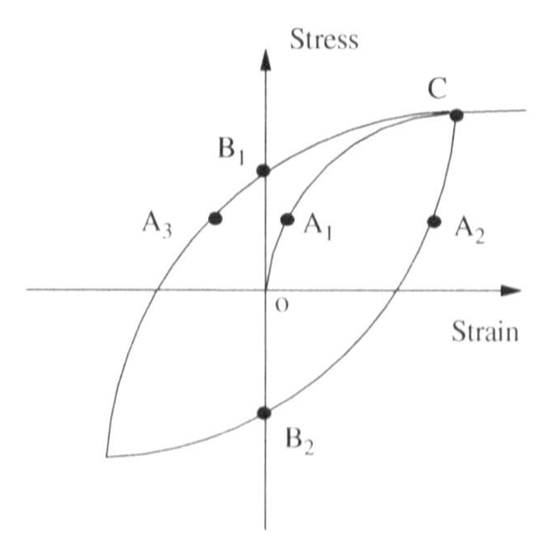

Figure 4.4 General stress–strain behavior

Taking a closer look at point C, it can be seen that, for a given point, different slopes are obtained in "loading" and "unloading," which implies a dependence on the direction of stress increment.

This dependence is only on the direction as **C** is a homogeneous function of degree zero on $\mathrm{d}\boldsymbol{\sigma}$.

Therefore, in this simple one-dimensional case, it is possible to write for loading

$$\mathrm{d}\boldsymbol{\varepsilon}_L = \mathbf{C}_L : \mathrm{d}\boldsymbol{\sigma} \tag{4.8}$$

and for unloading

$$\mathrm{d}\boldsymbol{\varepsilon}_U = \mathbf{C}_U : \mathrm{d}\boldsymbol{\sigma}$$

We observe that if we consider an infinitesimal cycle with $\mathrm{d}\boldsymbol{\sigma}$ followed by $-\mathrm{d}\boldsymbol{\sigma}$, the total change of strain is not zero as

$$\mathrm{d}\boldsymbol{\varepsilon} = \mathrm{d}\boldsymbol{\varepsilon}_L + \mathrm{d}\boldsymbol{\varepsilon}_U = (\mathbf{C}_L - \mathbf{C}_U) : \mathrm{d}\boldsymbol{\sigma} \neq 0 \tag{4.9}$$

This kind of constitutive law has been defined by Darve (1990) as incrementally nonlinear.

There are several alternatives to introduce the dependence on the direction of the stress increment, among which it is worth mentioning the multilinear laws proposed by Darve and coworkers in Grenoble (Darve and Labanieh 1982) or the hypoplastic laws of Dafalias (1986) or Kolymbas (1991). However, the simplest consists of defining in the stress space a normalized direction **n** for any given state of stress $\boldsymbol{\sigma}$ such that all possible increments of stress are separated into two classes, loading and unloading

$$\begin{aligned} \mathrm{d}\boldsymbol{\varepsilon}_L &= \mathbf{C}_L : \mathrm{d}\boldsymbol{\sigma} \quad \text{for} \quad \mathbf{n} : \mathrm{d}\boldsymbol{\sigma} > 0 \quad \text{(loading)} \\ \mathrm{d}\boldsymbol{\varepsilon}_U &= \mathbf{C}_U : \mathrm{d}\boldsymbol{\sigma} \quad \text{for} \quad \mathbf{n} : \mathrm{d}\boldsymbol{\sigma} < 0 \quad \text{(unloading)} \end{aligned} \tag{4.10}$$

Neutral loading corresponds to the limit case for which

$$\mathbf{n} : \mathrm{d}\boldsymbol{\sigma} = 0 \tag{4.11}$$

This is the starting point of the Generalized Theory of Plasticity, introduced by Zienkiewicz and Mroz (Mroz and Zienkiewicz 1985; Zienkiewicz and Mroz 1985) and later extended by Pastor and Zienkiewicz (Pastor et al. 1985, 1990; Zienkiewicz et al. 1985).

Introduction of this direction discriminating between loading and unloading defines a set of surfaces which is equivalent to those used in Classical Plasticity as will be shown later, but these surfaces need never be explicitly defined.

Continuity between loading and unloading states requires that constitutive tensors for loading and unloading are of the form

$$\mathbf{C}_L = \mathbf{C}^e + \frac{1}{H_L}\mathbf{n}_{gL} \otimes \mathbf{n} \tag{4.12}$$

and

$$\mathbf{C}_U = \mathbf{C}^e + \frac{1}{H_U}\mathbf{n}_{gU} \otimes \mathbf{n}$$

where $\mathbf{n}_{gL}$ and $\mathbf{n}_{gU}$ are arbitrary tensors of the unit norm and $H_{L/U}$ two scalar functions defined as loading and unloading plastic moduli. It can be very easily verified that both laws predict the same strain increment under neutral loading where both expressions are valid and hence nonuniqueness is avoided. As for such loading, the increments of strain using the expressions for loading and unloading are

$$d\boldsymbol{\varepsilon}_L = \mathbf{C}_L : d\boldsymbol{\sigma} = \mathbf{C}^e : d\boldsymbol{\sigma} \tag{4.13}$$

and

$$d\boldsymbol{\varepsilon}_U = \mathbf{C}_U : d\boldsymbol{\sigma} = \mathbf{C}^e : d\boldsymbol{\sigma}$$

It follows that material behavior under neutral loading is reversible and it can therefore be regarded as elastic. Indeed, the tensor $\mathbf{C}^{\mathbf{e}}$ characterizes elastic material behavior, and it can be very easily verified that for any infinitesimal cycle of stress $(d\boldsymbol{\sigma}, -d\boldsymbol{\sigma})$, where $d\boldsymbol{\sigma}$ corresponds to neutral loading conditions, the accumulated strain is zero.

This suggests that the strain increment can be decomposed into two parts

$$d\boldsymbol{\varepsilon} = d\boldsymbol{\varepsilon}^e + d\boldsymbol{\varepsilon}^p \tag{4.14}$$

where

$$d\boldsymbol{\varepsilon}^e = \mathbf{C}^e : d\boldsymbol{\sigma} \tag{4.15}$$

and

$$d\boldsymbol{\varepsilon}^p = \frac{1}{H_{L/U}} \left(\mathbf{n}_{gL/U} \otimes \mathbf{n}\right) : d\boldsymbol{\sigma} \tag{4.16}$$

We note that irreversible plastic deformations have been introduced without the need for specifying any yield or plastic potential surfaces, nor hardening rules. All that is necessary to specify are two scalar functions $H_{L/U}$ and three directions, $\mathbf{n}_{gL/U}$ and $\mathbf{n}$.

To account for softening behavior of the material, i.e. when H_L is negative, definitions of loading and unloading have to be modified as follows:

$$\begin{aligned} d\boldsymbol{\varepsilon}_L &= \mathbf{C}_L : d\boldsymbol{\sigma} \quad \text{for} \quad \mathbf{n} : d\boldsymbol{\sigma}^e > 0 \ \text{(loading)} \\ d\boldsymbol{\varepsilon}_U &= \mathbf{C}_U : d\boldsymbol{\sigma} \quad \text{for} \quad \mathbf{n} : d\boldsymbol{\sigma}^e < 0 \ \text{(unloading)} \end{aligned} \tag{4.17}$$

where $d\boldsymbol{\sigma}^e$ is given by

$$d\boldsymbol{\sigma}^e = \mathbf{C}^{e-1} : d\boldsymbol{\varepsilon} \tag{4.18}$$

We note here (and in what follows) that in matrix notation, the product forms are written simply as $d\boldsymbol{\varepsilon}_L = \mathbf{C}_L.d\boldsymbol{\sigma}$ for; $\mathbf{n}^{\mathrm{T}}.d\boldsymbol{\sigma}^e > 0$, etc.

4.2.2.2 Inversion of the Constitutive Tensor

Implementation of a constitutive model into finite element codes requires on many occasions an inversion of the constitutive tensor in order to express the increment of stress as a function of the strain increment. This inversion can only be automatically performed when the plastic modulus H is different from zero. Should this not be the case, the inversion would have to be carried out according to the procedure described below (Zienkiewicz and Mroz 1985).

First of all, a scalar λ is introduced

$$d\lambda = \frac{1}{H_{L/U}}\mathbf{n} : d\boldsymbol{\sigma} \tag{4.19}$$

and the increment of strain is written as

$$d\boldsymbol{\varepsilon} = \mathbf{C}^e : d\boldsymbol{\sigma} + d\lambda\mathbf{n}_{gL/U} \tag{4.20}$$

Both sides of the above equation are now multiplied by $\mathbf{n} : \mathbf{D}^e$

$$\mathbf{n} : \mathbf{D}^e : d\boldsymbol{\varepsilon} = (\mathbf{n} : \mathbf{D}^e) : (\mathbf{C}^e : d\boldsymbol{\sigma}) + (\mathbf{n} : \mathbf{D}^e) : d\lambda\mathbf{n}_{gL/U} \tag{4.21}$$

from which we obtain

$$\mathbf{n} : \mathbf{D}^e : d\boldsymbol{\varepsilon} = \mathbf{n} : d\boldsymbol{\sigma} + d\lambda(\mathbf{n} : \mathbf{D}^e : \mathbf{n}_{gL/U}) \tag{4.22}$$

where we have taken into account that the product $\mathbf{D}^e : \mathbf{C}^e$ is the fourth-order identity tensor.

Substituting now

$$\mathbf{n} : d\boldsymbol{\sigma} = d\lambda H_{L/U} \tag{4.23}$$

we obtain

$$\mathbf{n} : \mathbf{D}^e : d\boldsymbol{\varepsilon} = (H_{L/U} + \mathbf{n} : \mathbf{D}^e : \mathbf{n}_{gL/U})d\lambda \tag{4.24}$$

and

$$d\lambda = \frac{\mathbf{n} : \mathbf{D}^e : d\boldsymbol{\varepsilon}}{H_{L/U} + \mathbf{n} : \mathbf{D}^e : \mathbf{n}_{gL/U}} \tag{4.25}$$

If we now multiply by $\mathbf{D}^e$ both sides of

$$d\boldsymbol{\varepsilon} = \mathbf{C}^e : d\boldsymbol{\sigma} + d\lambda\mathbf{n}_{gL/U} \tag{4.26}$$

we have

$$d\boldsymbol{\sigma} = \mathbf{D}^e : d\boldsymbol{\varepsilon} - d\lambda\mathbf{D}^e : \mathbf{n}_{gL/U} \tag{4.27}$$

Substitution of the value of $d\lambda$ gives

$$d\boldsymbol{\sigma} = \mathbf{D}^e : d\boldsymbol{\varepsilon} - \frac{(\mathbf{D}^e : \mathbf{n}_{gL/U})(\mathbf{n} : \mathbf{D}^e : d\boldsymbol{\varepsilon})}{H_{L/U} + \mathbf{n} : \mathbf{D}^e : \mathbf{n}_{gL/U}} \tag{4.28}$$

which can be written as

$$\begin{aligned} d\boldsymbol{\sigma} &= \mathbf{D}^{ep} : d\boldsymbol{\varepsilon} \\ \mathbf{D}^{ep} &= \mathbf{D}^e - \frac{\mathbf{D}^e : \mathbf{n}_{gL/U} \otimes \mathbf{n} : \mathbf{D}^e}{H_{L/U} + \mathbf{n} : \mathbf{D}^e : \mathbf{n}_{gL/U}} \end{aligned} \tag{4.29}$$

If we make use of the vectorial formulation to represent tensors, the above expression can be written as

$$\mathbf{D}^{ep} = \mathbf{D}^e - \frac{\mathbf{D}^e \cdot \mathbf{n}_{gL/U} \cdot \mathbf{n}^T \cdot \mathbf{D}^e}{H_{L/U} + \mathbf{n}^T \cdot \mathbf{D}^e \cdot n_{gL/U}} \tag{4.30}$$

4.2.3 Classical Theory of Plasticity

4.2.3.1 Formulation as a Particular Case of Generalized Plasticity Theory

Classical Plasticity Theory can be considered as a particular case of the Generalized Theory described above by a suitable choice of the plastic modulus, directions $\mathbf{n}$ and $\mathbf{n}_{gL/U}$ and the elastic constitutive tensor.

A yield surface is first introduced as

$$f(\boldsymbol{\sigma}, \kappa) = 0 \tag{4.31}$$

where we have assumed that there is a set of scalar internal variables κ accounting for the material state and characterizing the size (and shape) of the yield surface. Sometimes, as will be discussed later, f depends also on the tensor variable α, as in the case of kinematic and anisotropic hardening models, for instance. Here we will restrict the discussion to the isotropic case stated above.

In the interior of the yield surface, there is no plastic deformation, and, consequently, the plastic modulus is $H = \infty$.

The loading–unloading direction is given by the normal to the surface

$$\mathbf{n}_{L/U} = \frac{\frac{\partial f}{\partial \sigma}}{\left|\frac{\partial f}{\partial \sigma}\right|} \tag{4.32}$$

where

$$\left|\frac{\partial f}{\partial \sigma}\right| = \left(\frac{\partial f}{\partial \sigma} : \frac{\partial f}{\partial \sigma}\right)^{\frac{1}{2}}$$

The direction of plastic flow is similarly derived from a plastic potential surface $g(\boldsymbol{\sigma}) = 0$ passing through the stress point considered,

$$\mathbf{n}_{gL/U} = \frac{\frac{\partial g}{\partial \sigma}}{\left|\frac{\partial g}{\partial \sigma}\right|} \tag{4.33}$$

Both surfaces can coincide, and the flow rule is then said to be associative or can be different in which case there is a nonassociative flow rule.

Therefore, the material behavior predicted by Classical Plasticity models presents a sharp transition from the elastic to the elastoplastic regime, with a discontinuity in the derivative of stress–strain curves.

The plastic modulus is obtained through the application of the so-called "consistency condition," i.e. the requirement that during yield, the stress point should always remain on the yield surface. A certain "hardening law" has to be introduced, relating $d\kappa$ to either incremental plastic work or to the increment of plastic strain.

4.2.3.2 Yield and Failure Surfaces

Following experimental evidence, plasticity theories postulate that irreversible or plastic strain appears whenever the stress reaches a surface $f(\sigma_{ij}, \kappa) = 0$. For all stress states in the interior of this surface, material behavior is elastic and $f(\sigma_{ij}, \kappa) < 0$. If κ is constant, the material cannot sustain higher stress and failure takes place. This is the reason why the yield surface is also known as the failure surface. Care should be taken, however, as in the case of materials with hardening, these surfaces can be different.

The scalar κ usually characterizes the size of the surface. This is, of course, a simplification, and more complex descriptions are available, such as

$$f\left(\sigma_{ij}, \alpha_{ij}, \kappa\right) = 0 \tag{4.34}$$

or

$$f\left(\sigma_{ij}, \alpha_{ij}, \kappa_1, \kappa_2, \kappa_3, \ldots\right) = 0$$

If the material is isotropic, the representation theorems of scalar functions of tensor variables allow a simpler expression for f

$$f(I_1, J_2, J_3, \kappa) = 0 \tag{4.35}$$

or

$$f(\sigma_1, \sigma_2, \sigma_3, \kappa) = 0$$

which can be further simplified to

$$f(I_1, J_2, J_3) - Y(\kappa) = 0 \tag{4.36}$$

or

$$f(\sigma_1, \sigma_2, \sigma_3) - Y(\kappa) = 0 \tag{4.37}$$

where $Y(\kappa)$ is generally some measure of strength.

I_1 is the first invariant of the stress tensor,

$$I_1 = \sigma_1 + \sigma_2 + \sigma_3 = \sigma_{\kappa\kappa} \tag{4.38}$$

J_2 and J_3 are the second and third invariants of the deviatoric stress tensor **s**,

$$s = \boldsymbol{\sigma} - \frac{1}{3} I_1 \mathbf{I} \tag{4.39}$$

$$J_2 = \frac{1}{2} tr\left(s^2\right) = \frac{1}{2} s_{ij} s_{ji} \tag{4.40}$$

$$J_3 = \frac{1}{3} tr\left(s^3\right) = \frac{1}{3} s_{ij} s_{j\kappa} s_{\kappa i} \tag{4.41}$$

and σ_1, σ_2, and σ_3 are the three principal stresses.

At this stage, it is convenient to define also the Lode's angle θ often used instead of J_3.

$$\theta = \frac{1}{3}\sin^{-1}\left(-\frac{3\sqrt{3}}{2}\frac{J_3}{J_2^{3/2}}\right) \tag{4.42}$$

with

$$-\frac{\pi}{6} \leq \theta \leq \frac{\pi}{6}$$

4.2.3.3 Hardening, Softening, and Failure

It is important to distinguish between the yield surface, inside which behavior of the material is elastic, and the failure surface, where failure takes place. To illustrate this, consider the example given in Figure 4.5 where a specimen of soft clay is being loaded from an initial state P_1 to failure at P_3. There, yield surfaces are the ellipses $f - (\kappa) = 0$.

The parameter κ in this case is associated with the (negative) plastic volumetric strain, i.e.

$$\mathrm{d}\kappa = -\mathrm{d}\varepsilon_v^p \tag{4.43}$$

and in Figure 4.5, we show the yield surface in the space of two stress invariants, the second, or deviatoric invariant and the first or mean, hydrostatic stress invariant. With each of these is associated an appropriate strain component $-n_v$ being the component in the direction of decreasing volumetric strain if plasticity is assumed associated. Thus, the three stages of loading P_1, P_2, and P_3 correspond to increasing values of κ as shown in Figure 4.5b.

It has to be noticed that plastic strain appears from the beginning of the test, as the initial stress is on the yield surface. If, for instance, we unload at P_2, there will exist a permanent deformation even when the stress has come back to the original state.

The process of increasing the size of the yield surface in this case is known as hardening. Comparing the conditions at P_1 and P_2, the elastic domain is bigger in the latter, and the material is harder in this sense. Notice that slopes of the stress–strain curves contradict this definition as the incremental response of the material is harder in the first case.

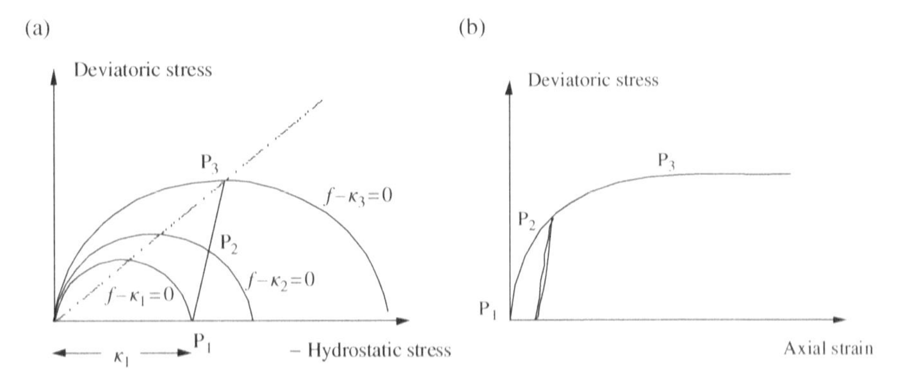

Figure 4.5 Typical hardening behavior of clay. (a) Yield surfaces (b) Stress–strain curve showing permanent strain upon unloading

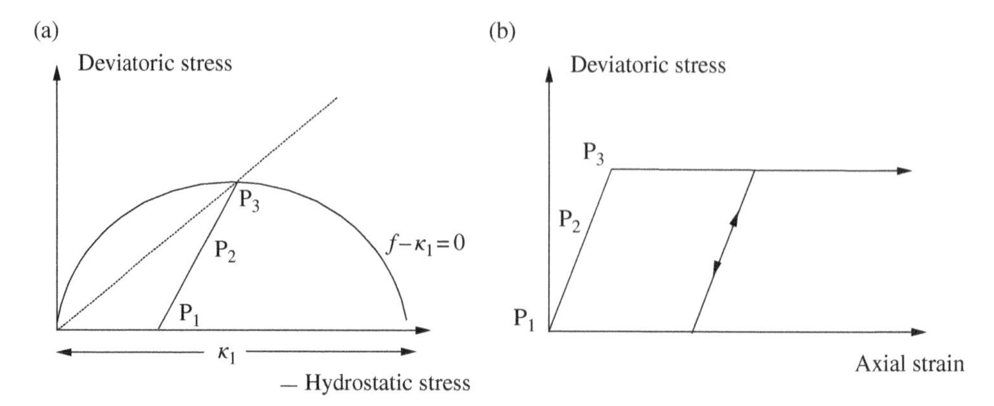

Figure 4.6 Ideal plasticity (κ = constant) (a) stress path; (b) stress–strain curve

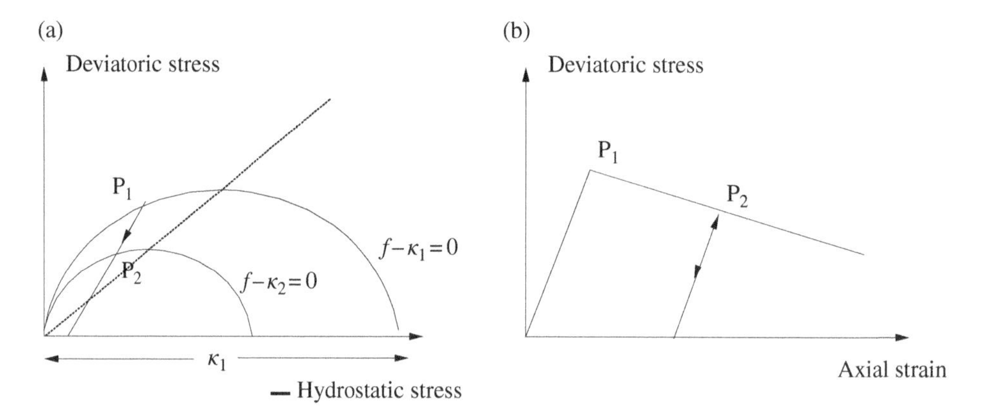

Figure 4.7 Softening behavior (a) stress path; (b) stress–strain curve

Hardening is not a common feature of all materials. Indeed, in the case shown in Figure 4.6, the size did not change and failure takes place as soon as the yield surface is reached.

In another loading case, the size of the yield surface may decrease, as shown in Figure 4.7, and softening behavior occurs.

4.2.3.4 Some Frequently Used Failure and Yield Criteria. Pressure-Independent Criteria: von Mises–Huber Yield Criterion

von Mises–Huber yield criterion assumes that plastic strain appears whenever the second invariant of the stress tensor reaches a critical value Y^2,

$$f = J_2 - Y^2 = 0 \tag{4.44}$$

where $Y(\kappa)$ is generally the tensile strength.

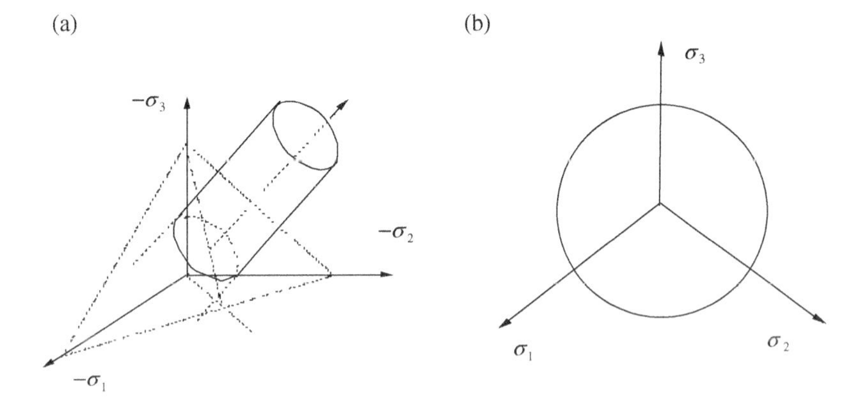

Figure 4.8 von Mises–Huber yield criterion. (a) In the principal stress space; (b) Section by Π plane

Alternative expressions are

i) In the principal stress space

$$f = (\sigma_1 - \sigma_2)^2 + (\sigma_2 - \sigma_3)^2 + (\sigma_1 - \sigma_3)^2 - 6Y^2 = 0 \tag{4.45}$$

ii) In general stress conditions

$$f = \left(\sigma_{xx} - \sigma_{yy}\right)^2 + \left(\sigma_{yy} - \sigma_{zz}\right)^2 + \left(\sigma_{zz} - \sigma_{xx}\right)^2 + 6\sigma_{xy}^2 + 6\sigma_{yz}^2 + 6\sigma_{zx}^2 - 6Y^2 = 0 \tag{4.46}$$

Taking into account that the condition $J_2 =$ constant corresponds to stress states σ_1, σ_2, and σ_3 such that the distance to the hydrostatic axis $\sigma_1 = \sigma_2 = \sigma_3$ is constant, von Mises criterion is represented in principal stress axes as a cylinder of radius $\sqrt{2J_2} = \sqrt{2}Y$ which is schematized in Figure 4.8a. In the same figure, we show a plane perpendicular to the hydrostatic axis, which is referred to as the Π plane. Its intersection with the von Mises cylinder is a circle, which is shown in Figure 4.8b.

A simple method of determining the constant Y is to perform a tension test $\sigma_2 = \sigma_3 = 0$ and to determine the instant at which plastic strain develops. If the value of limiting tensile stress is σ_Y, then we obtain

$$f = (\sigma_Y - 0)^2 + (0 - 0)^2 + (0 - \sigma_Y)^2 - 6Y^2 = 0 \tag{4.47}$$

from which

$$Y = \frac{1}{\sqrt{3}}\sigma_Y \tag{4.48}$$

In plane stress conditions, $\sigma_3 = 0$, and the expression of the criterion in principal stress axes is

$$f = (\sigma_1 - \sigma_2)^2 + \sigma_2^2 + \sigma_1^2 - 6Y^2 = 0 \tag{4.49}$$

which corresponds in the σ_1 and σ_2 axes to an ellipse with principal axes at 45° (Figure 4.9).

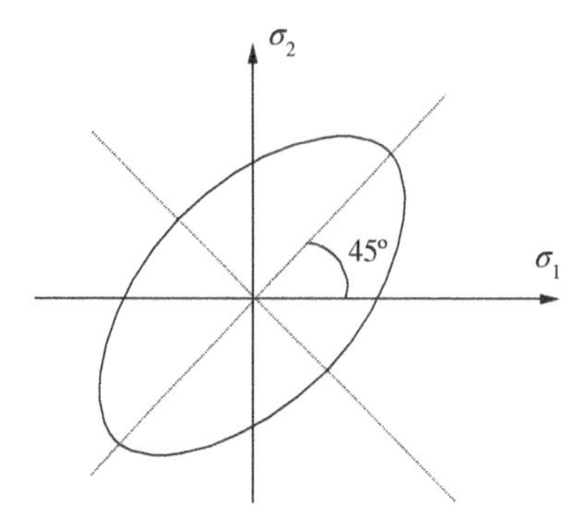

Figure 4.9 von Mises criterion for plane stress conditions

Tresca Criterion

The Tresca criterion, proposed in 1864, is based on the assumption that plastic straining of material appears when the maximum shear strain reaches a critical value Y. This condition, expressed in terms of the principal stresses, reads

$$(\sigma_{max} - \sigma_{min}) = Y \tag{4.50}$$

Substituting now the maximum and minimum principal stresses by their values in terms of the invariants I_1, J_2 and Lode's angle θ

$$\begin{pmatrix} \sigma_1 \\ \sigma_2 \\ \sigma_3 \end{pmatrix} = \frac{I_1}{3}\begin{pmatrix} 1 \\ 1 \\ 1 \end{pmatrix} + \frac{2\sqrt{J_2}}{\sqrt{3}}\begin{pmatrix} \sin\left(\theta + \frac{2\pi}{3}\right) \\ \sin(\theta) \\ \cos\left(\theta + \frac{4\pi}{3}\right) \end{pmatrix} \tag{4.51}$$

Noting that

$$\begin{aligned} \sigma_{max} &= \frac{I_1}{3} + \frac{2\sqrt{J_2}}{\sqrt{3}} \sin\left(\theta + \frac{2\pi}{3}\right) \\ \sigma_{min} &= I_1 + \frac{2\sqrt{J_2}}{\sqrt{3}} \cos\left(\theta + \frac{4\pi}{3}\right) \end{aligned} \tag{4.52}$$

we can write finally

$$2\sqrt{J_2}\cos\theta = Y \tag{4.53}$$

When plotted in the space of principal stresses, the Tresca yield criterion is a hexagonal prism, with its axis coincident with the hydrostatic axis $\sigma_1 = \sigma_2 = \sigma_3$ (Figure 4.10a). The section by the Π-plane is a regular hexagon as can be seen in Figure 4.10b.

Finally, the plane stress condition $\sigma_2 = 0$ is represented by

$$\begin{aligned} \frac{1}{2}\sigma_1 &= \pm Y \\ \frac{1}{2}\sigma_3 &= \pm Y \\ \frac{1}{2}(\sigma_1 - \sigma_3) &= \pm Y \end{aligned} \tag{4.54}$$

which are shown in Figure 4.11.

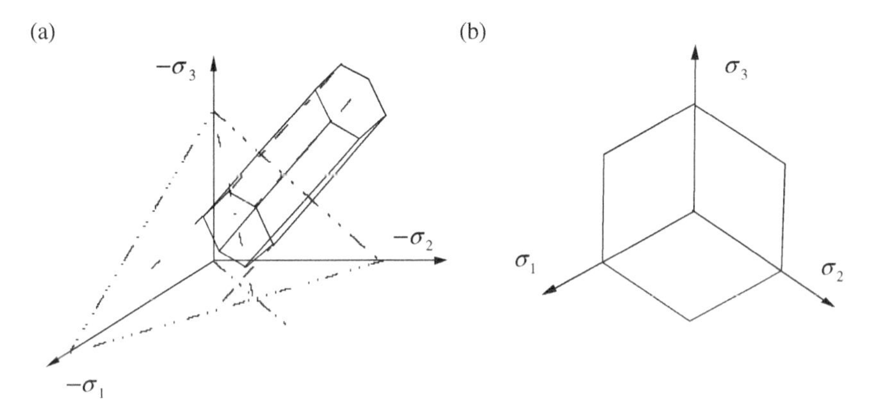

Figure 4.10 Tresca yield criterion. (a) In principal stress axes (b) in the Π-plane

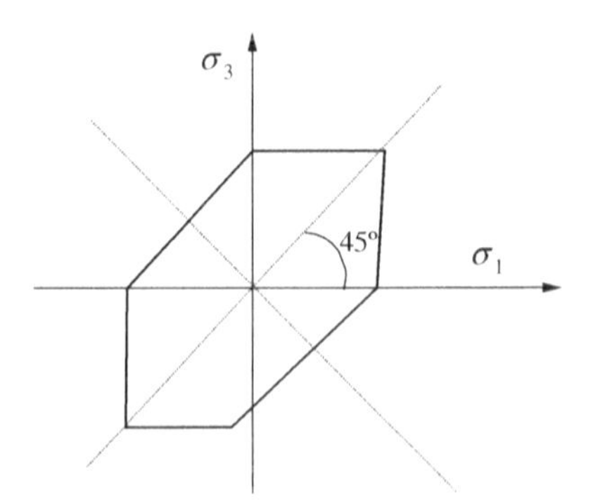

Figure 4.11 Tresca criterion for plane stress conditions

Pressure-Dependent Criteria: Mohr–Coulomb Surface

In 1773, Coulomb proposed the law

$$\tau = c - \sigma_n \tan \phi \tag{4.55}$$

to describe the conditions under which failure takes place in soils. He assumed that failure occurs on a plane on which the shear stress τ and the normal stress σ_n (compression negative) fulfill the above condition. Although it is not advisable to think of it as a yield surface, it has been used frequently in engineering practice and most finite element codes include it.

In terms of principal stresses or invariants, we will write

$$\tau = \frac{\sigma_1 - \sigma_3}{2} \cos \phi \tag{4.56}$$

and

$$\sigma_n = \frac{\sigma_1 + \sigma_3}{2} + \frac{\sigma_1 - \sigma_3}{2} \sin \phi \tag{4.57}$$

which can be obtained from geometrical considerations (Figure 4.12).

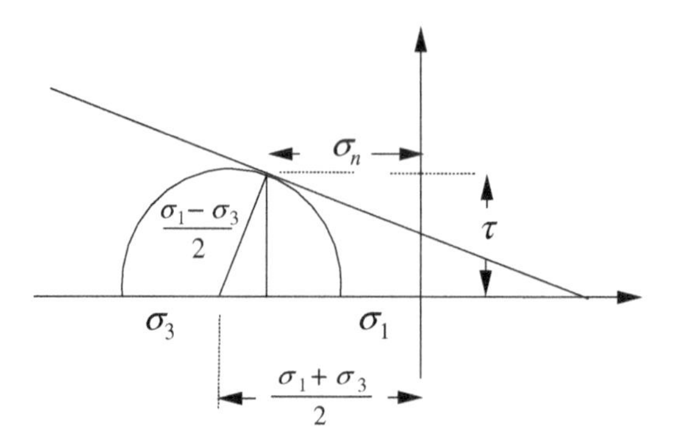

Figure 4.12 Mohr–Coulomb law

This results in

$$\frac{\sigma_1 - \sigma_3}{2}\cos\phi = c - \left\{\frac{\sigma_1 + \sigma_3}{2} + \frac{\sigma_1 - \sigma_3}{2}\sin\phi\right\}\tan\phi$$
$$(\sigma_1 - \sigma_3) = 2c\cos\phi - (\sigma_1 + \sigma_3)\sin\phi \tag{4.58}$$

and

$$(\sigma_1 + \sigma_3)\sin\phi + (\sigma_1 - \sigma_3) - 2c\cos\phi = 0$$

From above, using the relationships between principal stresses and invariants, it is easy to obtain

$$\frac{1}{3}I_1\sin\phi + \sqrt{J_2}\left\{\cos\theta - \frac{1}{\sqrt{3}}\sin\theta\sin\phi\right\} - c\cos\phi = 0 \tag{4.59}$$

The Mohr–Coulomb criterion is represented in the space of principal stresses as a hexagonal pyramid, which has been depicted in Figure 4.13.

Drucker–Prager Criterion

The Drucker–Prager criterion is an attempt to create a smooth approximation to the Mohr–Coulomb surface in the same manner as von Mises approximates Tresca. The surface is written as

$$\alpha I_1 + \sqrt{J_2} - Y = 0 \tag{4.60}$$

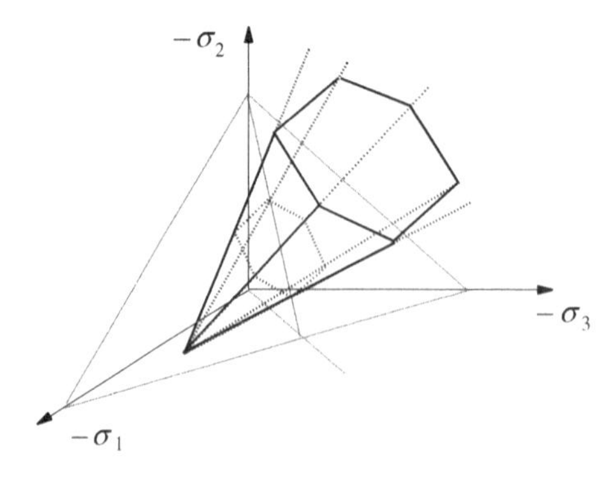

Figure 4.13 Mohr–Coulomb yield surface

and, when plotted in the space of principal stresses, consists of a cone in which the axis is coincident with the hydrostatic axis.

The section of this cone through the Π plane is a circle and when plotted in the mean hydrostatic pressure – deviatoric stress plane, the intersection with it consists of two lines with identical slope (compression and extension). Therefore, the friction angles corresponding to compression and extension are different, and, in fact, given the parameter α and a value of Lode's angle θ, if the intersections of the Drucker–Prager cone and the Mohr–Coulomb surfaces are to coincide for a certain value of Lode's angle θ, the relationship between the friction angle and α is

$$\alpha = \frac{1/3 \sin \phi}{\cos \theta - \frac{1}{\sqrt{3}} \sin \theta \sin \phi} \tag{4.61}$$

A similar relationship between cohesion and the parameter Y can easily be obtained as

$$Y = \frac{c \cos \phi}{\cos \theta - \frac{1}{\sqrt{3}} \sin \theta \sin \phi} \tag{4.62}$$

These relationships have to be taken into account when trying to use the Drucker–Prager criterion for plane strain conditions and what is known from experiments is cohesion and angle of friction.

Under cylindrical triaxial conditions, i.e. $\sigma^{\mathrm{T}} = (\sigma_1, \sigma_2 = \sigma_3)$, the angles of friction in compression and extension are different and can be obtained using the above relationship with $\theta = \pi/6$ and $\theta = -\pi/6$

$$\alpha_{\mathrm{c}} = \frac{2 \sin \phi}{\cos \frac{\pi}{6} - \frac{1}{\sqrt{3}} \sin \frac{\pi}{6} \sin \phi} \tag{4.63}$$

from where

$$\alpha_{\mathrm{c}} = \frac{2 \sin \phi}{3\sqrt{3} - \sqrt{3} \sin \phi} \tag{4.64}$$

and

$$\sin \phi = \frac{3\sqrt{3}\alpha_c}{2 + \sqrt{3}\alpha_c} \tag{4.65}$$

In a similar way,

$$\alpha_{\mathrm{e}} = \frac{2 \sin \phi}{3\sqrt{3} + \sqrt{3} \sin \phi} \tag{4.66}$$

$$\sin \phi = \frac{3\sqrt{3}\alpha_e}{2 - \sqrt{3}\alpha_e} \tag{4.67}$$

The values of Y for compression and extension are

$$Y_{\mathrm{c}} = \frac{6c \cos \phi}{\sqrt{3}(3 - \sin \phi)} \tag{4.68}$$

and

$$Y_e = \frac{6c\cos\phi}{\sqrt{3}(3+\sin\phi)} \tag{4.69}$$

Finally, it can be seen that for a given value of α, the relationship between angles of friction in extension and compression is

$$\frac{\sin\phi_e}{\sin\phi_c} = \frac{2+\sqrt{3}\alpha}{2-\sqrt{3}\alpha} \tag{4.70}$$

It is left to the reader as an exercise to demonstrate that there is a value of $\sin\phi_c$ for which the friction angle in extension reaches $\pi/2$.

It is interesting to mention that "rounded" Mohr–Coulomb surfaces have been proposed in the past (Zienkiewicz and Pande 1977) in which the slope M is assumed to vary as:

$$M = \frac{6\sin\phi}{3-\sin\phi\sin 3\theta}$$

In this way, the yield surface is smooth and coincides with the Mohr–Coulomb original surface in both triaxial compression and extension conditions.

4.2.3.5 Consistency Condition for Strain-Hardening Materials

If we assume that the material hardening is of "strain" type, there will exist a law relating to the increments of κ and ε

$$d\kappa = d\kappa(d\varepsilon^p) \tag{4.71}$$

Assuming that yielding occurs on the yield surface given by

$$df(\sigma,\kappa) = 0 \tag{4.72}$$

and hence

$$\frac{\partial f}{\partial \sigma} : d\boldsymbol{\sigma} + \frac{\partial f}{\partial \kappa} : d\kappa = 0 \tag{4.73}$$

This can be rewritten using (4.71) as

$$\frac{\partial f}{\partial \sigma} : d\boldsymbol{\sigma} + \frac{\partial f}{\partial \kappa} : \frac{\partial \kappa}{\partial \varepsilon^p} : d\varepsilon^p = 0 \tag{4.74}$$

If, in the above expression, we substitute $d\varepsilon^p$, it gives

$$\frac{\partial f}{\partial \sigma} : d\boldsymbol{\sigma} + \frac{\partial f}{\partial \kappa} : \frac{\partial \kappa}{\partial \varepsilon^p} : \mathrm{n}_{gL}\left(\frac{1}{H_L}\mathrm{n} : d\boldsymbol{\sigma}\right) = 0 \tag{4.75}$$

This expression can be further developed to

$$\left(\frac{\partial f}{\partial \sigma} : ds\right) + \frac{1}{\left|\frac{\partial f}{\partial \sigma}\right| \cdot \left|\frac{\partial g}{\partial \sigma}\right|}\left(\frac{\partial f}{\partial \kappa} : \frac{\partial \kappa}{\partial \varepsilon^p} : \frac{\partial g}{\partial \sigma}\right)\frac{1}{H_L}\left(\frac{\partial f}{\partial s} : ds\right) = 0 \tag{4.76}$$

from which the plastic modulus is finally obtained as

$$H_L = -\frac{\dfrac{\partial f}{\partial \kappa} : \dfrac{\partial \kappa}{\partial \varepsilon^p} : \dfrac{\partial g}{\partial \sigma}}{\left|\dfrac{\partial f}{\partial \sigma}\right| \cdot \left|\dfrac{\partial g}{\partial \sigma}\right|} \tag{4.77}$$

Using the alternative vector notation, the above expression is written as

$$H_L = -\frac{\dfrac{\partial f}{\partial \kappa} \cdot \dfrac{\partial \kappa}{\partial \varepsilon^p} \cdot \dfrac{\partial g}{\partial \sigma}}{\left|\dfrac{\partial f}{\partial \sigma}\right| \cdot \left|\dfrac{\partial g}{\partial \sigma}\right|} \tag{4.78}$$

where $\partial\kappa/\partial\boldsymbol{\varepsilon}^p$ is a square matrix.

Local failure conditions or continuing deformation at a constant stress state can happen whenever $H_L = 0$, which corresponds to:

$$\frac{\partial \kappa}{\partial \varepsilon^p} \cdot \frac{\partial g}{\partial \sigma} = 0 \tag{4.79}$$

for which either of the following conditions has to be fulfilled

$$(a)\frac{\partial \kappa}{\partial \varepsilon^p} = 0 \quad \text{(saturation)} \tag{4.80}$$

or

$$(b)\frac{\partial \kappa}{\partial \varepsilon^p} \cdot \frac{\partial g}{\partial \sigma} = 0 \quad \text{with} \quad \frac{\partial \kappa}{\partial \varepsilon^p} \neq 0$$

4.2.3.6 Computational Aspects

Most of the expressions given in the above sections simplify in the case of isotropic materials as all necessary items can be defined in terms of stress invariants. The yield surface, for instance, can be expressed as

$$f(I_1, J_2, J_3, \kappa) = 0 \tag{4.81}$$

or in terms of another alternative set of invariants.

The constitutive tensor $\mathbf{D}^{ep}$ which appears in nonlinear finite element computations can be expressed, as it was shown above, as a function of directions $\mathbf{n}$ and $\mathbf{n}_g$, and the scalar H, all of them dependent on the invariants.

However, what is needed in computations is the general three-dimensional form. Therefore, the expressions have to be transformed from the space of invariants to the general 3D space.

In the case of the Classical Theory of Plasticity, the constitutive tensor $\mathbf{D}^{ep}$ is written in vector notation as

$$\mathbf{D}^{ep} = \mathbf{D}^e - \frac{\mathbf{D}^e \cdot \dfrac{\partial g}{\partial \sigma} \cdot \dfrac{\partial f^T}{\partial \sigma} \cdot \mathbf{D}^e}{A + \dfrac{\partial f^T}{\partial \sigma} \cdot \mathbf{D}^e \cdot \dfrac{\partial g}{\partial \sigma}} \tag{4.82}$$

where we have introduced

$$A = H \cdot \left|\frac{\partial g}{\partial \sigma}\right| \left|\frac{\partial f}{\partial \sigma}\right| \tag{4.83}$$

having dropped, for simplicity, subindexes L/U referring to loading and unloading. This is precisely the Generalized Plasticity expression 4.28 with

$$\mathbf{n} = \frac{\partial f}{\partial \sigma} \quad \text{and} \quad \mathbf{n}_g = \frac{\partial g}{\partial \sigma} \tag{4.84}$$

A simple way to obtain either gradient in terms of invariants I_1, J_2, and θ is the following

$$\mathbf{n} = \frac{\partial f}{\partial \sigma} = \frac{\partial f}{\partial I_1}\frac{\partial I_1}{\partial \sigma} + \frac{\partial f}{\partial J_2}\frac{\partial J_2}{\partial \sigma} + \frac{\partial f}{\partial \theta}\frac{\partial \theta}{\partial \sigma} \tag{4.85}$$

where

$$\sin 3\theta = -\frac{3\sqrt{3}}{2}\frac{J_3}{J_2^{3/2}} \tag{4.86}$$

Differentiating this expression, we arrive at

$$\begin{aligned} 3\cos 3\theta \frac{\partial \theta}{\partial \sigma} &= -\frac{3\sqrt{3}}{2}\frac{1}{J_2^{3/2}}\frac{\partial J_3}{\partial \sigma} - \frac{3\sqrt{3}}{2}J_3\frac{\partial}{\partial \sigma}\left(\sqrt{J_2^{-3}}\right) \\ &= -\frac{3\sqrt{3}}{2}\frac{1}{J_2^{3/2}}\frac{\partial J_3}{\partial \sigma} - \frac{3\sqrt{3}}{2}J_3(-3)\frac{1}{J_2^2}\frac{\partial \sqrt{J_2}}{\partial \sigma} \end{aligned} \tag{4.87}$$

and, from here,

$$\frac{\partial \theta}{\partial \sigma} = -\frac{\sqrt{3}}{2\cos 3\theta}\left(\frac{1}{J_2^{3/2}}\frac{\partial J_3}{\partial \sigma} - \frac{3J_3}{J_2^2}\frac{\partial \sqrt{J_2}}{\partial \sigma}\right) \tag{4.88}$$

$$\mathbf{n} = \frac{\partial f}{\partial I_1}\frac{\partial I_1}{\partial \sigma} + \left(\frac{\partial f}{\partial \sqrt{J_2}} + \frac{3\sqrt{3}}{2\cos 3\theta}\frac{\partial f}{\partial \theta}\frac{J_3}{J_2^2}\right)\frac{\partial \sqrt{J_2}}{\partial \boldsymbol{\sigma}} - \frac{\sqrt{3}}{2\cos 3\theta}\frac{1}{J_2^{3/2}}\frac{\partial J_3}{\partial \sigma} \tag{4.89}$$

Taking into account now that

$$\frac{3\sqrt{3}}{2}\frac{J_3}{J_2^{3/2}}\frac{1}{\cos 3\theta}\frac{1}{\sqrt{J_2}} = -\frac{\sin 3\theta}{\cos 3\theta}\frac{1}{\sqrt{J_2}} = -\tan 3\theta \frac{1}{\sqrt{J_2}}$$

we have, finally,

$$\mathbf{n} = \frac{\partial f}{\partial I_1}\frac{\partial I_1}{\partial \sigma} + \left(\frac{\partial f}{\partial \sqrt{J_2}} + \frac{\tan 3\theta}{\sqrt{J_2}}\frac{\partial f}{\partial \theta}\right)\frac{\partial \sqrt{J_2}}{\partial \sigma} - \frac{\sqrt{3}}{2\cos 3\theta}\frac{1}{J_2^{3/2}}\frac{\partial J_3}{\partial \sigma}\frac{\partial f}{\partial \theta} \tag{4.90}$$

which can be written in a more compact way as:

$$\mathbf{n} = C_1\mathbf{n}_1 + C_2\mathbf{n}_2 + C_3\mathbf{n}_3 \tag{4.91}$$

where

$$C_1 = \frac{\partial f}{\partial I_1}$$
$$C_2 = \left(\frac{\partial f}{\partial \sqrt{J_2}} - \frac{\tan 3\theta}{\sqrt{J_2}}\frac{\partial f}{\partial \theta}\right) \tag{4.92}$$
$$C_3 = -\frac{\sqrt{3}}{2\cos 3\theta}\frac{1}{J_2^{3/2}}\frac{\partial f}{\partial \theta}$$

and

$$\mathbf{n}_1 = \frac{\partial I_1}{\partial \sigma}$$
$$\mathbf{n}_2 = \frac{\partial \sqrt{J_2}}{\partial \sigma} \tag{4.93}$$
$$\mathbf{n}_3 = \frac{\partial J_3}{\partial \sigma}$$

It can be seen that the set of constants $\{C_i\}$ depends on the yield criterion chosen, being independent of the vectors $\mathbf{n}_i$.

Next, we will obtain the explicit form of $\mathbf{n}_1$, $\mathbf{n}_2$, and $\mathbf{n}_3$.

Vector $\mathbf{n}_1$ is given by

$$\mathbf{n}_1 = \frac{\partial I_1}{\partial \sigma} = \frac{\partial \sigma_{\kappa\kappa}}{\partial \sigma_{ij}} = \frac{\partial(\sigma_{ik}\delta_{ki})}{\partial \sigma_{ij}} = \delta_{jk}\delta_{ki} = \delta_{ij} \tag{4.94}$$

or,

$$\frac{\partial}{\partial \sigma_x}\{\sigma_x + \sigma_y + \sigma_z\} = 1$$
$$\frac{\partial}{\partial \sigma_y}\{\sigma_x + \sigma_y + \sigma_z\} = 1$$
$$\frac{\partial}{\partial \sigma_z}\{\sigma_x + \sigma_y + \sigma_z\} = 1 \tag{4.95}$$
$$\frac{\partial I_1}{\partial \tau_{xy}} = \frac{\partial I_1}{\partial \tau_{yz}} = \frac{\partial I_1}{\partial \tau_{zx}} = 0$$

from which, in vector notation,

$$\mathbf{n}_1^{\mathrm{T}} = \{1, 1, 1, 0, 0, 0\} \tag{4.96}$$

To obtain vector $\mathbf{n}_2$, we will use tensor notation

$$\mathbf{n}_2 = \frac{\partial \sqrt{J_2}}{\partial \sigma} = \frac{1}{2\sqrt{J_2}}\frac{\partial J_2}{\partial \sigma} \tag{4.97}$$

Taking into account that

$$\mathbf{s} = \boldsymbol{\sigma} - \frac{I_1}{3}\boldsymbol{\delta} \tag{4.98}$$

$$J_2 = \frac{1}{2}\left(\boldsymbol{\sigma} - \frac{I_1}{3}\boldsymbol{\delta}\right) : \left(\boldsymbol{\sigma} - \frac{I_1}{3}\boldsymbol{\delta}\right) \tag{4.99}$$

we arrive at

$$J_2 = \frac{1}{2}\left(\boldsymbol{\sigma} : \boldsymbol{\sigma} - \frac{2}{3}I_1\boldsymbol{\delta} : \boldsymbol{\sigma} + \frac{I_1^2}{9}\boldsymbol{\delta} : \boldsymbol{\delta}\right) \tag{4.100}$$

and

$$J_2 = \frac{1}{2}\left(\boldsymbol{\sigma} : \boldsymbol{\sigma} - \frac{I_1^2}{3}\right) \tag{4.101}$$

from which it follows that

$$\begin{aligned}\frac{\partial J_2}{\partial \sigma_{ij}} &= \frac{1}{2}\left(\frac{\partial(\sigma_{kl}\sigma_{kl})}{\partial \sigma_{ij}} - \frac{1}{3}\frac{\partial\left(\sigma_{kk}^2\right)}{\partial \sigma_{ij}}\right) \\ &= \left(\sigma_{ij} - \frac{1}{3}\delta_{ij}\sigma_{kk}\right) = s_{ij}\end{aligned} \tag{4.102}$$

Therefore,

$$\mathbf{n}_2 = \frac{1}{2\sqrt{J_2}}\mathbf{s} \tag{4.103}$$

Vector $\mathbf{n}_3$ is given by

$$\mathbf{n}_3 = \frac{\partial J_3}{\partial \sigma} \tag{4.104}$$

where

$$J_3 = \frac{1}{3}s_{ij}s_{jk}s_{ki}$$

After some algebra, the final expression for $\mathbf{n}_3$ is

$$\mathbf{n}_3 = \begin{Bmatrix} s_{yy}s_{zz} - s_{yz}^2 + J_2/3 \\ s_{zz}s_{xx} - s_{zx}^2 + J_2/3 \\ s_{xx}s_{yy} - s_{xy}^2 + J_2/3 \\ 2\left(s_{xz}s_{xy} - s_{xx}s_{yz}\right) \\ 2\left(s_{xy}s_{yz} - s_{yy}s_{xz}\right) \\ 2\left(s_{yz}s_{zx} - s_{zz}s_{xy}\right) \end{Bmatrix} \tag{4.105}$$

Constants C_1, C_2, and C_3 depend on the yield criterion chosen. In the case of the von Mises yield criterion, which can be written as

$$f = J_2 - Y^2 = 0 \tag{4.106}$$

or

$$f = \sqrt{J_2} - Y = 0 \tag{4.107}$$

we find

$$C_1 = 0$$
$$C_2 = 1 \qquad (4.108)$$
$$C_3 = 0$$

Sometimes, an alternative expression using the 1D yield stress $\sigma_Y = \sqrt{3J_2}$ is used

$$f = \sqrt{3J_2} - \sigma_Y = 0 \qquad (4.109)$$

and then C_2 is $\sqrt{3}$.

An important aspect which we will not fully describe here is that of integration of the constitutive equations of classical plasticity, which can be found in the texts by Borja (2013) and Tamagnini et al. (2002).

4.3 Critical State Models

4.3.1 Introduction

Constitutive modelling of soil behavior is a keystone in the process of predicting the behavior of a geostructure. No finite element code will provide results of better quality than that of the constitutive equation implemented in it.

Today, there are a great variety of models able to deal with situations ranging from simple monotonic stress paths to cyclic loading, rotation of principal stress axes, and anisotropy.

This has been made possible because of extensive work developed in laboratories throughout the world. In the past years, coordination of effort between different groups has increased, and, as an important result, benchmark tests on some selected reference materials have been made available to constitutive modelers. Here, the initiatives of laboratories such as 3S in Grenoble, CERMES in Paris (both part of GRECO and GEO networks in France), and Case Western University in Cleveland (USA) have to be mentioned.

Most of the proposed benchmark tests deal with key aspects of granular and cohesive soil behavior:

1) Isotropic consolidation. Loading, unloading, and reloading. Memory effects.
2) Shear behavior in axisymmetric triaxial tests:
   - Drained and undrained tests
   - Effects of density and confining pressure
   - Liquefaction of loose sands
   - Memory effects: overconsolidation.
3) Unloading, reloading, and cyclic loading:
   - Densification
   - Pore pressure buildup. Liquefaction and cyclic mobility.
4) Three-dimensional effects

5) Anisotropy:
   - Material
   - Induced by loading.

These progressively more sophisticated tests have helped develop constitutive models of increasing complexity. There exists, however, a dramatic gap between these recently developed constitutive models and those used in day-to-day engineering practice. Several factors have contributed to it:

- Many industrial Finite Element codes do not implement, still, constitutive models suitable to realistic geotechnical analysis.
- To calibrate more advanced models, the engineer needs to be acquainted with them.
- Special laboratory tests are frequently required to obtain material parameters which cannot be obtained by direct observation from raw data.

The last section was devoted to the introduction of elastoplastic constitutive models in the framework of Generalized Plasticity, and it was shown there how Classical Plasticity models can be considered as particular cases of the theory. The simple models of Tresca and von Mises present severe limitations when applied to geomaterials in general and soils in particular. Drucker and Prager proposed, in 1952, an elastic-perfectly plastic constitutive model with an associated flow rule which could be applied to limit analysis problems. However, this model is not able to describe plastic deformations inside the yield surface cone as occur in common engineering situations. Moreover, the associated behavior is not valid as it would predict large dilatancy at failure.

Later, in 1957, Drucker, Gibson, and Henkel introduced an elastoplastic model including two fundamental ingredients, a closed yield surface which consisted of a cone and a circular cap, and a hardening law dependent on density, paving the way to modern plasticity.

At the same time, extensive research on the basic properties of soils in triaxial conditions was carried out at Cambridge University (Henkel 1956, 1960; Parry 1960) and these ideas were further elaborated, arriving not only at practical expressions describing volumetric hardening, but also at the concept of a line in the (e, p', q) space where the residual states lie. This line was referred to as the Critical State Line and is one of the basic ingredients of Critical State Theory introduced by the Cambridge group (Schofield and Wroth 1958; Roscoe and Burland 1968; Roscoe et al. 1958; Roscoe, Schofield and Thurairajah 1963).

The work on Cam Clay model continues, being worth mentioning the work done by Prof. R.I. Borja and coworkers Borja and Lee (1990), Borja (1991), Borja and Tamagnini (1998), Borja et al. (2001), Borja (2004), and Borja and Andrade (2006).

The purpose of this part is to describe classical elastoplastic models for soils, together with their limitations, which made it necessary to introduce new concepts.

4.3.2 Critical State Models for Normally Consolidated Clays

4.3.2.1 Hydrostatic Loading: Isotropic Compression Tests

One of the basic features of soil behavior is the importance of its density on its behavior. Both cohesive and granular soils exhibit changes of density caused by

- a change in effective confining pressure p' and
- changes of arrangement of grains in the structure induced by shearing of the material.

The simplest case in which the first mechanism occurs is hydrostatic loading of a soil specimen in which the confining pressure is varied. The process is carried out slowly enough to prevent the development of interstitial pore pressure (drained conditions).

If the initial state of stress is

$$\boldsymbol{\sigma}_0'^{\mathrm{T}} = -p'(1,1,1,0,0,0) = -p'\mathbf{m} \tag{4.110}$$

and the specimen is loaded according to

$$\Delta\boldsymbol{\sigma}'^{\mathrm{T}} = -\Delta p'(1,1,1,0,0,0) = -\Delta p'\mathbf{m} \tag{4.111}$$

the stress path will consist of a segment of a straight line along the hydrostatic axis $\sigma_1' = \sigma_2' = \sigma_3'$, or along the axis $q = 0$ if we are using the plane (p', q) (Figure 4.14). In the above, we have considered compressions as negative.

The change of volume can be described either by the volumetric strain,

$$\varepsilon_v = -tr(\boldsymbol{\varepsilon}) = -\mathbf{m}^{\mathrm{T}} \cdot \boldsymbol{\varepsilon} \tag{4.112}$$

where we have used the minus sign for consistency with the definition of p', or the change of voids ratio e

$$de = -d\varepsilon_v(1+e) \tag{4.113}$$

We will consider now a loading-unloading-reloading process 1-2-3-4-5-6-7-8 in which hydrostatic pressure p' is increased from p_1' to p_2', and then the specimen is unloaded to p_3' and reloaded again to p_4'. This cycle is followed by a loading branch 4–5, with a final pressure of p_5' (Figure 4.15a)

It can be seen that unloading and reloading branches differ, although volumetric strain developing in branch 2-3-4 can be considered to be reversible. However, irreversible plastic deformation occurs from 1 to 2 and from 4 to 5. This behavior is typical of soft clays and can be sketched as shown in Figure 4.15b.

If time effects can be neglected, the response of soft cohesive soil can be idealized in the $\ln(p') - e$ plot as a line of slope λ (Points 1-2-5-8)

$$de = -\lambda \frac{dp'}{p'} \tag{4.114}$$

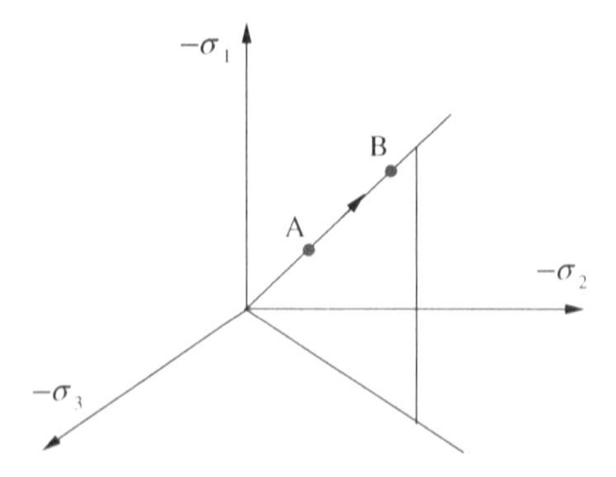

Figure 4.14 Hydrostatic compression stress path

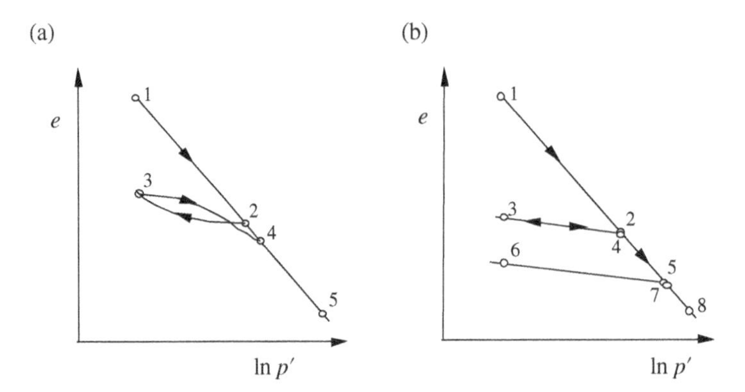

Figure 4.15 Hydrostatic compression test on a normally consolidated clay. (a) Experimental results; (b) Idealized behavior

or, alternatively

$$d\varepsilon_v = \frac{\lambda}{1+e}\frac{dp'}{p'} \tag{4.115}$$

This line is often referred to as the "Normal Consolidation Line," (NCL) and is one of the basic ingredients of modern plasticity models for soils. The parameter λ depends on the type of soil and it can be related to the Plasticity Index (PI) by the empirical relation (Atkinson and Bransby 1978)

$$\lambda = \mathrm{PI}/171$$

It is important to note that if a Mohr–Coulomb or a Drucker–Prager yield surface had been used, no plastic deformation would have been produced, and, therefore, this is a severe limitation of all finite element codes implementing, as unique options, yield criteria which are open in the hydrostatic axis. If such plastic deformation needs to be reproduced, closed yield criteria should be used instead. Figure 4.16 illustrates this fact. If the stress increases from point 1 to 2, both states are inside yield surface f_1, and therefore, no plastic strain is produced in the process. The solution is to use yield surfaces intersecting the hydrostatic axis $q = 0$. Loading from 1 to 2 expands the yield surface which can be assumed to harden as the soil densifies.

The observed volumetric strain can be decomposed into elastic and plastic parts according to

$$d\varepsilon_v = d\varepsilon_v^e + d\varepsilon_v^p = \frac{\lambda}{1+e}\frac{dp'}{p'} \tag{4.116}$$

During unloading from 2 to 3 and subsequent reloading to 4, the behavior will be purely elastic

$$d\varepsilon_v = d\varepsilon_v^e = \frac{\kappa}{1+e}\frac{dp'}{p'} \tag{4.117}$$

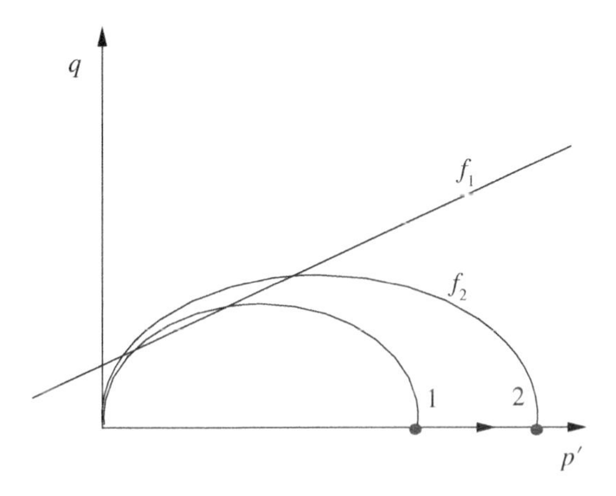

Figure 4.16 Open and closed yield surfaces

where κ is a new constant characterizing the elastic volumetric response. It can be related to the bulk modulus by

$$K_v = \frac{1+e}{\kappa} p' \tag{4.118}$$

Once the stress point reaches the yield surface again, a plastic strain develops, the yield surface expands and the soil continues hardening.

A simple law relating to the size of the yield surface, which will be denoted by p_c to the plastic volumetric strain, is obtained as

$$d\varepsilon_v^p = d\varepsilon_v - d\varepsilon_v^e = \frac{(\lambda - \kappa)}{1+e} \frac{dp'_c}{p'_c} \tag{4.119}$$

from where

$$\frac{dp'_c}{d\varepsilon_v^p} = \left(\frac{1+e}{\lambda - \kappa}\right) p'_c \tag{4.120}$$

The subscript "c" refers to consolidation and the process 1-2-5 is referred to as "isotropic consolidation."

If a soil specimen that has been subjected in the past to a consolidation pressure of $p'_c = p'_2$ is tested at a lower pressure p'_3, it will be possible to observe in the curve $e - \ln p'$ a change of slope such as depicted in Figure 4.15a in the branch 3-4-5. This soil is referred to as "overconsolidated," while soils at the NCL are called "normally consolidated." Both the concepts can be easily understood in the framework of plasticity, as overconsolidated soils are characterized by the stress state being inside the yield surface at the initial state.

The "Overconsolidation Ratio," or OCR, is a parameter measuring the degree of overconsolidation,

$$\text{OCR} = \frac{p'_{c\,\max}}{p'_c} \tag{4.121}$$

Of course, these definitions apply only to simple hydrostatic loading conditions, but can be generalized to more complex situations where the OCR will be the ratio of the measures of two stress states.

4.3.2.2 Triaxial Rest

So far, we have considered only stress paths where no shear strain is induced unless the soil is anisotropic. It was seen that isotropic compression results in densification and hardening of soil and it was mentioned that another mechanism causing densification was shear. Here we will concentrate in the shear behavior of normally consolidated clays subjected to symmetric or cylindrical triaxial stress conditions

$$\sigma = (\sigma_1, \sigma_2 = \sigma_3, \sigma_3)^{\mathrm{T}} \tag{4.122}$$

which hereafter will be referred to as "triaxial."

The stress conditions applied in a triaxial cell are sketched in Figure 4.17, and consist of a cell pressure σ_3 applied through a fluid, usually water, and a vertical additional load $(\sigma_1 - \sigma_3)$ referred to as a "deviator" applied with a ram.

The triaxial test is commonly used in the laboratory to determine soil properties as the desired stress paths can be reproduced quite accurately. There are several problems like membrane penetration, inhomogeneities caused by the development of narrow zones where the strain localizes, known as "shear bands," and friction with the upper and lower rigid caps. In addition to these, it is worth mentioning those problems related to accurate measurement of vertical loads, changes in specimen cross section, homogenization of pore pressures inside the specimen, and measurements of axial, radial, and volumetric strain.

Two main types of tests are currently used: (i) consolidated drained; and (ii) consolidated undrained. In the first case, a saturated soil specimen is brought to an initial state of hydrostatic stress

$$\boldsymbol{\sigma}' = (\sigma_0', \sigma_0', \sigma_0')^{\mathrm{T}} = -p'(1,1,1) \tag{4.123}$$

where the pore pressure is zero.

The load is applied slowly to avoid pore pressure buildup and once the initial conditions are reached, the vertical load applied through the ram is increased. The stress path is

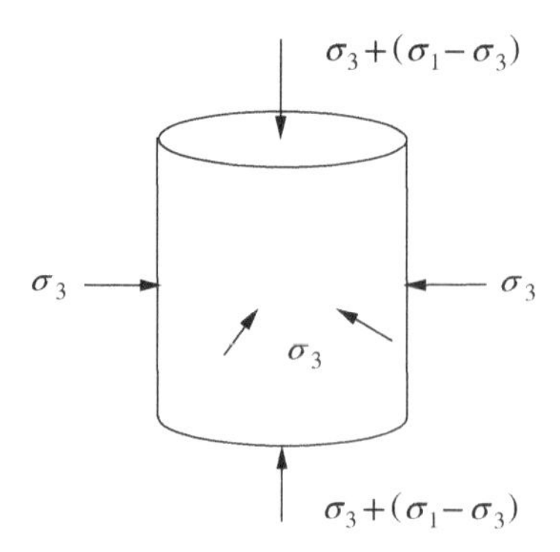

Figure 4.17 Triaxial stress conditions

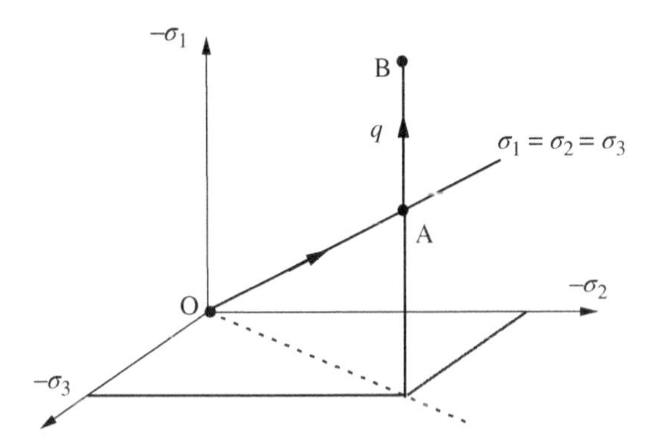

Figure 4.18 Consolidated drained stress path

depicted in Figure 4.18. As pore pressures are zero (drained conditions), the total and effective stresses coincide.

In this figure, it can be seen that both the hydrostatic pressure p' and the deviatoric stress are increasing. The stress path can be studied either in the space of principal stresses, or, alternatively, in the spaces (I_1', J_2, θ) or (p', q, θ). The last is very convenient as these invariants are given by

$$p' = -\frac{1}{3}(\sigma_1' + \sigma_2' + \sigma_3') = -\frac{1}{3}(\sigma_1' + 2\sigma_3') \tag{4.124}$$

and

$$q = \sqrt{3J_2} \tag{4.125}$$

with

$$J_2 = \frac{1}{6}\left[(\sigma_1' - \sigma_3')^2 + (\sigma_1' - \sigma_3')^2 + (0)^2\right] = \frac{1}{3}(\sigma_1' - \sigma_3')^2 \tag{4.126}$$

from which we have

$$q = -(\sigma_1' - \sigma_3') \tag{4.127}$$

which is precisely the stress induced by the vertical load applied through the ram. In the above, both stresses are negative and we have supposed that the absolute value of σ_1' is higher than that of σ_3'.

The measures of strain are those work-associated to p' and q

$$\varepsilon_v = -(\varepsilon_1 + 2\varepsilon_3) \tag{4.128}$$

$$\varepsilon_s = -\frac{2}{3}(\varepsilon_1 - \varepsilon_3) \tag{4.129}$$

Concerning Lode's angle, it is kept constant during the test, provided that $(\sigma_1' - \sigma_3')$ does not change sign during the test. This fact occurs when applying compression–extension cycles.

The stress path in the (p', q) plane is shown in Figure 4.19, where it can be seen that due to its positive slope, and for a given absolute value of the deviator q, the angle AOB is smaller

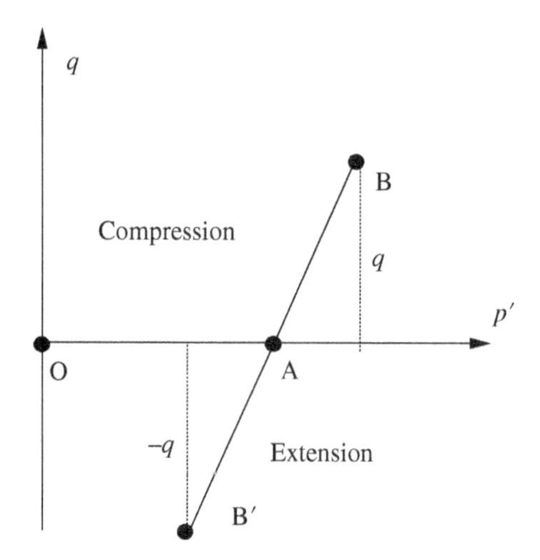

Figure 4.19 Consolidated drained stress path in $p' - q$ plane

than AOB′. Therefore, if the failure surface is of Mohr–Coulomb type, $M_c \geq M_e$, the soil will fail earlier in extension. The slope of the stress path can be easily obtained as

$$\Delta p' = -\frac{1}{3}\left(\Delta\sigma_1' + 2\Delta\sigma_3'\right) \tag{4.130}$$

and

$$\Delta q = -\left(\Delta\sigma_1' - \Delta\sigma_3'\right) \tag{4.131}$$

from where, taking into account that σ_3' is constant,

$$\frac{\Delta q}{\Delta p'} = 3 \tag{4.132}$$

The results obtained in compression tests on normally consolidated clays are similar to those depicted in Figure 4.20. The main features are the following:

- There is a tendency of the soil to compact as the test proceeds, caused by the increase of p' and a rearrangement of soil particles.
- Failure takes place at a certain value of stress ratio $\eta = M$ for tests performed at different confining pressures.
- Soil strength and compaction depend on confining pressure and increase with it.

The second type of triaxial test is the Consolidated Undrained (CU) test, where, after consolidation, the drainage valve is closed to prevent dissipation of pore pressure. The test has to be carried out slowly enough for the pore pressure to be homogeneous through the specimen. During the test, measurements of pore pressure, axial strain, vertical stress, and cell pressure are taken to monitorize the stress path. Figure 4.21 shows typical results obtained in drained consolidated clays.

It can be seen how the effective stress path bends toward the origin as a consequence of pore pressure increase caused by the tendency of soil to compact. Again the failure takes place at a line of slope M_c, which coincides with that obtained in drained tests. This test

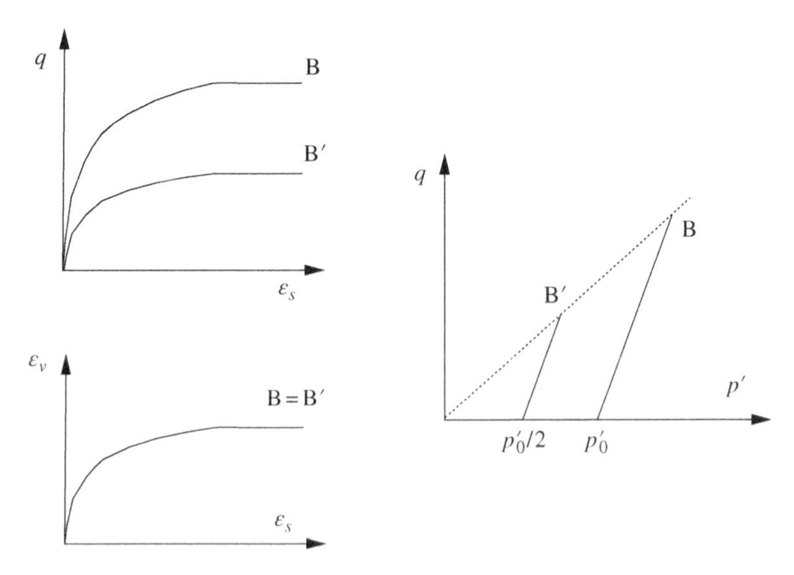

Figure 4.20 Typical results of CD tests on normally consolidated clays

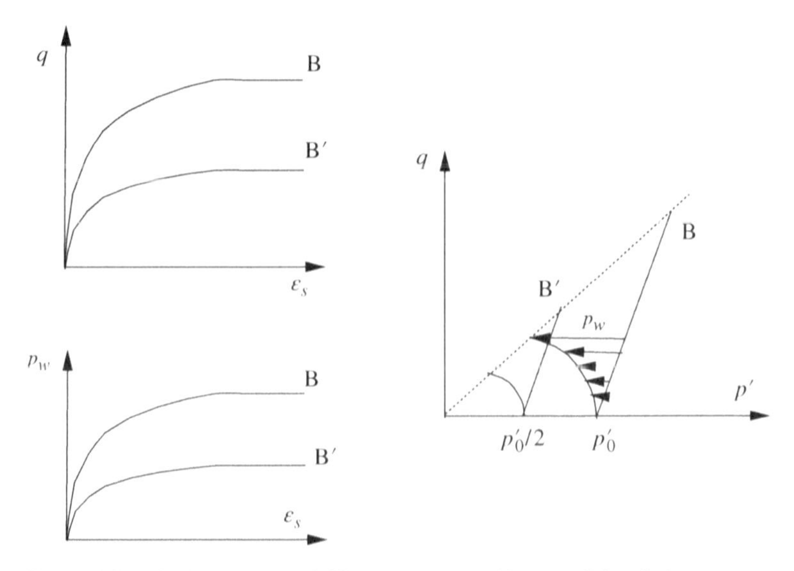

Figure 4.21 Typical results of CU tests on normally consolidated clays

has been classically used to characterize soil behavior under "fast" loading, where pore pressures do not have time to dissipate, in short-term stability analysis. Of course, this is a simplification and a complete coupled analysis should be performed instead.

It has to be noticed that soil strength is lower in undrained than in drained conditions because of the generated pore pressures.

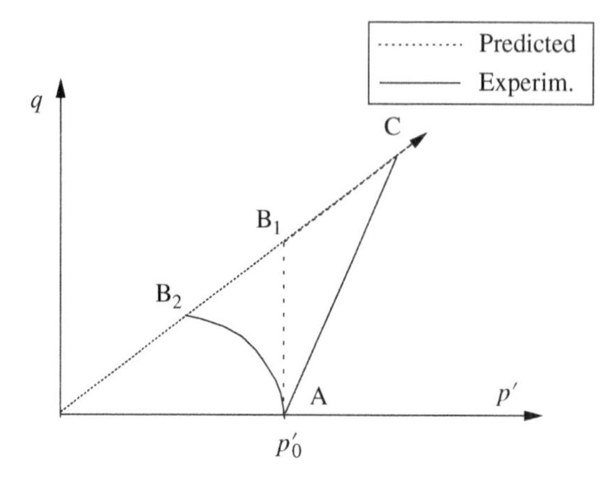

Figure 4.22 Predicted (Mohr–Coulomb) and observed behavior in CU tests

Undrained behavior of normally consolidated clays provides an interesting illustration of the shortcomings of the Mohr–Coulomb criterion when used as a yield surface.

Figure 4.22 compares the predicted behavior of such a model with that observed in the laboratory. It can be seen how the model overestimates soil strength because it cannot predict the pore pressures caused by plastic volumetric strain which develop during the test. In the Mohr–Coulomb model, no plastic strain is produced until the yield surface is reached. In addition to that, if the flow rule is associative, dilation and negative pore pressures will develop at failure, and the stress path will turn to the right following the yield surface (B1-C). This process will be endless and the deviatoric stress will keep increasing continuously. In reality, the process is stopped by cavitation of the pore fluid.

4.3.2.3 Critical State

It can be said that modern plasticity models for soils are based both on the pioneering work of Drucker et al. (1957), who first introduced the ideas of volumetric hardening and a closed yield surface, and on the theoretical and experimental work of researchers from the University of Cambridge, who provided the framework of Critical State Soil Mechanics, in which elastoplastic models for soils could be developed.

The basic ingredients of Critical State Soil Mechanics are the following:

- There exists a line in the $(e, \ln p')$ plane in which all stress paths in normally consolidated clays lie, which is referred to as the "Normal Consolidation Line" (NCL). This line was depicted in Figure 4.15b (1-2-5-8). The interest of this line is that it provides a volumetric hardening rule which can be generalized to general stress conditions (Roscoe et al. 1963).
- There exists a line in the space $(e, \ln p', q)$ where all residual states lie, independently of the type of test and initial conditions (Parry 1960). The projection of this line on to the $(e, \ln p')$ plane is parallel to the NCL, and divides initial states into "wet" and "dry," depending on whether they lie in the space between both lines or not (Figure 4.23). At this line, shear deformation takes place without a change of volume.
- The stress paths resulting either from consolidated drained and undrained tests lie on a unique state surface referred to as the "Roscoe Surface." This fact was found

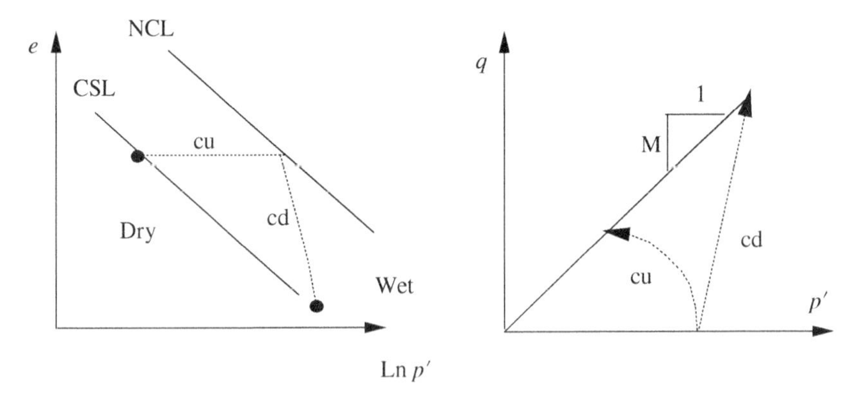

Figure 4.23 Normal consolidation and Critical State lines

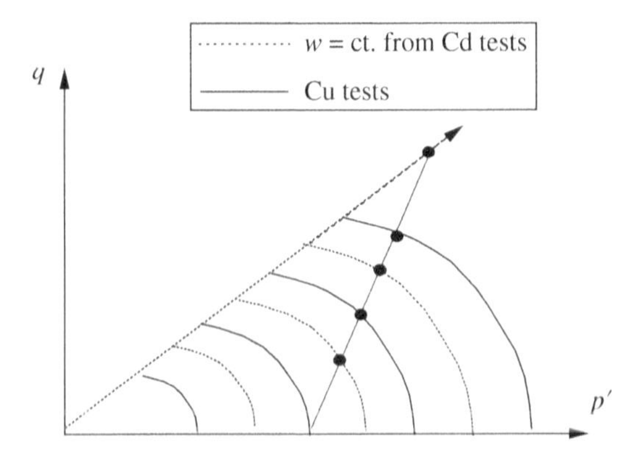

Figure 4.24 Constant water content lines as obtained from CD and CU tests (sketched)

experimentally by Henkel (1960) who plotted the water content contours obtained in drained tests and found that undrained test paths followed these lines as well (Figure 4.24). This fact is not directly applicable to elastoplastic models, as these isolines are not yield surfaces corresponding to constant values of the hardening parameter. In fact, during undrained paths, the soil hardens as plastic volumetric strain is produced, while the sum of the plastic and elastic increments of volumetric strain is kept constant. What is useful, however, is that it gives a hint of the kind of yield surface.

From here, simple elastoplastic models can be derived. The first step, as mentioned above, is to assume as a hardening rule

$$\frac{\mathrm{d}p'_c}{\mathrm{d}\varepsilon_v^{\mathrm{p}}} = \left(\frac{1+e}{\lambda-\kappa}\right)p'_c \tag{4.133}$$

where p'_c is a parameter characterizing the size of the yield surface.

Next, the yield surface is determined. Roscoe et al. (1963) assumed that incremental plastic work

$$\delta W^{\mathrm{p}} = \boldsymbol{\sigma}' : \mathbf{d}\boldsymbol{\epsilon}^{\mathbf{p}} = p' \cdot \mathrm{d}\varepsilon_{v}^{\mathrm{p}} + q \cdot \mathrm{d}\varepsilon_{\mathrm{s}}^{\mathrm{p}} \tag{4.134}$$

was given by

$$\delta W^{\mathrm{p}} = Mp'\mathrm{d}\varepsilon_{\mathrm{s}}^{\mathrm{p}} \tag{4.135}$$

from where

$$Mp'\mathrm{d}\varepsilon_{\mathrm{s}}^{\mathrm{p}} = p' \cdot \mathrm{d}\varepsilon_{v}^{\mathrm{p}} + q \cdot \mathrm{d}\varepsilon_{\mathrm{s}}^{\mathrm{p}} \tag{4.136}$$

Using the above expression, dilatancy $d_g = \mathrm{d}\varepsilon_{\mathrm{v}}^{\mathrm{p}}/\mathrm{d}\varepsilon_{\mathrm{s}}^{\mathrm{p}}$ is obtained as

$$d_g = M - \frac{q}{p'} \tag{4.137}$$

where it is interesting to note that dilatancy is zero at the Critical State Line.

The normal to the plastic potential surface is proportional to

$$\left(\mathrm{d}\varepsilon_{\mathrm{v}}^{\mathrm{p}}, \mathrm{d}\varepsilon_{\mathrm{s}}^{\mathrm{p}}\right)^{\mathrm{T}} \tag{4.138}$$

and

$$\left(\frac{\partial g}{\partial p'}, \frac{\partial g}{\partial q}\right)^{\mathrm{T}} \tag{4.139}$$

Therefore, dilatancy is given by

$$d_g = \frac{\frac{\partial g}{\partial p'}}{\frac{\partial g}{\partial q}} = M - \frac{q}{p'} \tag{4.140}$$

If we take into account that, along the surface

$$d_g = \frac{\partial g}{\partial p'}\mathrm{d}p' + \frac{\partial g}{\partial q}\mathrm{d}q = 0 \tag{4.141}$$

$$\frac{\frac{\partial g}{\partial p'}}{\frac{\partial g}{\partial q}} = -\frac{\mathrm{d}q}{\mathrm{d}p'} \tag{4.142}$$

we obtain

$$M - \frac{q}{p'} = -\frac{\mathrm{d}q}{\mathrm{d}p'} \tag{4.143}$$

which can be integrated to obtain the plastic potential

$$g \equiv q + Mp' \ln\left(\frac{p'}{p'_c}\right) = 0 \tag{4.144}$$

where p'_c is the abscissa at which the surface intersects the hydrostatic axis $q = 0$.

This surface has been depicted in Figure 4.25, where it can be seen that the normal to the surface at $p' = p'_c$ is not directed along the axis. Therefore, the normal will not be uniquely defined in three-dimensional stress conditions, although it can be assumed that the surface is rounded off in the proximity of the axis so that the normal is directed along it.

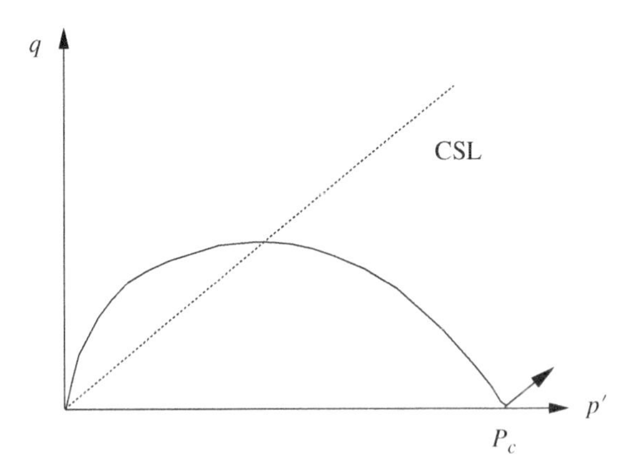

Figure 4.25 Yield and plastic potential surfaces of Cam-Clay model

If the flow rule is assumed to be associated, the yield surface coincides with g.

This model was further elaborated by Burland (1965), who suggested an ellipse as a yield surface. The work dissipation was given by

$$\delta W^{p2} = \left(p' d\varepsilon_{\mathrm{v}}^{\mathrm{p}}\right)^2 + \left(q d\varepsilon_{\mathrm{s}}^{\mathrm{p}}\right)^2 \tag{4.145}$$

from which dilatancy is obtained as

$$d_{\mathrm{g}} = \frac{(M^2 - \eta^2)}{2\eta} \tag{4.146}$$

with

$$\eta = \frac{q}{p'} \tag{4.147}$$

The yield surface can easily be obtained by integration of the above and is given by

$$f \equiv q^2 + M^2 p'(p' - p_c) = 0 \tag{4.148}$$

which is depicted in Figure 4.26.

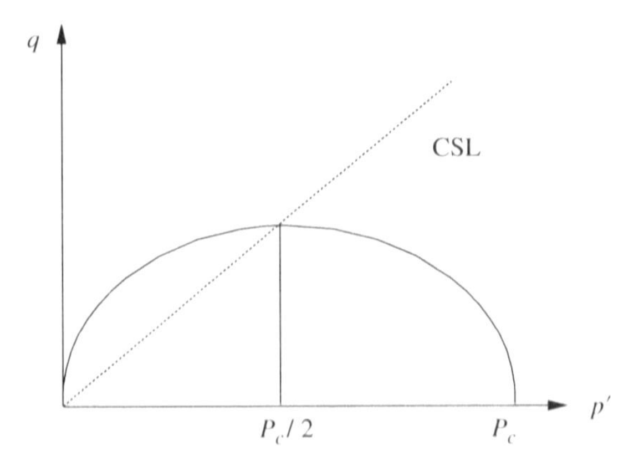

Figure 4.26 Yield surface of the modified Cam-Clay model

4.3.3 Critical State Models for Sands

The models described so far are able to predict with reasonable accuracy the behavior of normally consolidated clays. They depart from reality (i) when applied to overconsolidated soils, as it is not possible to reproduce inelastic strain which develops inside the yield surface, and (ii) when applied to sands. The main differences are the following:

- Critical State Models for clays are based on the idea that hardening, which depends on volumetric plastic strain can be determined from isotropic compression tests with paths following the NCL. Normally, consolidated clays have a unique relationship between void ratio and pressure under isotropic compression. Sands, on the contrary, can be prepared in a laboratory at different densities under a given initial confining pressure, which results in different NCLs. Indeed, it can be said that there exist infinite NCLs for sand. Moreover, the NCLs depart from the classical straight lines on the plane $(e, \ln p')$ and present a curved shape.
- Behavior of sands depends on density. Constitutive modelers have often considered the same sand in dense and loose states as different materials. The problem is that confining pressure affects the loose or dense nature of sand. Given sand, prepared in a laboratory in a dense state, will behave as dense at low-medium confining pressures. As pressure is increased, its behavior will change to loose progressively.
- Very loose sands under undrained shearing present liquefaction. The phenomenon consists of a sudden drop of resistance of the soil, which behaves as a viscous fluid.

In order to develop critical state-based constitutive models for sands, it is important to take into account the above-listed peculiarities. The next sections will be devoted to explore them in detail.

4.3.3.1 Hydrostatic Compression

There exist many tests reported in the literature describing the behavior of sands under a wide range of confining pressures (Seed and Lee 1967; Vesic and Clough 1968; Castro 1969; Pestana and Whittle 1995; Jefferies and Been 2000).

Figure 4.27 (Seed and Lee 1967) depicts the behavior of different specimens of sand prepared at different densities. At high pressures, all the four paths approach a common curve, for which the name "Limiting Compression Curve" has been proposed (Pestana and Whittle 1995). Along all these paths, both elastic and plastic strains are produced.

Figure 4.28 shows a loading-reloading cycle for two different initial densities. It is important to notice the difference between the unloading curve of the lower density and the loading of the higher density sand, which do not coincide even if they have a common point.

It can be concluded that given sand has an infinite number of NCLs, corresponding to all different densities at which it can be prepared. Pestana and Whittle (1995) proposed to divide confining pressures into different segments, and name them as: (i) Very low confining pressures, up to 10^1 kPa, (ii) Low, from 10^1 to 10^3 kPa, (iii) Elevated, up to 10^4, (iv) High, up to 10^5, (v) Very high, up to 10^6, and (vi) Ultra high, up to kPa. They provided some engineering examples where the isotropic compression behavior at high pressures is important

At high confining pressures, in addition to classical plastic deformation mechanisms such as rearrangements of grains, sliding, and plastification of contacts, crushing of grains can

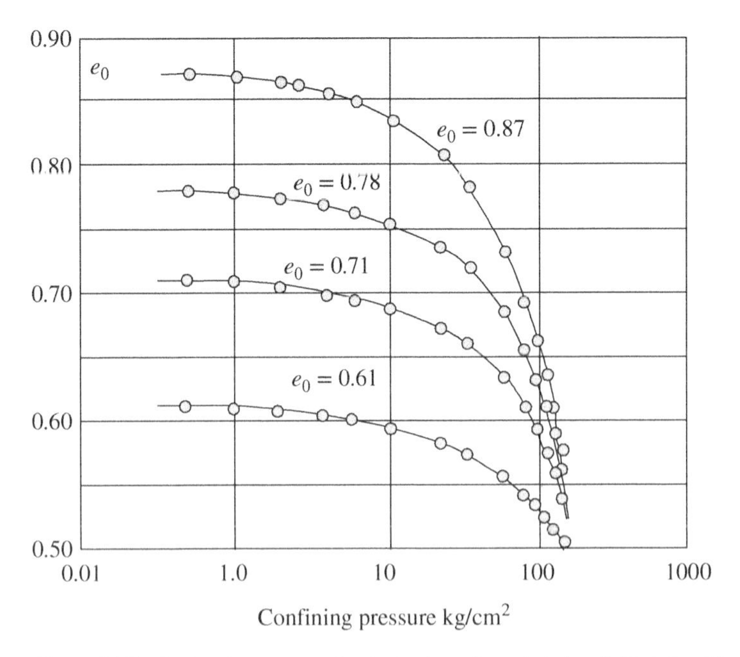

Figure 4.27 Isotropic compression behavior of sand at four initial densities. *Source:* From Seed and Lee (1967)

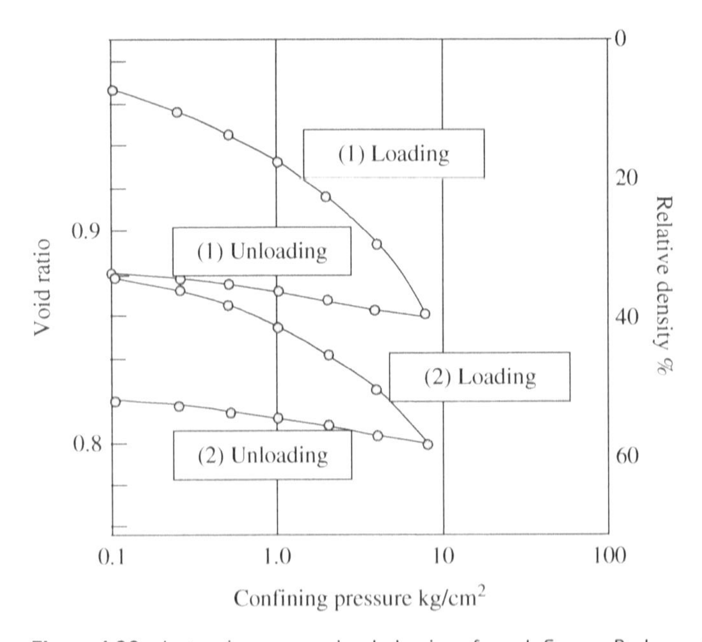

Figure 4.28 Isotropic compression behavior of sand. *Source:* Redrawn from Castro (1969)

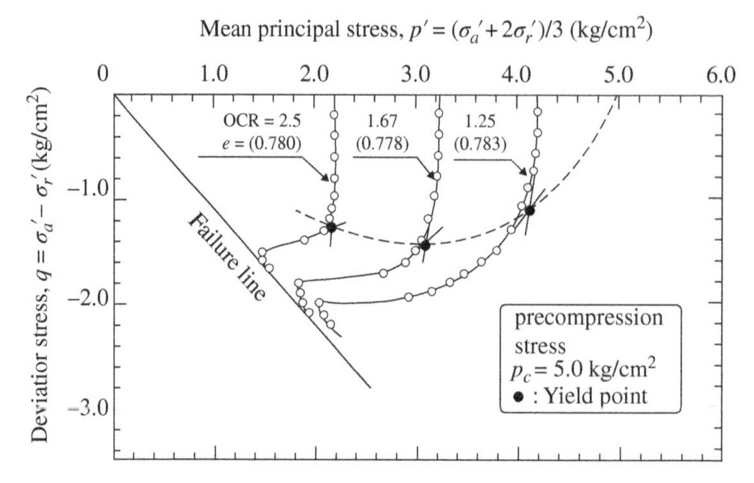

Figure 4.29 Influence of isotropic overconsolidation on shear behavior. *Source:* Based on Ishihara and Okada (1982)

take place. Breakage initiates at fractures produced at regions close to contacts (Sowers et al. 1965; Miura and Yamanouchi 1975; and, more recently, by Alonso and Oldecop 2003).

Analytical expressions for the NCLs have been proposed by Pestana and Whittle (1995) and Jefferies and Been (2000).

An important fact to point out is the experimental evidence of coupling between volumetric plastic strains produced under isotropic compression and the shear behavior. Figure 4.29 (Ishihara and Okada 1982) shows the stress paths in CU tests of different specimens of sand which have been overconsolidated under pressure of 500 kPa. By inspecting the stress paths, it is possible to find the position of the yield surface.

Coop et al. (2004) observed in carbonate sands stable gradings at high strains in the ring shear apparatus. Breakage appeared during shear, probably coupled to the soil dilatancy.

4.3.3.2 Dense and Loose Behavior

The behavior of granular soils depends mainly on density and two extreme classes of behavior can be identified. Dense sands show, when sheared in triaxial conditions, a behavior is similar to that depicted in Figure 4.30. In the first part of the test, the sand contracts, reaching a minimum voids ratio, and, from there, it dilates. Concerning the deviatoric stress, it increases until a peak is reached, and then it softens. Finally, it stabilizes at residual conditions, where the plastic flow takes place at constant volume. The results sketched in the figure follow the ideas of Taylor (1948) who suggested that the moment at which the stress ratio (deviatoric stress in the plot) reaches the value at which residual conditions will take place later, the volumetric strain presents a peak. An important difficulty encountered is that the specimen is no longer homogeneous long before residual conditions, as strain concentrates in shear bands. Therefore, the observed softening is rather of structural than of material nature. If the sand is prepared in a very loose state, the results are those sketched in Figure 4.30. There is no peak in the deviatoric stress and the sand compacts through the

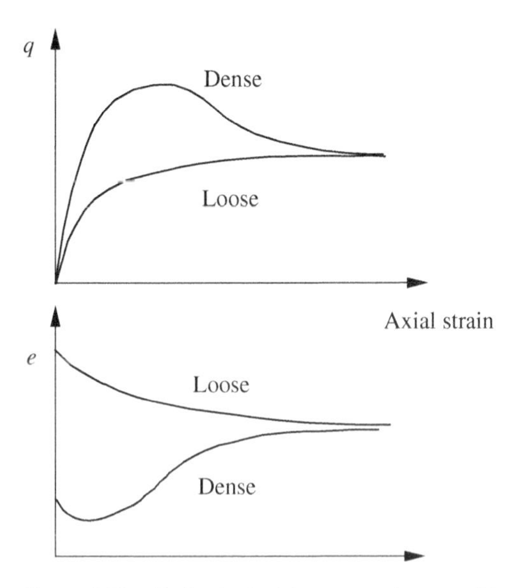

Figure 4.30 Drained triaxial tests on dense and loose sand

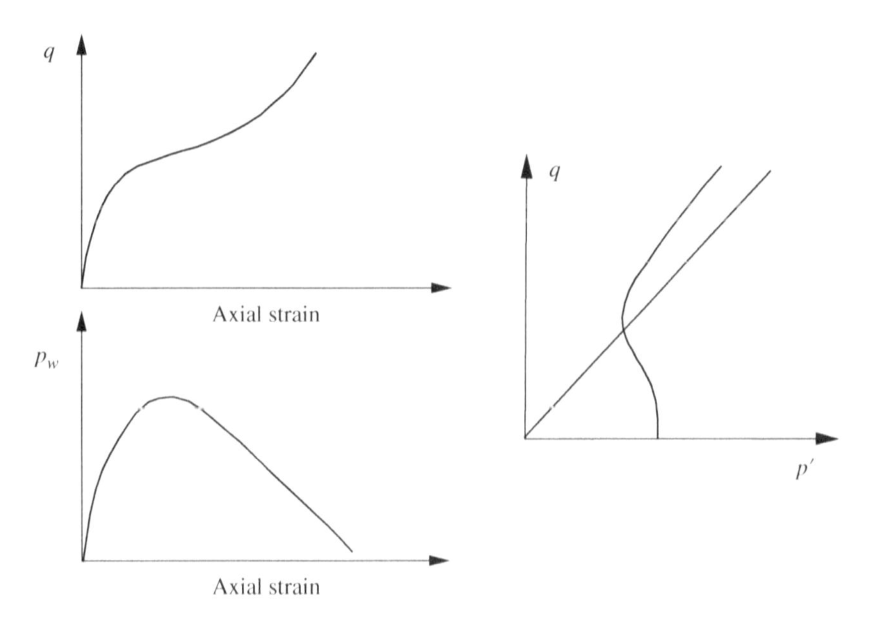

Figure 4.31 Undrained behavior of dense sand in CU triaxial test

test. The residual conditions in the planes $(e, \ln p')$ and (p, q') coincide for both densities, which indicate the existence of a Critical State at which residual conditions take place.

Under undrained conditions, dense sand's tendency to dilate results on a decrease of the pore pressure, which can be observed in Figure 4.31. The deviatoric stress q does not exhibit a peak and it continuously increases until the pore pressure falls below cavitation and the test has to be stopped.

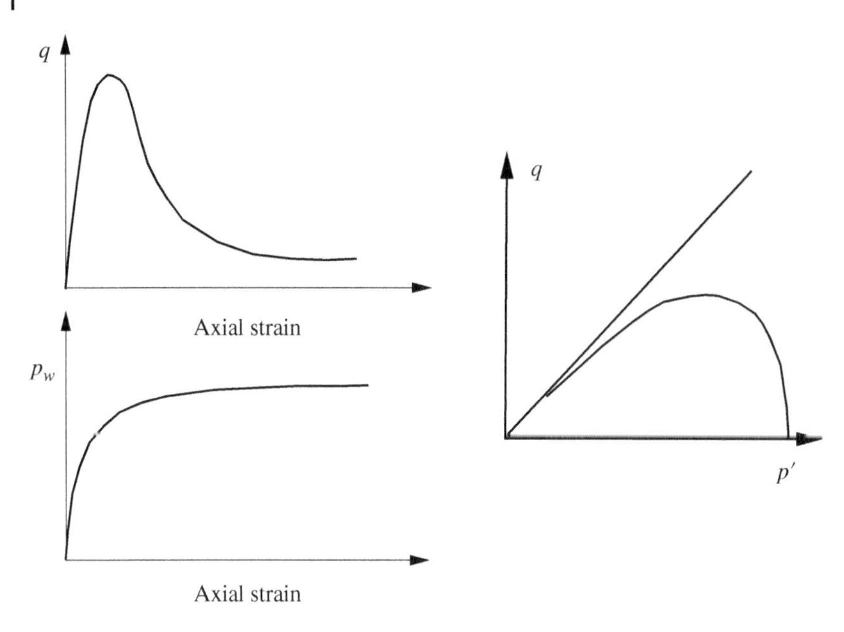

Figure 4.32 Liquefaction of very loose sand in CU triaxial test

On the other side of the density spectrum, very loose sands present liquefaction under undrained conditions (Figure 4.32). The phenomenon consists of a sudden drop of resistance of the soil, which behaves as a viscous fluid. It is important to note that the behavior in the descending branch corresponds to increasing values of the stress ratio, and, therefore, it is not sound to assume this behavior as softening. The separation from the vertical of the stress path shows that plasticity is present from the beginning of the test.

So far, we have considered only the influence of density. One paramount aspect of sand behavior is that "loose" and "dense" behavior depends not only on density but also on confining pressure. Indeed, sand that exhibits dense behavior at low confining pressures behaves as loose when the initial confining pressure is increased. Figure 4.33 (Ishihara 1973) provides an interesting example. Four samples having the same initial void ratio are consolidated at four different confining pressures 0.1, 1, 2, and 3 MPa) and then sheared. It is interesting to note how the sample consolidated at 0.1 MPa behaves as dense, while the sample consolidated at 3 MPa exhibits the "quasi-liquefaction" behavior typical of loose sands.

4.3.3.3 Critical State Line

The existence of a Critical State Line for sands in the (e, p', q) space is widely admitted today. Experimental results on CD triaxial tests show that samples with different initial void ratios tested under the same confining pressure have a common residual state with the same values of the void ratio, deviatoric stress, and effective hydrostatic pressure (Figure 4.30). If specimens of sand with the same void ratio but different initial confining pressures are tested under CU conditions, they will arrive at a common residual state (see

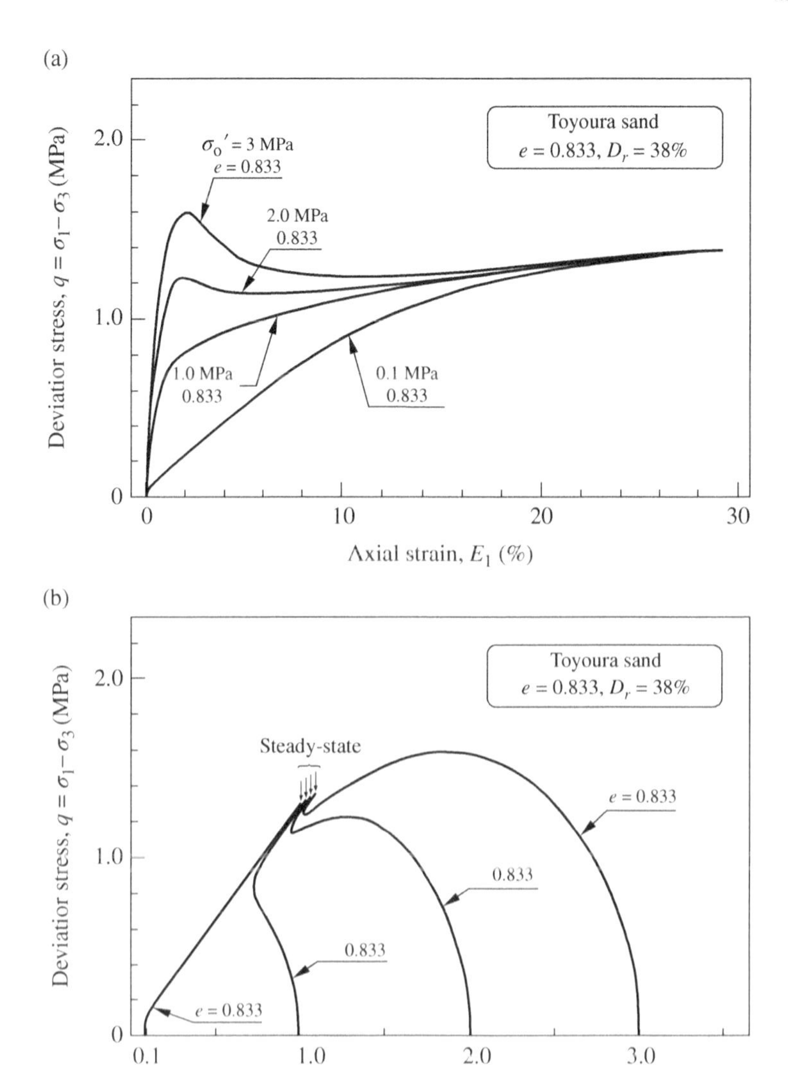

Figure 4.33 Behavior of Toyoura sand showing the influence of confining pressure. *Source:* From Ishihara (1973)

Figure 4.33). As a first approximation, it is assumed that the CSL is the line defined by $q - M_g p' = 0$ and $e - e_0 = \lambda \ln \left(p'/p_0'\right)$.

The concept of a critical void ratio at which a granular soil can undergo any amount of deformation was first proposed by Casagrande in 1935 and 1936, who stated that shear deformation of dense and loose specimens tends to produce the same "critical density." Castro (1969) provides a detailed description of Casagrande's findings which include a comment on the possible effects of shear bands where deformation concentrates. Casagrande investigated the effect of the confining pressure on the critical void ratio, and concluded that it decreases as the confining pressure increases (Casagrande 1938).

Roscoe et al. (1958) performed shear tests on several granular materials including steel balls and glass beads, confirming the findings of Casagrande. The projection on the (p', q) plane was a straight line. It is curious to note that they kept the name "critical void ratio line" (CVRL). This line is indeed the CSL.

Castro (1969) performed a complete series of triaxial tests on three different sands in order to gain an insight into the liquefaction phenomenon. He studied the locus where the residual conditions lied at liquefaction, and concluded that it was a straight line on the $(e, \ln p')$ plane. When comparing it with the residual conditions in drained tests, he found they did not coincide, and introduced the concept of the "Steady State Line" (SSL) where loose granular soils under undrained shear developed what he called a "flow structure." SSL line was parallel to the CSL, but lied below it.

In the case of specimens showing limited liquefaction, Castro found the locus where these conditions lied and obtained yet a third straight line parallel to SSL and CSL. This line was called by Ishihara (1973) the "quasi steady state line" or QSSL. Castro explained that this line corresponds to points in the stress paths where the path toward a flow structure reversed, returning toward the CSL.

Concerning the shape of the CSL in the (e, p') plane, it has been assumed to be a straight line when plotting e vs. $\ln p'$. Experiments at high confining pressures done by Verdugo (1992) showed separation from the straight line. He proposed a linear representation in (e, p') plane. Li and Wang (1998) have proposed a linear relation between e and $(p'/p_{atm})^{\xi}$ where p_{atm} is the atmospheric pressure and ξ an exponent to be obtained experimentally. (See also Li 1997). Figure 4.34 shows for Toyoura sand the CSL plotted in both $(e, \ln p')$ and $(e, (p')^{\xi})$ using $\xi = 0.7$ (Li 1997).

4.3.3.4 Dilatancy

Dilatancy is an important feature of soil behavior as it provides information on the plastic potential. The study of dilatancy is closely related to that of the critical state. The first study of sand dilatancy is due to Reynolds (1985) who pointed out that dense sands tend to dilate when sheared. Taylor (1948) studied the coupling between volume change vs. strain and stress vs. strain curves in sands, as explained in Section 4.3.3.2. See also Rowe (1962). From the point of view of elastoplasticity, it can be assumed that dilatancy is always zero at the line $\eta = M_g$, either before reaching the critical state or there. In fact, several investigators proposed a separate denomination of this line referring to it as Characteristic State Line (Habib and Luong 1978) or Line of Phase Transformation (Ishihara et al. 1975).

An important question is the dependence of dilatancy on density and confining pressure. Some experimental results such as those of Tatsuoka and Ishihara (1974) hint that the stress ratio at which the behavior changes from contractive to dilatant is not constant. Indeed, experiments reported by them showed that it decreases from a value close to $\eta = q/p' = 1.5$ to $\eta = 1$ as the soil densifies.

This effect has been studied by Li and Dafalias (2000) who proposed a dilatancy law depending on density and confining pressure which will be described in Section 4.3.3.6. The simplest law is that of Nova (1982), which assumes a linear dependency of the form

$$d_g = (1 + \alpha)\left(M_g - \eta\right) \tag{4.149}$$

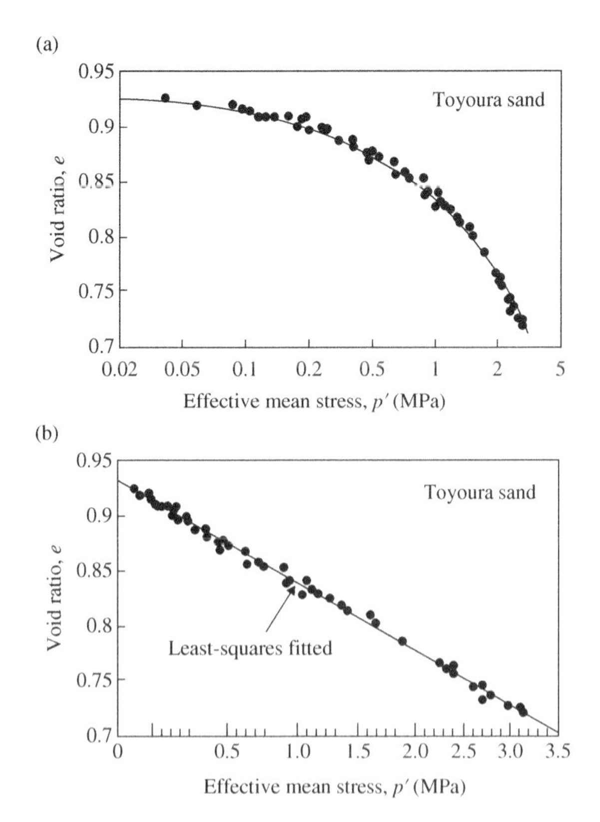

Figure 4.34 CSL plotted in $(e, \ln p')$ (a) and (e, p'^{ξ}) (b) planes.
Source: From Li and Wang (1998)

4.3.3.5 A Unified Approach to Density and Pressure Dependency of Sand Behavior: The State Parameter

We have already seen how the loose or dense behavior of given sand depends both on density and confining pressure. This means that neither density nor confining pressure alone can fully characterize sand behavior, but a combination of both.

We will introduce here the concept of state parameter which plays a fundamental role both in understanding and behavior and in modeling it.

The idea of a joint dependency of sand behavior on density and pressure has attracted the attention of researchers since the early works of Roscoe and Poorooshasb (1963), Wroth and Basset (1965), Cole (1967), and Seed and Lee (1967). The latter authors performed two series of tests on sands of different densities at constant voids ratios and constant confining pressure. By analyzing the volume change tendency at failure, they concluded that (i) for a giving confining pressure, there is a critical voids ratio – or a discriminant density – which separates dilative from contractive behavior, (ii) for a given voids ratio, there is a critical confining pressure separating contractive from dilative specimens, and (iii) the relationships obtained in both cases between confining pressure and void ratio coincide.

Uriel (1973) and Uriel and Merino (1979) proposed a simple constitutive model able to describe with a single set of parameters the behavior of dense and loose sands, and the effect of the confining pressure. The model is based on a state parameter ψ defined by Uriel as a function of the discriminant density ρ_d and the density at the critical state ρ_C

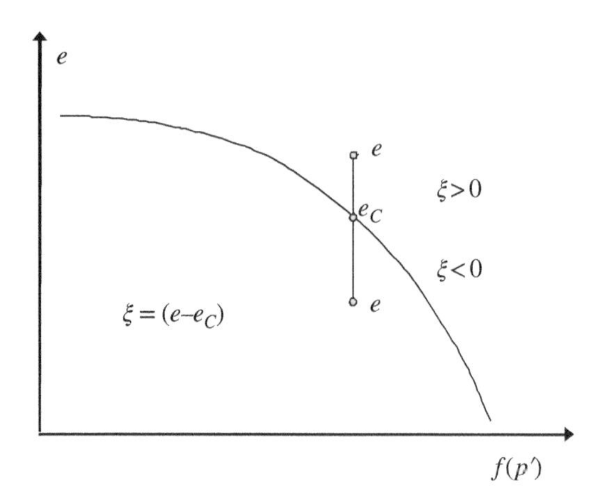

Figure 4.35 Definition of the state parameter. *Source:* Based on Been and Jefferies (1985)

$$\psi = \frac{\rho - \rho_d}{\rho_C - \rho_d} \tag{4.150}$$

The most widely accepted state parameter today is that proposed by Been and Jefferies (1985). It is defined as illustrated in Figure 4.35 as the difference between the actual voids ratio and the voids ratio at critical state under the same confining pressure

$$\psi_s = e - e_C \tag{4151}$$

This state parameter is positive for loose states and negative for denses. Another definition provided by Wan and Guo (1998) is

$$\psi_q = \frac{e}{e_c} \tag{4.152}$$

and presents values larger than 1 for loose specimens and smaller for dense.

As an alternative, Ishihara (1973) introduced a more refined definition. Since then, many researchers have introduced a state parameter in their constitutive models. It is worth mentioning the work of Jefferies (1993), Saitta (1994), Bahda (1997), Bahda et al. (1997), Manzari and Dafalias (1997), Li et al. (1999), Gajo and Wood 1999, Ling and Yang (2006), Ling and Liu (2003), Tonni et al. (2006), Manzanal (2008), and Manzanal et al. (2011a).

Definitions of the state parameter based on the CSL present limitations in the case of very loose soils, soil with fines or when crushing occurs. Javanmardi et al. (2018) have proposed a modification accounting for this effect, by introducing an alternative reference state curve. The dependence of CSL on crushing has been studied by Douadji and Hicher (2010), being worth mentioning the contributions of Dafalias (2016) and Theocharis et al. (2017).

4.3.3.6 Constitutive Modelling of Sand Within Critical State Framework

So far, we have presented in a succinct way the most important aspects of sand behavior, which we have to take into account when modeling:

i) Behavior of sand depends both on density and confining pressure. As the latter increases, the sand behaves more in a "loose" manner.

ii) Noval (virgin) hydrostatic compression produces plastic volumetric strain and there exists an infinity of NCLs depending on the initial void ratio at which the sand specimen has been built (Figure 4.27, Lee and Seed 1967).
iii) Residual states lie on a unique CSL in the (e, p', q) space. The projection on the plane (p', q) can be assumed to be a straight line, but for the projection on the (e, p') space, several alternative forms have been proposed.
iv) Analysis of dilatancy suggests the existence of a line on the (p', q) space where dilatancy is zero. This is the Characteristic State Line (Habib and Luong 1978) or Line of Phase Transformation (Ishihara et al. 1975), and can be different from the CSL (see Figure 4.35).
v) If we consider the drained behavior of dense sands, we can see that the deviatoric stress reaches a peak and then decreases to a residual value on the CSL. Before reaching the peak, the stress path crosses the projection of the CSL on the (p', q) plane but the material does not fail at this moment. After the peak, the decrease of deviatoric stress is associated with an increase in voids ratio (softening).
vi) Liquefaction of very loose sands during undrained loading shows a peak in the deviatoric stress which does not correspond to softening, as the stress ratio $\eta = p'/q$ which is a measure of mobilized friction is continuously increasing and this has to be interpreted as hardening.

Following these experimental facts, we can conclude that constitutive models for sands within the CS framework should implement the following features:

- Hardening laws should depend both on volumetric and deviatoric plastic strains to avoid failure at the Characteristic State Line (v), and should ensure that plastic modulus is zero at CSL (iii)
- Be nonassociative, otherwise liquefaction could not be modeled in a consistent way (vi)
- In case a single-yield surface is used, it should be closed, otherwise the model will not be able to reproduce plastic volumetric strain during hydrostatic loading (ii)
- If the model is to be used under a wide range of confining pressures, it is necessary to incorporate a state parameter (i, iv)

We will comment next on these requirements.

Concerning *hardening laws*, a dependence on deviatoric plastic strain is required to cross the line $\eta = M_g$, as the plastic modulus is zero there otherwise. Deviatoric hardening was introduced by Nova (1977) and Wilde (1977), who assumed a hardening parameter of the type

$$Y = \varepsilon_v^p + D\xi \tag{4.153}$$

where ξ is the accumulated deviatoric shear strain

$$\xi = \int \| \mathrm{d}\varepsilon_s^p \| \tag{4.154}$$

The size of the yield surface was made to depend on Y, and the plastic modulus in triaxial conditions was found to be proportional to

$$\frac{\partial g}{\partial p'} + D\frac{\partial g}{\partial q} \tag{4.155}$$

Therefore, the Critical State Line at which

$$\frac{\partial g}{\partial p'} = 0 \tag{4.156}$$

can be crossed without the plastic modulus is zero. Failure will occur when

$$\frac{\partial g}{\partial p'} + D\frac{\partial g}{\partial q} = 0 \tag{4.157}$$

which happens at a stress ratio higher than M_g. Once there, if D is kept constant, the path will not return to the CSL and failure will take place with dilation. Another possibility, proposed by Nova (1982), consists of making them drop D to zero, which results in a discontinuity of slope but with the desired result of coming back to CS conditions. Finally, a hardening law with saturation can be assumed to hold for D

$$\mathbf{D} = \beta_0 \exp\left(-\beta_1 \xi\right) \tag{4.158}$$

This law was suggested by Wilde (1977) and applied to a bounding surface model by Pastor et al. (1985).

If a negative value of D is assumed, then the plastic modulus becomes zero at a stress ratio lower than critical, and liquefaction-like behavior can be modelled in the softening regime. As discussed above, it is more sound to assume this process to be of the hardening type as the stress ratio is continuously increasing.

The second ingredient is a nonassociative flow rule as suggested by Poorooshasb et al. (1966, 1967), Nova and Wood (1979), Nova (1982), Zienkiewicz et al. (1975), and Pastor et al. (1985).

The plastic potential and flow rules can be determined from the experiment, as shown in Nova and Wood (1978), where surfaces were defined by different analytical expressions valid for different ranges of the stress ratio.

Pastor et al. (1985) used as plastic potential a simplification of that proposed by Nova and Wood (1979), assuming now that a single expression was valid for the full range of stress ratios

$$g \equiv q - M_g p' \cdot \left(1 + \frac{1}{\alpha}\right)\left[1 - \left(\frac{p'}{p'_g}\right)^{\alpha}\right] \tag{4.159}$$

where p'_g is the abscissa at which it intersects the p' axis.

This surface can be obtained from the dilatancy rule

$$d_g = (1 + \alpha) \cdot (M_g - \eta) \tag{4.160}$$

As yield surfaces, they proposed curves belonging to the same family

$$f \equiv q - M_f \cdot p'\left(1 + \frac{1}{\alpha}\right)\left[1 - \left(\frac{p'}{p'_c}\right)^{\alpha}\right] \tag{4.161}$$

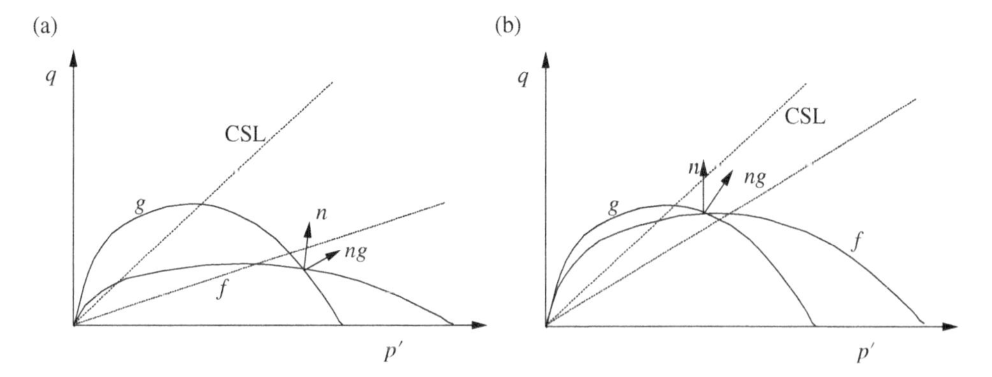

Figure 4.36 Plastic potential and yield surfaces for (a) loose sands (b) dense sands. *Source:* From Balasubramanian and Chaudhry (1978)

where $M_f \neq M_g$ in general. An interesting fact reported by them is that the ratio M_f/M_g depends on the relative density and indeed it can be assumed to be the same as D_r.

Figure 4.36a and b shows plastic potential and yield surfaces for very loose and dense sands.

Concerning the **state parameter**, it can be assumed that it influences any (or all) the basic ingredients of a plasticity model, i.e. the dilatancy law, the hardening law, and the geometry of the yield surface.

The dilatancy law depends on density for a given confining pressure. Among the laws proposed, we can mention those of Manzari and Dafalias (1997) and Li and Dafalias (2000), which can be written as

$$d_g = A\{M \ f_1(\psi_s) - \eta\} \tag{4.162}$$

where A is a constant, M is the slope of the CSL on the (p', q) plane, and $f_1(\psi_s)$ is a function of the state parameter $\psi_s = e - e_C$ which is equal to one at the CSL. Therefore, at CSL $e = e_C$, $\psi_s = 0$, $\psi_s = 0$, $f_1(\psi_s) = 1$, and $d_g = 0$.

Li and Dafalias (2000) proposed

$$A = \frac{d_0}{M} \quad f_1(\psi_s) = \exp(m\psi_s) \tag{4.163}$$

Saitta (1994) proposed a hardening law of the type

$$\frac{dp_c}{p_c} = \frac{1 + e_0}{\lambda - \kappa}\left(d\varepsilon_v^p + Dd\varepsilon_s^p\right) \tag{4.164}$$

where the deviatoric component D was given by $D = \max(0, \psi_s)$

Li and Dafalias (2000) introduced a plastic modulus proportional to

$$\left(M_g \exp(-n\psi_s) - \eta\right) \tag{4.165}$$

In this way, the plastic modulus is not zero when crossing the Characteristic State Line, as $\exp(-n\psi) \neq 1$ but reaches zero at CSL, where $\psi_s = 0$.

The state parameter can also be introduced in the yield function. Saitta (1994) proposed a yield surface of the form $M_f = M_g - \psi_s M'_f$ where M'_f is a new parameter. It is interesting to note that at CSL, $M_f = M_g$.

4.4 Generalized Plasticity Modeling

4.4.1 Introduction

Section 4.2 was devoted to introducing Generalized Plasticity, studying how Classical Plasticity was a particular case. The models presented there were not suitable to describe many features of geomaterial behavior. Then, in Section 4.3, we studied the main features of the behavior of cohesive and frictional soils within the framework of modern Critical State Soil Mechanics theories. The purpose of this section is to describe Generalized Plasticity Theory Models which can be applied to monotonic and cyclic loading of soils, starting with the most basic model for clays. Since the time the first GPT model proposed by Pastor and Zienkiewicz and Chan in 1986 and 1987, many developments have taken part. The model was extended to include anisotropy (Pastor 1991), the state parameter (Saitta 1994; Bahda 1997; Bahda et al. 1997; Ling and Yang 2006; Manzanal 2008; Manzanal et al. 2011a), damage (Fernández Merodo et al. 2004, 2005) and unsaturated soils (Bolzon et al. 1996; Tamagnini and Pastor 2004; Santagiuliana and Schrefler 2006; Manzanal et al. 2011b).

This section will be devoted to describing some of the GPT models mentioned above, focusing on models for saturated soils, which have been applied for both monotonic and cyclic loading conditions.

4.4.2 A Generalized Plasticity Model for Clays

4.4.2.1 Normally Consolidated Clays

The simplest case of normally consolidated clay under virgin loading will be considered first. We will begin by assuming that a residual critical state exists in the space $e - p' - q$, where e is the voids ratio and p', q the effective confining pressure and a measure of deviatoric stress.

To obtain a flow rule, it is possible to use values of dilatancy measured in laboratory tests

$$d_g = \frac{d\varepsilon_v^p}{d\varepsilon_s^p} \frac{d\varepsilon_v}{d\varepsilon_s} \tag{4.166}$$

where ε_v and $d\varepsilon_s$ are strain measures work-conjugated to p' and q

$$\delta W = \sigma' : d\varepsilon = (p'q).\begin{pmatrix} d\varepsilon_v \\ d\varepsilon_s \end{pmatrix} \tag{4.167}$$

defined as

$$\begin{aligned} d\varepsilon_v &= -tr(d\varepsilon) \\ d\varepsilon_s &= \frac{2}{\sqrt{3}}\left(\frac{1}{2}\mathbf{de} : \mathbf{de}\right)^{\frac{1}{2}} \mathbf{de} = \text{dev}(d\varepsilon) = d\varepsilon - \frac{1}{3} d\varepsilon_v \end{aligned} \tag{4.168}$$

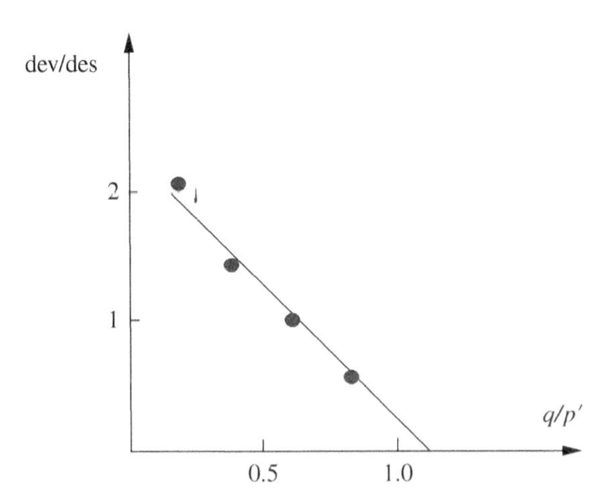

Figure 4.37 Dilatancy of soft Bangkok clay. *Source:* From Balasubramanian and Chaudry (1978)

If the ratio between the plastic increments of volumetric and deviatoric strain is assumed to be the same, then the ratio of total (elastic plus plastic) increments observed in laboratory tests, direction $\mathbf{n}_g$ can be immediately obtained.

To this end, experiments carried out by Balasubramanian and Chaudhry (1978) using constant p'/q stress paths suggest that dilatancy can be approximated by a straight line in the p'–q plane (Figure 4.37).

Dilatancy is therefore expressed as

$$d_g = \frac{\mathrm{d}\varepsilon_v^p}{\mathrm{d}\varepsilon_s^p} = (1+\alpha)(M_g - \eta) \tag{4.169}$$

where M_g is the slope of the Critical State Line in the p'–q plane, α a material constant and η is the so-called stress ratio defined as

$$\eta = \frac{g}{p'} \tag{4.170}$$

Direction $\mathbf{n}_g$ is now given by

$$\mathbf{n}_g^{\mathrm{T}} = (n_{gv}, n_{gs}) \tag{4.171}$$

with

$$n_{gv} = \frac{d_g}{\sqrt{1+d_g^2}} \tag{4.172}$$

$$n_{gs} = \frac{1}{\sqrt{1+d_g^2}} \tag{4.173}$$

This law can also be used to describe the dilatancy of granular materials, as it was suggested in Pastor et al. (1985) according to test results reported by Frossard (1983).

Concerning direction $\mathbf{n}$, we will assume that the flow rule is associative, following Atkinson and Richardson, who performed experiments in three cohesive soils and found little discrepancies from plastic potential and yield surfaces (Atkinson and Richardson 1985).

Therefore, we will have

$$\mathbf{n}^{\mathrm{T}} = (n_v, n_s) \tag{4.174}$$

with

$$n_v = \frac{d}{\sqrt{1+d^2}} \tag{4.175}$$

$$n_s = \frac{1}{\sqrt{1+d^2}} \tag{4.176}$$

and

$$d = d_g \tag{4.177}$$

In what follows, we will drop the subindex "g" referring to the plastic potential surface as it coincides with the yield surface.

To obtain the plastic modulus for virgin loading, we will consider an isotropic consolidation test of a normally consolidated specimen, for which the increments of volumetric elastic and total strain are given by

$$\mathrm{d}\varepsilon_v^{\mathrm{e}} = \frac{\kappa}{1+e}\frac{\mathrm{d}p'}{p'} \tag{4.178}$$

and

$$\mathrm{d}\varepsilon_v = \frac{\lambda}{1+e}\frac{\mathrm{d}p'}{p'} \tag{4.179}$$

from which the plastic volumetric strain increment is

$$\mathrm{d}\varepsilon_v^{\mathrm{p}} = \frac{(\lambda-\kappa)}{1+e}\frac{\mathrm{d}p'}{p'} \tag{4.180}$$

Comparing now the above equation to the general expression for the plastic strain increment

$$\mathrm{d}\varepsilon^{p} = \frac{1}{H_L}\mathbf{n}.(\mathbf{n} : \mathrm{d}\boldsymbol{\sigma}') \tag{4.181}$$

which particularizes to

$$\mathrm{d}\varepsilon_v^{\mathrm{p}} = \frac{1}{H_L}\mathrm{d}p' \tag{4.182}$$

for the stress path considered here, it can be concluded that the plastic modulus H_L is given by

$$H_L = p'\frac{1+e}{(\lambda-\kappa)} = H_0 p' \tag{4.183}$$

The parameters λ and κ are the slopes of the normal consolidation and elastic unloading lines in the $(e, \ln p')$ plane and H_0 is a material constant.

To generalize this expression of the plastic modulus to other conditions than isotropic compression paths, we will make the assumption that plastic modulus depends on the mobilized stress ratio, decreasing as the latter increases until reaching a value of zero at the critical line ($\eta = q/p' = M$).

Therefore,

$$H_L = H_0 p' f(\eta) \tag{4.184}$$

where $f(\eta)$ is such that

$$f(\eta) = 1 \quad \eta = 0$$

and

$$f(\eta) = 0 \quad \eta = M$$

A suitable form was proposed in Pastor et al. (1990)

$$f(\eta) = \left(1 - \frac{\eta}{M}\right)^{\mu} \frac{\left(1 + d_0^2\right)}{\left(1 + d^2\right)} \operatorname{sign}\left[1 - \frac{\eta}{M}\right] \tag{4.185}$$

where $d_0 = (1 + \alpha)M$ and μ can be taken as two for most clays.

So far, we have analyzed only the behavior in the triaxial plane, but the above expressions can be immediately generalized to all three-dimensional conditions by assuming that M depends on Lode's angle θ according to a suitable law. Below we define a smoothed version of Mohr's criterion widely used in practice (Zienkiewicz and Pande 1977).

$$M = \frac{18M_c}{18 + 3(1 - \sin 3\theta)} \tag{4.186}$$

where M_c is the slope of the Critical State Line obtained in standard compression triaxial tests and θ Lode's angle

$$-\frac{\pi}{6} < \theta = \frac{1}{3}\sin^{-1}\left[\frac{3\sqrt{3}}{2}\frac{J_2'}{J_2'^{3/2}}\right] < \frac{\pi}{6} \tag{4.187}$$

Elastic constants are assumed to depend on p' according to the laws

$$d\varepsilon_s^e = \frac{1}{G_0}\frac{p_0'}{p'}dp' \tag{4.188}$$

and

$$\begin{aligned} d\varepsilon_v^e &= \frac{\kappa}{1 + e}\frac{1}{p'}dp' \\ d\varepsilon_v^e &= \frac{\kappa}{1 + e}\frac{1}{p_0'}\frac{p_0'}{p'}dp' = \frac{1}{K_{ev0}}\frac{p_0'}{p'}dp' \end{aligned} \tag{4.189}$$

where

$$K_{ev0} = \frac{1 + e}{\kappa}p_0'$$

The model presented so far concerns normally consolidated clays under virgin loading. In order to assess model performance, their predictions will be compared against a full set of tests carried out by Balasubramanian and Chaudhry in 1978 for soft Bangkok clay.

The proposed model has five parameters, i.e. two elastic constants, the slope of the Critical State Line M, the constant α characterizing dilatancy and H_0, which appears in the plastic modulus. To determine them, the following procedure may be followed:

- The elastic constants can be easily determined from unloading-reloading tests. Here, they were found from the constant p' tests reported.
- The slope M of the Critical State Line on the $(p' - q)$ plane is found from drained, undrained, or constant p' tests.
- The parameter α controlling dilatancy can be found from dilatancy plots and it is given by

$$\alpha = \frac{d}{M - \eta} - 1 \tag{4.190}$$

- Finally, constant H_0 can be found as a function of λ, κ, and e as described above.

The results obtained with the proposed model are shown in Figures 4.38–4.40. First of all, Figure 4.38 shows the constant p' tests together with the experimental results. Next, the model is applied to simulate the consolidated undrained behavior of Bangkok clay (Figure 4.39), and, finally, we present the results obtained for consolidated drained tests (Figure 4.40), using for all cases the same values of the constitutive parameters.

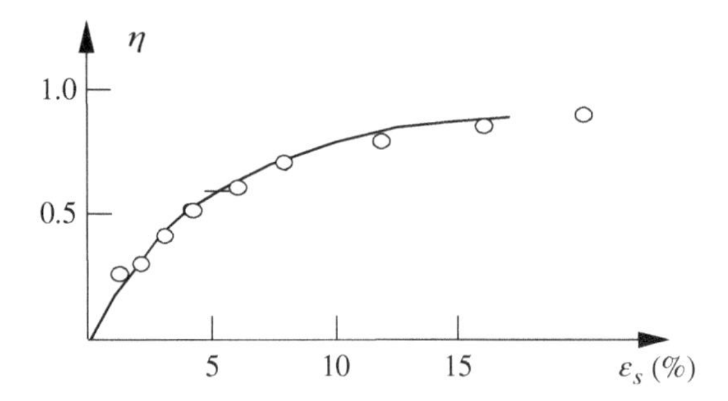

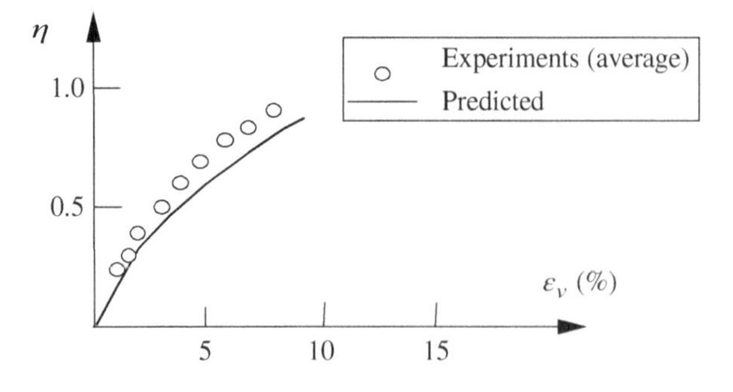

Figure 4.38 Constant p' test on Bangkok clay. *Source:* From Balasubramanian and Chaudhry (1978)

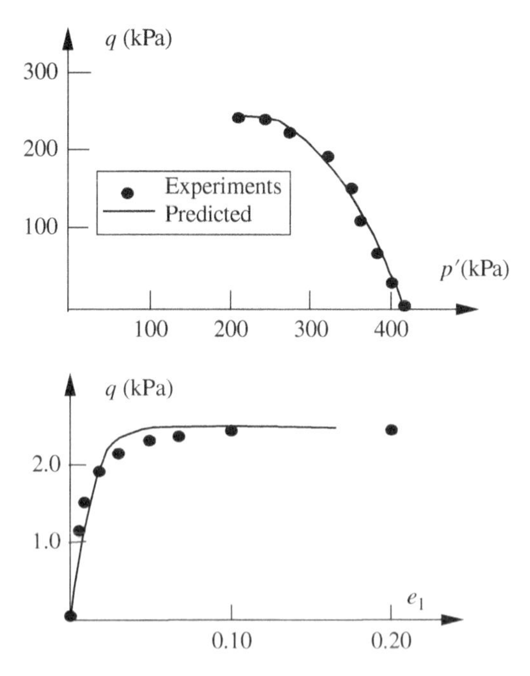

Figure 4.39 Consolidated undrained tests on Bangkok clay. *Source:* From Balasubramanian and Chaudhry (1978)

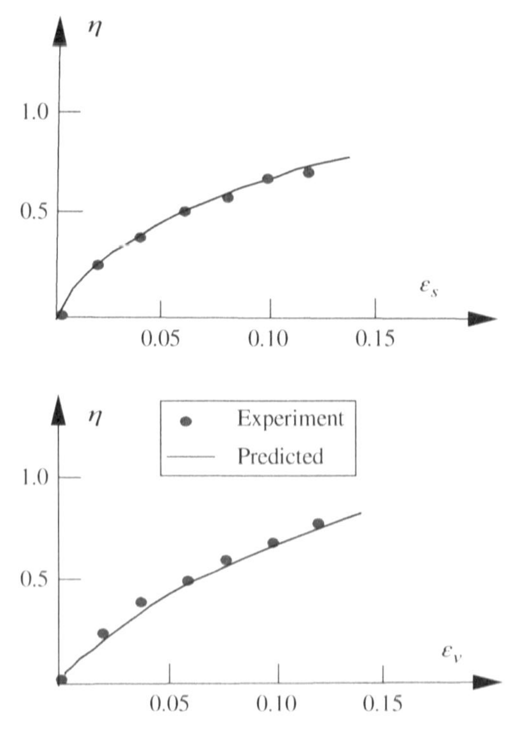

Figure 4.40 Consolidated drained tests on Bangkok clay. *Source:* From Balasubramanian and Chaudhry (1978)

The elastic behavior has been approximated in a simple manner. Hyperelastic models, such as the ones proposed by Borja et al. (1997), not only provide a more consistent description but are necessary when integrating the constitutive equation using implicit methods. The reader will find in Borja (2013) and Tamagnini et al. (2002) the details.

4.4.2.2 Overconsolidated Clays

The model described in Section 4.4.2.1 can be extended to describe the behavior of overconsolidated clays. To this end, we will introduce a function accounting for the memory of past history, which will consist of storing the past event of maximum intensity. The mobilized stress function proposed in Pastor et al. (1990) is

$$\zeta = p' \cdot \left\{1 - \left(\frac{1+\alpha}{\alpha}\right)\frac{\eta}{M}\right\}^{1/\alpha} \tag{4.191}$$

which will be used in the plastic modulus

$$H_L = H_0 p' \{f(\eta) + g(\xi)\} \left(\frac{\zeta_{\mathrm{MAX}}}{\zeta}\right)^{\gamma} \tag{4.192}$$

where $f(\eta)$ has been given in the previous section and ζ_{MAX} is the maximum value previously reached by the mobilized stress function.

In the above, we have introduced a deviatoric strain-hardening function $g(\xi)$ (Wilde 1977),

$$g(\xi) = \beta \exp(-\beta\xi) \tag{4.193}$$

where

$$\xi = \int \mathrm{d}\xi \tag{4.194}$$

and

$$\mathrm{d}\xi = (\mathbf{de}^p : \mathbf{de}^p)^{1/2} \tag{4.195}$$

$$\beta = \beta_0 \left(1 - \frac{\zeta}{\zeta_{\mathrm{MAX}}}\right) \tag{4.196}$$

Therefore, two additional parameters γ and β_0 are needed to extend the range of application of the model to overconsolidated clays.

It should be noted that for the first or virgin loading of clays, the above expressions reduce to those previously proposed for normally consolidated clays.

Figures 4.41 and 4.42 show the behavior of normally and heavily consolidated Weald clay reported by Henkel (1956), together with the model predictions.

It is important to note that in overconsolidated clays, the peak value of the stress ratio may be higher than M, then decreasing to reach it as a residual state.

For cyclic loading, it is possible to obtain quite satisfactory results for clays using simple elastic unloading and thus avoid the introduction of additional parameters. We show in Figure 4.43 the performance with the above assumption for cyclic tests with constant stress amplitude carried out by Taylor and Bacchus (1969).

Finally, Table 4.1 gives the parameters used in the simulations described above.

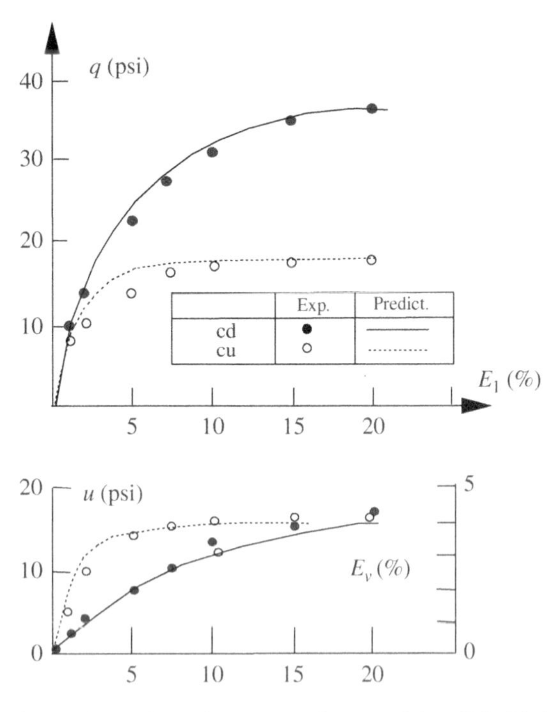

Figure 4.41 Behavior of normally consolidated Weald clay. *Source:* Based on Henkel (1956)

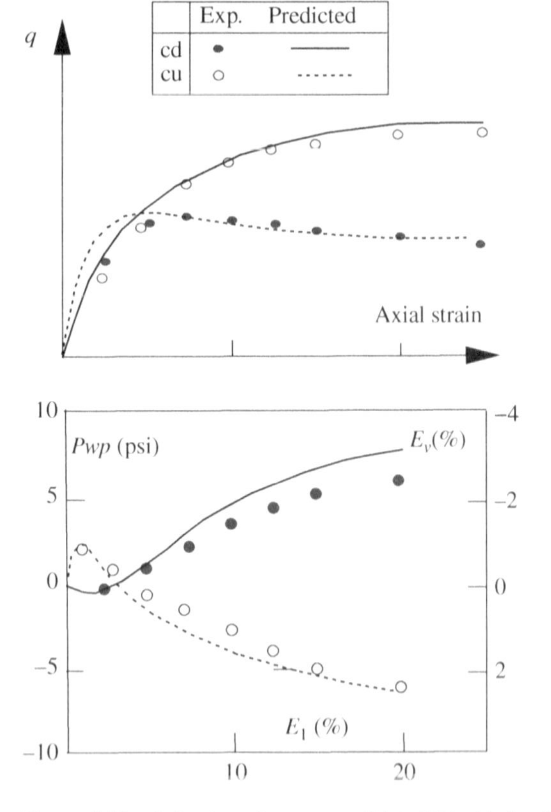

Figure 4.42 Behavior of overconsolidated Weald clay (OCR = 24)

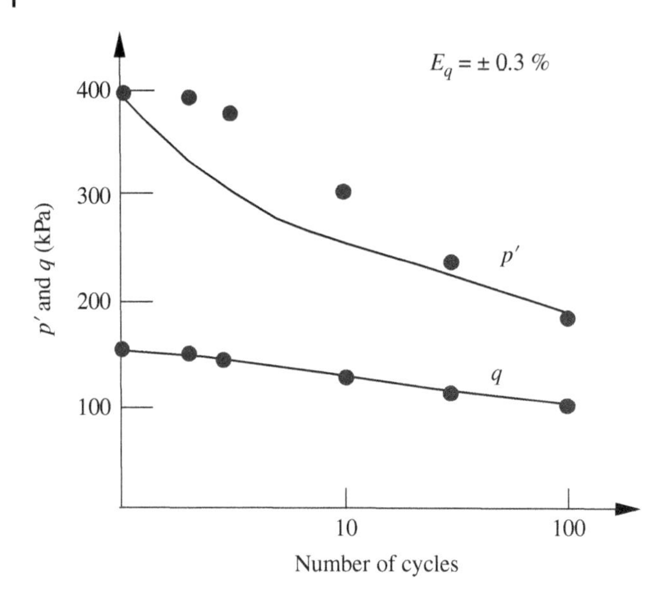

Figure 4.43 Behavior of clay under two-way strain-controlled triaxial loading. *Source:* From Taylor and Bacchus (1969)

Table 4.1 Parameters used in the simulations

	Bangkok clay	Weald clay	Taylor and Bacchus
G_0 (kg/cm^2)	124.2	766.0	440.0
K_{evo} (kg/cm^2)	150.0	800.0	640.0
M	1.10	0.90	1.50
H_0	6.60	165.0	25.0
μ	2.0	3.0	2.5
β	—	0.10	0.17
γ	—	0.40	8.0

4.4.3 The Basic Generalized Plasticity Model for Sands

4.4.3.1 Monotonic Loading

Following experimental results reported by Frossard in drained triaxial tests, dilatancy can be approximated by a linear function of the stress ratio η similar to that used in the preceding section for normally consolidated clays.

$$d_g = (1 + \alpha)(M_g - \eta) \tag{4.197}$$

using M_g instead of M as we will assume that the flow rule is nonassociated. Dilatancy is zero at the line

$$\eta = M_g$$

which coincides with the projection of the Critical State Line on the plane $(p'-q)$.

This line has also been referred to as the "characteristic state line" (Habib and Luong 1978) or the "line of phase transformation" (Ishihara et al. 1975) and plays an important role in modeling sand behavior as will be shown later. It has to be noted that this line is not the Critical State Line, which will be reached at residual conditions. Whether the Critical State Line existed or not has been a matter of discussion during past years due to the difficulty of obtaining homogeneous specimens at failure after shear bands have developed. However, recent experiments carried out at Grenoble by Desrues have shown that inside the shear band, a critical void ratio is reached.

During a test, this line can be crossed the first time, with the specimen still far from the residual state. If shearing continues, the stress path will finally approach the Critical State Line.

Therefore, the condition $\eta = M_g$ represents two different states at which dilatancy is zero, the "characteristic state" and the critical state.

The direction of plastic flow $\mathbf{n}_{gL}$ can be determined in the triaxial space by similar procedures used in cohesive soils, giving

$$\mathrm{n}_g^{\mathrm{T}} = \left(n_{gv}, n_{gs}\right) \tag{4.198}$$

with

$$n_{gv} = \frac{d_g}{\sqrt{1 + d_g^2}} \tag{4.199}$$

$$n_{gs} = \frac{1}{\sqrt{1 + d_g^2}} \tag{4.200}$$

So far, the behavior of granular and cohesive soils coincides.

However, use of nonassociative flow rules is necessary for modeling unstable behavior within the hardening region and the direction $\mathbf{n}$ should be specified as different from $\mathbf{n}_{gL}$.

We do this by writing

$$\mathbf{n}^{\mathrm{T}} = (n_v, n_s) \tag{4.201}$$

with

$$n_v = \frac{d_f}{\sqrt{1 + d_f^2}}$$

$$n_s = \frac{1}{\sqrt{1 + d_f^2}}$$

where

$$d_f = (1 + \alpha)(M_f - \eta) \tag{4.202}$$

Again, both M_f and M_g depend on Lode's angle in the manner suggested in Zienkiewicz and Pande (1977).

It has to be remarked that both directions have been defined without reference to any yield or plastic potential surfaces, though, of course, these can be established *a posteriori*.

In fact, it is possible to integrate the above directions to obtain both plastic potential and yield surfaces

$$f = \left\{ q - M_f p' \left(1 + \frac{1}{\alpha}\right) \left[1 - \left(\frac{p'}{p'_c}\right)^{\alpha}\right] \right\} \tag{4.203}$$

$$g = \left\{ q - M_g p' \left(1 + \frac{1}{\alpha}\right) \left[1 - \left(\frac{p'}{p'_g}\right)^{\alpha}\right] \right\} \tag{4.204}$$

where the size of both surfaces is characterized by the integration constants p'_c and p'_g.

Both surfaces were depicted in Figure 4.36a and b for medium-loose sand and agree well with experimental data obtained from acoustic emission (Tanimoto and Tanaka 1986). Similar yield surfaces were proposed by Nova (1982).

To derive a suitable expression for the plastic modulus H_L, it is necessary to take into account several well-established experimental facts:

i) Residual conditions take place at the Critical State Line

$$\left(\frac{q}{p'}\right)_{\text{res}} = M_g \tag{4.205}$$

ii) Failure does not necessarily occur when this line is first crossed.
iii) The frictional nature of material response requires the establishment of a boundary separating impossible states from those which are permissible.

A convenient law was introduced in Pastor and Zienkiewicz (1986) in the form

$$H_L = H_0 p' H_f \{H_v + H_s\} \tag{4.206}$$

where

$$H_f = \left(1 - \frac{\eta}{\eta_f}\right)^4 \tag{4.207}$$

together with

$$\eta_f = \left(1 + \frac{1}{\alpha}\right) M_f \tag{4.208}$$

limit the possible states, and where

$$H_v = \left(1 - \frac{\eta}{M_g}\right) \tag{4.209}$$

$$H_s = \beta_0 \beta_1 \exp\left(-\beta_0 \xi\right) \tag{4.210}$$

are of a similar form to the expressions proposed for clays.

To illustrate the predictive capability of the proposed model, we will consider the next several sets of experiments reported in the literature (Taylor 1948; Castro 1969; Saada and Bianchini 1989) and which cover the basic features of granular soil behavior under monotonic loading.

i) Very loose sands exhibit liquefaction under undrained shearing as has been described in Chapter 3. Considering the qualitative results shown there, it is important to remember that the material densifies during the whole process, which is shown by a continuous increase in pore water pressure, suggesting that the soil is hardening.

   This seems to contradict the fact that a peak exists and the material can be thought of as softening. However, in a frictional material, strength has to be analyzed in terms of mobilized stress ratios rather than deviatoric stress, and no peak is presented by this parameter.

   This behavior can be considered unstable in the sense of Drucker (1956, 1959)

$$\mathrm{d}\boldsymbol{\sigma}^{\mathrm{T}}\mathrm{d}\boldsymbol{\varepsilon}^{\mathrm{p}} < 0 \tag{4.211}$$

   having thus

$$\mathrm{d}\boldsymbol{\sigma}^{\mathrm{T}}\left(\frac{1}{H}\mathbf{n}_g \cdot \mathbf{n}^{\mathrm{T}}\right)\mathrm{d}\boldsymbol{\sigma} < 0 \tag{4.212}$$

   If such a feature is to be modelled with a positive plastic modulus, the associated plasticity has to be abandoned, choosing

$$\mathbf{n}_g \neq \mathbf{n} \tag{4.213}$$

   Figure 4.44a–c shows, respectively, the stress paths, deviatoric stress vs. axial strain and pore pressures obtained by Castro, together with the model predictions, which agree well with them.

ii) At the other end of the density range, peaks exist in deviatoric stress during drained shear of very dense sands, this effect developing progressively as density is increased.

   The factor H_s is introduced in the expression giving a plastic modulus to account for:

   - crossing of the Characteristic State Line ($\eta = M_g$) without immediately producing failure,
   - reproduction of softening,
   - residual conditions taking place at the Critical State Line.

   To illustrate the role of the plastic modulus in the transition from softening to hardening regimes, let us consider a drained triaxial test (Figure 4.45). During the first part of the path, both H_v and H_s are positive and decrease in a monotonous way. At $\eta = M_g$, i.e. when crossing the Characteristic State Line, H_v becomes zero, while H_s is still positive. If the process continues, a moment arrives at η_p where

$$H_v + H_s = 0 \tag{4.214}$$

   with

$$\eta_p > M_g \tag{4.215}$$

   If the test is run under displacement control, the deviatoric stress does not change for an infinitesimal variation of the strain

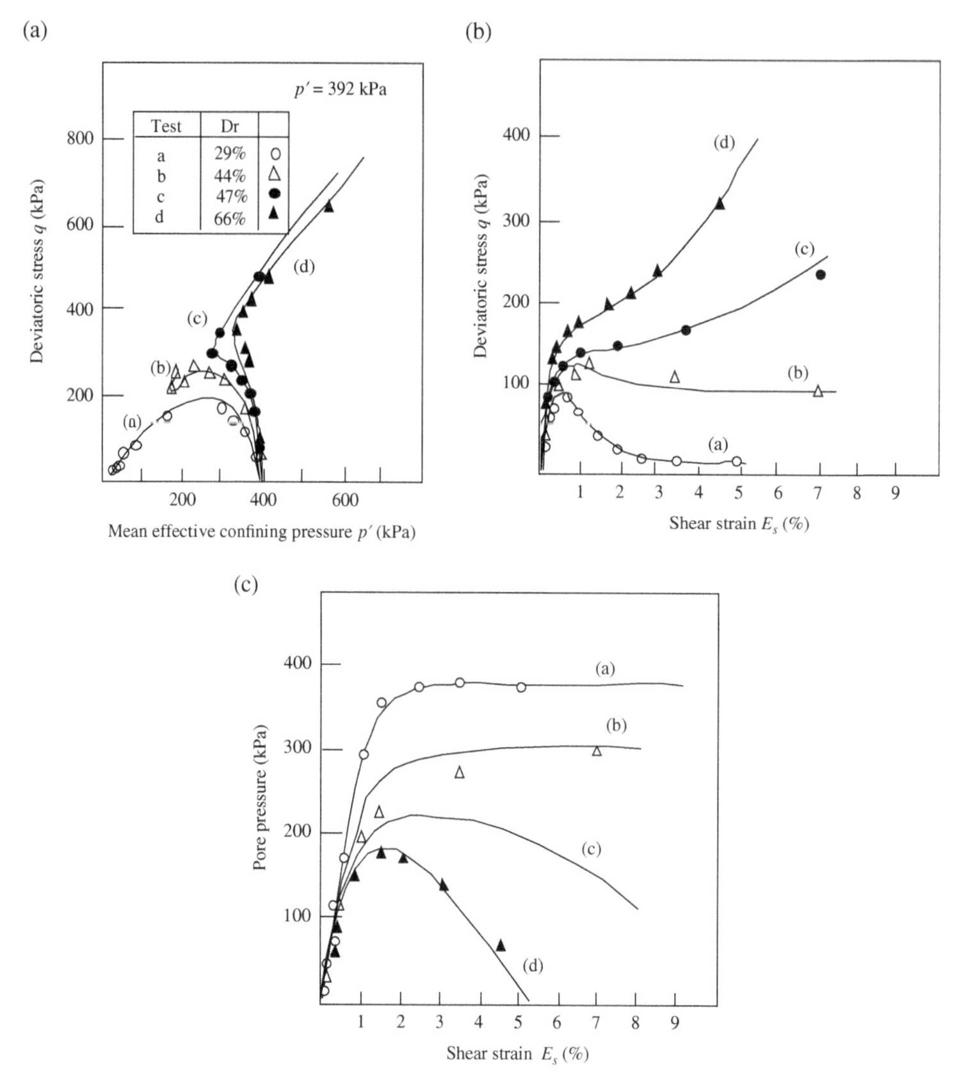

Figure 4.44 Undrained behavior of Banding sand. *Source:* Data from Castro (1969). Computed results shown by the solid line. (a) $p'-q$ plot; (b) deviatoric stress vs. shear strain plot; (c) pore pressure vs. shear strain plot

$$
\begin{aligned}
&\mathrm{d}p' = \mathrm{d}q = 0 \\
&\mathrm{d}\varepsilon_s \neq 0 \\
&\mathrm{d}\varepsilon_v \neq 0
\end{aligned}
$$

Meanwhile, H_s has decreased, and consequently, the plastic modulus becomes negative. The soil has entered the softening regime, and from this moment, the deviatoric stress will present a descending branch.

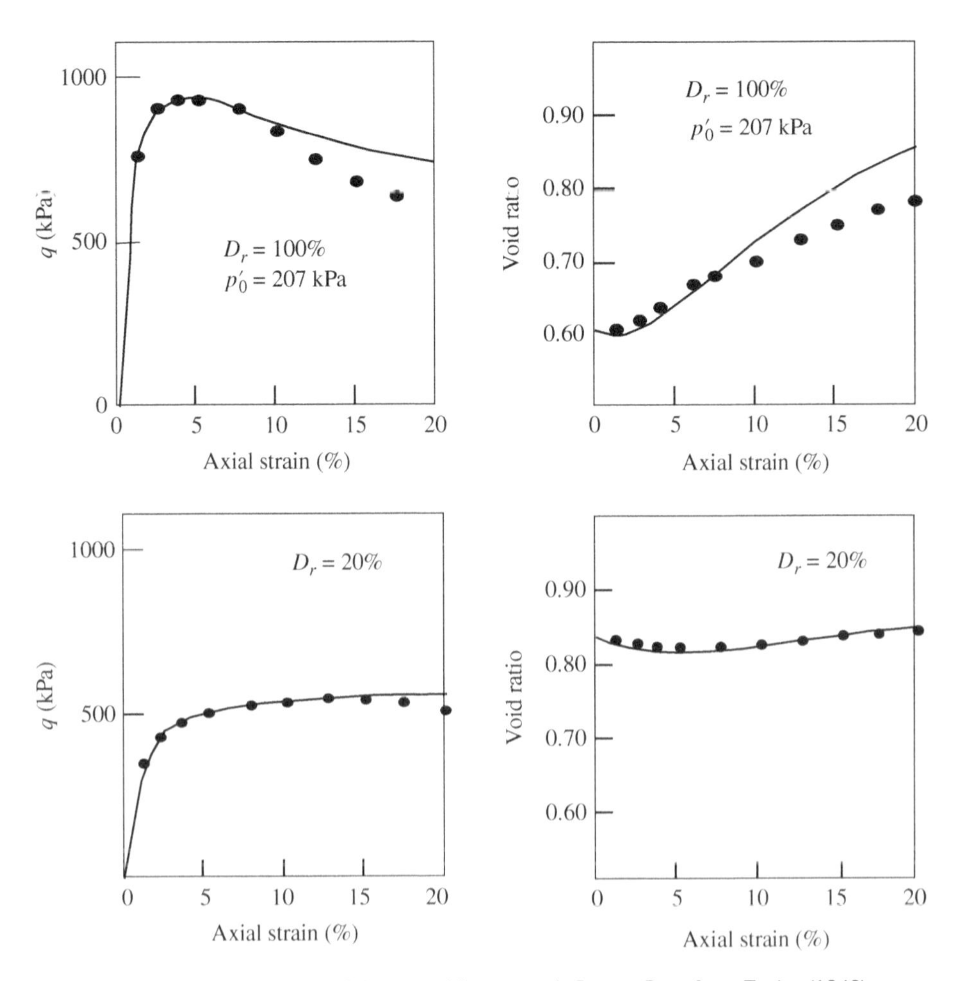

Figure 4.45 Drained behavior of dense and loose sand. *Source:* Data from Taylor (1948)

The deviatoric strain-hardening function, H_s will vanish as deformation progresses, reaching a final asymptotic value of zero at $\eta = M_g$, this time at the Critical State Line. During the softening process,

$$d\boldsymbol{\sigma}^{\mathrm{T}} d\boldsymbol{\varepsilon}^{\mathrm{p}} < 0 \tag{4.216}$$

and

$$d\boldsymbol{\sigma}^{\mathrm{T}} \left(\frac{1}{H} \mathbf{n}_g \mathbf{n}^{\mathrm{T}} \right) d\sigma < 0 \tag{4.217}$$

It can be seen that there is no need on this occasion for nonassociativeness to ensure the existence of peaks as H is negative, and, in fact, very dense sands may exhibit the limiting associative behavior with

$$M_f = M_g$$

The ratio M_f/M_g seems to be dependent on relative density, and in Pastor et al. (1985), a suitable relation was proposed as

$$\frac{M_f}{M_g} = D_r \tag{4.218}$$

where D_r is the relative density.

Figures 4.45 and 4.46a and b show model predictions for dense and loose sand response in drained conditions (Taylor 1948; Saada and Bianchini 1989).

Care should be taken when analyzing the results of tests in general, and in the case of dense sands in particular, as failure localizes along narrow zones referred to as shear bands. From the moment of their inception, the specimen is no longer homogeneous and the experimental results correspond to a boundary value problem rather than a homogeneous body. However, we have to stress the following facts:

- Even if the specimen is not homogeneous, softening must exist for the sample to exhibit a peak.
- The overall response is governed by the ratio between the width of the shear band and the length of the specimen. This effect is similar to what can be observed in numerical computations and which has been referred to as mesh-dependence.
- Experimental evidence seems to indicate the existence of a residual critical state. To obtain it, it is simpler to use loose rather than very dense specimens.

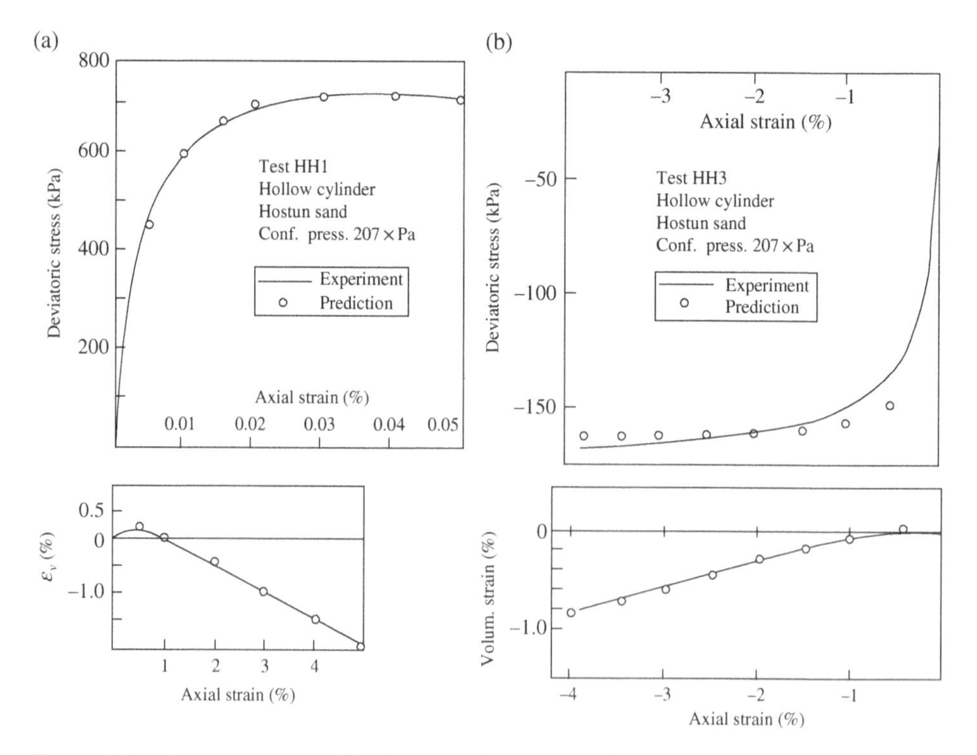

Figure 4.46 Drained behavior of Hostun sand. *Source:* From Saada and Bianchini (1989). (a) Compression test; (b) Extension test

iii) Undrained shearing of medium-loose to dense sands shows the intermediate characteristics to be discussed. Once the characteristic state line is reached, an upturn in the stress path is produced as the soil changes from contractive to dilative behavior. If the material is isotropic, determination of the CSL position can easily be performed from a point at which the undrained stress path has a vertical tangent in $(p' - q)$ space, as then,

$$d\varepsilon_v^p = 0$$

and

$$d\varepsilon_v^e = 0$$

as

$$dp' = 0$$

Figure 4.44a–c shows how relative density influences the undrained behavior of sand, together with predictions of the proposed model.

At this point, a model has been produced such that:

i) it reproduces the most salient features of sand under monotonic shearing;
ii) it is very simple as no surfaces are involved and consistency conditions do not have to be fulfilled;
iii) it is computationally efficient in FE codes as the stress point does not have to be brought back to the yield surface and tangent moduli are easily established.

4.4.3.2 Three-Dimensional Behavior

So far, we have considered the triaxial response of soil (compression and extension). However, proposed relations have been made dependent not only on I_1' and J_2 (or, alternatively, p' and q), but also on the third invariant or equivalently on Lode's angle θ.

Hence, the soil response can be generalized to any path out of the triaxial plane.

Denoting triaxial stress parameters p', q, and θ by $\boldsymbol{\sigma}*$

$$\boldsymbol{\sigma}^* = \begin{pmatrix} p' \\ q \\ \theta \end{pmatrix} \tag{4.219}$$

and the Cartesian stress tensor by $\boldsymbol{\sigma}$, invariance of the contracted product $\boldsymbol{\sigma} : \mathbf{n}$ results in (Chan et al. 1988)

$$d\boldsymbol{\sigma} : \mathbf{n} = d\boldsymbol{\sigma}^* : \mathbf{n}^* \tag{4.220}$$

then, substituting in the above

$$d\boldsymbol{\sigma}^* = \frac{\partial \boldsymbol{\sigma}^*}{\partial \boldsymbol{\sigma}} : d\boldsymbol{\sigma} \tag{4.221}$$

results in

$$\mathbf{n} = \mathbf{n}* \frac{\partial \boldsymbol{\sigma}^*}{\partial \boldsymbol{\sigma}} \tag{4.222}$$

and similarly

$$\mathbf{n}_g = \mathbf{n}_g^* \frac{\partial \boldsymbol{\sigma}^*}{\partial \boldsymbol{\sigma}} \tag{4.223}$$

Finally, the increment of plastic strain is given by

$$d\varepsilon^p = \frac{1}{H}\left(\mathbf{n}_g \otimes \mathbf{n}\right) : d\boldsymbol{\sigma} \tag{4.224}$$

and the constitutive tensor $\mathbf{C}^{ep}$ in Cartesian coordinates can be written as

$$C_{ijkl}^{ep} = \frac{\partial \sigma_m}{\partial \sigma_{ij}} C_{mn}^{ep} \frac{\partial \sigma_n}{\partial \sigma_{kl}} \tag{4.225}$$

This, of course, allows any specified stress or strain path to be followed, as illustrated in the following examples.

Figure 4.47 shows the predicted behavior for p' and θ constant tests against results obtained in the hollow cylinder device (Pastor et al. 1989). The model was calibrated with conventional triaxial compression and extension tests only.

b	EXP.	PRED.
0.0	○	———
0.277	●	—·—
0.666	Δ	—··—·
1.000	▲	- - - -

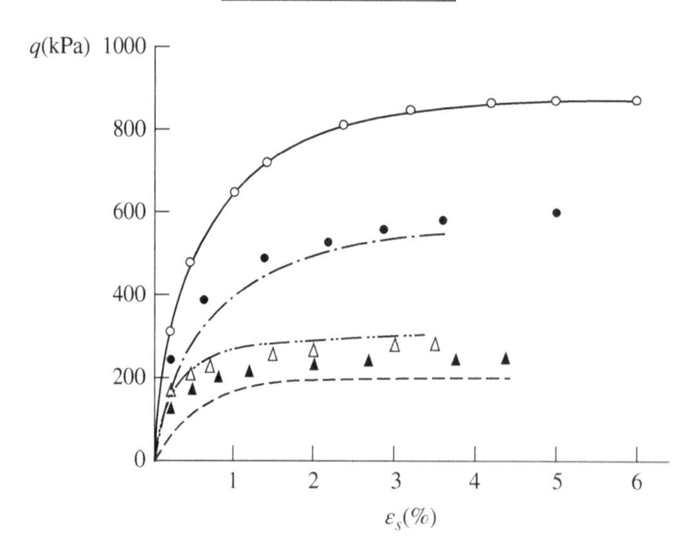

Figure 4.47 Constant b tests on Reid sand. *Source:* From Saada and Bianchini (1989)

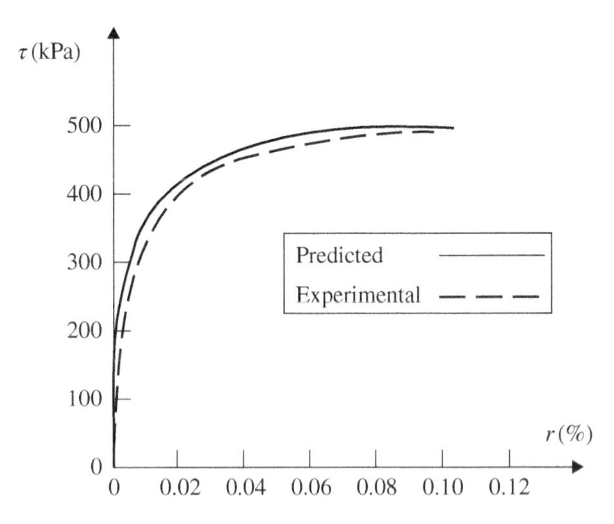

Figure 4.48 Shear of sand with rotation of principal stress axes. *Source:* From Saada and Bianchini (1989)

The hollow cylinder device is able to perform tests with rotation of principal stress axes by combining axial and radial stresses with a variable shear stress τ.

If, from a triaxial state (σ_1, $\sigma_2 = \sigma_3$, $\tau = 0$), the shear stress is increased, leaving the specimen to drain freely, q will increase while p' is kept constant, and, in addition to this, the principal stress axis will change.

In Figure 4.48 (Pastor et al. 1989), it is shown how the proposed model can reproduce this special path. In this case, the effect of increasing q prevails over rotation of principal stress axes.

However, a pure rotation of principal stress directions will not induce any response from the material as the model is defined in terms of stress and strain invariants.

Experiments with pure rotation paths show a plastic volumetric strain under drained conditions and pore pressure generation under undrained loading.

Generalization of the proposed model for these phenomena can be made by considering several mechanisms, as it was proposed in Pastor et al. (1990) using models able to introduce strain and load-induced anisotropy (Pastor 1991).

4.4.3.3 Unloading and Cyclic Loading

The first step toward modeling cyclic behavior of sands is to understand what happens when unloading and reloading. Concerning the former, the response is characterized as isotropic and elastic in most classical plasticity models, which is not always very accurate.

In fact, it can be observed from experiments that higher pore pressures than those correspondent to elastic unloading appear. Figure 4.49 depicts results obtained by Ishihara and Okada (1982) on undrained shearing of loose sands under reversal of stress. Isotropic elastic unloading is characterized by zero volumetric plastic strain and, as under undrained conditions the volume is constant, the volumetric elastic strain should also be zero and, therefore, p' should not change (a variation of p' causes a change in volumetric elastic

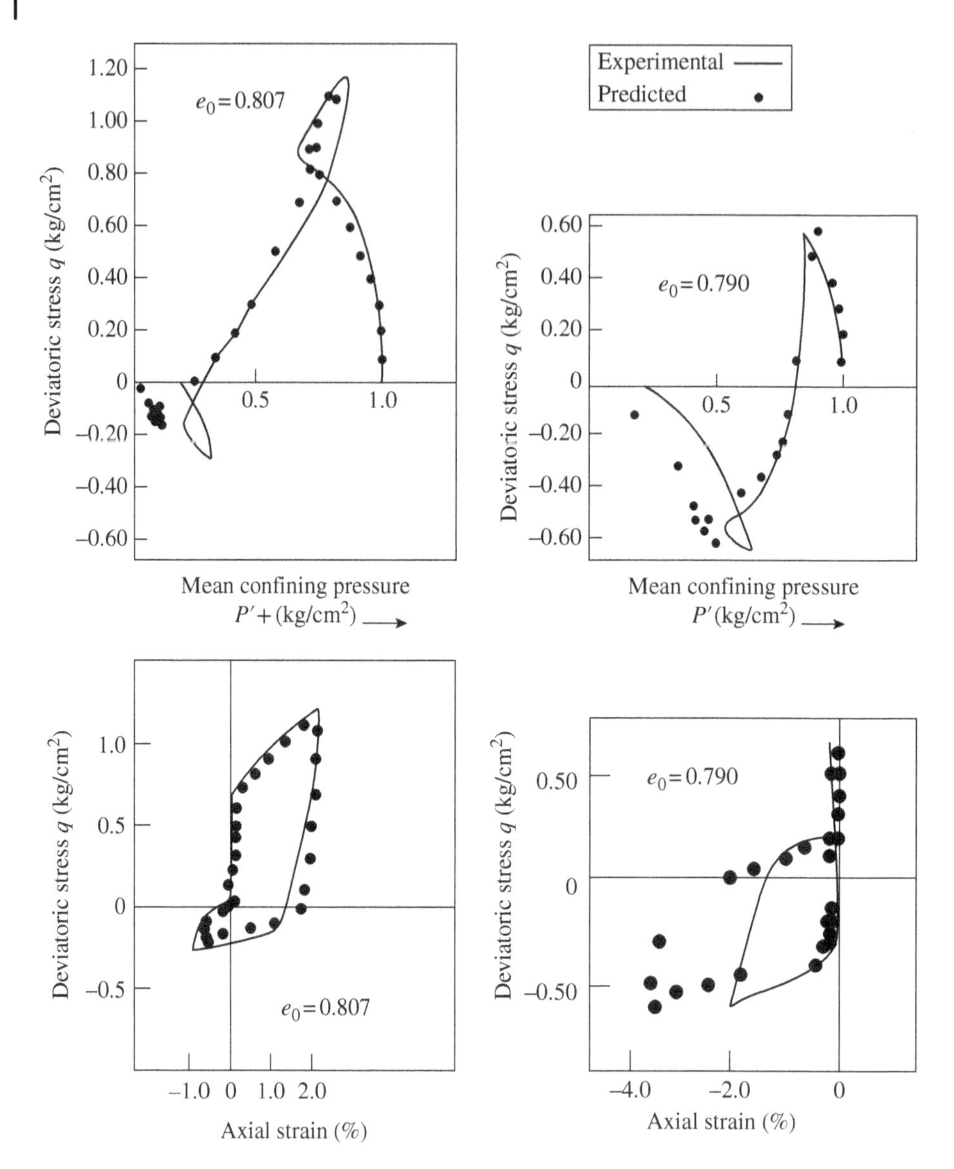

Figure 4.49 Undrained behavior of loose sand under reversal of stress. *Source:* Based on Ishihara and Okada (1982)

strain). Instead of unloading along a vertical line, the stress path turns toward the origin which indicates higher pore pressures than isotropic elastic. This phenomenon depends on the stress ratio η_u from which unloading takes place, its importance increasing with it.

Two possible explanations are possible:

1) Either the material structure has changed after having crossed the characteristic state line, and the new distribution of contacts makes the specimen anisotropic (Bahda 1997), or

2) Plastic deformations develop during unloading.

If we assume that plastic strains appear upon unloading, and that they are of a contractive nature, a simple expression for the plastic modulus fulfilling these requirements was proposed in Pastor et al. (1990)

$$\begin{aligned} H_u &= H_{u0}\left(\frac{M_g}{\eta_u}\right)^{\gamma_u} && \text{for } \left|\frac{M_g}{\eta_u}\right| > 1 \\ &= H_{u0} && \text{for } \left|\frac{M_g}{\eta_u}\right| \leq 1 \end{aligned} \tag{4.226}$$

and extends the range of the model so far proposed hierarchically.

To determine the direction of plastic flow produced upon unloading, we note that irreversible strains are of a contractive (densifying) nature.

The direction $\mathbf{n}_{gU}$ can thus be provided by

$$\mathbf{n}_{gU} = \left(n_{guv}, n_{gus}\right)^{\mathrm{T}}$$

where

$$n_{guv} = -abs\left(n_{gv}\right) \tag{4.227}$$

and

$$n_{gus} = +n_{gs} \tag{4.228}$$

Concerning reloading, it is necessary, as it was done in Section 4.4.31 for clays, to take into account the history of past events. Here, we will modify the plastic modulus introducing a *discrete memory factor* H_{DM} as

$$H_{DM} = \left(\frac{\zeta_{\max}}{\zeta}\right)^{\gamma} \tag{4.229}$$

where ζ was defined above as

$$\zeta = p' \cdot \left\{1 - \left(\frac{1+\alpha}{\alpha}\right)\frac{\eta}{M}\right\}^{1/\alpha}$$

and γ is a new material constant.

Finally, theplastic modulus is given by

$$H_L = H_0 \cdot p' \cdot H_f(H_v + H_s)H_{DM} \tag{4.230}$$

Figure 4.49 shows a prediction of this model extension for the experimental results of Ishihara and Okada.

It is possible now to model cyclic phenomena as liquefaction and cyclic mobility which appear in loose and medium sands under cyclic loading, and which are responsible for the catastrophic failure of structures subjected to earthquakes.

Both phenomena are largely caused by the overall tendency of medium and loose sands to densify when subjected to drained cyclic shearing. If the load is applied fast enough or the

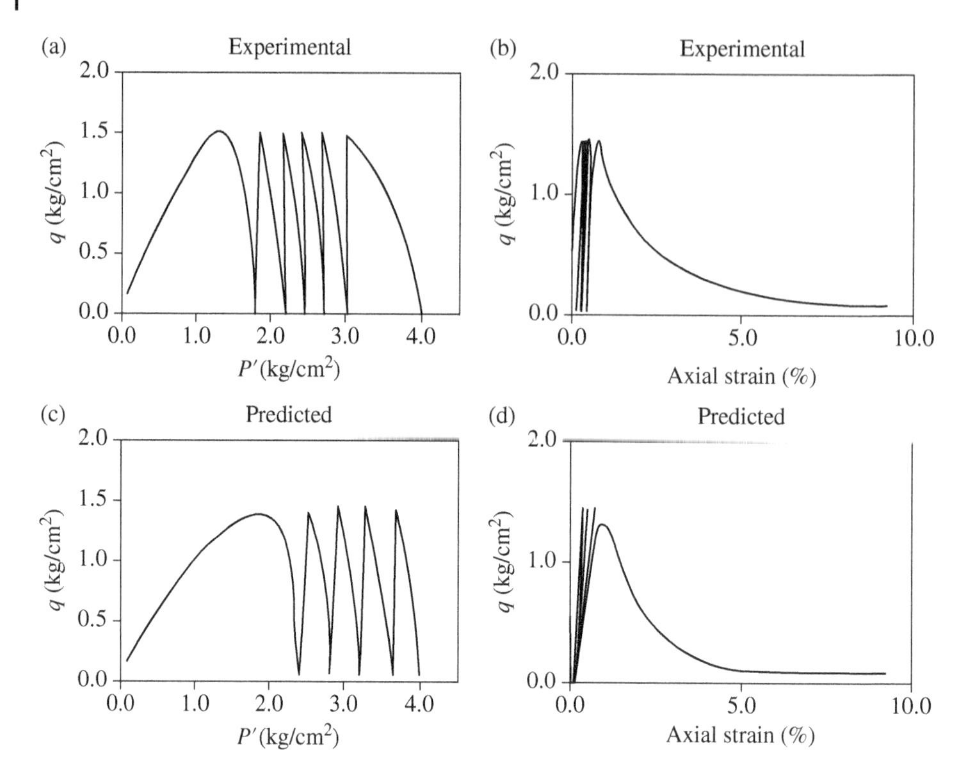

Figure 4.50 Liquefaction of loose banding sand under cyclic loading. (a, b) Experimental data. (c, d) Predictions. *Source:* Data from Castro (1969)

permeability is relatively small, this mechanism causes progressive pore pressure buildup leading to failure.

In the case of very loose sands, liquefaction takes place following a series of cycles in which the stress path migrates toward lower confining pressures. Figure 4.50 shows the results obtained by Castro (1969) in his pioneering work.

Denser sands do not exhibit liquefaction but *cyclic mobility*. Failure here is progressive since the stress path approaches the characteristic state line by its shift caused by pore pressure buildup. Deformations during unloading cause the stress path to turn toward the origin and strains produced during the next loading branch are of higher amplitude.

Figure 4.51 shows both the experimental results obtained by Tatsuoka on Fuji river sand.

Table 4.2 above gives the model parameters used in the preceding simulations.

The model can be further elaborated as shown in Pastor et al. (1987) and Pastor et al. (1993) by improving the way in which the history of past events is taken into account. To this end, two elements are introduced:

1) a surface defining the maximum level of stress reached, and
2) the point at which the last reversal took place.

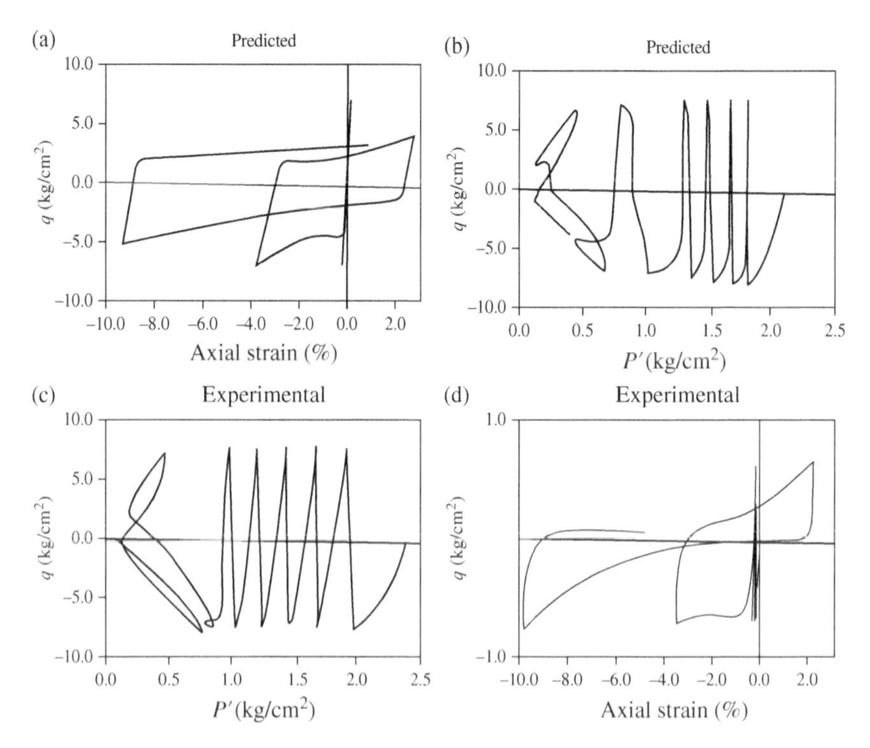

Figure 4.51 Cyclic mobility of loose Niigata sand. (a, b) Experimental data. (c, d) Predictions. *Source:* Data from Tatsuoka (1976)

Directions $\mathbf{n}$ and $\mathbf{n}_g$ and the plastic modulus H_L depend on the relative position of the stress C with respect to the point at which the load was reversed, B, and an image point, D, defined on the same mobilized stress surface as B.

To obtain the values of H_L, $\mathbf{n}$, and $\mathbf{n}_g$, suitable interpolation rules are used. In particular, $\mathbf{n}$ is interpolated from $-\mathbf{n}$ to $\mathbf{n}$ using a linear law. The direction of plastic flow is obtained again by defining a suitable dilatancy at C, d_{Cg} which is interpolated from an initial value d_{g0} to

$$d_{gD} = (1+\alpha)\left(M_g - \eta_D\right) \tag{4.231}$$

The initial value of the dilatancy at the reversal point d_{g0} is given by

$$d_{g0} = (1+\alpha)\left(M_g - C_g \eta_B\right) \tag{4.232}$$

where the constant $C_g (0 < C_g < 1)$ varies with the density, being close to zero for medium-loose sands.

The plastic modulus is interpolated between an initial value H_{U0} and its final value at the image point on the mobilized stress surface H_D. The initial value can be assumed to be infinite to decrease a possible accumulation of plastic strain under very low amplitude cycles.

$$H = H_{U0} + f \cdot (H_D - H_{U0}) \tag{4.233}$$

Table 4.2 Model parameters used in simulations.

	Figure 4.40a	Figure 4.40b	Figure 4.40c		Figure 4.40d	Figure 4.41 (dense)
K_{eu0}	35 000	35 000	35 000		35 000	30 000
G_0	52 500	52 500	52 500		52 500	50 000
M_f	0.4	0.545	0.570		0.72	0.72
M_g	1.5	1.32	1.12		1.03	1.28
H_0	350	350	350		350	16 000
β_0	4.2	4.2	4.2		4.2	2.25
β_1	0.2	0.2	0.2		0.2	0.2
γ	—	—	—		—	—
H_{u0}	—	—	—		—	—
γ_u	—	—	—		—	—
	Figure 4.41 (loose)	**Figure 4.42**	**Figure 4.45**	**Figure 4.46**	**Figure 4.47**	**Figures 4.43 and 4.44**
K_{ev0}	30 000	43 000	35 000	35 000	65 000	10 5000.
G_0	50 000	37 000	65 000	52 500	30 000	20 0000
M_f	0.50	1.2	0.80	0.40	0.71	1.17
M_g	1.33	1.26	1.30	1.50	1.5	0.90
H_0	4 000	1 000	1 600	350	800	1 750
β_0	2.25	2.0	1.5	4.2	3.8	1.2
β_1	0.2	0.13	0.10	0.2	0.16	0.14
γ	—	—	1.0	4	1	4
H_{u0}	—	—	200	600	250	—
γ_u	—	—	2	2	5	—

where f is an interpolation function depending on the relative position of the points B, C, and D and which is 1 when C and D coincide.

Concerning the rule to obtain the image stress point D, there are several alternative possibilities. For instance, it can be obtained as the intersection of the straight line joining the reversal and the stress point with the mobilized stress surface, as depicted in Figure 4.52.

This interpolation law provides a smooth transition between unloading to reloading. In fact, unloading may be considered a new loading process. It is important to remark that direction of plastic flow and unit vector **n** will not be functions of the stress state only, but of the past history as well.

Finally, the influence of sand densification under cyclic loading can be taken into account by introducing into the plastic modulus a factor H_d

$$H_d = \exp\left(-\gamma_d \varepsilon_v^{\mathrm{p}}\right) \tag{4.234}$$

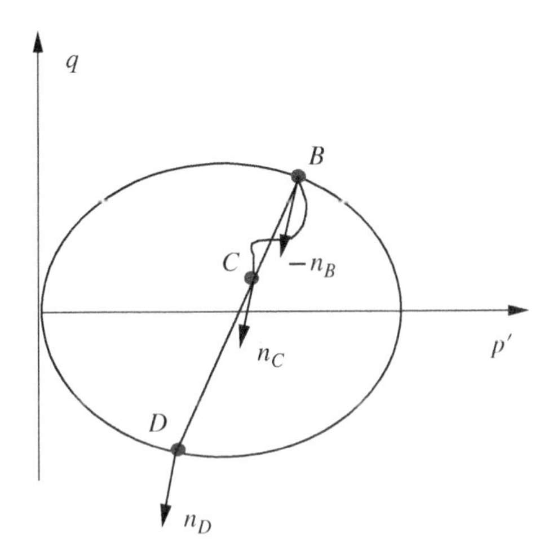

Figure 4.52 Interpolation rule

Figure 4.53 shows the densification of loose sand under cyclic loading and it can be observed how the volumetric and deviatoric plastic strain produced decreases with the number of cycles.

It should be mentioned here that since the simple models we have described here were proposed, several improvements and modifications have been introduced, particularly at CERMES (Paris) (Saitta 1994), where research recently finished has succeeded to include state parameters describing, in a consistent way, the behavior of sand under different conditions of confining pressure and relative density (Bahda 1997; Bahda et al. 1997).

4.4.4 Anisotropy

4.4.4.1 Introductory Remarks

Anisotropy in geomaterials is caused either by the arrangement of particles such as occurring in natural deposits in which the grains may have their major axes on the bedding planes, or by the spatial distribution of contacts and contact forces. In the first case, it is found that the strength is higher when tested along the deposition direction. This effect can introduce important errors if not taken into account. For instance, if the number of cycles to liquefaction is determined using a standard triaxial testing machine to evaluate the liquefaction potential of a natural sand deposit, the value obtained will be greater, and, therefore, the strength will be underestimated.

Several theories have been proposed within the framework of plasticity to describe both initial and stress-induced anisotropy. Basically, anisotropy has been approached most of the times by changing the position, orientation and shape of isotropic yield, loading or plastic potential surfaces, in such a way that those changes were dependent on tensors such as stress or strain and not only on their invariants.

Initial or fabric anisotropy could be reproduced as well by introducing initial movements and distortions on the surfaces.

The combination of kinematic and anisotropic hardening laws proposed by Mroz (1967) have provided a suitable way to model the anisotropic behavior of soils (Mroz et al. 1979; Hashiguchi 1980; Ghaboussi and Momer 1982; Hirai 1987; Liang and Shaw 1991; Cambou

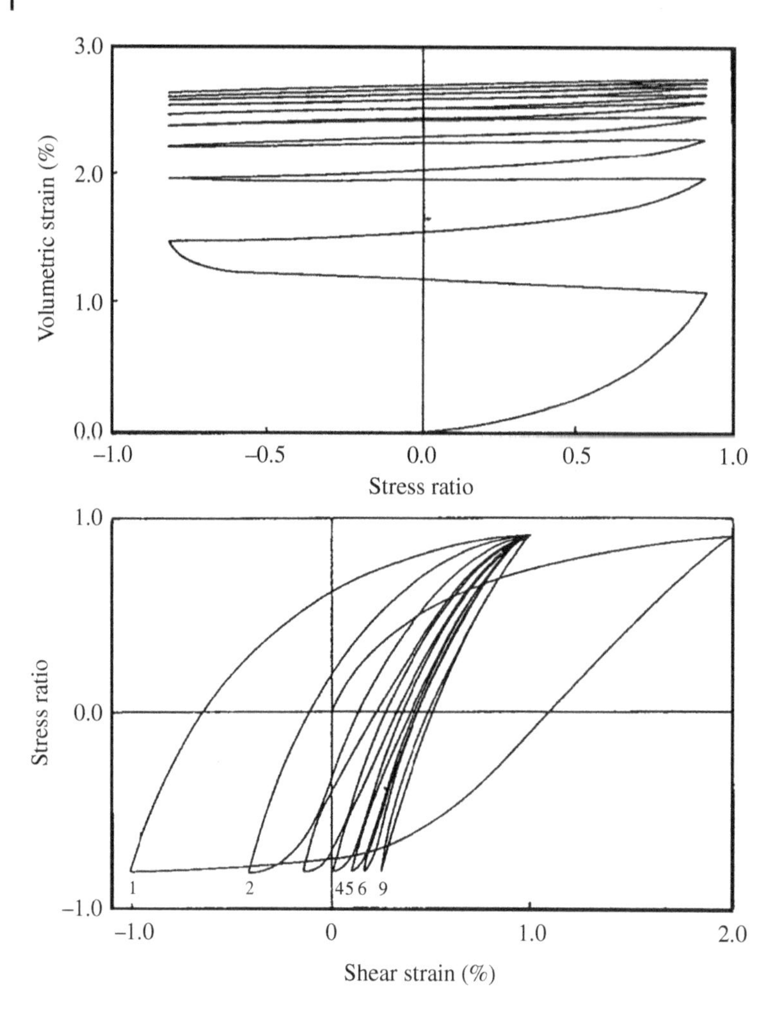

Figure 4.53 Densification of medium-loose sand under drained cyclic loading

and Lanier 1988; di Prisco et al. 1993). Surfaces can be allowed to expand following isotropic hardening rules and to translate and rotate under anisotropic hardening laws.

If only initial or fabric anisotropy is to be considered, a simple way to introduce anisotropic surfaces is to define a modified second invariant of the deviatoric stress tensor

$$\overline{J}_2' = \frac{1}{2} s_{ij} A_{ijkl} s_{kl} = \frac{1}{2} \mathbf{s} : A : \mathbf{s} \tag{4.235}$$

where A_{ijkl} is a fourth-order tensor characterizing material anisotropy.

This method was initially suggested by Hill (1950) and extended by Nova and Sacchi (1982) and Nova (1986) to soils and soft rocks. If J_2' is substituted by $\overline{J}_2'$ in any isotropic plasticity model (such as the Cam–Clay, for instance), one finds yield surfaces that have been distorted and rotated. This effect can be introduced also by directly formulating the surfaces on the stress space (Banerjee and Youssif 1986) or by deriving them from modified anisotropic flow rules, as proposed by Anandarajah and Dafalias (1986).

Baker and Desai (1984) suggested to include the effect of stress-induced anisotropy via joint stress and stress invariants. Surfaces were made dependent on

$$K_1 = \mathrm{tr}(\boldsymbol{\sigma} \cdot \boldsymbol{\varepsilon}^p)$$
$$K_2 = \mathrm{tr}\left(\boldsymbol{\sigma} \cdot \boldsymbol{\varepsilon}^{p2}\right)$$
$$K_3 = \mathrm{tr}\left(\boldsymbol{\sigma}^2 \cdot \boldsymbol{\varepsilon}^p\right)$$
$$K_4 = \mathrm{tr}\left(\boldsymbol{\sigma}^2 \cdot \boldsymbol{\varepsilon}^{p2}\right)$$

in addition to invariants of stress and strain tensors.

This approach is based on the representation theorem of scalar functions depending upon two symmetric second-order tensors $\boldsymbol{\sigma}$ and $\boldsymbol{\varepsilon}^p$ in this case). All theories mentioned above are able to introduce the anisotropic response of geomaterials even when there is a single mechanism of deformation. Multi-mechanism theories can also describe anisotropic behavior, provided that they are not formulated in terms of the three stress invariants only. Multilaminate models as introduced by Pande and Sharma (1983) consider that deformation is caused by dilation and slip taking place at all possible contact planes within the material. Of all possible active planes, only a reduced number of sampling planes is considered. The overall response is obtained by a process of numerical integration extended to sampling planes.

Alternatively, if attention is focused only on planes normal to XY, YZ, and ZX, and their responses are grouped together, one finally arrives at a three-mechanism model (Matsuoka 1974; Aubry et al. 1982).

Both multi-laminate and multi-mechanism models of the type described above can produce plastic strain under pure rotation of principal stress axes (Pande and Sharma 1983; Pastor et al. 1990).

Alternatively, the behavior of the material can be assumed to be caused by the superposition of responses to variations in σ_1', σ_2', and σ_3', and can then be generalized to more general stress conditions. This has been proposed by Darve and Labanieh (1982) and applied to complex stress paths, including anisotropy effects.

Finally, it should be mentioned that material fabric plays a paramount role on geomaterials' anisotropic response and it is, in turn, modified by the deformation process. The fabric may be approximated by a second-order tensor, which can be incorporated into the constitutive equations. An interesting way has been recently proposed by Pietruszczak and Krucinski (1989) and consists of adding two components to obtain the increment of plastic strain. The first one corresponds to an isotropic-hardening mechanism, and the second one accounts for deviations of isotropy, which are made dependent on fabric tensor.

4.4.4.2 Proposed Approach

It has been mentioned above that material structure or fabric has to be incorporated in the constitutive equations to account for both initial and induced anisotropy. Here, it will be assumed that fabric can be described by a second-order structure tensor **A**, which will determine its type of symmetry. If **Q** is a rotation or reflection tensor, the class of symmetry will be defined by the set of operators **Q** which fulfill

$$A = \mathbf{Q}^T \cdot \mathbf{A} \cdot \mathbf{Q} \tag{4.236}$$

For instance, transversely isotropic materials will be described by **A** invariant under **Q** given by

$$\mathbf{Q} = \begin{pmatrix} 1 & 0 & 0 \\ 0 & \cos\theta & -\sin\theta \\ 0 & \sin\theta & \cos\theta \end{pmatrix} \tag{4.237}$$

where it has been assumed that the plane of isotropy is XY.

The structure tensor will have the form

$$A = \begin{pmatrix} A_{11} & 0 & 0 \\ 0 & A_{22} & 0 \\ 0 & 0 & A_{33} = A_{22} \end{pmatrix} \tag{4.238}$$

and it can be easily checked that the invariance relation $\mathbf{A} = \mathbf{Q}^{\mathrm{T}} \cdot \mathbf{A} \cdot \mathbf{Q}$ is verified.

If the initial "structure" of the material is described by $\mathbf{A}^0$, $\mathbf{A}$ will vary along the loading process, according to

$$\mathbf{A} = \mathbf{A}^0 + \int \mathrm{d}\mathbf{A} \tag{4.239}$$

where $\mathrm{d}\mathbf{A}$ will depend on the plastic strain

$$\mathrm{dA} = \mathrm{dA}(\mathrm{d}\boldsymbol{\varepsilon}^{\mathrm{p}}) \tag{4.240}$$

Now, the structure tensor can be used to define a fourth-order anisotropy tensor **B** from which a modified second invariant $\overline{J}_2'$ can be derived as suggested by Hill (1950); Nova and Sacchi (1982) and Nova (1986). Following Cowin (1985), **B** can be expressed as a combination of terms listed below.

$$\begin{array}{ll} (i) & \boldsymbol{\delta} \otimes \boldsymbol{\delta} \\ (ii) & \boldsymbol{\delta} \otimes \mathbf{A} \quad \mathbf{A} \otimes \boldsymbol{\delta} \\ (iii) & \boldsymbol{\delta} \otimes \mathbf{A}^2 \quad \mathbf{A}^2 \otimes \boldsymbol{\delta} \\ (iv) & \mathbf{A} \otimes \mathbf{A} \\ (v) & \mathbf{A} \otimes \mathbf{A}^2 \quad \mathbf{A}^2 \otimes \mathbf{A} \\ (vi) & \mathbf{A}^2 \otimes \mathbf{A}^2 \end{array} \tag{4.241}$$

where compact notation has been used.

If a transverse isotropic material is considered, the tensor **B** referred to principal axes is given by

$$\begin{pmatrix} B_1 & B_2 & B_3 & & & & & & \\ B_3 & B_2 & B_4 & & & & & & \\ B_3 & B_4 & B_2 & & & & 0 & & \\ & & & B_5 & & & & & \\ & & & & B_5 & & & & \\ & & & & & B_5 & & & \\ & & & & & & B_5 & & \\ & & & & & & & B_6 & \\ & 0 & & & & & & & B_6 \end{pmatrix} \tag{4.242}$$

where rows and columns include components of the tensor with their components ordered as

11 22 33 12 21 13 31 23 32

and, therefore, component 1133 is located in the first row and third column.

In the above,

$$2B_6 = B_2 - B_4 \tag{4.243}$$

It can be seen that the resulting anisotropy tensor depends only on five constants and that the form proposed by Nova and Sacchi (1982) is a particular case of the above expression in which B_1 and B_4 have been made one and zero, respectively.

So far, only J_2' has been extended to account for anisotropy. However, geomaterial behavior is also dependent on first and third invariants, and anisotropy should also be reflected on them. New invariants $\overline{I}_2'$ and $\overline{J}_3'$ can be introduced in a similar manner, by defining $\mathbf{B}^I$ and $\mathbf{B}^{III}$ which are tensors of orders two and six.

The first anisotropy tensor $\mathbf{B}^I$ would be dependent on $\boldsymbol{\delta}$ and A, and $\mathbf{B}^{III}$ on double tensorial products of $\boldsymbol{\delta}$, A, and $\mathbf{A}^2$ such as $\boldsymbol{\delta} \otimes \boldsymbol{\delta} \otimes \boldsymbol{\delta}$, $\mathbf{A} \otimes \boldsymbol{\delta} \otimes \boldsymbol{\delta}$, etc.

Therefore, the extended set of invariants is given by

$$\begin{aligned}
\overline{I}_1' &= \mathrm{B}^I : \sigma' \\
\overline{J}_2' &= \sigma' : \mathrm{B}^{II} : \sigma' \\
\overline{J}_3' &= \sigma' : \left(\sigma' : \mathrm{B}^{III} : \sigma'\right)
\end{aligned} \tag{4.244}$$

Finally, constitutive laws derived for isotropic materials in terms of $\overline{I}_1'$, $\overline{J}_2'$, and $\overline{J}_3'$ can be generalized to anisotropic situations by substituting them by modified forms given above.

An interesting particular case is obtained when $\mathbf{A}$ is taken as

$$A = \boldsymbol{\varepsilon}^{\mathrm{p}} \tag{4.245}$$

Then, the constitutive law can be seen to be dependent on joint stress–strain invariants, as proposed by Baker and Desai (1984).

As mentioned above, the result of substituting stress invariants by their modified forms can be viewed as introducing a rotation and a distortion of yield and plastic potential surfaces, and indeed some kinematic hardening models in which back stress is introduced can be considered as particular cases of the theory outlined above. They present, however, the advantage of being simpler to develop.

If the three tensors introduced to produce the modified invariants are defined as:

$$\begin{aligned}
\mathrm{B}^I &= \mathrm{A} \\
\mathrm{B}^{II} &= (\delta \otimes \delta - \mathrm{A} \otimes \mathrm{A})\delta\mathrm{A} \\
\mathrm{B}^{III} &= \frac{1}{6}\boldsymbol{\delta} \otimes \mathrm{C} \otimes \boldsymbol{\delta} - \frac{1}{9}(\mathrm{A} \otimes \boldsymbol{\delta} \otimes \boldsymbol{\delta} + \boldsymbol{\delta} \otimes \mathrm{A} \otimes \boldsymbol{\delta} + \boldsymbol{\delta} \otimes \boldsymbol{\delta} \otimes \mathrm{A}) \\
&\quad + \frac{1}{9}(\boldsymbol{\delta} \otimes \mathrm{A} \otimes \mathrm{A} + \mathrm{A} \otimes \boldsymbol{\delta} \otimes \mathrm{A} + \mathrm{A} \otimes \mathrm{A} \otimes \boldsymbol{\delta}) - \frac{1}{3}\mathrm{tr}\left(\mathrm{A}^3\right)\mathrm{A} \otimes \mathrm{A} \otimes \mathrm{A}
\end{aligned}$$

It can be checked that this choice corresponds to a pure rotation of the yield and plastic potential surfaces. In the case of generalized plasticity, where surfaces are not introduced, it will be a rotation of directions $\mathbf{n}$ and $\mathbf{n}_{gL/U}$ (Pastor 1991).

The model described in Pastor (1991) and Pastor et al. (1993) introduced a rotation characterized by a new direction of the plastic potential and yield surface axes given by a unit tensor **a** given by:

$$\mathbf{a} = \frac{\mathrm{A}}{\mathrm{tr}(\mathrm{A}^2)^{1/2}} \tag{4.246}$$

The first modified invariant was obtained as

$$\overline{I}'_1 = \frac{1}{\sqrt{3}}\mathrm{tr}(\boldsymbol{\sigma}' \cdot \mathbf{a}) \tag{4.247}$$

which is proportional to the projection of stress tensor along **a**.

To obtain the second invariant J'_2, modified deviatoric stress was defined as

$$\overline{\mathbf{s}} = \boldsymbol{\sigma}' - \sqrt{3}\mathbf{a} \cdot \overline{I}'_1 \tag{4.248}$$

Alternatively, a modified form of the classical set of invariants $(\overline{p}', \overline{q}, \overline{\theta})$ can be obtained. Care should be taken when defining a new Lode's angle as the trace of the new modified deviatoric stress is not zero. A possible solution consists of defining a new deviatoric stress **t** as

$$\mathbf{t} = \mathbf{s} - \frac{1}{3}\mathrm{tr}(\mathbf{s})\boldsymbol{\delta} \tag{4.249}$$

from which

$$\begin{aligned} G_2 &= \frac{1}{2}\mathrm{tr}\left(\mathbf{t}^2\right) \\ G_3 &= \frac{1}{2}\mathrm{tr}\left(\mathbf{t}^3\right) \end{aligned} \tag{4.250}$$

and

$$\overline{\theta} = -\frac{1}{3}\sin^{-1}\left(\frac{3^{3/2}}{2}\frac{G_3}{G_2^{3/2}}\right) \tag{4.251}$$

The modified invariants

$$\overline{p}' = \frac{1}{3}\overline{I}'_1 \quad \overline{q} = \sqrt{3G_2} \tag{4.252}$$

and $\overline{\theta}$ can now be introduced into the generalized plasticity model described in Section 4.4.3, to describe the anisotropic behavior of sand.

4.4.4.3 A Generalized Plasticity Model for the Anisotropic Behavior of Sand

Sand deposits exhibit an anisotropic response caused by the alignment of sand grains on horizontal planes. This initial or inherent anisotropy may be modified by subsequent strains developed as the material is loaded.

If a specimen of such material is brought to failure, grains will be reorganized as deformation increases, changing the initial structure.

It has been shown above how the material response can be described by providing suitable expressions for tensors $\mathbf{n}_{gL/U}$ and plastic modulus $H_{L/U}$. Following this approach, simple models have been derived for isotropic materials (Pastor et al. 1985, 1990; Zienkiewicz et al. 1985) in terms of invariants I'_1, J'_2, and J'_3 or p', q, and θ. It will be shown next how to obtain a simple generalized plasticity model for anisotropic sands.

Concerning the plastic flow rule, two experimental facts will be recalled:

- If experimental data obtained on granular soils with initial anisotropy such as is given in Yamada and Ishihara (1979) are analyzed, it can be found that the zero volumetric incremental strain surface may be described by the same simple relation proposed in the preceding section for isotropic sand

$$\eta' = \frac{q}{p'} = M_g(\theta) \tag{4.253}$$

where M_g depends only on Lode's angle θ and η' is the stress ratio. The tests performed by Yamada and Ishihara (1979) consisted of proportional, radial paths performed at constant p' and θ. A detailed description of both the testing procedures and the results obtained is given there. The samples exhibited a strong anisotropy as grains were arranged such that their long axes were horizontal. Therefore, different behavior was observed along paths such as ZC and YC which have the same value of Lode's angle. However, no such dependence was found for M_g, which will be assumed to be independent of anisotropy.
- A similar analysis may be carried out to study how soil dilatancy is affected by soil anisotropy. Miura and Toki (1984) analyzed the experiments of Yamada and Ishihara (1979) and concluded that soil anisotropy did not greatly affect the dilatancy behavior of sand, confirming the linear relation between dilatancy and stress ratio proposed above for isotropic soils.

$$d_g = (1 + \alpha)\cdots(M_g - \eta) \tag{4.254}$$

It was found that the parameter α depended on θ and the following relationship was proposed

$$\alpha(\theta) = (1 + \alpha_c)\frac{M_{gc}}{M_g(\theta)} - 1 \tag{4.255}$$

where subindex "c" refers to values obtained for $\theta = 30^\circ$.

So far, it has not been necessary to introduce the modified invariants described above and expressions giving $M_g(\theta)$ and dilatancy holds for both isotropic and anisotropic materials.

The modified invariants $(\overline{p}', \overline{q}, \overline{\theta})$ will be used, however, in the definitions of direction $\mathbf{n}$ and plastic modulus $H_{L/U}$ directly.

Microstructure tensor $\mathbf{A}$ was assumed to have an initial value $\mathbf{A}^0$ corresponding to the initial fabric. Therefore, it should reflect material symmetries. Naturally deposited sands exhibiting transverse isotropy will have

$$\mathbf{A}^0 = \begin{pmatrix} A_1 & & \\ & A_2 = A_3 & \\ & & A_3 \end{pmatrix} \tag{4.256}$$

with axis X_1 coinciding with the direction of deposition, while isotropic materials will be characterized by

$$\mathbf{A}^0 = \begin{pmatrix} A_1 & & \\ & A_2 = A_1 & \\ & & A_3 = A_1 \end{pmatrix} \tag{4.257}$$

As the material is loaded, deformation will produce rotation of grains and rearrangement of microstructure. Therefore, **A** will change and d**A** will be a function of the increment of strain.

General expressions for this function have been suggested by Pietruszczak and Krucinski (1989). Here we will assume that d**A** may be expressed as

$$\mathrm{d}\mathbf{A} = \mathrm{d}\underline{\epsilon}^{\mathrm{p}}\{C_{a1}\exp\left(-C_{u0}\xi\right)\} \tag{4.258}$$

where C_{a0} and C_{a1} are two material parameters and ξ is the accumulated deviatoric strain.

If C_{a0} and C_{a1} are taken as zero and one, respectively, and no initial anisotropy exists, **A** will coincide with ϵ^{p}, and the modified invariants will be functions of both the stress and mixed stress–strain invariants.

To show the performance of the proposed model, Figure 4.50 reproduces both the results obtained by Yamada and Ishihara (1979) and the model predictions (Pastor 1991). The tests were run on the Fuji River, with specimens constructed by pluviation of sand through the water to simulate the natural deposition process. This resulted in a highly anisotropic structure which was modified by subsequent loading. Cubic samples of sand were tested on a true triaxial apparatus along different paths. Those depicted in Figure 4.54 correspond to compression and extension along the vertical.

It is interesting to notice how the anisotropic structure induced by the deposition process resulted in a higher overall stiffness for the specimen tested in compression along the vertical and how the situation was reversed when loaded in extension.

4.4.5 A State Parameter-Based Generalized Plasticity Model for Granular Soils

One important limitation of the basic model for sands described in Section 4.4.3 is that specimens of given sand with different densities require a different set of parameters to reproduce the observed behavior. Since the basic model for sands was proposed (Pastor et al. 1989), efforts were devoted to improving the model to include the double dependence on density and pressure. It is worth mentioning here the work of Saitta (1994), Bahda (1997), Bahda et al. (1997), Ling and Liu (2003), Ling and Yang (2006), Tonni et al. (2006), Manzanal (2008), and Manzanal et al. (2011a, 2011b).

The extensions are based on introducing changes on the main ingredients of the basic model, i.e. the direction of plastic flow, the loading-unloading discriminating direction **n**, the plastic modulus, including history and unloading, and the elastic constants.

The models implement the state parameter ψ as a key ingredient. In most cases, the definition is that of Been and Jefferies (1985), $\psi = e - e_c$ which was illustrated in Figure 4.35. Therefore, a suitable expression for the CSL is needed. Manzanal (2008) and Manzanal et al. (2011a) used the form proposed by Li (1997), while Ling and Yang (2006) used a variant of it, based on the work of Li and Wang (1998)

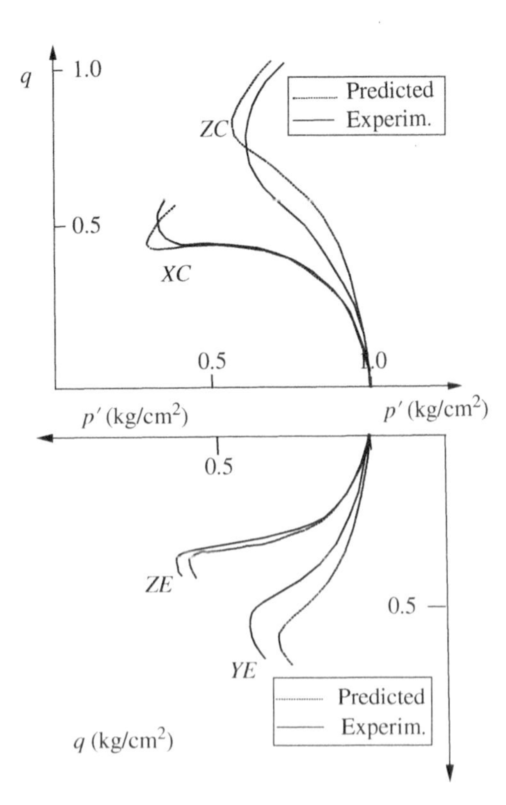

Figure 4.54 Anisotropic behavior of Fuji River sand in triaxial compression and extension

$$e = e_{atm} + \lambda \left(\frac{p'}{p'_{atm}} \right)^{\xi} \tag{4.259}$$

Regarding the **dilatancy**, most authors use the law proposed by Li and Dafalias (2000):

$$d = \frac{d_0}{M_g} \left(\eta_{PTS} - \eta \right) \tag{4.260}$$

where η_{PTS} is the stress ratio of the Phase Transformation Line where dilatancy is zero (Ishihara et al. 1975; Habib and Luong 1978). We will assume that η_{PTS} is given by

$$\eta_{PTS} = M_g \exp(m\psi) \tag{4.261}$$

where d_0 is a model parameter which can be related to the dilatancy law used in the basic model by:

$$d_0 = (1 + \alpha) M_g \tag{4.262}$$

It is interesting to note that in the case of loose sands, $\psi > 0$ and $\psi_{PTS} > M_g$ while in the case of dense sands, $\psi < 0$ and $\psi_{PTS} < M_g$. We have depicted in Figure 4.55 the dilatancy laws of dense and loose sand together with the original dilatancy law of the basic model. The results agree well with experimental data (Verdugo and Ishihara 1996).

From here, we can derive the components of the unit tensor $\mathbf{n}_{gL}$.

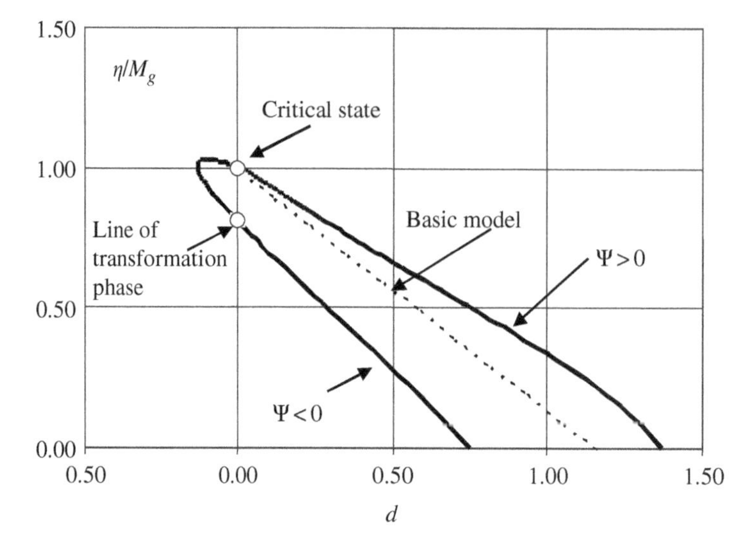

Figure 4.55 State parameter-based dilatancy laws for dense and loose sand, together with the dilatancy law of the basic model

The **loading-unloading** discriminating direction **n** is assumed to be of the form

$$n = (n_v, n_s)$$

with

$$n_v = \frac{d_f}{\sqrt{1 + d_f^2}} \qquad n_s = \frac{1}{\sqrt{1 + d_f^2}}$$

with d_f given by

$$d_f = \frac{d_{0f}}{M_f}\left(M_f \exp\left(m\psi\right) - \eta\right)$$

In above, d_{0f} and M_f are parameters of the model. The structure is similar to that of the plastic flow direction vector $\boldsymbol{n}_g$. Concerning M_f, the ratio $\frac{M_f}{M_g}$ it is related to relative density. Pastor et al. (1990) suggested using the relation

$$\frac{M_f}{M_g} \approx D_r$$

Alternatively, Manzanal (2008) and Manzanal et al. (2011a) proposed the law

$$\frac{M_f}{M_g} = h_1 - h_2\psi_q$$

where h_1, h_2, and ψ_q are parameters. The latter can be estimated as

$$\left(\frac{e_{\min}}{e_\Gamma}\right) \le \psi_q \le \left(\frac{e_{\max}}{e_{\min}}\right)$$

In above, e_{min}, e_{max} are the minimum and maximum void ratios, and e_Γ is the intercept of the CSL with $p' = 1$ on the plane $(p' \;, e)$.

The plastic modulus is assumed to be given by

$$\begin{aligned} H &= H_0 \sqrt{p' p'_0} H_{DM} \cdot H_f (H_v + H_s) \quad & \eta > 0 \\ &= H_0 \sqrt{p' p'_0} H_{DM} & \eta = 0 \end{aligned} \tag{4.263}$$

where H_{DM}, H_f, and H_s are kept the same as in the basic model, p'_0 is a reference pressure and

$$\begin{aligned} H_0 &= H'_0 \exp\left(-\beta'_0 \psi_q \right) \\ H_v &= H_{v0} (\eta_P - \eta) \end{aligned} \tag{4.264}$$

The stress ratio η_P corresponds to the peak value of the stress path, $\eta_P = M_g \exp(-\beta_v \psi)$ (Li and Dafalias 2000), and H_{v0} and β_v are model parameters.

Figure 4.56 shows the performance of the enhanced, state parameter-based version, when applied to Toyoura sand (Verdugo and Ishihara 1996). The figure depicts a series of consolidated undrained tests performed on medium-dense sand (left) and a very loose sand (right). In both cases, four different confining pressures of 100, 1000, 2000, and 3000 kPa have been chosen. All the tests have been modelled with the same set of parameters and the model predictions capture the essential features of sand behavior observed in the experiments.

Figure 4.57 compares the model predictions against the experimental results of drained triaxial tests on Toyoura sand (Verdugo and Ishihara 1996). Two different confining

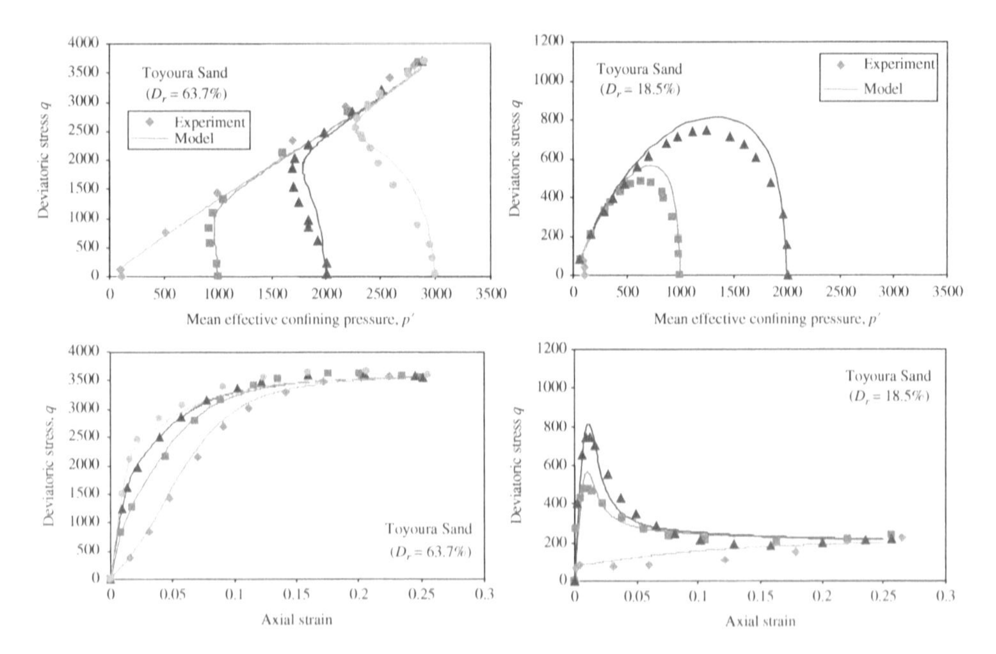

Figure 4.56 Consolidated undrained tests in Toyoura sand (Verdugo and Ishihara 1996): experiments and model predictions. *Source:* Based on Verdugo and Ishihara (1996)

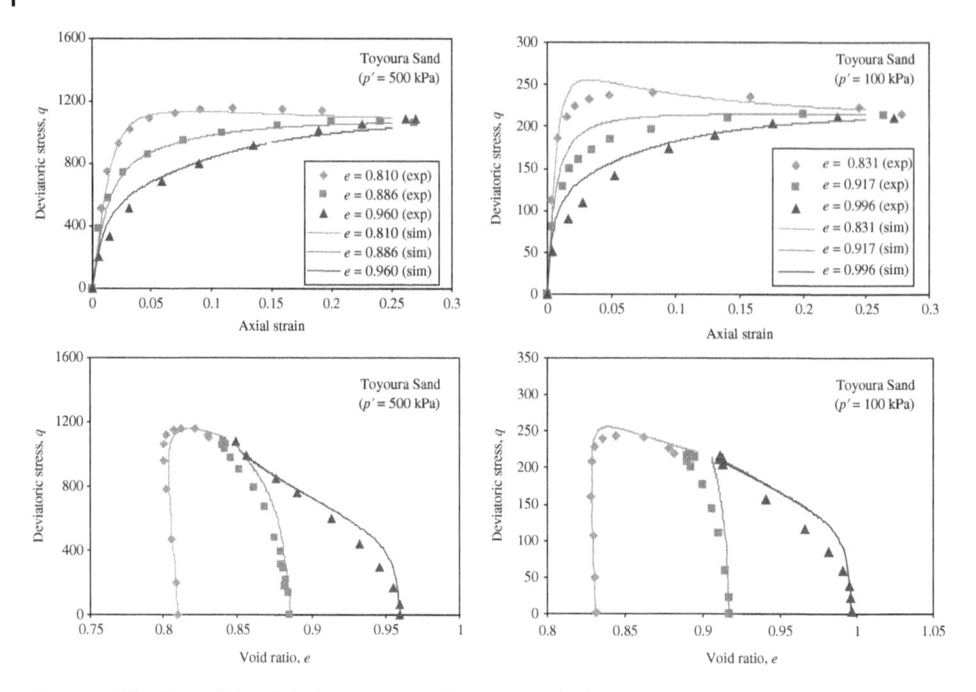

Figure 4.57 Consolidated drained tests on Toyoura sand (Verdugo and Ishihara 1996): experiments and model predictions. *Source:* Based on Verdugo and Ishihara (1996)

pressures of 100 and 500 kPa have been used, and in each case, three different densities have been tested. The set of model parameters is the same used for the undrained case shown in Figure 4.55. Again, the model is able to reproduce well the observed behavior.

Regarding unloading and reloading plastic moduli, the reader is addressed to the paper by Ling and Yang (2006), where the authors propose improvements that are successfully tested against cyclic loading tests.

4.4.6 Generalized Plasticity Modeling of Bonded Soils

Fernández Merodo et al. (2004, 2005) proposed an extension of the basic generalized plasticity model of Pastor et al. (1989) for cohesive frictional soils with debonding and applied it to both experimental tests and the fast catastrophic flowslide of Las Colinas (El Salvador, February 2001), induced by an earthquake. The authors interpreted the mechanism as a dry liquefaction problem caused by a sudden collapse of the soil loose, metastable structure (see also Pastor et al. 2002).

Progressive destruction of bonds was implemented in the basic generalized plasticity model following the work of Gens and Nova (1993) and Lagioia and Nova (1995), who based their model on assuming that the size of the yield surface on the p'–q plane, characterized by its intersections with the p' axis p_t and p_c respectively, depended on the degree of debonding (Figure 4.58)

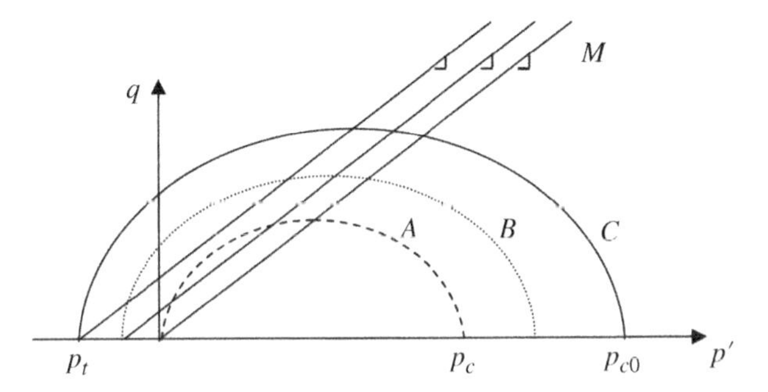

Figure 4.58 Successive yield surfaces for increasing degrees of bounding. Surface A corresponds to unbounded material. *Source:* Baseed on Lagioia and Nova (1995)

According to Lagioia and Nova (1995), p_t varied as:

$$p_t = p_{t0} \exp\left(-\rho_t \varepsilon^d\right) \tag{4.265}$$

where ε^d is the accumulated volumetric strain and ρ_t a model parameter.

Fernández Merodo et al. (2004, 2005) kept the same structure of the model, introducing some simple modifications.

They defined a new hydrostatic stress as

$$p* = p' + p_t \tag{4.266}$$

from which the modified stress ratio $\eta* = q/p*$ was introduced.

The key ingredients are:

$$\begin{aligned}
&H_L = [H_0 p* - H_b] H_f^* (H_v^* H_s) H_{DM}^* \\
&H_f^* = \left(1 - \frac{\eta^*}{\eta_f^*}\right)^4 \quad H_v^* = \left(1 - \frac{\eta^*}{M_g}\right) \\
&H_{DM} = \left(\frac{\zeta_{\max}^*}{\zeta^*}\right)^\gamma \quad \zeta^* = p^*\left[1 - \frac{\alpha}{1+\alpha M_f}\frac{\eta^*}{}\right] \\
&H_b = b_1 \varepsilon^d \exp\left(-b_2 \varepsilon^d\right)
\end{aligned} \tag{4.267}$$

where b_1, b_2 are new material constants.

The model predicted well the observations by Lagioia and Nova (1995). We include in Figure 4.59 the results of an isotropic consolidation test

4.4.7 Generalized Plasticity Models for Unsaturated Soils

So far, we have dealt with saturated soils, for which there is a framework within which we can understand and explain many fundamental aspects of their behavior. This framework is based on the effective stress principle, which allows us to understand the different behavior under fully drained or undrained conditions. In the majority of testing devices, pore pressures, when generated, are measured and recorded, so the effective stress can be obtained. The situation in unsaturated soils is not the same.

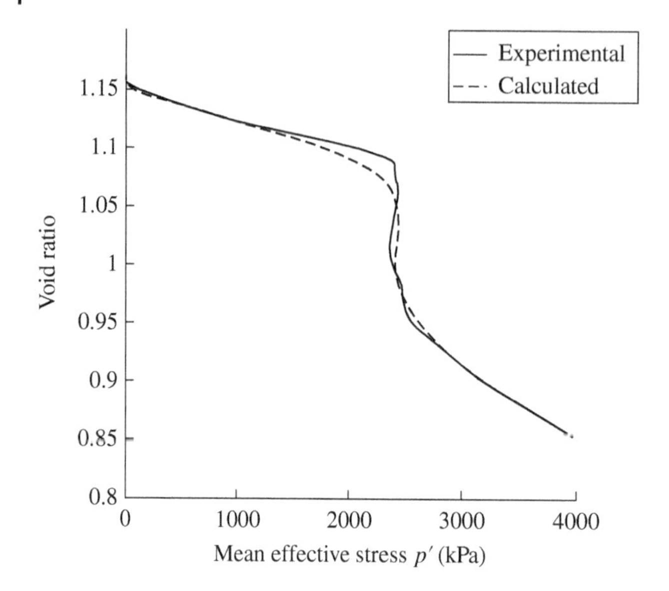

Figure 4.59 Isotropic compression test. *Source:* Redrawn from Lagioia and Nova (1995)

Unsaturated soils present an apparent cohesion due to the negative pressure in the pore water. The term "suction" or "matric suction" is introduced to define the difference between air and water pressure:

$$s = p_a - p_w \tag{4.268}$$

In many engineering applications, the pore air will be the at atmospheric pressure, i.e. $p_a = 0$ but in laboratory tests, researchers conduct tests with nonzero p_a.

Just to illustrate the effect of suction, consider that building a sandcastle requires sand with an apparent cohesion that does not exist either on fully dry or fully saturated sands.

Suction was considered by Bishop (1959) who proposed a generalization of the effective stress principle as:

$$\sigma' = \sigma - \chi s I + p_a I \tag{4.269}$$

where χ is a parameter varying between zero and one, often referred to as "Bishop effective stress parameter" (see also Section 2.5 of Chapter 2). The above expression can be rewritten as

$$\sigma' = (\sigma + p_a I) - \chi s I = \overline{\sigma} - \chi s I \tag{4.270}$$

where the stress tensor is decomposed into a "net stress tensor" $\overline{\sigma}$ and the suction term. Please note the use of negative values for compression components. Jennings and Burland (1962) showed that the effective stress proposed by Bishop is not able to reproduce collapse in wetting paths where suction decreases and the hydrostatic component of the net stress increases in such a way that Bishop-effective stress is constant. This led Bishop and Blight (1963) to state that "The principle of effective stress can be applied to saturated soils only if account is taken of the effective stress path. In the case of partly saturated soils, it is not the effective stress path only, but the paths of the two components $\bar{\sigma}$ and s which have to be taken into account," introducing the so-called "bi-tensorial" formulations based on the net stress and the matric suction tensors which have been widely used later on.

Even if it was widely accepted that Bishop stress alone could not explain the behavior of unsaturated soils, some researchers continued working on the effective stress tensor approach. Schrefler (1984) and Lewis and Schrefler (1987) derived the effective stress tensor averaging the contributions of the mixture components, arriving to

$$\sigma' = \sigma + p_a I - S_r s I \tag{4.271}$$

which coincides with Bishop stress if we assume $\chi = S_r$. We will refer to it as effective stress, and concentrate on cases where $p_a = 0$. Also Schrefler et al. (1990) dealt with two-phase flow. Hassanizadeh and Gray (1990) and Gray and Hassanizadeh (1991) arrived at the same expression following an alternative way, based on averaging and including the effect of interfaces. Alternative forms have been proposed by Coussy (1995, 2004), the former being the differential of the latter. However, the limitation described by Jennings and Burland (1962) remained unanswered in models using only the effective stress tensor.

Houlsby (1997) analyzed the work input to an unsaturated granular material and obtained its rate as

$$\delta W = \{\sigma + p_a I - S_r(p_a - p_w)I\}\dot{\varepsilon} + (p_a - p_w)n\dot{S}_r \tag{4.272}$$

plus three additional terms related to friction forces between soil skeleton and air and water, and the compressibility of air. He confirmed the thermodynamic consistency of the effective stress tensor proposed by Schrefler (1984) and insisted on the necessity of describing soil behavior using the suction and the work-conjugated strain-like variable $n\dot{S}_r$.

Based on Houlsby work, a new generation of models for unsaturated soils based both on the effective stress and suction was produced. After Jommi (2000) proposed a general framework for Critical State models, attention first focused on modeling isotropic compression tests (Gallipoli et al. 2003; Wheeler et al. 2003; Sheng et al. 2004), then shearing behavior was modeled (Tamagnini 2004; Tamagnini and Pastor 2004; Fernández Merodo et al. 2005; Santagiuliana and Schrefler 2006; who enhanced a previous model by Bolzon et al. 1996; Manzanal 2008). It is also worth mentioning the work of Nuth and Laloui (2008). Please note that in the example of Section 8.6, the models of Bolzon et al. (1996) and Santagiuliana and Schrefler (2006) have been adopted. They adopt already the format of generalized plasticity.

This is a key issue: unsaturated soil modeling requires two stress measures and two associated, work-conjugated strain variables. The strain increment will be decomposed into:

$$d\varepsilon = d\varepsilon^e + d\varepsilon^p + d\varepsilon^s \tag{4.273}$$

where the third component is the strain increment caused by the increment of suction.

Within the framework of generalized plasticity, we can therefore write the increment of strain as (Tamagnini 2004; Fernández Merodo et al. 2005; Manzanal et al. 2011b):

$$d\varepsilon = \boldsymbol{C}^e d\sigma' + \frac{1}{H_{L/U}} \boldsymbol{n}_{gL/U} \otimes \boldsymbol{n} d\sigma' + \frac{1}{H_s} \boldsymbol{n}_{gL/U} ds \tag{4.274}$$

where we have introduced an additional plastic modulus for suction-driven strain. This is the basic structure of generalized plasticity models for unsaturated soils.

As pore pressure changes and, consequently, effective stress changes can be large, it is important to keep the state parameter in the formulation, which means to study and modify accordingly the influence of suction in the CSL. Concerning CSL projection on the ln $p' - e$ plane, the problem is similar to that found when studying the NCL. Gallipoli et al. (2003) proposed a normalization method which made all CSLs corresponding to different suctions to collapse into a single curve. Manzanal et al. (2011b) proposed the relation between pressures at a given state and saturation:

$$\frac{p'}{p'_S} = \exp\{a[\exp(b\xi) - 1]\} \tag{4.275}$$

where a, b are model parameters and ξ is a "cementation variable" introduced by Gallipoli et al. (2003), given by

$$\xi = f(s)\,(1 - S_r) \tag{4.276}$$

The function $f(s)$ proposed by them is a relation obtained by Haines (1925) and Fisher (1926) who considered an assembly of spheres with the same size (monodisperse), and the stabilizing interparticle force exerted by water menisci. The function $f(s)$ is the relation between the increments of the stabilizing hydrostatic stress at a given suction and at zero suction.

Regarding the CSL line on the (p'–q) plane, Manzanal et al. (2011b) proposed to change the classical equation $q - M p' = 0$ by using a "corrected" definition of the effective degree of saturation in the effective stress as

$$S_{re} = \frac{S_r - S_{r0}}{1 - S_{r0}} \tag{4.277}$$

where S_{r0} is the minimum degree of saturation at high suctions, accounting for adsorbed water.

The plastic moduli are also modified. A suitable alternative is the following:

$$\begin{aligned} H_L &= H_0\sqrt{p' \cdot p_0} H_{DM}\{H_f(H_v + H_s)\} \\ H_b &= \omega(\xi) H_0\sqrt{p' \cdot p_0} H_{DM} H_v \end{aligned} \tag{4.278}$$

where the discrete memory factor is now

$$H_{DM} = \left[\frac{\zeta_{\max} J_s}{\zeta}\right]^{\gamma} \tag{4.279}$$

In the above equation, the same measure of the maximum mobilized stress is defined for the saturated soils, and J_s is chosen as

$$J_s = \exp[c g(\xi)] \tag{4.280}$$

where ξ was defined in Equation 4.274.

So far, we have analyzed the increments of strain induced by changes of effective stresses and suction. We still need to describe the relation between changes in the degree of

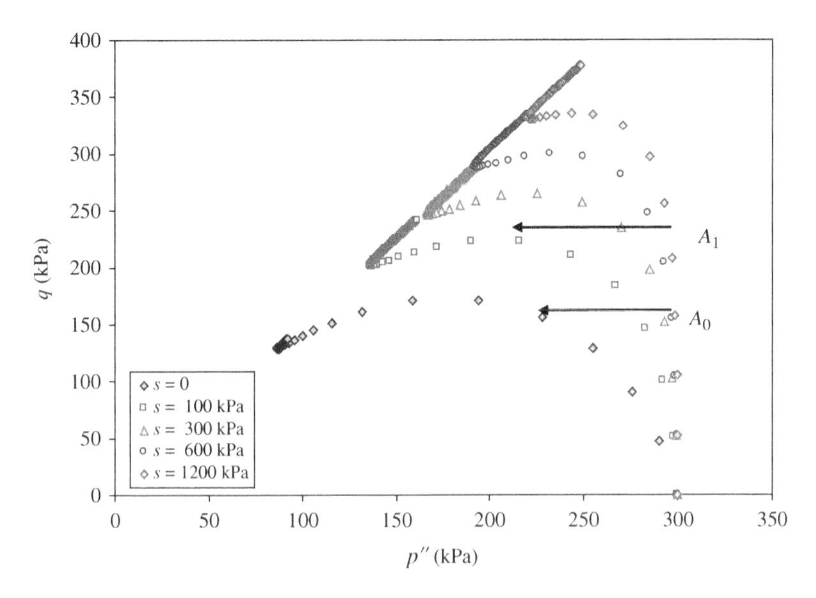

Figure 4.60 Collapse of very loose granular sands when suction decreases. *Source:* From Manzanal (2008)

saturation and effective stress and suction. This can be done in a simple manner by adding the contributions of volumetric strains and the water retention curves, which, for constant net stress, relate changes of degree of saturation and suction.

We will include here a simple yet interesting example case which is that of the undrained test of type A where the specimen is loaded up to certain deviatoric stress and then the suction is decreased. Figure 4.60 shows simulations of undrained tests (Manzanal 2008), where we have represented two decreasing suction paths. Depending on the magnitude of deviatoric stress, failure can be of catastrophic type (A_1) while, in others, the specimen will reach equilibrium (path starting at A_0).

4.4.8 Recent Developments of Generalized Plasticity Models

In the preceding sections, we have described some extensions and improvements of the original basic generalized plasticity model proposed by Pastor et al. (1990), covering work done by the authors in the past years. Here we will describe some interesting enhancements proposed in the last years, covering alternative definitions of the state parameter, breakage of grains, sand-reinforcement interaction, modeling of gravelly soils and sand-gravel mixtures, sandstone, and applications to rockfill structures.

Regarding the **state parameter**, Sadeghian and Namin (2013) investigated how the implementation of the state parameter could improve the quality of the numerical model predictions. They compared different alternatives for defining the state parameter, and applied the model to one of the VELACS project cases, model No. 3 of VELACS project (Arulanandan and Scott 1993). For more details on VELACS, see Chapter 7.

In all cases, both pore pressures and vertical displacements using the state parameter showed a better agreement with the experimental data, though it is difficult to decide which definition of the state parameter is better. The same conclusion applies to the laboratory tests on Toyoura sand done for the same purpose.

Cen et al. (2018a) proposed a new generalized plasticity model for sands within the framework of Critical State Soil Mechanics using the state parameter in order to extend the range of application of the model to high pressures, where grain breakage can generate additional volumetric strain. The key modification is the definition of a Critical State Line able to describe the behavior of sands at higher pressures and an alternative definition of the state parameter.

The state parameter was defined as $\psi = e/e_c$, where e_c is the voids ratio at the current effective confining pressure. In order to avoid negative values, they propose to use the relation $e_c = a \exp\left(-b\frac{p'}{p_{atm}}\right)$ which is shown to be valid at relative effective confining pressures $\frac{p'}{p_{atm}}$ up to 36.

Dilatancy law was chosen as $d_g = d_{g0}\left(\psi - \frac{\eta}{M_g}\right)$ with the state parameter definition given above. The loading-unloading parameter was also modified by choosing $d_f = d_{f0}\left(\psi - \frac{\eta}{M_f}\right)$ with $M_f = M_g(\chi_1 e - \chi_2 - \lambda\psi)$.

Finally, plastic modulus expression keeps its original structure, with components changed according to laws proposed by Li and Dafalias (2000). The resulting model is not complex and agrees reasonably well with experimental data.

Regarding **breakage of grains**, the model by Cen et al. (2018b) was applied to describe the behavior of granular materials including breakage of grains in Cen et al. (2018b). The model was calibrated and applied to cases where the confining pressures reached 3 MPa.

Breakage of grains was also taken into account in a model proposed by Liu and Ling (2008), which extended the previous work of Liu et al. (2006) for behavior of interfaces under both monotonic and cyclic loading. Interfaces of foundations with granular soils present an important accumulation of volumetric strain – referred to as cyclic degradation – leading to a reduced strength of the interface and grain crushing. The model pays particular attention to the discrete memory and loading/unloading components. It is based on the similarity of formulations for sand materials and interfaces using Critical State Soil Mechanics concepts noticed by Boulon and Nova (1990). The key ingredients are: (i) to define a new state parameter based on a definition of void ratio for the interface, and (ii) to allow CSL to translate when grains break. Breakage is made dependent on the work dissipated by shear stresses on the interface.

It seems clear that the amount of breakage will depend on the interface materials. Farzaneh and Iraji (2015) and Iraji and Farzaneh (2017) studied the case of **soil-reinforcement interaction** using a two-phase model in which generalized plasticity described the behavior of the soil mass. According to the authors, in this way, it is possible to account for phenomena such as softening of the reinforced soil mass. The reinforcement material was described by a classic elastoplastic Tresca law.

Another reinforced material, the fiber-reinforced cemented sand, has been studied by Abioghli and Hamidi (2019). The generalized plasticity model follows the approach of Manzanal et al. (2011a) which has been described in this chapter.

Gravelly and rockfill materials are used in geostructures such as rockfill dams and railroad platform layers. In the latter case, they are subjected to a very high number of cycles, requiring expensive maintenance work to ensure serviceability. Liu et al. (2014) analyze the cyclic behavior of dense gravelly soils including grain breakage using generalized plasticity. The model combines concepts from bounding surface and generalized plasticity. One key ingredient is the stress state at load reversal, $\boldsymbol{\sigma}^r$ from which a new stress measure is defined as $\overline{\boldsymbol{\sigma}} = \boldsymbol{\sigma} - \boldsymbol{\sigma}^r$. The following surfaces are introduced: (i) stress reversal, (ii) maximum stress ratio, (iii) phase transformation, and (iv) peak strength. From here, relative distances involving the current position of the stress state and the surfaces are used to characterize the constitutive behavior of the soil. This idea can be related to the continuous plasticity extension described in this text. A new interesting contribution consists of defining a unique plastic modulus for loading/unloading/reloading paths. The model is calibrated using laboratory tests from a rockfill dam that was severely damaged by the Great Wenchuan earthquake in 2008 and granite ballast used in railroad platforms.

Rockfill materials and their applications have been studied by Xu et al. (2012), Zou et al. (2013), Wei and Zhu (2013), Dong et al. (2013), Fu et al. (2014), Liu and Zhou (2013), Liu et al. (2014), Wang et al. (2015), Liu and Gao (2017), and Bian et al. (2019). In the paper by Xu et al. (2012), a detailed study of a Chinese dam, the Zingpu concrete face rockfill dam is presented.

The proposed generalized plasticity models aim to better describe the behavior of rockfill materials and structures made of them, such as dams of large heights, where phenomena such as settlements at the time of filling the reservoirs and seismic response are important.

In less involved situations, **sand-gravel mixtures** were modeled using generalized plasticity by Goorani and Hamidi (2015). The model incorporates new Critical State Lines for the mixture, using the CSL for sand as a reference and considering the composition of the mixture. It is based on the model by Manzanal et al. (2011a) and applied to poorly graded sand-gravel mixtures. The main elements of the model – state parameter, dilatancy, flow and loading directions, and plastic modulus – take into account the gravel content in a simple yet effective manner.

Regarding **sandstone**, Weng and Ling (2013) have proposed a generalized plasticity model. Hyperelastic behavior is characterized by an energy density function from which the elastic constitutive tensor is derived, which allows to describe different behaviors under pure isotropic compression and shear according to observations. The stress ratio definition is modified, introducing $\eta = q/q_f$ where q_f is the value of q at the failure surface, which, depending on the observed nonlinearity can be of Drucker–Prager or Hoek–Brown types. Regarding the plastic modulus, it is assumed to be of the form

$$H_L = H_0 (p'/p_{atm})^{1/2} H_f H_s$$

with $H_f = (1 - \eta^2)$ $H_s = \exp(-\beta_0 \xi_s)$ $\xi_s = \int d\xi_s$ the integral characterizing the accumulated plastic shear strain. The model was calibrated using data from four different sandstones. It is worth mentioning the ability of the model to reproduce cyclic behavior.

4.4.9 A Note on Implicit Integration of Generalized Plasticity Models

Constitutive equations in elastoplasticity are expressed as differential relations between the increments of stress and strain. In generalized plasticity, they are written as

$$d\boldsymbol{\varepsilon} = d\boldsymbol{\varepsilon}^e + d\boldsymbol{\varepsilon}^p = \boldsymbol{C}^e : d\boldsymbol{\sigma}' + \frac{1}{H_{U/L}} \boldsymbol{n}_{gL/U} \otimes \boldsymbol{n} : d\boldsymbol{\sigma}' = \boldsymbol{C}^{ep} : d\boldsymbol{\sigma}' \tag{4.281}$$

which can be inverted, leading to:

$$d\boldsymbol{\sigma}' = \boldsymbol{D}^{ep} : d\boldsymbol{\varepsilon} \tag{4.282}$$

When solving boundary value problems with finite elements, the mathematical model consists of a set of differential equations including balance of mass, the balance of linear momentum, and constitutive relations. In cases where the system is nonlinear, implicit methods using techniques, such as Newton–Raphson method, are used to solve the discretized equations in every increment. Regarding the constitutive equation, integration has to be performed in a consistent manner if we wish to achieve second-order accuracy.

Explicit techniques are not optimal for many nonlinear problems, and indeed they can lead to inaccurate results. Implicit methods have been used since the early work of Ortiz and Popov (1985), being worth mentioning the contributions of Simo and Taylor (1986), Simo and Hughes (1986), Borja and Lee (1990), Borja (1991), Borja and Tamagnini (1998), and Borja et al. (2001). In all the cases mentioned above, the constitutive equation included a yield surface and the consistency condition was used in the algorithm.

In generalized plasticity and in other theories such as hypoplasticity, the situation is different as no surfaces are defined. Heeres (2001), Zhang et al. (2001), and de Borst and Heeres (2002) proposed alternative methods to perform an implicit integration of generalized plasticity equations. This work was further refined by Mira et al. (2009), who proposed to use a generalized midpoint algorithm.

The key ingredients were:

1) The algorithm was formulated in the elastic strain space instead of in the stress space.
2) Hyperelasticity was used to characterize the elastic behavior of the material, following the work of Houlsby et al. (2005).

The stress update equation was

$$\boldsymbol{\sigma}^{n+1} = \boldsymbol{\sigma}^n + \boldsymbol{D}^e : \left(\Delta\boldsymbol{\varepsilon}^{n+1} - \lambda^{n+1}\boldsymbol{n}_g^{n+\alpha}\right) \tag{4.283}$$

where λ is the plastic multiplier, and $\boldsymbol{n}_g^{n+\alpha} = \boldsymbol{n}_g^{n+\alpha}((1-\alpha)\boldsymbol{\sigma}^n + \alpha\boldsymbol{\sigma}^{n+1})\ \ \alpha \in [0, 1]$.

The algorithm is second-order accurate for $\alpha = 1/2$ and the first order for all other cases (Ortiz and Popov 1985).

As described in Mira et al. (2009), in the case of elastoplasticity, in addition to the stress integration equation, there are additional equations describing the evolution of hardening parameters, and other relevant internal variables. In the generalized plasticity case, instead of enforcing consistency on the yield surface, this condition is replaced for another concerning the consistency parameter $d\lambda$

$$d\lambda = \frac{\boldsymbol{n} : \boldsymbol{D}^e : d\boldsymbol{\varepsilon}}{H_{L/U} + n : \boldsymbol{D}^e : \boldsymbol{n}_{gL/U}} \tag{4.284}$$

The variables used in the iterations are the elastic strain, the plastic modulus, and the consistency parameter. The residuals are:

$$
\begin{aligned}
r_{\varepsilon}^{n+1} &= \boldsymbol{\varepsilon}^{e,n+1} - \left(\boldsymbol{\varepsilon}^{e,n} + \Delta\boldsymbol{\varepsilon}_{\text{trial}}^{e,n+1} - \Delta\lambda^{n+1}\boldsymbol{n}_{g}^{n+\alpha}\right) \\
r_{H}^{n+1} &= H^{n+1} - H\left(\boldsymbol{\sigma}(\boldsymbol{\varepsilon}^{e,n+\alpha}), \boldsymbol{\varepsilon}^{p}\left(\Delta\lambda^{n+1}\right)\right) \\
r_{\lambda}^{n+1} &= \Delta\lambda^{n+1}.\left(H^{n+1} + \boldsymbol{n}^{n+\alpha} : \boldsymbol{D}^{n+\alpha} : \boldsymbol{n}_{g}^{n+\alpha}\right) - \boldsymbol{n}^{n+\alpha} : \boldsymbol{D}^{n+\alpha} : \Delta\boldsymbol{\varepsilon}^{n+1}
\end{aligned} \tag{4.285}
$$

The Newton–Raphson scheme is written in a compact manner as:

$$
\boldsymbol{r}^{i+1,n+1} = \boldsymbol{r}^{i,n+1} + \boldsymbol{J}_{r,a}^{i+1} \cdot d\boldsymbol{a}^{i+1,n+1} \tag{4.286}
$$

where $\boldsymbol{r}^{i+1,\,n+1}$ is the residual vector at iteration $i + 1$ and time step $n + 1$, $d\boldsymbol{a}^{i+1,\,n+1}$ the increment of the variable vector $\boldsymbol{a}$, and $\boldsymbol{J}_{r,a}^{i+1}$ is the Jacobian matrix

$$
\begin{aligned}
\boldsymbol{r}^{n+1T} &= \left(r_{\varepsilon}^{n+1}, r_{H}^{n+1}, r_{\lambda}^{n+1}\right) \\
\boldsymbol{a}^{n+1T} &= \left(\varepsilon^{e,n+1}, H^{n+1}, \Delta\lambda^{n+1}\right) \\
\boldsymbol{J}_{r,a}^{i+1} &= \frac{\partial \boldsymbol{r}}{\partial \boldsymbol{a}}
\end{aligned} \tag{4.287}
$$

The consistent tangent operator $\hat{\boldsymbol{D}}^{n+1}$ which will be used in Newton–Raphson iterations in order to ensure second-order convergence is obtained as:

$$
\hat{\boldsymbol{D}}^{n+1} = \frac{d\boldsymbol{\sigma}^{n+1}}{d\boldsymbol{\varepsilon}^{n+1}} = \frac{d\boldsymbol{\sigma}^{n+1}}{d\boldsymbol{\varepsilon}^{e,n+1}}\frac{d\boldsymbol{\varepsilon}^{e,n+1}}{d\boldsymbol{\varepsilon}^{n+1}} = \boldsymbol{D}^{e,n+1}\frac{d\boldsymbol{\varepsilon}^{e,n+1}}{d\boldsymbol{\varepsilon}^{n+1}} \tag{4.288}
$$

Details of the method are detailed in the paper by Mira et al. (2009).

4.5 Alternative Advanced Models

4.5.1 Introduction

So far, we have discussed in the previous sections some simple classical plasticity models for soils that have proven to reproduce accurately enough the behavior of soil under monotonic loading. They incorporated as the basic ingredients a plastic potential and a yield surface, the latter being allowed to expand or contract depending on whether the material was hardening or softening. However, material remained elastic within the yield surface, where no plastic deformation can develop. The immediate consequence is that these models are unable to reproduce either the behavior of overconsolidated soils, or phenomena occurring during cyclic loading, such as pore pressure generation in fast processes or densification. Indeed, both phenomena are related, as the latter is a direct consequence of the tendency of soil to compact. Generalized plasticity models, as we have seen, are able to provide such an accumulation of volumetric strain or pore water pressure buildup. In addition to generalized plasticity, there exist other interesting alternatives, which started with kinematic hardening models. Here, we will describe bounding surface plasticity and hypoplasticity-based models in a succinct way, giving the reader and idea of the theory on which they are based and possible applications.

4.5.2 Kinematic Hardening Models

The most interesting approach consisted in extending the theory of plasticity beyond the limits imposed in the classical formulation. The first successful theory was the multi-surface kinematic-hardening model proposed by Mroz (1967), where a set of "loading surfaces" within an outer "boundary" surface was postulated. Since then, further developments and improvements have taken place (Prévost 1977; Mroz et al. 1978; Hirai 1987). The number of surfaces allows us to keep track of loading events such as the maximum stress level reached or points at which stress has reversed. Large-intensity loading events erase lower-intensity events. An elastic domain may also be postulated, corresponding to the volume enclosed by the inner surface. As the stress is increased from an initial state, the surfaces reached by the stress path translate until a new loading surface is attained. This movement must comply with a rule which ensures that surfaces never intersect each other.

An improvement of this "multi-surface model" was provided by Mroz et al. (1981), who introduced an infinite number of nested loading surfaces, making the hardening modulus depend on the ratio of the sizes of active loading and outer or consolidation surfaces. In this way, the field of plastic moduli was made continuous in the whole domain enclosed by the outer surface. Memory of loading events was kept through the position and size of the surfaces at which stress reversal took place. The elastic domain was assumed to shrink to a point.

It can be seen that both the "multisurface" and the "infinite number of surfaces" models are able to reproduce most of the basic features of soils under cyclic loading, such as memory of past events and plastic deformation during unloading.

Mroz and Norris (1982) showed an application of the model to the cyclic behavior of normally consolidated and overconsolidated clays, and found that the model was able to predict final states lying on an "equilibrium line," as observed by Sangrey et al. (1969).

Other elastoplastic, kinematic, or anisotropic hardening models have been shown to perform well in modeling liquefaction and other cyclic loading phenomena (Aubry et al. 1982; Ghaboussi and Momen 1982; Hirai 1987). However, the price to pay in numerical computations is high, and simplified versions were sought. We will describe next the bounding surface plasticity, which provided such simplification.

4.5.3 Bounding Surface Models and Generalized Plasticity

If the number of surfaces is reduced to two, i.e. the outer or consolidation and the inner or yield, a field of hardening moduli can still be described by prescribing the variation between both surfaces. This model – one of the most effective – was independently proposed by Krieg (1975) and Dafalias and Popov (1975), and evolved to what is known today as "Bounding Surface Theory" (Dafalias and Herrmann 1982; Dafalias 1986; Bardet 1989; Kaliakin and Dafalias 1989; Wang et al. 1990).

A similar approach, the "subloading surface model" was proposed by Hashiguchi and Ueno (1977) and Hashiguchi et al. (1989).

On the bounding surface, plastic strain develops according to classical plasticity theory, with directions $\mathbf{n}$ and $\mathbf{n}_g$ given by the normals to the bounding and plastic potential surfaces, and the plastic modulus obtained through application of the consistency condition

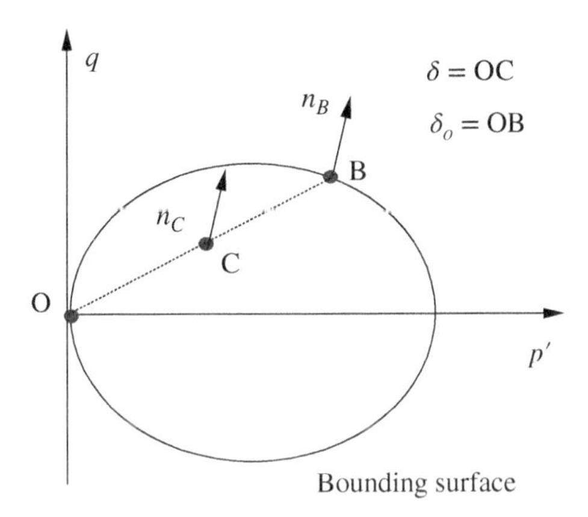

Figure 4.61 Bounding surface with radial interpolation

describing material hardening or softening properties. In the case of loading processes beginning at the bounding surface, the results coincide with those of classical plasticity. However, for loading processes inside it, such as may occur in cyclic loading, the difference is that bounding surface models are able to introduce plastic deformations by using some interpolation rules relating the stress point P (C in Figure 4.61) to an image of it on the BS, P_{BS} (B in Figure 4.61). Simple interpolation rules were proposed by Dafalias and Herrmann (1982), and by Zienkiewicz et al. (1985). There, to obtain the image point P_{BS}, a line was drawn passing through the origin and point P, its intersection with the bounding surface being taken as the image point. Directions $\mathbf{n}$ and $\mathbf{n}_g$ in P were assumed to be those at P_{BS} and the plastic modulus was interpolated according to a simple law

$$H_L = H_L^{BS}\left(\frac{\delta_0}{\delta}\right)^{\gamma} \tag{4.289}$$

where δ is the distance from the origin to the stress point P, and δ_0 the distance between the origin and the image point P_{BS}, γ being a parameter of the model (Figure 4.59).

The main shortcoming of early BS models was their inability to reproduce plastic deformations that develop when unloading, and it was overcome within the more general framework of generalized plasticity (Pastor et al. 1985). Here, the model was of bounding surface type for loading, but plastic deformations during unloading were introduced within the more general framework of generalized plasticity. A further step was given by Pastor, Zienkiewicz, and Chan, introducing a full generalized plasticity model in Pastor and Zienkiewicz (1986) and Pastor et al. (1990), which was applied by the authors to reproduce the behavior of both cohesive and frictional soils under monotonic and cyclic loading.

Since the pioneering work of Dafalias, many researchers have extended the original theory to cope with granular soils, state parameter, unsaturated soils, grain crushing, and anisotropy. Indeed, bounding surface plasticity was soon applied to sands by Bardet (1986), Wang et al. (1990), and Manzari and Dafalias (1997) among others.

Bardet (1986) based his model in Critical State Theory. Characteristic state line was used for the plastic potential providing the direction of plastic flow. Stress ratios at characteristic state and peak conditions were used in the model. As for interpolation, he chose a radial mapping to obtain the image stress on the bounding surface. The model incorporated nine constants, of which two described the elastic behavior, two the Critical State Line in e-ln p plane, one characterized the shape of the ellipse used as bounding surface, then the residual friction angle (compression and extension), the peak failure angle and the plastic modulus.

Wang et al. (1990) set their model in the framework of bounding surface hypoplasticity, within which the direction of the plastic strain increment depends on that of the stress. Moreover, the effects of rotational shear were accounted for. A new loading surface was introduced at each stress reversal point. The number of constants increased up to 15. The model was applied to the undrained behavior of sand under monotonic and cyclic loading, describing well the cyclic mobility exhibited by medium-loose sands.

So far, the bounding surface models describing sand behavior used different sets of parameters depending on the density of the material. The introduction of the state parameter allowed constitutive modelers to cope with sands of different initial densities using a single set of constants. Manzari and Dafalias' (1997) contribution incorporated the state parameter, their model being applied to sands in a large range of states under both monotonic and cyclic loading. The model incorporated two surfaces and was based on Critical State Theory. In addition to yield and bounding surfaces, the model used a critical state and a dilatancy surface. As for hardening, it implemented both an isotropic and a kinematic hardening.

The concept of the state parameter and its applications to constitutive modeling was revisited by Wang et al. (2002) who proposed an alternative state parameter based on the ratio of current mean confining hydrostatic pressure and that at critical state for the same voids ratio.

One important feature of granular soils is the anisotropy, either structural or induced by the stress–strain history of the material. Bounding plasticity modelers started as soon as 1986 (Dafalias 1986) to extend the theory to cope with such phenomena. Again, the model was set within the Critical State Framework. The approach consisted in rotating the bounding surface according to work dissipation during loading. It is interesting to note that rotation of a surface in the stress space is a particular case of using mixed invariants in its formulation (Pastor 1991).

Li and Dafalias (2004), following Li and Dafalias (2002), extended the model to describe the inherent anisotropy of sand, caused, for instance, by deposition of nonspherical grains. They introduced a fabric tensor from which a scalar variable describing anisotropy was derived. The results showed that sand behavior was very much sensitive to the relation between shear and the anisotropy axes.

Improving bounding surface models for anisotropic soils has continued since then, being worth mentioning the work of Taiebat and Dafaias (2007), Ling et al. (2016), and Jian et al. (2017a, 2017b).

Regarding extensions to crushable materials such as rockfill materials, the strategy is the same explained for generalized plasticity models. It is based on introducing (i) a state parameter, which depends on the position of the Critical State Line, and (ii) dependence of the Critical State Line on the amount of crushing. Examples of bounding surface models for crushable grained materials are those of Kan and Taiebat (2014) and Xiao et al. (2020).

Modelling of unsaturated soils, in general, is based, as explained when describing generalized plasticity applications on a suitable measure of the stress, which includes both the effective stress tensor and the suction. Models are based on the dependence of the Critical

State Line on the suction. As examples of applications of bounding surface models to unsaturated soils, it is worth mentioning the works of Russell and Khalili (2006), and the more recent ones of Zhao et al. (2019), Bruno et al. (2020), and Zhou et al. (2020).

4.5.4 Hypoplasticity and Incrementally Nonlinear Models

One of the basic characteristics of the elastoplastic tensors $\mathbf{D}_{ep}$ and $\mathbf{C}_{ep}$ is their dependence on the direction of loading $\mathbf{u} = \mathrm{d}\boldsymbol{\sigma}/\|\mathrm{d}\boldsymbol{\sigma}\|$, which is taken into account in a simple way by introducing a direction $\mathbf{n}$ for each mechanism of deformation considered.

Alternatively, it is possible to provide general expressions for the constitutive tensors satisfying all the necessary requirements, such as dependence on $\mathbf{u}$. Hypoplasticity is, in this sense, the most promising framework in which this goal can be achieved. Among all the constitutive models of this kind, it is worth mentioning those proposed by Darve and Labanieh (1982), Dafalias (1986), and Kolymbas (1991).

One of the first models which can be considered of this kind was introduced in Grenoble by Darve and Labanieh (1982), and since the early days of the theory, it has been considerably improved (Desrues and Chambon 1993).

They are referred to as incrementally nonlinear models as they are based on the assumption that the incremental nonlinearity in the constitutive tensor may be approximated by suitable interpolation laws once material behavior along different stress paths has been established.

Darve and Labanieh (1982) suggested that these paths could correspond to positive and negative directions along principal stress axes 1, 2, and 3. Consequently, six values of $\mathbf{C}$ were required for the interpolation along a particular direction.

The constitutive tensor was assumed to be given by

$$\mathbf{C} = \mathbf{C}^{I} + \mathbf{C}^{II} \cdot \mathbf{u} \tag{4.290}$$

where $\mathbf{u}$ is the unit tensor along the direction of loading increment. If constitutive tensors along positive and negative directions of the principal directions are $\mathbf{N}^{+}$ and $\mathbf{N}^{-}$, a simple incrementally nonlinear law could be given by

$$\mathrm{d}\boldsymbol{\varepsilon} = \left\{\frac{\mathbf{N}^{+}\mathbf{N}^{-}}{2}\right\}\mathrm{d}\boldsymbol{\sigma} + \left\{\frac{\mathbf{N}^{+} + \mathbf{N}^{-}}{2\|\mathrm{d}\boldsymbol{\sigma}\|}\right\}\mathrm{d}s^{2} \tag{4.291}$$

where

$$\begin{aligned} \mathrm{d}\boldsymbol{\sigma} &= (\mathrm{d}\sigma_1, \mathrm{d}\sigma_2, \mathrm{d}\sigma_3)^{\mathrm{T}} \\ \mathrm{d}\boldsymbol{\varepsilon} &= (\mathrm{d}\varepsilon_1, \mathrm{d}\varepsilon_2, \mathrm{d}\varepsilon_3)^{\mathrm{T}} \end{aligned} \tag{4.292}$$

The model proved to reproduce well the behavior of soils under both monotonic and cyclic loading (Darve et al. 1986).

Dafalias and coworkers presented extensions of the bounding surface model within the framework of hypoplasticity (Dafalias 1986; Wang et al. 1990).

Hypoplastic models have been introduced also in Karlsruhe by Kolymbas and coworkers (Wu and Kolymbas 1990; Kolymbas and Wu 1993), producing general expressions for the constitutive tensor.

4.6 Conclusion

This chapter has presented some rate-independent models which are able to predict the most salient features of soil behavior under monotonic and cyclic loading. We have done a selection, as otherwise, this chapter would have required a full book for the task. Along the chapter, we have seen how more and more complex aspects of soil behavior could be included, the price being a high number of constitutive parameters. During the last years, a different approach has attracted the attention of modelers: direct simulations with discrete elements. Laboratory specimens of sand subjected to complex stress paths have been successfully reproduced just by using a very short number of parameters, such as friction between particles. The high computational effort which could be required to solve realistic boundary value problems makes this approach difficult for most applications. As the computer power keeps increasing, it is possible to forecast that such an approach could be used in the not-so-far future. A promising example is given in a recent paper by Sibille et al. (2019).

References

Abioghli, H. and Hamidi, A. (2019). A constitutive model for evaluation of mechanical behavior of fiber-reinforced cemented sand, *J. Rock Mech. Geotech. Eng.*, **11**, 2, 349–360.

Alonso, E. and Oldecop, L. (2003). Comportement des remblais en enrochement, *Rev. Fr. Géotech.*, **102**, 3–19.

Anandarajah, A. and Dafalias, Y. F. (1986). Bounding surface plasticity III: application to anisotropic cohesive soils *J. Eng. Mech. ASCE*, **112**, 12, 1292–1318.

Arulanandan, K. and Scott, R. F. (1993). Project VELACS—control test results, *J. Geotech. Eng.*, **119**(8), https://doi.org/10.1061/(ASCE)0733-9410(1993)119:8(1276).

Atkinson, J. H. and Bransby, P. L. (1978). *The Mechanics of Soils: An Introduction to Critical State Soil Mechanics*, McGraw Hill, London.

Atkinson, J. H. and Richardson, D. (1985). Elasticity and normality in soil: experimental examinations, *Géotechnique*, **35**, 443–449.

Aubry, D. Hujeux, J. C. Lassoudière, F. and Meimon, Y. (1982). A double memory model with multiple mechanisms for cyclic soil behaviour, in *Numerical Models in Geomechanics. International Symposium*, R. Dungar, G. N. Pande and J. A. Studer (Eds), pp. 3–13, Balkema, Rotterdam.

Bahda, F. (1997). Etude du comportement du sable au triaxial. PhD thesis, ENPC-CERMES, Paris.

Bahda, F. Pastor, M. and Saitta, A. (1997). A double hardening model based on generalized plasticity and state parameters for cyclic loading of sands, in *NUMOG VI: International Symposium on Numerical Models in Geomechanics*, G. N. Pande and S. Pietruszczak (Eds), Montreal, Quebec, 2 July 1997, pp. 33–38, A.A. Balkema, Brookfield Publication, Netherlands.

Baker, R. and Desai, C. S. (1984). Induced anisotropy during plastic straining, *Int. J. Num. Anal. Methods Geomech.*, **8**, 167–185.

Balasubramanian, A. S. and Chaudhry, A. R. (1978). Deformation and strength characteristics of soft Bangkok clay, *J. Geotech. Eng. Div. ASCE*, **104**, 9, 1153–1167.

Banerjee, P. K. and Yousif, N. B. (1986). A plasticity model for the mechanical behaviour of anisotropically consolidated clay, *Int. J. Num. Anal. Methods Geomech*, **10**, 521–541.

Bardet, J. P. (1986). Bounding surface plasticity model for sands, *J. Eng. Mech.*, **112**, (11), 1198–1217.

Bardet, J. P. (1989). Prediction of deformations of Hostun and Reid Bedford sands with a simple bounding surface plasticity model, in *Constitutive Equations for Granular Non-cohesive Soils*, A. Saada and G. Bianchini (Eds), 131–148, Balkema.

Been, K. and Jefferies, M. (1985), A state parameter for sands, *Géotechnique*, **35**, 99–112

Bian, S., Wu, B and Ma, Y. (2019), Modeling static behavior of rockfill materials based on generalized plasticity model, *Adv. Civil Eng.*, **2019**, 2371709, doi: 10.1155/2019/2371709

Bishop, A. W. (1959). The principle of effective stress, *Tek. Ukeblas*, **39**, 859–863.

Bishop, A. W. and Blight, G. E. (1963). Some aspects of effective stress in saturated and partly saturated soils, *Géotechnique*, **13**, (3), 177–197

Bolzon, G., Schrefler, B. A. and Zienkiewicz, O. C. (1996). Elasto-plastic constitutive laws generalised to partially saturated states, *Géotechnique*, **46**, (2), 279–289.

Borja, R. I. (1991). Cam-Clay plasticity, Part II: Implicit integration of constitutive equation based on a nonlinear elastic stress predictor, *Comput. Methods Appl. Mech. Engrg.*, **88**, 225–240.

Borja, R. I. (2004). Cam-Clay plasticity. Part V: a mathematical framework for three-phase deformation and strain localization analyses of partially saturated porous media, *Comput. Methods Appl. Mech. Engrg.*, **193**, 5301–5338.

Borja, R. I. (2013). *Plasticity: Modeling and Computation*. Springer Science & Business Media.

Borja, R. I. and Andrade, J. E. (2006). Critical state plasticity. Part VI: Meso-scale finite element simulation of strain localization in discrete granular materials, *Comput. Methods Appl. Mech. Engrg.*, **195**, 5115–5140.

Borja, R. I. and Lee, S. R. (1990). Cam-Clay plasticity, Part I: implicit integration of elasto-plastic constitutive relations, *Comput. Methods Appl. Mech. Engrg.*, **78**, 49–72.

Borja, R. I. and Tamagnini, C. (1998). Cam-Clay plasticity, Part III: extension of the infinitesimal model to include finite strains, *Comput. Methods Appl. Mech. Engrg.*, **155**, 73–95.

Borja, R. I., Tamagnini, C. and Amorosi, A. (1997). Coupling plasticity and energy-conserving elasticity models for clays, *J. Geotech. Geoenviron. Eng.*, **123**(10), 948–957.

Borja, R. I., Lin, C.-H. and Montans, F. J. (2001). Cam-Clay plasticity, Part IV Implicit integration of anisotropic bounding surface model with nonlinear hyperelasticity and ellipsoidal loading function, *Comput. Methods Appl. Mech. Engrg.*, **190**, 3293–3323.

Boulon, M. and Nova, R. (1990). Modelling of soil-structure interface behaviour a comparison between elastoplastic and rate type laws, *Comput. Geotechnics*, **9**, (1–2), 21–46.

Bruno, A. W., Gallipoli, D., Rouainia, M., and Lloret-Cabot, M. (2020). A bounding surface mechanical model for unsaturated cemented soils under isotropic stresses, *Comput. Geotech.*, **125**, 103673. doi: 10.1016/j.compgeo.2020.103673

Burland, J. B. (1965). Correspondence on 'The yielding and dilatation of clay', *Géotechnique*, **15**, 211–214.

Cambou, B. and Lanier J. (1988), Induced anisotropy in cohesionless soil: experiments and modelling, *Comput. Geotech.* **6**, 291–311.

Casagrande (1938). The shearing resistance of soils and its relation to the stability of earth dams. *Proc. Soils Found. Conf.*, 1938 US Engineering Department.

Castro, G. (1969). Liquefaction of sands. PhD thesis, Harvard University, Harvard Soil Mech. Series no. 81.

Coussy, O. (2004). *Poromechanics*, John Wiley and Sons, Chichester, ISBN 0-470-84920-7.

Cen, W. J, Luo, J. R., Bauer, E. and Zhang, W. D. (2018a). Generalized plasticity model for sand with enhanced state parameters, *J. Eng. Mech.*, **144**, (12), 04018108, doi: 10.1061/(ASCE) EM.1943-7889.0001534

Cen, W. J., Luo, J. Zhang, W. and Rahman, M. S. (2018b). An enhanced generalized plasticity model for coarse granular material considering particle breakage, *Adv. Civil Eng.*, **2018**, 7242936 doi: 10.1155/2018/7242936

Chan, A. H. C. Zienkiewicz, O. C. and Pastor, M. (1988). Transformation of incremental plasticity relation from defining space to general cartesian stress space, *Commun. Appl. Num. Meth.*, **4**, 577–580.

Cole, E. R. (1967). The behaviour of soils in the simple shear apparatus. PhD thesis, Cambridge University.

Coop, M. R., Sorensen, K. K., Bodas Freitas, T. and Georgoutsos, G. (2004). Particle breakage during shearing of a carbonate sand, *Geotechnique*, **54**, 3, 157–163.

Coulomb, C. A. (1773) Essai sur une application (8) des regles de maxlmis et de minimis A quelques problemea de statique relatifs A l'architecture. Memoires de mathematiques et de physique présentes A l'Academie, *Royale des Sciences par divers savants, et lus dans ses Assemblées* **7**, 343, reimprimé dans la Theorle des machines simples, Paris 1809 et 1821.

Coussy, O. (1995). *Mechanics of Porous Media*, Wiley, Chichester.

Cowin, S. C. (1985). The relationship between the elasticity tensor and the fabric tensor, *Mech. Materials*, **4**, 137–147.

Dafalias, Y. F. (1986). Bounding surface plasticity. I: mathematical foundation and hypoplasticity, *J. Eng. Mech. ASCE*, **112**, 966–987.

Dafalias, Y. F.(2016). Must critical state theory be revisited to include fabric effects? *Acta Geotech.*, **11**, (3), 479–491.

Dafalias, Y. F. and Herrmann, L. R. (1982). Bounding surface formulation of soil plasticity, in *Soil Mechanics—Transient and Cyclic Loads*, G. N. Pande and O. C. Zienkiewicz (Eds), Ch. 10, 253–282, Wiley.

Dafalias, Y. F. and Popov, E. P. (1975). A model of non-linearly hardening materials for complex loadings, *Acta Mech.* **21**, 173–192.

Darve, F. (1990). (Ed) *Geomaterials. Constitutive Equations and Modelling*. Elsevier Applied Science.

Darve, F. and Labanieh, S. (1982). Incremental constitutive law for sands and clays: simulation of monotonic and cyclic tests, *Int. J. Numer. Anal. Meth. Geomechs.*, **6**, 243–275.

Darve, F. Flavigny, E. and Rojas, E. (1986). A class of incrementally non-linear constitutive relations and applications to clays, *Comput. Geotech.*, **2**, 43–66.

de Borst, R. and Heeres, O. M.(2002). A unified approach to the implicit integration of standard, non-standard and viscous plasticity models, *Int. J. Numer. Anal. Methods Geomech.*, **26**, 1059–1070

de Saint-Venant, B. (1870). Memoire sur la torsion des prismes. Memoire sur la flexion des prismes, *Comptes Rendus Acad. Sci. Paris*, **70**, 473.

Desrues, J. and Chambon, R. (1993). A new rate type constitutive model for geomaterials: CloE, in *Modern Approaches to Plasticity*, D. Kolymbas (Ed), pp. 309–324, Elsevier.

di Prisco, C. Nova, R. and Lanier, J. (1993). A mixed isotropic kinematic hardening constitutive law for sand, in *Modern Approaches to Plasticity*, D. Kolymbas (Ed), pp. 83–124, Balkema.

Dong, W., Hu, L., Yu, Y. Z. and Lv, H. (2013). Comparison between Duncan and Chang's EB model and the generalized plasticity model in the analysis of a high earth-rockfill dam, *J. Appl. Math.*, **2013**, 709430, 12 pages.

Douadji, A. and Hicher, P. Y. (2010). An enhanced constitutive model for crushable granular materials, *Int. J. Numer. Anal. Meth. Geomech.*, **34**, 555–580.

Drucker, D. C. (1956). On uniqueness in the theory of plasticity, *Quart. Appl. Math.*, **14**, 35–42.

Drucker, D. C., (1959). A definition of unstable inelastic material, *J. Appl. Mech.*, **26**, 101–106.

Drucker, D. C. and Prager, W. (1952). Soil mechanics and plastic analysis or limit design, *Quart. Appl. Math.*, **10**, 157–165.

Drucker, D. C., Gibson, R. E. and Henkel, D. J. (1957). Soil mechanics and workhardening theories of plasticity, *Trans. ASCE* **122**, 338–346.

Farzaneh, O. and Iraji, A. (2015). Two-phase model for nonlinear dynamic simulation of reinforced soil walls based on a modified Pastor-Zienkiewicz-Chan model for granular soil, *J. Eng. Mech.*, **142**, 10.1061/(ASCE)EM.1943-7889.0000985, 04015072

Fernández-Merodo, J. A., Pastor, M., Mira, P., Tonni, L., Herreros, M. I., Gonzalez, E. and Tamagnini, R. (2004). Modelling of diffuse failure mechanisms of catastrophic landslides, *Comp. Methods Appl. Mech. Engrg.*, **193**, 2911–2939.

Fernández Merodo, J. A., Tamagnini, R., Pastor, M. and Mira, P. (2005). Modelling damage with generalized plasticity, *Riv. Ital. Geotecn.*, **4**, 32–42.

Fisher, R. A. (1926) On the capillary forces in an ideal soil; correction of formulas by W.B. Haines, Journal of Agricultural Science, 16, 492–505.

Frossard, E. (1983) Une équation d'écoulement simple pour les materiaux granulaires, *Géotechnique*, **33**, 1, 21–29.

Fu, Z., Chen, S. and Peng, C. (2014) Modeling cyclic behavior of rockfill materials in a framework of generalized plasticity, *Int. J. Geomech.*, **14**, 2, 191–204

Gajo, A. and Muir Wood, D. (1999). Severn-Trent sand: a kinematic-hardening constitutive model: the q-p formulation, *Géotechnique*, **49**, 595–614.

Gallipoli, D., Gens, A.; Sharma, R. and Vaunat, J. (2003). An elastoplastic model for unsaturated soil incorporating the effects of suction and degree of sturation on mechanical behaviour, *Géotechnique*, **53**, (1), 123–135.

Gens, A. and Nova, R. (1993). Conceptual bases for a constitutive model for bonded soils and weak rocks. *Proceedings of International Symposium on Geotechnical Engineering of Hard Soils-Soft Rocks*, Athens, Greece, 20–23 September 1993, Rotterdam: Balkema, pp. 485–494.

Ghaboussi, J. and Momer, H. (1982). Modelling and analysis of cyclic behaviour of sands, in *Soil Mechanics-Transient and Cyclic Loads*, G. N. Pande and O. C. Zienkiewicz (Eds), pp. 313–342 Wiley.

Goorani, M. and Hamidi, H. (2015). A generalized plasticity constitutive model for sand-gravel mixtures, *Int. J. Civil Eng. Trans. B Geotech. Eng.*, **13**, (2), 133–145.

Gray, W. G. and Hassanizadeh, S. M. (1991). Unsaturated flow theory including interfacial phenomena, *Water Resour. Res.*, **27**, 8, 1855–1991.

Habib, P. and Luong, M. P. (1978). Sols pulvurulents sous chargement cyclique. *Materiaux and Structures Sous Chargement Cyclique*, Ass. Amicale des Ingenieurs Anciens Eléves de l'Ecole Nationale des Ponts et Chaussées, Palaiseau (28–29 September 1978), 49–79.

Haines W. B. (1925). Studies in the physical properties of soils: II. A note on the cohesion developed by capillary forces in an ideal soil, *J. Agric. Sci.*, **15**, (4), 529–535.

Hashiguchi, K. (1980). Constitutive equations of elastoplastic materials with elastic-plastic transition. *J. Appl. Mech. ASME*, **47**, 266–272.

Hashiguchi, K. and Ueno, M. (1977). Elastoplastic Constitutive laws of granular materials. in *Constitutive Equations of Soils*, 9th. Int. Congr. Soil Mech. Found. Engng., S. Murayama and A. N. Schofield (Eds), pp. 73–82, JSSMFE.

Hashiguchi, K. Imamura, T. and Ueno, M. (1989). Prediction of deformation behaviour of sands by the subloading surface model, in *Constitutive Equations for Granular Non-cohesive Soils*, A. Saada and G. Bianchini (Eds), 131–148, Balkema.

Hassanizadeh, S. M. and Gray, W. G. (1990). Mechancs and thermodynamics of multiphase flow in porous media including interphase boundaries, *Adv. Water Resour.*, **13**, (4), 169–186.

Heeres, O. M. (2001). Modern strategies for the numerical modelling of the cyclic and transient behavior of soils. Dissertation, Delft University of Technology.

Henkel, D. J. (1956). The effect of overconsolidation on the behaviour of clays during shear, *Géotechnique*, **6**, 139–150.

Henkel, D. J. (1960). The shear strength of saturated remoulded clay. *Proceedings of Research Conference on Shear Strength of Cohesive Soils*, Boulder, Colorado, United States, June, pp. 533–540.

Hill, R. (1950). *The Mathematical Theory of Plasticity*, Oxford, Clarendon Press.

Hirai, H. (1987). An elastoplastic constitutive model for cyclic behaviour of sands, *Int. J. Num. Anal. Meth. Geomech.*, **11**, 503–520.

Houlsby, G. T. (1997). The work input to an unsaturated granular material, *Géotechnique*, **47**, (1), 193–196.

Houlsby, G. T., Amorosi, A. and Rojas, E. (2005). Elastic moduli of soils dependent on pressure: a hyperelastic formulation, *Geotechnique*, **55**, (5), 383–392.

Huber, M. T. (1904). *Czasopismo techniczne, Lemberg*, **22**, 81.

Iraji, A. and Farzaneh, O. (2017). Two-phase model for nonlinear elasto-plastic behavior of reinforced soil structures using Pastor-Zienkiewicz-Chan model for matrix phase, *Soils Found.*, **57**, (2017), 1014–1029.

Ishihara, K. (1973). Liquefaction and flow failure during earthquakes, *Géotechnique*, **43**, 315–415.

Ishihara, K. and Okada, S. (1982). Effects of large preshearing on cyclic behaviour of sand, *Soils Found.*, **22**, 109–125.

Ishihara, K. Tatsuoka, F. and Yasuda, S. (1975). Undrained deformation and liquefaction of sand under cyclic stress, *Soils Found.*, **15**, 29–44.

Javanmardi, Y., Imam, S. M. R., Pastor, M., and Manzanal, D. (2018). A reference state curve to define the state of soils over a wide range, *Geotechnique*, **68**, (2), 95–106. doi: 10.1680/jgeot.16.P.136.

Jefferies, M. G. (1993). Nor.Sand: a simple critical state model for sand, *Géotechnique*, **43**, 91–103.

Jefferies, M. G. and Been, K. (2000). Implication for critical state from isotropic compression of sand, *Géotechnique*, **50**, 419–429.

Jennings, J. E. B. and Burland, J. B. (1962). Limitations to the use of effective stresses in partly saturated soils, *Géotechnique*, **12**, (2), 125–144.

Jiang, J., Ling, H. I., Kaliakin, V. N. (2017a). Evaluation of an anisotropic elastoplastic–viscoplastic bounding surface model for clays, *Acta Geotech.*, **12**, 335–348 doi: 10.1007/s11440-016-0471-7.

Jiang, J., Ling, H. I and Yang, L. (2017b). Approximate simulation of natural structured soft clays using a simplified bounding surface model, *Int. J. Geomech.*, **17**, (7), 06016044 doi:10.1061/(ASCE)GM.1943-5622.0000864

Jommi, C. (2000). Remarks on the constitutive modelling of unsaturated soils, in *Experimental Evidence and Theoretical Approaches in Unsaturated Soils*, Tarantino, A. and Mancuso, C. (Eds), pp. 139–153, Trento, Italy, Balkema, Rotterdam.

Kaliakin, V. N. and Dafalias, Y. F. (1989). Simplifications to the bounding surface model for cohesive soils, *Int. J. Num. Anal. Meth. Geomech.*, **13**, 91–100.

Kan, M.E. and Taiebat, H.A. (2014). A bounding surface plasticity model for highly crushable granular materials. *Soils Found.* **54**(6):1188–1201

Kolymbas, D. (1991). An outline of hypoplasticity, *Arch. Appl. Mech.*, **61**, 143–151.

Kolymbas, D. and Wu, W. (1993). Introduction to hypoplasticity, in *Modern Approaches to Plasticity*, Kolymbas, D. (Ed), 213–224, Elsevier.

Krieg, R. D. (1975). A practical two-surface plasticity theory, *J. Appl. Mech., Trans. ASME*, **E42**, 641–646.

Lagioia, R.and Nova, R. (1995). An experimental and theoretical study of the behaviour of a calcarenite in triaxial compression, *Géotechnique*, **45**, 4, 633–648.

Lewis, R. W. and Schrefler, B. A.(1987). *The Finite Element Method in the Deformation and Consolidation of Porous Media*, Wiley, Chichester.

Li, X. S. (1997). Modeling of dilative shear failure, *J. Geotech. Geoenviron. Eng.*, **123**, (7), 609–616.

Li, X. L. and Dafalias, Y. F. (2000). Dilatancy for cohesionless soils, *Geotechnique*, **50**, 449–460.

Li, X. S. and Dafalias, Y. F. (2002). Constitutive modeling of inherently anisotropic sand behavior, *J. Geotech. Geoenviron. Eng.*, **128**, 10, 868–880

Li, X. S. and Dafalias, Y. F. (2004). A constitutive framework for anisotropic sand including nonproportional loading, *Géotechnique*, **54**, 1, 41–55

Li, X. S. and Wang, Y. (1998). Linear representation of steady-state line for sand, *J. Geotech. Geoenviron. Eng.*, **124**, (12), 1215–1217.

Li, X. S., Dafalias, Y. F. and Wang, Z. L. (1999). State dependent dilatancy in critical state constitutive modelling of sand, *Can. Geotech. J.*, **36**, 599–611

Liang, R. L. and Shaw, H. L. (1991). Anisotropic plasticity model for sands, *J. Geotech. Eng. ASCE*, **117**, 6, 913–933.

Ling, H. I. and Liu, H. (2003). Pressure-level dependency and densification behavior of sand through generalized plasticity model, *J. Eng. Mech. ASCE*, **129**, (8), 851–860.

Ling, H. I. and Yang, S. (2006). Unified sand model based on the critical state and generalized plasticity, *J. Eng. Mech.*, **132**, (12), 1380–1391

Ling, H. I., Hung, C. and Kaliakin, V. N. (2016). Application of an enhanced anisotropic bounding surface model in simulating deep excavations in clays, *J. Geotech. Geoenviron. Eng.*, **142**, (11), 04016065.

Liu, M. and Gao, Y. (2017). Constitutive modeling of coarse-grained materials incorporating the effect of particle breakage on critical state behavior in a framework of generalized plasticity, *Int. J. Geomech.*, **17**, 5, 04016113.

Liu H. and Ling H. I. (2008). Constitutive description of interface behavior including cyclic loading and particle breakage within the framework of critical state soil mechanics, *Int. J. Numer. Anal. Methods Geomech.*, 2008, **32**, (12), 1495–1514.

Liu, H. and Zou, D. (2013). An associated generalized plasticity framework for modeling gravelly soils considering particle breakage, *J. Eng. Mech. ASCE*, **139**, (5), 606–615.

Liu, H., Ling, H. and Song, E. (2006). Constitutive modeling of soil–structure interface through the concept of critical state soil mechanics, *Mech. Res. Commun.*, **33**, 515–531.

Liu, H., Zou, D. and Liu, J. (2014). Constitutive modeling of dense gravelly soils subjected to cyclic loading, *Int. J. Numer. Anal. Methods Geomech.*, **38**, 1503–1518, doi: 10.1002/nag.2269.

Manzanal, D. (2008). Constitutive model based on generalized plasticity incorporating state parameter for saturated and unsaturated sand (in Spanish). PhD thesis, School of Civil Engineering, Polytechnic University of Madrid.

Manzanal, D., Fernandez Merodo, J. A and Pastor, M. (2011a). Generalized plasticity state parameter-based model for saturated and unsaturated soils. Part 1: Saturated state, *Int. J. Numer. Anal. Meth. Geomech*, **35**, 12, 1347–1362

Manzanal, D., Pastor, M. and Fernández Merodo, J. A. (2011b). Generalized plasticity state parameter-based model for saturated and unsaturated soils. Part II: Unsaturated soil modeling, *Int. J. Numer. Anal. Meth. Geomech*, **35**, 18, 1899–1917.

Manzari, M. and Dafalias, Y. F. (1997). A critical state two-surface plasticity model for sands, *Géotechnique*, **47**, 255–272.

Matsuoka, H. (1974). Stress-strain relationships of sands based on the mobilized plane, *Soils Found.*, **14**, 2, 47–61.

Melan, E. (1938). Zur Plastizität des räumlichen Kontinuums, *Ing. Arch.*, **9**, 116–126. https://doi.org/10.1007/BF02084409.

Mira, P., Tonni, L., Pastor, M., and Fernández Merodo, J. A. (2009). A generalized midpoint algorithm for the integration of a generalized plasticity model for sands, *Int. J. Numer. Meth. Eng.*, **77**, 1201–1223

Miura, S. and Toki, S. (1984). Elastoplastic stress-strain relationship for loose sands with anisotropic fabric under three-dimensional stress conditions, *Soils Found.*, **24**, 43–57.

Miura, N. and Yamanouchi, T.(1975). Effect of water on the behavior of a quartz-rich sand under high stresses, *Soils Found.*, **15**, (4), 23–34.

Mroz, Z. (1967). On the description of anisotropic work-hardening, *J. Mech. Phys. Solids*, **15**, 163–175.

Mroz, Z. and Norris, V. A. (1982). Elastoplastic and viscoplastic constitutive models for soils with applications to cyclic loading, in *Soil Mechanics—Transient and Cyclic Loads*, G. N. Pande and O. C. Zienkiewicz (Eds), pp. 173–217, Wiley.

Mroz, Z. and Zienkiewicz, O. C. (1985). Uniform formulation of constitutive equations for clay and sand, in *Mechanics of Engineering Materials*, C. S. Desai and R. H. Gallaher (Eds), Ch. 22, pp. 415–450 Wiley.

Mroz, Z. Norris, V. A. and Zienkiewicz, O. C. (1978). An anisotropic hardening model for soils and its application to cyclic loading, *Int. J. Num. Anal. Meth. Geomech.*, **2**, 203–221.

Mroz, Z. Norris, V. A. and Zienkiewicz, O. C. (1979). Application of an anisotropic hardening model in the analysis of elastic-plastic deformation of soils, *Geotechnique*, **29**, 1–34.

Mroz, Z. Norris, V. A. and Zienkiewicz, O. C. (1981). An anisotropic critical state model for soils subjected to cyclic loading, *Géotechnique*, **31**, 451–469.

Nova, R. (1977). On the hardening of soils, *Arch. Mech. Stos.*, **29**, 3, 445–458.

Nova, R. (1982). A constitutive model for soil under monotonic and cyclic loading, in *Soil Mechanics-Transient and Cyclic Loads*, G. N. Pande and O. C. Zienkiewicz (Eds), pp. 343–374, Wiley.

Nova, R. (1986). An extended cam clay model for soft anisotropic rocks, *Comp. Geotech.*, **2**, 69–88.

Nova R. and Sacchi G. (1982). A model of the stress-strain relationship of orthotropic geological media, *J. Mech. Theor. Appl.*, **1**, 6, 927–949.

Nova, R. and Wood, D. M. (1978). An experimental program to define yield function for sand, *Soils Found.*, **18**, 77–86.

Nova, R. and Wood, D. M. (1979). A constitutive model for sand, *Int. J. Num. Anal. Meth. Geomech.* **3**, 255–278.

Nuth, M. and Laloui, L. (2008). Effective stress concept in unsaturated soils: clarification and validation of a unified framework *Int. J. Num. Anal. Meth. Geomech*, **32**, 771–801

Ortiz, M., Popov, E. P. (1985). Accuracy and stability of integration algorithms for elastoplastic constitutive equations, *Int. J. Numer. Methods Eng.*, **2**, 1561–1576.

Pande, G. N. and Sharma, K. G. (1983). Multi-laminate model for clays: a numerical evaluation of the influence of principal stress axes, *Int. J. Num. Anal. Meth. Geomech.*, **7**, 397–418.

Parry, R. H. G. (1960). Triaxial compression and extension tests on remoulded saturated clay, *Géotechnique*, **10**, 166–180.

Pastor, M. (1991). Modelling of anisotropic soil behaviour, *Comput. Geotech.*, **11**, 173–208.

Pastor, M. and Zienkiewicz, O. C. (1986). A generalized plasticity hierarchical model for sand under monotonic and cyclic loading, in *Proceedings of the 2nd International Conference on Numerical Models in Geomechanics, Ghent (Belgium), 31st March–4 April*, G. N. Pande and W. F. Van Impe (Eds), pp. 131–150, M. Jackson and Son Publications.

Pastor, M. Zienkiewicz, O. C. and Leung, K. H. (1985). Simple model for transient soil loading in earthquake analysis. II Non associative models for sands, *Int. J. Num. Anal. Meth. Geomech.*, **9**, 477–498.

Pastor, M. Zienkiewicz, O. C. and Chan, A. H. C. (1987). A generalized plasticity continuous loading model for geomaterials, in *Numerical Methods in Engineering: Theory and Approximation*, O. C. Zienkiewicz, G. N. Pande and J. Middleton (Eds), pp. 1–11, Martinus Nijhoff Publications.

Pastor, M. Zienkiewicz, O. C. and Chan, A. H. C. (1989) Generalized plasticity model for three-dimensional sand behaviour, in *Constitutive Equations for Granular Non-Cohesive Soils*, A. Saada and G. Bianchini (Eds), 535–549, Balkema.

Pastor, M. Zienkiewicz, O. C. and Chan, A. H. C. (1990). Generalized plasticity and the modelling of soil behaviour, *Int. J. Num. Anal. Meth. Geomech.*, **14**, 151–190.

Pastor, M., Zienkiewicz, O., Guang-Dou, X. and Peraire, J. (1993). Modelling of sand behaviour: cyclic loading, anisotropy and localization, in *Modern Approaches to Plasticity* D. Kolymbas (Ed), pp. 469–492, Elsevier.

Pastor, M., Quecedo, M., Fernández Merodo, J. A., Herreros, M. I., Gonzalez, E., and Mira, P. (2002). Modelling tailings dams and mine waste dumps failures, *Géotechnique*, **52**, (8), 579–591.

Pestana, J. M. and Whittle, A. J. (1995). Compression model of cohesionless soils, *Géotechnique*, **45**, 611–631.

Pietruszczak, S. and Krucinski, S. (1989). Description of anisotropic response of clays using tensorial measure of structural disorder, *Mech. Mater.*, **8**, 237–249.

Poorooshasb, H. B. Holubec, I. and Sherbourne, A. N. (1966). Yielding and flow of sand in triaxial compression (Parts II and III), *Can. Geotech. J.*, **3**, 179–190.

Poorooshasb, H. B. Holubec, I. and Sherbourne, A. N. (1967). Yielding and flow of sand in triaxial compression (Parts II and III), *Can. Geotech. J.*, **4**, 376–397.

Prandtl, L. (1924). Spannungsverteilung in plastischen Körpern, *Proc. 1st Int. Cong. Appl. Mech.*, pp. 43, Delft.

Prévost, J. H. (1977). Mathematical modelling of monotonic and cyclic undrained clay behaviour, *Int. J. Num. Anal. Meth. Geomech.*, **1**, 195–216.

Reuss, A. (1930). Berücksichtigung der elastischen Formänderung in der Plastizitätstheorie. *Zeits. Ang. Math. Mech.*, **10**, 266.

Reynolds, O. (1985). The dilation of media composed of rigid particles in contact, *Philos. Mag.*, **S5**, 20, 469–481.

Roscoe, K. H. and Burland, J. B. (1968). On the generalized stress-strain behaviour of wet clay, in *Engineering Plasticity*, J. Heyman and F. A. Leckie (Eds), pp. 535–610, Cambridge University Press.

Roscoe, K. and Poorooshasb, H. (1963). A theoretical and experimental study of strain in triaxial compression tests on normally consolidated clays, *Geotechnique.*, **13**, 12–38, 10.1680/geot.1963.13.1.12.

Roscoe, K. H. Schofield, A. N. and Wroth, C. P. (1958). On the yielding of soils, *Géotechnique*, **8**, 22–53.

Roscoe, K. H. Schofield, A. N. and Thurairajah, A. (1963). Yielding of clays in states wetter than critical, *Géotechnique*, **13**, 211–240.

Rowe, P. W. (1962). The stress-dilatancy relation for static equilibrium of an assembly of particles in contact, *Proc. Roy. Soc.*, **A269**, 500–527.

Russell, A. and Khalili, N. (2006). A unified bounding surface plasticity model for unsaturated soils, *Int. J. Numer. Anal. Meth. Geomech.*, **30**, 181–212. doi:10.1002/nag.475

Saada, A. and Bianchini, G. (1989). (Eds) *Constitutive Equations for Granular Non-Cohesive Soils*, Balkema.

Sadeghian, S. and Namin, M. L.(2013). Using state parameter to improve numerical prediction of a generalized plasticity constitutive model, *Comput. Geosci.*, **51**, 255–268.

Saitta, A. (1994). Modélisaton élastoplastique du comportement mécanique des sols application à la liquefaction des sables et à la sollicitation d'expansion de cavité. PhD thesis, ENPC-CERMES, Paris.

Sangrey, D. A. Henkel, D. J. and Esrig, M. I. (1969). The effective stress response of a saturated clay soil to repeated loading, *Can. Geotech. J.*, **6**, 241–252.

Santagiuliana, R. and Schrefler B. A. (2006). Enhancing the Bolzon-Schrefler-Zienkiewicz constitutive model for partially saturated soil, *Transp. Porous Media*, **65**, 1–30.

Schofield, A. N. and Wroth, C. P. (1958). *Critical State Soil Mechanics*, McGraw-Hill.

Schrefler, B. A. (1984). The finite element method in soil consolidation (with applications to surface subsidence). PhD thesis, University College of Swansea C/Ph/76/84.

Schrefler, B. A., Simoni, L., Li, X. and Zienkiewicz, O. C. (1990). Mechanics of partially saturated porous media, in *Numerical Methods and Constitutive Modelling in Geomechanics*, CISM Courses and Lectures, No 311, C.S. Desai, G. Gioda (Eds), 169–209, Springer Verlag.

Seed, H. B. Lee, K. H. (1967). Undained strength characteristics of cohesionless soils, *J. Soil Mech. Found. Div. Proc. ASCE*, **93**, 6, 333–360.

Sheng, D., Sloan, S. W. and Gens, A. (2004). A constitutive model for unsaturated soils: thermomechanical and computational aspects, *Comput. Mech.*, **33**, (6), 453–465.

Sibille, L., Villard, P., Darve, F. and Aboul Hosn, R. (2019). Quantitative prediction of discrete element models on complex loading paths, *Int. J. Numer. Anal. Methods Geomech.*, **43**, 1–30, 10.1002/nag.2911

Simo, J. C. and Hughes, T. J. R. (1986). General return mapping algorithms for rate independent plasticity, in *Constitutive Laws for Engineering Materials*, C. S. Desai, E. Krempl, P. D. Kiousis and T. Kundu (Eds), pp. 221–231, Elsevier, Amsterdam.

Simo, J. C. and Taylor, R. L. (1986). A return mapping algorithm for plane stress elastoplasticity, *Int. J. Numer. Methods Eng.*, **22**, 649–670.

Sowers, G. F., Williams, R. C. and Wallace, T. S. (1965). Compressibility of broken rock and settlement of rockfills, *Proc. 6th Int. Conf. Soil Mech. Montreal*, **2**, 561–565.

Taiebat, M. and Dafalias, Y. F. (2007). SANISAND: simple anisotropic sand plasticity model, *Int. J. Numer. Analyt. Methods Geomech.*, **32**, 8, 915–948.

Tamagnini, R. (2004). An extended Cam-Clay model for unsaturated soils with hydraulic hysteresis, *Géotechnique*, **54**, (2), 223–228.

Tamagnini, R. and Pastor, M. (2004). A thermodynamically based model for unsaturated soils: a new framework for generalized plasticity. *Proceedings of the 2nd International Workshop on Unsaturated Soils*, Mancuso (ed.), Naples, Italy, June, 1–14

Tamagnini, C, Castellanza, R and Nova, R 2002. Implicit integration of constitutive equations in computational plasticity, *Revue française de génie civil.*, **6**, (6), 1051–67.

Tanimoto, K. and Tanaka, Y. (1986). Yielding of soils as determined from acoustic emission, *Soils Found.*, **26**, 69–80.

Tatsuoka, F. (1976). Stress-dilatancy relations of anisotropic sands in three dimensional stress condition, *Soils Founda*, **16**, (2), 1–18.

Tatsuoka, F. and Ishihara, K. (1974). Drained deformation of sand under cyclic stresses reversing direction, *Soils Found.*, **14**, 51–65.

Taylor, D. W. (1948). *Fundamentals of Soil Mechanics*, Wiley.

Taylor, P. W. and Bacchus, D. R. (1969). Dynamic cyclic strain tests on a clay. *Proceedings of the 7th International Conference on Soil Mechanics and Foundation Engineering I*, Mexico, 401–409.

Theocharis, A. I., Vairaktaris, E., Dafalias, Y. F. and Papadimitriou, A. G. (2017). Proof of incompleteness of critical state theory in granular mechanics and its remedy. *J. Eng. Mechanics, ASCE*, **143**, (2), 04016117.

Tonni, L., Cola, S. and Pastor, M. (2006). A generalized plasticity approach for describing the behaviour of silty soils forming the Venetian lagoon basin, in *Numerical Methods in Geotechnical Engineering*, Schweiger (Ed), pp. 93–99. Taylor & Francis.

Tresca, H. (1864). Sur l'Ecoulement des Corps Solides Soumis à des Fortes Pressions, *Comptes Rendus Acad. Sci. Paris*, **59**, 754.

Uriel, A. O. (1973). Discussion to Spec. Session 2, Problems of non linear soil mechanics, *Proceedings of the 8th International Conference on Soil Mechanics and Foundation Engineering Moscow*, **4.3**, 78–80.

Uriel, A. O and Merino, M. (1979). Harmonic response of sands in shear. *Third Int. Conf. on Num. Methods Geomechanics*, Aachen, 2–6 April 1979.

Verdugo, R. (1992). Characterization of sandy soil behavior under large deformation ci.nii.ac.jp.

Verdugo, R., and Ishihara, K (1996). The steady state of sandy soils, *Soils Found.*, **36**, 81–91.

Vesic, A. S. and Clough, G. W. (1968). Behaviour of granular materials under high stresses, *J. Soil Mech. Found. Div. Proc. ASCE*, **94**, 661–688.

von Mises, R. (1913). Göttinger Nachrichten, math.-phys. Klasse, 582.

Wan, R. G. and Guo P. J. (1998). A simple constitutive model for granular soils: modified stress-dilatancy approach, *Comput. Geotech.*, **22**(2), 109–133.

Wang, Z. L. Dafalias, Y. F. and Shen, C. K. (1990). Bounding surface hypoplasticity model for sand, *J. Eng. Mech. ASCE*, **116**, 983–1001.

Wang, Z. L., Dafalias, Y. F., Li, X. S., Makdisi, F. I. (2002). State pressure index for modeling sand behavior, *J. Geotech. Geoenviron. Eng.*, **128**, (6), 511–519.

Wang, Z. J., Chen, S. S. and Fu, Z. Z. (2015). Dilatancy behaviors and generalized plasticity constitutive model of rockfill materials, *Rock Soil Mech.*, **36**, (7), 1931–1938

Wei, K. and Zhu, S.(2013). A generalized plasticity model to predict behaviors of the concrete-faced rock-fill dam under complex loading conditions, *Eur. J. Environ. Civ. Eng.*, **17**, 7, 579–597.

Weng, M. C. and Ling, H. I. (2013). Modeling the behavior of sandstone based on generalized plasticity concept, *Int. J. Numer. Anal. Methods Geomech.*, **37**, 2154–2169.

Wheeler, S. J., Sharma, M. K. and Buisson, M. S. R. (2003). Coupling of hydraulic hysteresis and stress-strain behaviour in unsaturated soils, *Géotechnique*, **53**,(1), 41–54.

Wilde, P. (1977). Two-invariant dependent model of granular media, *Arch. Mech. Stos*, **29**, 799–809.

Wroth, C. P. and Bassett, N. (1965). A stress-strain relationship for the shearing behaviour of sand, *Géotechnique*, **15**, 32–56.

Wu, W. and Kolymbas, D. (1990). Numerical testing of the stability criterion for hypoplastic constitutive equations, *Mech. Mater.*, **29**, 195–201.

Xiao, Y., Sun, Z., Stuedlein, A. W. Wang, C., Wu, Z., Zhang, Z. (2020), Bounding surface plasticity model for stress-strain and grain-crushing behaviors of rockfill materials, *Geosci. Front.*, **11**, 495–510.

Xu, B., Zou, D. and Liu, H (2012). Three-dimensional simulation of the construction process of the Zipingpu concrete face rockfill dam based on a generalized plasticity model, *Comput. Geotech.*, **43**, 143–154.

Yamada, Y. and Ishihara, K. (1979). Anisotropic deformation characteristics of sand under three-dimensional stress conditions, *Soils Found.*, **19**, 2, 79–94.

Zhang, H. W., Heeres, O. M., de Borst, R., Schrefler, B. A.(2001). Implicit integration of a generalized plasticity constitutive model for partially saturated soil, *Eng. Comput.*, **18**, 314–336.

Zhao, N. F., Ye, W. M., Wang, Q., Chen, B. and Cui, Y. G. (2019). A bounding surface model for unsaturated compacted bentonite, *Eur. J. Environ. Civil Eng.*, 1–5. doi: 10.1080/19648189.2019.1651222.

Zhou, C, Tai, P, Yin, J-H. (2020). A bounding surface model for saturated and unsaturated soil-structure interfaces, *Int. J. Numer. Anal. Methods Geomech.*, **44**, (18), 2393–2592. doi: 10.1002/nag.3123.

Zienkiewicz, O. C. and Mroz Z. (1985) Generalized Plasticity formulation and application to Geomechanics, in *Mechanics of Engineering Materials.* C. S. Desai and R. H. Gallagher (Eds), Ch. 33, pp, 691, Wiley.

Zienkiewicz, O. C. and Pande, G. N. (1977). Some useful forms of isotropic yield surfaces for soil and rock mechanics, in *Finite Elements in Geomechanics*, G. Gudehus (Ed), Chapter 5, pp. 179–190, Wiley.

Zienkiewicz, O. C. Humpheson C. and Lewis, R. W. (1975). Associated and non-associated viscoplasticity and plasticity in soil mechanics, *Géotechnique*, **25**, 671–689.

Zienkiewicz, O. C. Leung, K. H. and Pastor, M. (1985). Simple model for transient soil loading in earthquake analysis. I Basic model and its application, *Int. J. Num. Anal. Meth. Geomech.*, **9**, 453–476.

Zou, D., Xu, B., Kong, X., Liu, H. and Zhou, Y. (2013). Numerical simulation of the seismic response of the Zipingpu concrete face rockfill dam during the Wenchuan.

5

Special Aspects of Analysis and Formulation

Radiation Boundaries, Adaptive Finite Element Requirement, and Incompressible Behavior

5.1 Introduction

In the presentation of the essential theory and the finite element discretization procedures, we have deliberately omitted some "finer points" which, on occasion, might be essential to obtain more accurate or more generally applicable solutions to realistic engineering problems. We shall introduce these "finer points" in the present chapter in sufficient detail to allow the reader to follow the current literature and to devise his or her own program modifications.

The chapter will be divided into four sections corresponding to the topics discussed and each section can be studied independently. These sections are:

§5.2 Far-field solutions in quasi-static problems.
§5.3 Input for earthquake analysis and the radiation boundary.
§5.4 Adaptive refinement for improved accuracy and the capture of localized phenomena.
§5.5 Stabilization of computation for nearly incompressible behavior with equal interpolation.

5.2 Far-Field Solutions in Quasi-Static Problems[1]

Sometimes geomechanical problems are defined over very large domains which may be considered as unbounded. Typical examples are problems of regional scale such as land subsidence due to fluid extraction from the underground or simulation of natural subsidence.

Numerical modeling of such problems sometimes produces inaccurate answers due to the difficulty of properly fixing the boundary conditions at large distances from the center of the model.

The usual technique within the finite element method is the truncation of the mesh, introducing a far-field boundary at some distance from the central area. Conditions valid at infinity are then usually imposed at this fictitious boundary. The accuracy of the simulation depends then on the location of the far-field boundary, and this dependence may differ for each interacting field. Obviously, more accurate results are obtained by increasing

Computational Geomechanics: Theory and Applications, Second Edition. Andrew H. C. Chan, Manuel Pastor, Bernhard A. Schrefler, Tadahiko Shiomi and O. C. Zienkiewicz.

the size of the modeled domain. This means an increased number of degrees of freedom (dofs) in problems sometimes defined over very long time spans. Hence, the ensuing numerical models are very large, both in space and in time. An analyst often attempts to reduce the costs by truncating the mesh as close as possible to the area of interest. The correctness of the outcome is not easy to judge and trial-and-error analyses should be performed to assess the influence of the boundary and its assumed conditions on the final results. General criteria do not exist due to the fact that the coupled solution is strongly dependent on the response characteristics of each interacting field. Usually, the matrices relative to the different interactive fields present eigenvalues which differ greatly in magnitude from each other. For this reason, the accuracy of the solution depends on the natural time scale of each field and on the time span of interest for the analysis. It has been observed (Simoni and Schrefler 1989) that errors do not always affect all the state variables in the same way and not necessarily even for the state variables of the field where the forcing function is applied. This fact is more difficult to detect because errors may occur only after a certain time span, depending on the intrinsic velocity of each field. Sometimes, the time span of the trial-and-error analysis is too short to reveal them.

Hence, certain care must be exercised in the truncation approach to avoid unjustified confidence in the results. Appropriate modeling techniques are therefore mandatory for obtaining acceptable results.

Within the framework of the finite element method, infinite elements (IE) seem to offer particular advantages in solving this type of problems. Here we use mapped infinite elements (MIE), which were first applied to consolidation problems by Simoni and Schrefler (1987). MIE are at the same time efficient and simple to use in quasi-static problems. Their use allows for a more realistic assessment of the boundary conditions and reduces the number of dofs required for accurate numerical modeling.

There are essentially two infinite element formulations currently in use for quasi-static problems: decay function infinite elements and mapped infinite elements. In the first case, a finite element is extended so that it stretches to infinity in the direction of one local axis (say ξ). The shape functions $M_i(\xi,\eta)$ of the original finite element in the η-direction are retained, whereas, in the ξ-direction, they are simply multiplied by a decay function $f_i(\xi)$:

$$N_i(\xi,\eta) = M_i(\xi,\eta) f_i(\xi) \tag{5.1}$$

under the conditions that at node i, the shape function N_i must be equal to unity and tend to the far-field value at infinity. The decay function ensures the finiteness condition of the solution and has to be chosen to obtain a reasonable description of the physics of the problem. One of the most widely used decay functions is the exponential function:

$$f_i(\xi) = \exp\frac{\xi_i - \xi}{L} \tag{5.2}$$

where the presence of ξ_i ensures the unit value at node i and L is a length that determines the speed of the decay. Another popular choice is a reciprocal function of the type

$$f_i(\xi) = \left[\frac{\xi_i - \xi_0}{\xi - \xi_0}\right]^n \tag{5.3}$$

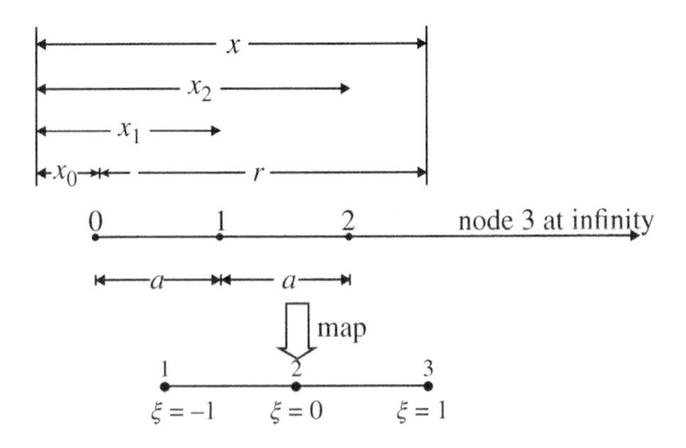

Figure 5.1 Global to local mapping of a one-dimensional infinite element.

where ξ_0 is the distance from some origin. This must be outside the infinite element, hence opposite to the side that extends to infinity. Extension to the case where a finite element is stretched to infinity in more than one direction is straightforward. The use of these types of elements requires particular quadrature formulae, e.g. Gauss-Laguerre or Gauss-Hermite, because the domain of the element is $[-1,\infty]$. A general presentation of the method is available in the literature (Bettess 1992).

As far as the mapped type of element is concerned, we consider the one-dimensional case shown in Figure 5.1, where the infinite element, ranging from node 1 to node 3 (at infinity) through node 2, is mapped onto the parent element defined by the local co-ordinate system $-1 \leq \xi \leq 1$. Point 0 is called the *pole* of the transformation and is positioned arbitrarily, with the only restriction that once it is positioned, the location of node 2 is defined by

$$x_2 = 2x_1 - x_0 \tag{5.4}$$

The interpolation from local to global coordinates is obtained by means of the standard finite-element method for isoparametric elements:

$$x(\xi) = \sum_{i=1}^{2} M_i(\xi)\, x_i \tag{5.5}$$

where the summation extends to finite nodes only and the mapping functions are

$$\begin{aligned} M_1(\xi) &= -2\xi/(1-\xi) \\ M_2(\xi) &= (1+\xi)/(1-\xi) \end{aligned} \tag{5.6}$$

From these relations, it follows that in global coordinates, the positions $x = x_1, x_2, \infty$ correspond to the points $\xi = -1, 0, 1$. Moreover, as

$$M_1(\xi) + M_2(\xi) = 1 \tag{5.7}$$

the transformation does not depend on the adopted reference system.

The field variables Φ (displacements and pressure) are now interpolated using the standard shape function N_i as follows:

$$\Phi(\xi) = \sum_{i=1}^{n+1} N_i(\xi)\, \Phi_i(\xi) = \alpha_0 + \alpha_1\xi + \alpha_2\xi^2 + \alpha_3\xi^3 + \ldots.. + \alpha_n\xi^n \tag{5.8}$$

where α_i contains the nodal values of Φ. Solving equation (5.5) for ξ yields

$$\xi = 1 - \frac{2a}{x-(x_1-a)} = 1 - \frac{2a}{r} \tag{5.9}$$

where r indicates the distance from the pole 0 to a general point belonging to the element and parameter a=x_2-x_1=x_1-x_0. By introducing equation (5.9) in equation (5.8), the expression of the field variable ξ as a function of the global coordinate r is obtained:

$$\Phi(r) = \beta_0 + \frac{\beta_1}{r} + \frac{\beta_2}{r^2} + \frac{\beta_3}{r^3} + \ldots\ldots + \frac{\beta_n}{r^n} \tag{5.10}$$

which represents the polynomial form defined in the unlimited domain x corresponding to a polynomial of degree n defined in the limited domain ξ.

From the above relations, the role of pole 0 and of its position within the obtained approximation can easily be seen: the type of decay depends on the choice of the pole and on the shape functions in (5.8). The vanishing of the variable Φ at infinity is obtained by setting $\beta_0 = 0$ that is by restricting the summation in (5.8) to the *finite* nodes. In such a way, we lose the possibility of representing constant values of the field variables and it is impossible to represent a constant displacement field for the infinite elements.

The procedure can be easily extended to the case of two or three dimensions, by simply using shape function products. Elements may be generated which are either singly or doubly infinite. For example, the infinite element in Figure 5.2a can be considered as a 9-node Lagrangian element in which the 3 nodes corresponding to $\xi = 1$ are at infinity. The relative mapping functions $\hat{M}_i(\xi,\eta)$ are given by equation (5.6) in the infinite direction multiplied by the usual quadratic shape function N_i in the η direction:

$$\begin{array}{lll} \hat{M}_1(\xi,\eta) = M_1(\xi)N_1(\eta), & \hat{M}_2(\xi,\eta) = M_1(\xi)N_2(\eta), & \hat{M}_3(\xi,\eta) = M_1(\xi)N_3(\eta) \\ \hat{M}_4(\xi,\eta) = M_2(\xi)N_3(\eta), & \hat{M}_5(\xi,\eta) = M_2(\xi)N_2(\eta), & \hat{M}_6(\xi,\eta) = M_2(\xi)N_1(\eta) \end{array} \tag{5.11}$$

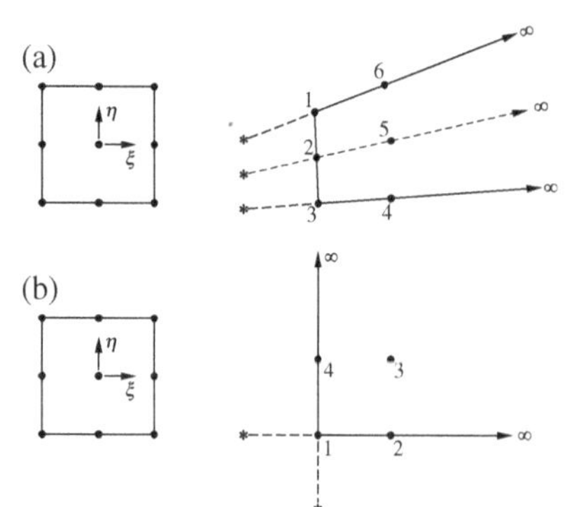

Figure 5.2 Two-dimensional mapped infinite elements: (a) Lagrangian biquadratic singly infinite element; (b) biquadratic doubly infinite element.

Similar expressions can be obtained for the element in Figure 5.2b, which is an infinite element extending to infinity in two directions. In this element, all nodes corresponding to $\xi = 1$ or $\eta = 1$ are at infinity. The mapping functions are the product of equation (5.5) in the ξ and η directions, and the shape functions are properly chosen among those of the 9-noded Lagrangian finite element. Elements like these (doubly infinite elements) can be used in corner areas between elements extending to infinity in different directions.

For a detailed treatment of the question, see Bettess (1992) in which the shape and mapping functions for other two- and three-dimensional elements are also listed.

Both the geometry and the expansion of the field variable Φ of the mapped infinite element depend on the same pole or group of poles in the case of more than one dimension. When choosing the pole(s), we must, then, take into account both the geometry and the physical characteristics of the problem, in particular:

- the pole(s) must be external to the infinite element
- the element sides, extending toward infinity, must be parallel or divergent to avoid the overlapping of elements and preserve mapping uniqueness
- the numbers and positions of the connecting nodes between infinite and finite elements must coincide to ensure continuity across common sides, or adjacent infinite elements.

Once these restrictions have been taken into account, the shifting of the pole(s) involves a variation in the type of decay for the field variables only. As far as the numerical integration is concerned, the usual formulae for finite element codes can be used (e.g. Gauss–Legendre).

The choice of the decay parameters of the approximations is the main concern of the analyst, who must appreciate the physics of the problem to be solved. This choice, however, is easier than truncating the mesh and setting the proper boundary conditions. Hence, infinite elements, in particular, the mapped ones, represent a powerful tool in dealing with unbounded domain problems because the numerical procedure for the formation of the pertinent matrices is very similar to the one used for the standard finite elements. The only adjustment concerns the existence of mapping functions different from the shape functions. The similarity of the numerical procedure allows the use of the standard time-stepping techniques for integration in time even in the presence of infinite elements.

Generally speaking, unbounded domains may be discretized with infinite elements only. However, the best results are obtained by using finite and infinite elements at the same time. The area of greater interest, where nonlinearities may also arise, is discretized by means of the former, whereas the surrounding area is modeled through infinite elements (preferably biquadratic or higher order). The only care must be paid to keeping the continuity on common sides of finite and infinite elements.

The use of shape functions of the same order (parabolic) to approximate pore pressures and displacements in finite elements often results in pressure oscillations both in time and space because the Babuska–Brezzi condition is violated (Section 3.2.3). The same has been observed in the use of the infinite elements. To avoid space oscillations, smoothing techniques as suggested in (Reed 1984) can be used. However, an effective way to eliminate these oscillations is given either by the split algorithm of Section 5.5.5 or by a Discontinuous Galerkin procedure in time shown in Section 5.4.2 and 6.4.3. Both come at the expense of additional degrees of freedom.

Mapped infinite elements of this section can also be used in dynamic problems as shown in Heider et al. (2012). The outgoing dynamic waves toward infinity are absorbed at the FE-IE interface by means of a viscous damping boundary. In such a way, wave reflection is prevented and the dynamic response of the unbounded domain is imitated.

The solution to a problem of compaction due to the withdrawal of water through a well from a single aquifer will be shown in Section 6.4.2 as an example of the application of infinite elements.

5.3 Input for Earthquake Analysis and Radiation Boundary

5.3.1 Specified Earthquake Motion: Absolute and Relative Displacements

The input for earthquake analysis is based on measured recorded data of actual earthquakes and is generally presented as the values of the displacement $\mathbf{u}$ and/or of the acceleration $\ddot{\mathbf{u}}$ at the time interval of 0.02 seconds given for the duration of the earthquake.[2]

If the time history of the input can be specified, we can proceed as outlined in this book to obtain the solution by time integration of the discretized form of the equations of motion such as those given by equations (2.11) and (2.13) of Chapter 2.

The simplest case for the specification of input is illustrated in Figure 5.3 which attempts to model a structure resting on a stratified soil foundation of unlimited extent by specifying the input motion at some arbitrary internal boundary shown.

Such a model corresponds well with such physical models as those of the shaking table or centrifuge where the specified boundary represents a "box" into which the model is fitted and which moves in a specified manner. In Chapter 7, we shall show several calculations which correspond to such physical experiments and which model the real phenomena of practice reasonably well.

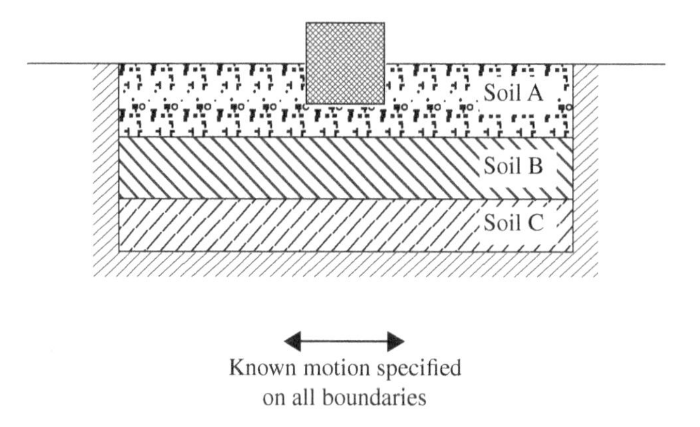

Figure 5.3 Specified motion on the boundaries of a "shaking table box" modeling of an infinite foundation.

With all displacements or tractions at the boundaries specified, we can use the discretization of Chapter 3 and proceed with the solution of any transient problem. It should, however, be remarked that if only uniform motion is specified on the boundaries, it is sometimes convenient to recast the equations of motion in terms of the relative displacement $\mathbf{u}_R$ which we define as

$$\mathbf{u}_R = \mathbf{u} - \mathbf{u}_E \tag{5.12}$$

where $\mathbf{u}_E = \mathbf{u}_E(t)$ is the prescribed earthquake motion that does not depend on the position. The governing equations (2.20) and (2.21) of Chapter 2 now become (neglecting the source terms and putting $\alpha = 1$)

$$\mathbf{S}^T\boldsymbol{\sigma} - \rho\ddot{\mathbf{u}}_R + \rho(\mathbf{b} + \ddot{\mathbf{u}}_E) = \mathbf{0} \tag{5.13a}$$

$$\nabla^T\mathbf{k}(-\nabla p - \rho_f\ddot{\mathbf{u}}_R + \rho_f(\mathbf{b} + \ddot{\mathbf{u}}_E)) + \mathbf{m}^T\dot{\boldsymbol{\varepsilon}} + \frac{\dot{p}}{Q} = 0 \tag{5.13b}$$

with the boundary condition on the input boundary being

$$\mathbf{u} = \mathbf{u}_E \tag{5.14a}$$

which is replaced by

$$\mathbf{u}_R = 0 \tag{5.14b}$$

If the relative velocity is used in the finite element discretization of the problem, the numerical computations are identical to those of the absolute displacement if the same initial conditions (e.g. $\mathbf{u} = 0$) are assumed. However, the input is now the acceleration $\ddot{\mathbf{u}}_E$ giving a prescribed body force and this is often more accurately known.

In a more realistic treatment of the foundation problem, we shall impose somewhat different boundary conditions recognizing the fact that in the input, only the incoming wave motion is specified and that outgoing waves must leave the problem domain unimpeded. Figure 5.4 again shows the problem, initially suggested in Figure 5.3, indicating the position

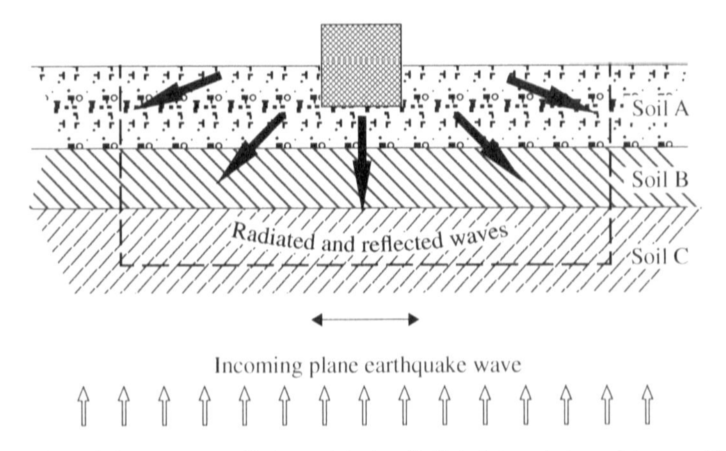

Figure 5.4 A more realistic model of an "infinite" foundation with a specified incoming wave. *Source:* Based on Zienkiewicz and Huang (1990).

of the same limiting boundary but on which the motion will now not be directly specified. We shall discuss this problem in the next section in more detail and suggest how such problems can be dealt with. First, however, a few words about the way knowledge of the seismic input wave is obtained.

The seismic signal is usually measured at or near the free surface and it represents the modification of the original seismic wave which is initiated at the earthquake source caused by passing through different material zones and involving a number of internal reflections and refractions at the interfaces between layers of a different material.

The geological conditions at the site will very often be such that the so-called "bedrock" exists as a zone of a significantly more rigid material underneath softer soil layers. Any incoming seismic wave passing from the bedrock to the softer soil layers will be selectively amplified depending on the material properties of both bedrock and the soil layers. The significant consequence of the presence of the bedrock lies in the fact that all of the reflected waves are practically trapped inside the soft soil layers, as only a small fraction of these can be transmitted back to the bedrock through the interface with the softer soil. If the bedrock is significantly more rigid, the transmitted wave is smaller than the reflected one back toward the soil surface. In such a case, the simple fixed-base approach is valid and no transmitting boundary conditions need to be imposed on the bedrock level as practically no waves get transmitted into the bedrock.

The need for an arbitrary model truncation emerges in cases where no distinct base rock exists or when the extent of the softer soil layers is so great that it would be prohibitive to include the whole zone in a mathematical model. Such a situation may also arise when the nonlinear material behavior can be expected only near the surface and deeper layers (with material properties still far from bedrock-like characteristics) are expected to remain elastic.

To model such a case correctly it is necessary to reconstruct the incoming seismic wave at the model truncation boundary. In the simplest case of a one-dimensional elastic, homogeneous, isotropic wave propagation problem involving the free surface, it is very well known that the free surface displacement wave equals the double of the incoming displacement wave. Here the incoming signal can be easily extracted from the recorded total signal on the undisturbed surface. Even in the case of the elastic nonhomogeneous domain, the incoming signal can again be extracted from the total signal recorded on the surface. Therefore, in the following, it will be assumed that the incoming wave (displacement, velocity, or acceleration) is known at a position corresponding to the model truncation boundary, and that outside of this homogeneous elastic conditions pertain.

5.3.2 The Radiation Boundary Condition: Formulation of a One-Dimensional Problem

We return here once again to the problem of a stratified, horizontal foundation such as we have considered in the previous section but now without a superposed structure. Clearly, if we consider a vertical slice shown in Figure 5.5, isolated by cut sections AA and BB, we note immediately that the problem is one-dimensional, i.e. that the displacements, stresses, etc., do not vary with the horizontal coordinate $\mathbf{x}$.

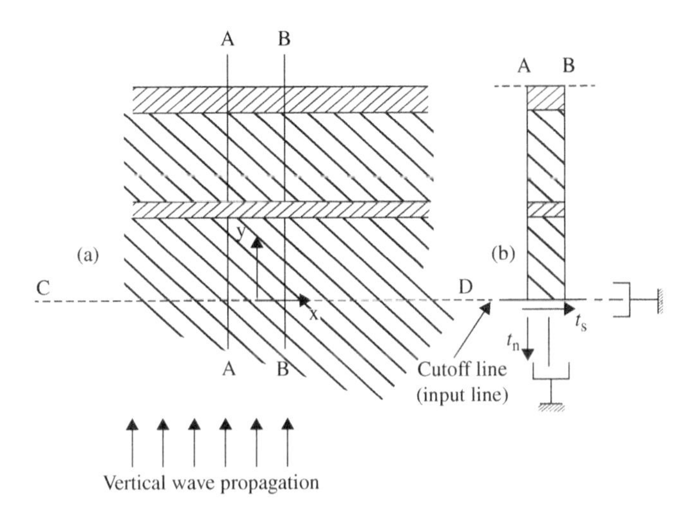

Figure 5.5 A horizontally stratified foundation subject to vertically propagating compression or shear waves: (a) the corresponding 1-D problem with (b) arbitrary cutoff. *Source:* Based on Zienkiewicz and Huang (1990).

The equations governing the problem are still (5.13a) and (5.13b) with the unknown variables remaining as $\mathbf{u}_R$ which is now, however, a function of the coordinates y and time t. Thus

$$\mathbf{u}_R = \mathbf{u}_R(y, t) \tag{5.15}$$

Thus, all the derivatives with respect to the x-axis are made identically zero.

To demonstrate the wave nature of the problem, we shall assume that in the vicinity of the arbitrary, "input" surface CD (and below this surface), the conditions are such that:

i) only isotropic elastic behavior exists;
ii) the body forces have been taken into account independently; and finally that
iii) the dynamic phenomena are sufficiently rapid so that the flow in the porous medium can be neglected and $k = 0$ is assumed.

Writing the total relative velocity in terms of its components

$$\mathbf{u}_R = \begin{Bmatrix} u_x \\ u_y \end{Bmatrix} \tag{5.16}$$

the system (5.13) reduces, in the absence of internal flow, to

$$\frac{\partial \sigma_{xy}}{\partial y} - \rho \ddot{u}_x = 0 \tag{5.17a}$$

$$\frac{\partial \sigma_{yy}}{\partial y} - \rho \ddot{u}_y = 0 \tag{5.17b}$$

where only total stresses are considered. The elastic constitutive relation under isotropic undrained conditions gives

$$\sigma_{xy} = G\frac{\partial u_x}{\partial y} \tag{5.18a}$$

and

$$\sigma_{yy} = \widetilde{K}\frac{\partial u_y}{\partial y} \tag{5.18b}$$

where

$$G = \frac{E}{2(1+\nu)} \tag{5.19a}$$

is the shear modulus and

$$\widetilde{K} = \frac{E(1-\nu)}{(1+\nu)(1-2\nu)} \tag{5.19b}$$

is the one dimensional constrained modulus or restrained axial modulus. E and ν are Young's modulus and Poisson's ratio.

Equation (5.17a) becomes on insertion of the above

$$\frac{\partial^2 u_x}{\partial y^2} - \frac{\rho}{G}\ddot{u}_x = 0 \tag{5.20a}$$

and (5.17b) becomes

$$\frac{\partial^2 u_y}{\partial y^2} - \frac{\rho}{\widetilde{K}}\ddot{u}_y = 0 \tag{5.20b}$$

Each of the above equations corresponds to the well-known scalar wave equation

$$\frac{\partial^2 \phi}{\partial y^2} - \frac{1}{c^2}\frac{\partial^2 \phi}{\partial t^2} = 0 \tag{5.21}$$

which has the solution

$$\phi = \phi_{\mathrm{I}}(y - ct) + \phi_{\mathrm{o}}(y + ct) \tag{5.22}$$

in which c is the wave velocity and ϕ_{I} and ϕ_{o} represent two waves traveling in the positive and negative directions of y respectively (incoming and outgoing waves).

Thus, $\phi = u_x$ represents shear waves traveling with velocity

$$c_{\mathrm{s}} = \sqrt{\frac{G}{\rho}} \tag{5.23}$$

and $\phi = u_y$ represents compressive waves traveling with velocity

$$c_{\mathrm{c}} = \sqrt{\frac{\widetilde{K}}{\rho}} \tag{5.24}$$

We observe that c_c tends to infinity for fully incompressible solid and fluid situations.

To obtain the radiation condition, we observe the solution sought at the "cutoff" line CD should represent only an outgoing wave, i.e.

$$\phi = \phi_o(y + ct) \tag{5.25}$$

We observe immediately that

$$\frac{\partial \phi}{\partial t} = \phi' c \quad \text{and} \quad \frac{\partial \phi}{\partial y} = \phi' \tag{5.26}$$

where

$$\phi' \equiv \frac{\mathrm{d}\phi(y + ct)}{\mathrm{d}(y + ct)}$$

or that on the boundary

$$\frac{\partial \phi}{\partial y} = \frac{1}{c}\frac{\partial \phi}{\partial t} \tag{5.27}$$

to ensure the existence of outgoing waves alone. Using the relationships (5.18) and noting the definitions of (5.23) and (5.24), we will observe that on the boundary CD, the tangential traction becomes

$$t_x \equiv \sigma_{xy} = \frac{G}{c_s}\frac{\partial u_x}{\partial t} \tag{5.28a}$$

and the normal traction becomes

$$t_y \equiv \sigma_{yy} = \frac{\widetilde{K}}{c_c}\frac{\partial u_y}{\partial t} \tag{5.28b}$$

This is equivalent to the requirement that on the boundary, "dashpots" of suitable strength are imposed in tangential and normal directions. Representation of such radiation (or quiet) boundary conditions in the manner presented above was suggested almost simultaneously by Zienkiewicz and Newton (1969) and Lysmer and Kuhlemeyer (1969).

In the one-dimensional case presented here, the radiation condition is exact. However, on many occasions, it has been used effectively on two- or three-dimensional boundary shapes where the conditions of Equations (5.28a) and (5.28b) imply dashpots in the tangential and normal direction at any position of the boundary. The numerical tests of the effectiveness of such a radiation boundary condition are presented by Zienkiewicz et al. (1987b) where it was shown that for a given wave input form, identical results are obtained independently of the arbitrary cutoff position.

Many alternative forms of radiation boundary conditions have been developed. Here the early work of Smith (1973) which was then generalized by Zienkiewicz et al. (1987a) is one possibility. Alternative methods are discussed by Wolf and Song (1996), White et al. (1977), Kunar and Marti (1981), and Zienkiewicz and Taylor (1991) (see also Section 5.3.4).

It is usual to conduct an analysis in terms of the relative displacement $\mathbf{u}_R$ defined by (5.3) and to apply the radiation condition to this relative displacement only (see Zienkiewicz et al. 1987a, Clough and Penzien 1993).

5.3.3 The Radiation Boundary Condition: Treatment of Two-Dimensional Problems

A more general situation of engineering interest is the one illustrated in Figure 5.6 where a structure "perturbs" the simple one-dimensional solution of the layered foundation. Once again the horizontal boundary on which the vertically propagating waves enter the problem domain is treated identically to that of the one-dimensional case. Indeed, identical "dampers" are placed on this boundary to ensure transmission of the exiting waves (but now these are approximate only as the transmission conditions do not apply exactly to waves exiting obliquely to the boundary).

More serious difficulties are, however, posed on the two vertical boundaries AA and BB and the boundary condition which needs to be imposed on these. Clearly, at points that are far away from the superposed structure, the solution must be asymptotic to the previously discussed one-dimensional one. A possible way of dealing with the boundary conditions on these sections is therefore to impose the radiation damper between the interior region and the one-dimensional, free-field, solution. Such a treatment is suggested by Zienkiewicz et al. (1988) but a simpler alternative is that of repeatable boundary conditions, which is also given there. The latter achieves identical results more simply.

In the repeatable boundary condition, which is illustrated in Figure 5.7, it is assumed that a sequence of structures is placed on the foundation at regular intervals B. The treatment of such repeatable conditions is simple in the finite element context (see Zienkiewicz et al. 2005a) as clearly the values of displacement, stresses, etc., are identical on such a section as A or B due to periodicity and the assembly of nodal values at these boundaries is ensured by suitable node numbering.

In Figure 5.8, we illustrate a test problem where different depths and widths of the analyzed domain are used. A homogeneous elastic material is here assumed throughout the space and Figure 5.9 shows the time histories of displacement, acceleration, and stress

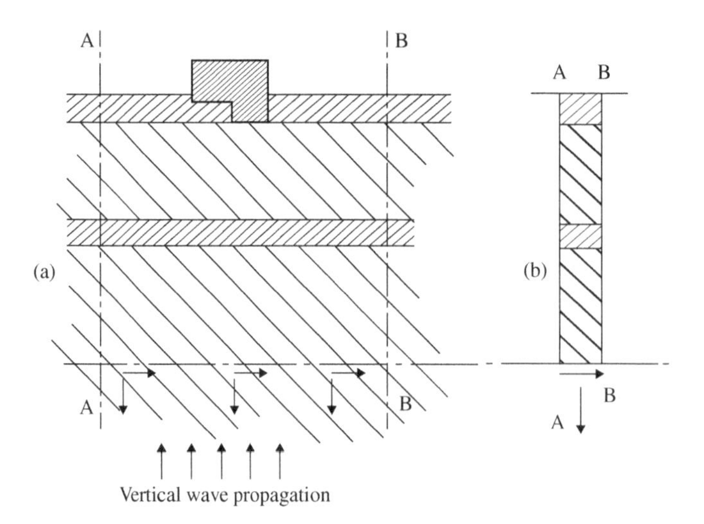

Figure 5.6 Foundation of Figure 5.5 perturbed by the imposition of a structure (a) and the 1D problem (b). *Source:* Based on Zienkiewicz and Huang (1990).

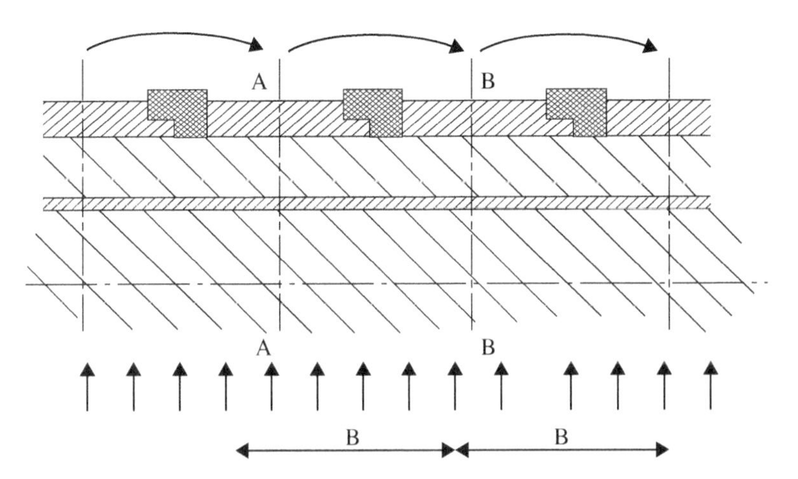

Figure 5.7 Repeatable boundary conditions. Displacement at A = displacement at B. *Source:* Based on Zienkiewicz and Huang (1990).

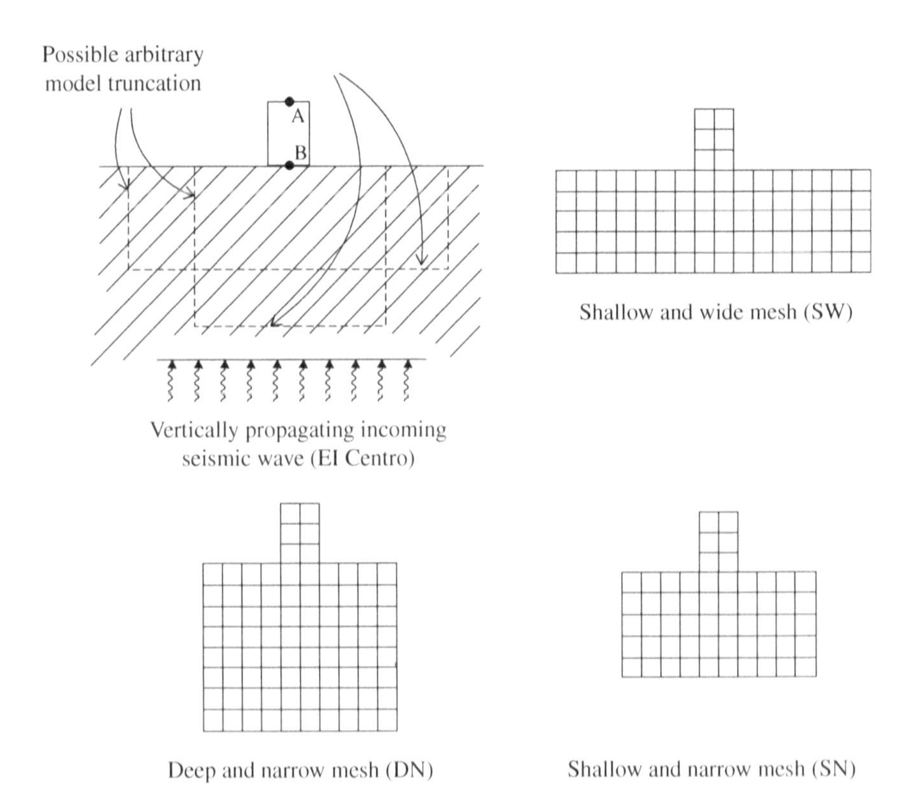

Figure 5.8 Two-dimensional model problem and three meshes (SN, DN, and SW). *Source:* Based on Zienkiewicz and Huang (1990).

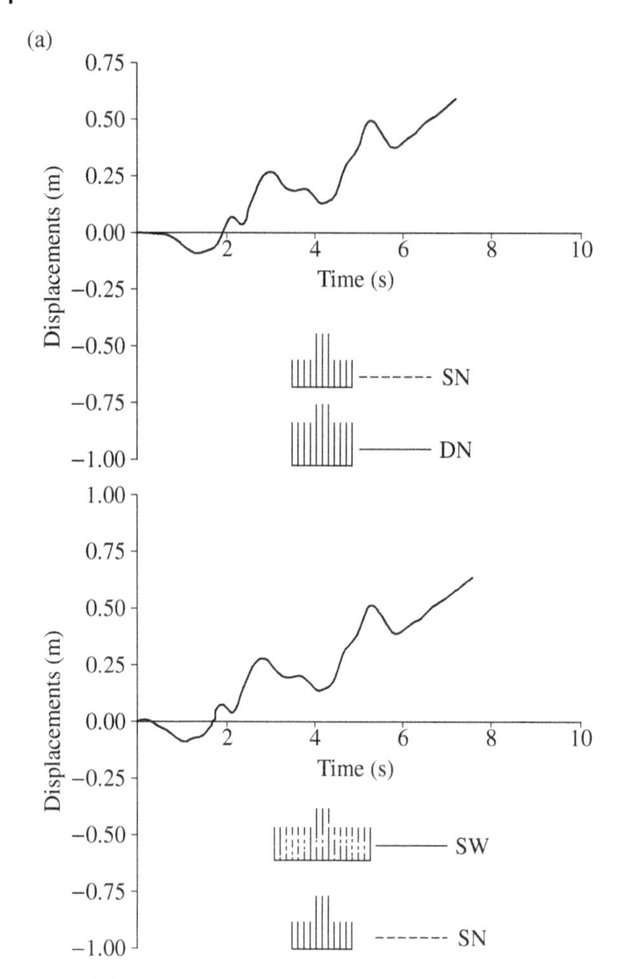

Figure 5.9 The problem of Figure 5.8. (a) Time history of horizontal displacement at point B; (b) Time history of horizontal acceleration at point B; (c) Time history of shear stress at point B for the input of the El Centro earthquake at the base of the mesh. *Source:* Based on Zienkiewicz and Huang (1990).

for a typical point at the base of the structure and with different domains of computation. It is surprising to note how little the results are affected by the extent of the domain assumed.

5.3.4 The Radiation Boundary Condition: Scaled Boundary-Finite Element Method[3]

To satisfy the radiation condition for general two- and three-dimensional situations of engineering problems is a challenging research topic. This is evidenced by the lack of such a capacity in commercial software packages. One development toward achieving this goal

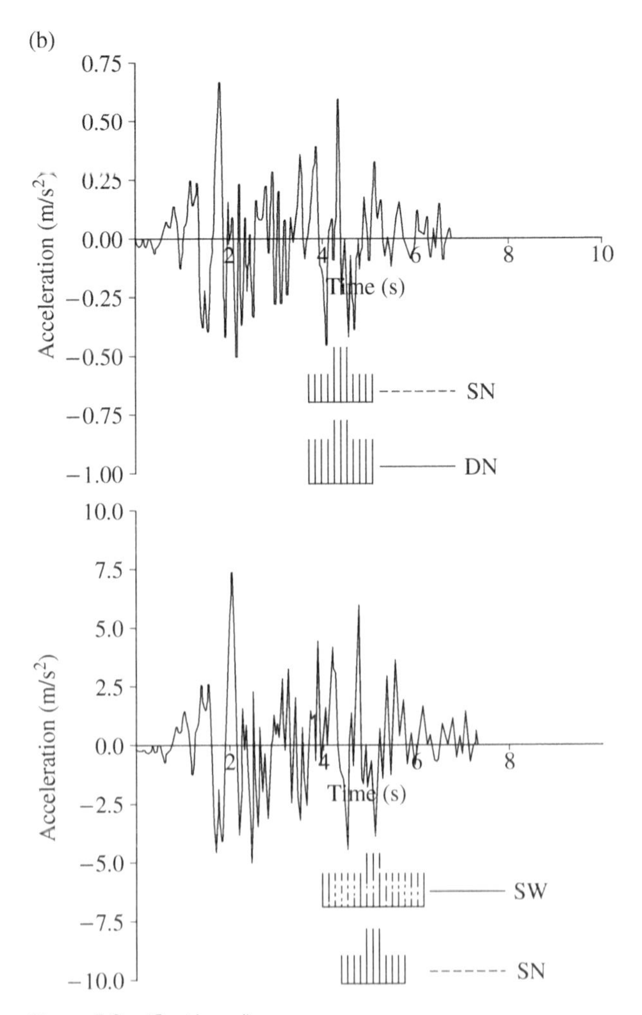

Figure 5.9 (Continued)

is the scaled boundary finite-element method, which was also referred to as the consistent infinitesimal finite-element method in its early stage of development (Wolf and Song 1996).

Following the substructure technique as illustrated in Figure 5.10, the structure and a part of soil adjacent to the structure are regarded as a substructure. The remaining part of soil of unlimited extent is another substructure which exhibits linear behavior. The two substructures interact at their common boundary called the "structure-soil interface". It is well known (Wolf, 1988) that the substructure of soil can be represented in the frequency domain by its dynamic stiffness matrix $\mathbf{S}^{\infty}(\omega)$ (ω denotes excitation frequency). The nodal force $\mathbf{R}(\omega)$-displacement $\mathbf{u}(\omega)$ relationship on the structure-soil interface is decomposed as

$$\mathbf{R}(\omega) = \mathbf{S}^{\infty}(\omega)\mathbf{u}(\omega) = i\omega\mathbf{C}_{\infty}\mathbf{u}(\omega) + \mathbf{K}_{\infty}\mathbf{u}(\omega) + \mathbf{S}_{r}^{\infty}(\omega)\mathbf{u}(\omega) \tag{5.29}$$

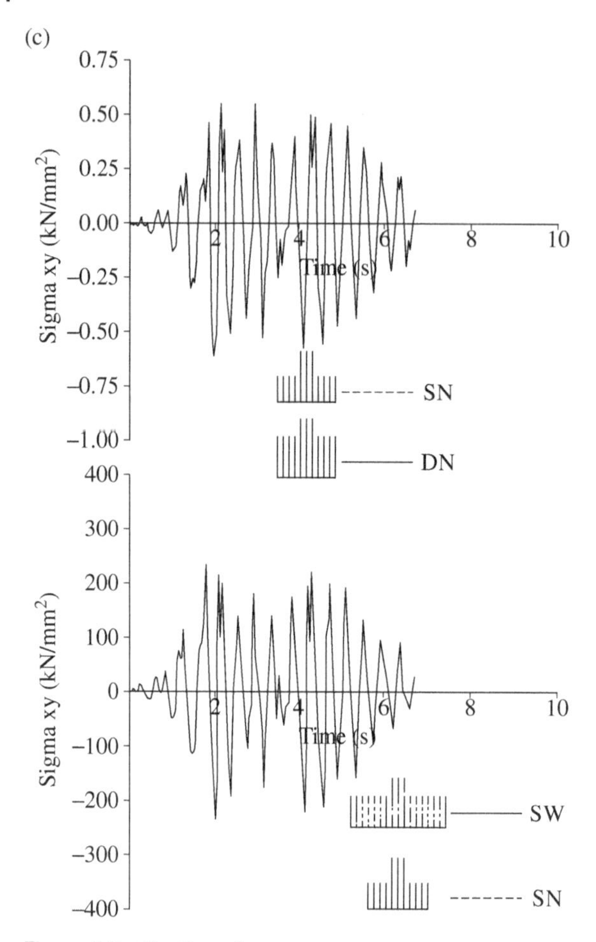

Figure 5.9 (Continued)

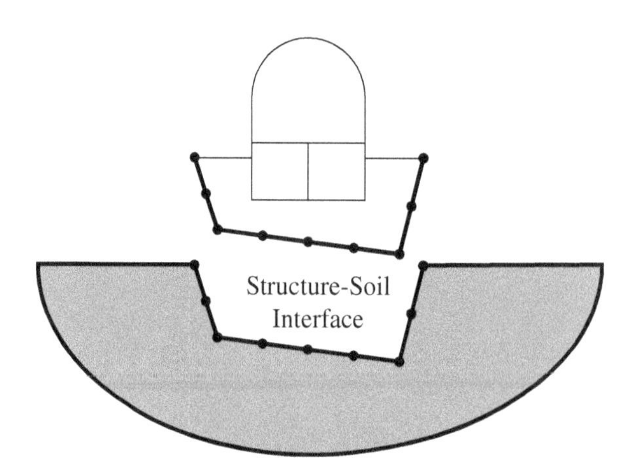

Figure 5.10 Substructure technique for seismic analysis of structures

where $\mathbf{C}_\infty$ and $\mathbf{K}_\infty$ represent the "dampers" and "springs" distributed on the structure–soil interface. $\mathbf{S}_r^\infty(\omega)$ is the remaining regular term. In the time domain, the nodal force–displacement relationship is expressed as

$$\mathbf{R}(t) = \int_0^t \mathbf{S}^\infty(t-\tau)\mathbf{u}(\tau)d\tau = \mathbf{C}_\infty \dot{\mathbf{u}}(t) + \mathbf{K}_\infty \mathbf{u}(t) + \int_0^t \mathbf{S}_r^\infty(t-\tau)\mathbf{u}(\tau)d\tau \tag{5.30}$$

The first two terms represent the instantaneous response and the convolution integral represents the lingering response. When only the first term is retained, the formulation is that of the viscous boundary consisting of "dampers" (Lysmer and Kuhlemeyer 1969). The viscous-elastic boundary corresponds to the first two terms.

The scaled boundary finite-element method provides a tool to evaluate the dynamic stiffness matrix $\mathbf{S}^\infty(\omega)$ and the unit-impulse response matrix $\mathbf{S}^\infty(t)$ at the structure–soil interface (Wolf and Song 1996). In this method, only the structure–soil interface is discretized. The radiation condition is satisfied rigorously at infinity. No fundamental solution is required. Anisotropic materials can be treated without additional computational efforts. It can be seamlessly coupled with finite elements.

Same as other rigorous methods, the scaled boundary finite-element formulation is spatially and temporally global. The computational effort increases rapidly with the size of the problem and the number of time steps. It is desirable in engineering applications to introduce approximations to reduce the spatial and temporal coupling so that the computational cost is minimized without significantly affecting accuracy. Several such approximation schemes of the rigorous scaled boundary finite-element approach have been proposed.

Zhang et al. (1999b) introduced approximations in both time and space. The structure–soil interface is divided into patches of smaller sizes, which is equivalent to reduce a matrix to sub-matrices of smaller dimensions. A unit-impulse matrix is approximated with a few linear segments by making use of its asymptotic properties. This significantly reduces the number of operations in evaluating the convolution. Yan et al. (2004) applied linear system theory to approximate the convolution integral by a system of linear equations. Lehmann (2007) and Schauer and Lehmann (2009) replaced the fully populated unit-impulse response matrix with hierarchical matrices. A substantial improvement in efficiency for matrix–vector products in the convolution integral was reported.

Radmanovic and Katz (2010) proposed an implicit algorithm to evaluate the unit-impulse response matrix. Instead of a stepwise constant approximation, a linear interpolation is adopted. This allows the use of a larger size of time step in evaluating the unit-impulse matrix. The asymptotic behavior of the unit-impulse response is considered to further reduce the computational effort. A speed-up of more than an order of magnitude is reported by using the rigid prismatic foundation shown in Figure 5.11. This development has been incorporated into a commercial software package.

Bazyar and Song (2008) developed a high-order local transmitting boundary condition based on the scaled boundary finite-element method. The dynamic-stiffness matrix is determined recursively as continued fractions. After introducing auxiliary variables, the force–displacement relationship is formulated as an equation of motion with symmetric frequency-independent static stiffness and damping matrices. The computationally expensive

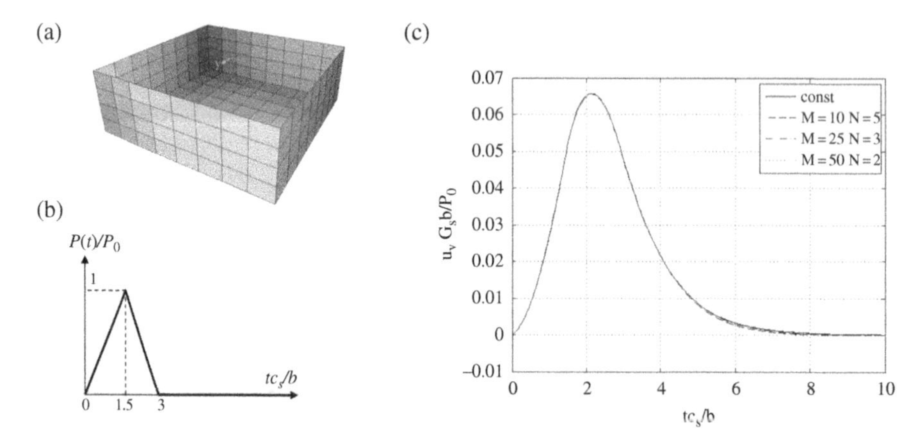

Figure 5.11 A rigid prismatic foundation embedded in a half-space subjected to a triangular concentrated load: (a) scaled boundary finite-element mesh; (b) loading time history; (c) vertical displacement response (const: piecewise constant discretization of convolution integral; M: linear segments of unit-impulse response; N: number of time steps within one linear segment when evaluating the convolution integral).

task of evaluating the unit-impulse response matrix and the subsequent computation of convolution integrals are circumvented. Song and Bazyar (2008) applied the technique of the reduced set of base functions to minimizing the number of spatial degrees of freedom. Taking advantage of the sparsity of the coefficients of the scaled boundary finite-element equation, the reduced set of base functions can be constructed efficiently. The vertical response of a strip footing under an impulse load is illustrated in Figure 5.12. The result obtained with a reduced set of eight modes and a continued fraction solution of seven terms is very close to the results of the rigorous approach (Wolf and Song 1996) and of the extended mesh at an early time.

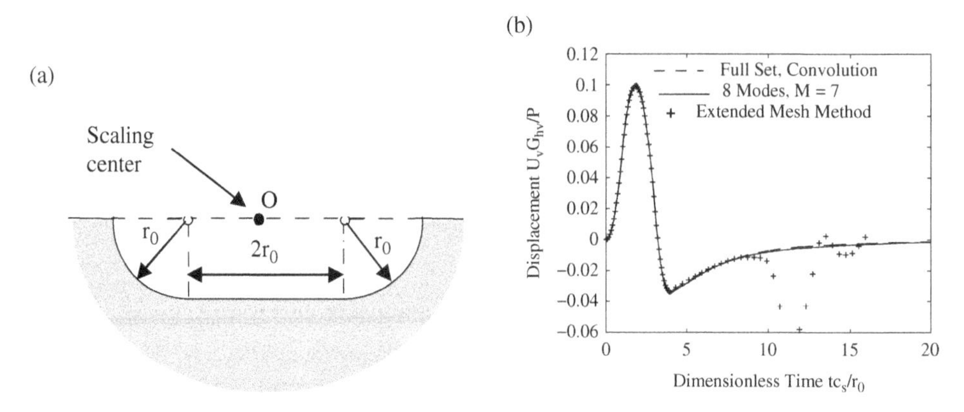

Figure 5.12 A rigid strip footing embedded in a transversely isotropic half-plane: (a) geometry; (b) vertical displacement.

5.3.5 Earthquake Input and the Radiation Boundary Condition – Concluding Remarks

Except for the case of Section 5.3.4, we have limited our discussion to that of the behavior of the two-dimensional foundation layer problems with a vertically propagating wave input. Extension of the problem to three dimensions for the same wave input is trivial but, of course, three dampers will now be necessary on the radiation boundaries. Greater difficulties are presented by problems in which the earthquake (or shock) waves enter the boundary obliquely or indeed horizontally. Here, of course, the input motion history will be dependent on the position and the determination of this in itself is a major problem. However, once such motion is established, it is possible to apply radiation boundary conditions throughout. We shall not discuss this difficult problem further as it is not frequently encountered in practice.

5.4 Adaptive Refinement for Improved Accuracy and the Capture of Localized Phenomena

5.4.1 Introduction to Adaptive Refinement

Accuracy control and adaptive finite element refinement are, of course, of importance in all analysis problems, even if the material behavior is linearly elastic. However, the need for adaptive refinement is even greater when plastic deformations are pronounced as here often very sharp gradients of displacements can occur, leading in the limit to localized displacement discontinuities. In Figure 5.13, we show a typical plastic deformation pattern

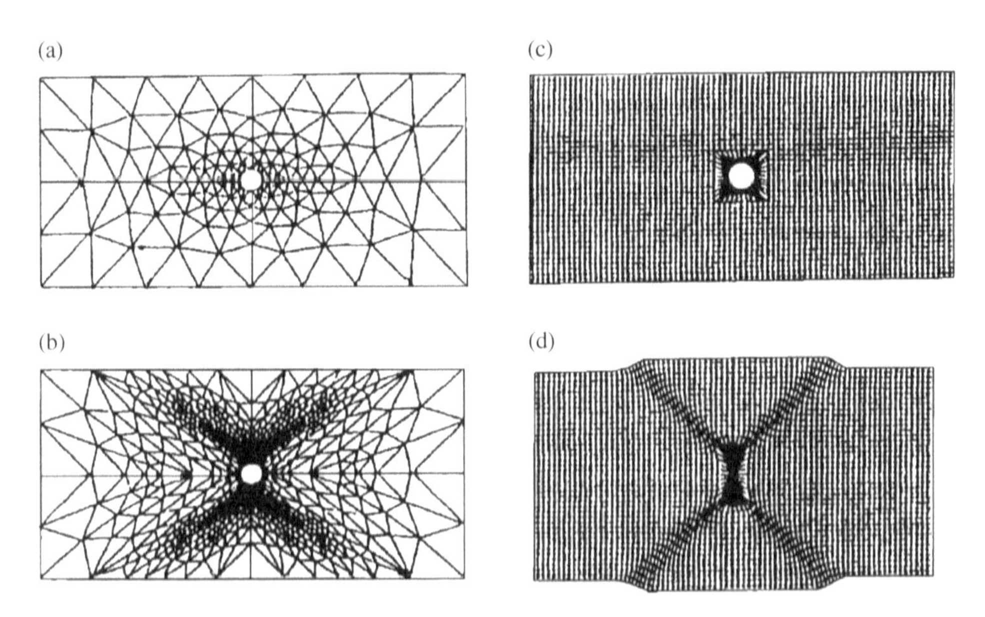

Figure 5.13 First adaptive solution of a purely plastic deformation problem. A perforated bar (a) initial mesh; (b) final adapted mesh with elongation DOF 1039; (c) initial material configuration; (d) final material deformation. *Source:* Based on Zienkiewicz and Huang (1990).

occurring in a uniformly stretched tensile specimen with a small perforation (viz. Zienkiewicz and Huang 1990). In this analysis, the mesh was adaptively refined with elements not only being reduced in size near the displacement discontinuity, but also stretched in the direction of this discontinuity which is indicated by the material deformation pattern.

The "capture" of this discontinuity, as it actually develops, can be achieved approximately with finite elements using a continuous interpolation. However, by sufficient refinement, the exact solution can be approached as closely as possible. How far the refinement should proceed is a question that is difficult to answer precisely and here we will need to revert to the notion of error tolerance.

In an adaptive refinement of many engineering problems in which the errors are distributed throughout the domain, it is convenient to introduce error norms (such as the frequently-used energy error norm) and to require that the error of that norm be kept below a certain value, usually as a fixed percentage of the total value of energy norm in the domain. Such an approach is not recommended in studies of plastic deformation or localization as our interest cannot, in general, be described by a single "number". We may, for instance, wish to find the maximum values of the loads carried by the structure of a particular size at a particular stage of deformation or indeed the maximum loads occurring throughout the deformation history. Alternatively, the interest may lie in determining precisely the position of the region where large strains or discontinuities occur.

For such problems, we can separate the process of error determination and of refinement of the mesh. Thus, the latter can be guided, for instance, by requiring that such a quantity (or indicator) as

$$h_{\min}\frac{\partial \phi}{\partial s} = C \tag{5.31}$$

where C is constant between all the elements, $h_{\min}$ is their minimum size, and s is the direction of the maximum gradient of ϕ, the function of interest. This quantity can be interpreted as the maximum value of the first term of the Taylor expansion defining the local error of the scalar quantity ϕ. By ensuring that the mesh is generated so that the quantity C is constant throughout all elements, we achieve a solution that captures well all local discontinuities and which is efficient in achieving the progression which gives overall accuracy. At any stage of refinement, estimates of error are possible by using various recovery procedures (see Zienkiewicz and Zhu 1992), but, alternatively, convergence to an exact solution can be studied by simply reducing the constant C in refinement.

The use of the indicator defined by (5.31) allows element elongation to be included in the refinement, as, on many occasions, the feature occurring at high gradients is almost one-dimensional. Indeed, in a truly one-dimensional feature, the maximum sizes of an element along its direction would be arbitrary and any reasonable value would be fixed on the maximum element size $h_{\max}$. However, if the contours of the function ϕ diverge or are curved, the upper limit $h_{\max}$ can be specified more closely. Thus, for instance, if the contours separated by the value of $h_{\min}$ diverge by an angle θ, then the limit on $h_{\max}$ could be replaced as

$$h_{\max} < \frac{h_{\min}}{\theta} \tag{5.32a}$$

as a bound based on the variation of the smallest dimension of the element.

With curved contours

$$h_{max} < \alpha R \tag{5.32b}$$

is often specified where α is circa 0.1 and R is the radius of curvature. These types of procedure are discussed in detail by Zienkiewicz and Wu (1994) in the context of fluid mechanics.

An alternative refinement indicator has been used for a longer time in fluid mechanics. This is a requirement that

$$h_{min}^2 \left| \frac{\partial^2 \phi}{\partial s^2} \right|_{max} = \tilde{C} \tag{5.33}$$

specifies the minimum size of elements. This specification was first formulated by Peraire et al. (1987) and is very effective in the capture of shocks. Here, the elongation of elements can be computed directly in terms of principal curvatures.

$$\frac{h_{max}}{h_{min}} = \frac{\left| \frac{\partial^2 \phi}{\partial s^2} \right|_{min}}{\left| \frac{\partial^2 \phi}{\partial s^2} \right|_{max}} \tag{5.34}$$

It appears that the first indicator (i.e. that of (5.31)) is most efficient in the capture of narrow discontinuities, but both provide a remeshing that gives a rapid convergence and reduction of both local and global errors. Figure 5.14 shows how an adaptive analysis based on the first indicator can model discontinuity developed during the failure of the foundation under an eccentric load. Here a von Mises type of yield surface is used with ideal plasticity assumptions (Zienkiewicz et al. 1995a, 1995b).

In Chapter 3, we have already mentioned that special conditions have to be satisfied by mixed finite element forms for incompressible, or nearly incompressible, behavior such as is encountered under undrained conditions. Indeed, such behavior will occur in many applications of plasticity using von Mises or Tresca yield surfaces. For an adequate solution, it is always necessary to use here special mixed forms of elements which are outlined in Chapter 3. In the two examples quoted already, we used the T6C/3C triangle where six nodes define a quadratic variation of continuous displacement and three nodes interpolate pressures in a continuous manner.

In Figure 5.15, we show again an analysis of an ideal elastoplastic problem in which a strong localization occurs. Here two forms of regular mesh are compared – one named "lucky" mesh in which the triangle subdivision lines follow approximately the slip surface, and the other, the "bad" mesh in which these lines are orthogonal to the slip surface. It is clearly noted that for the same subdivision, the "bad" mesh gives answers which are always inferior to those of the "lucky" mesh. However, the adaptive solution starting from either refinement shows nearly exact values of the collapse load.

A serious problem with the adaptive analysis of nonlinear problems of plasticity in which the results are path-dependent is that of data transfer between the various stages of analysis. In principle, the control of the error should be achieved at each load increment separately and this, of course, necessitates the transfer of history-dependent data such as stresses, strains, etc., from the mesh of the previous step to that used in the next increment. To avoid

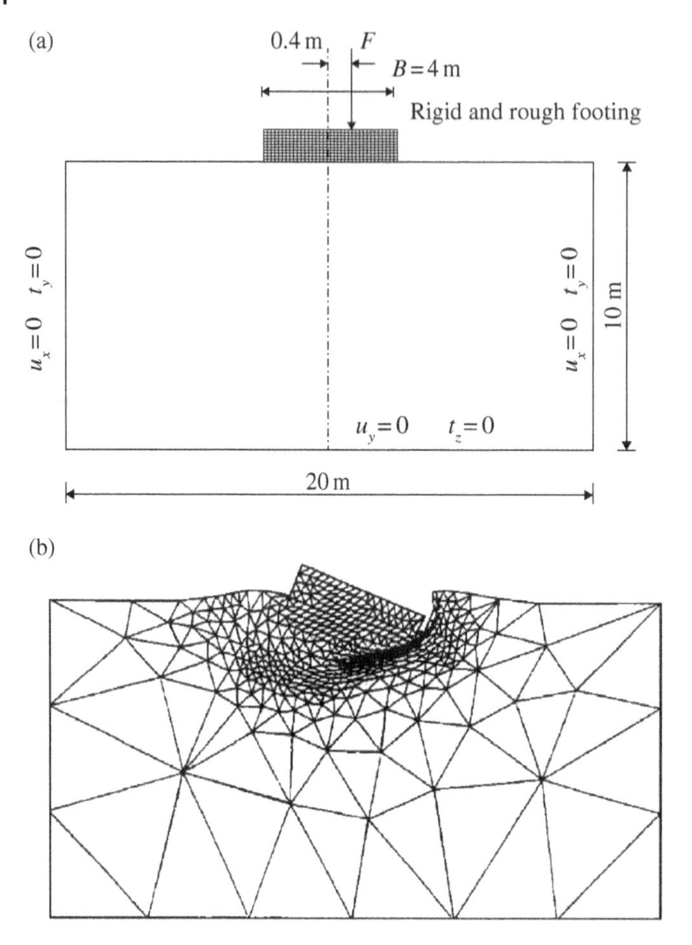

Figure 5.14 Adaptive solution of the problem of foundation collapse with an ideally plastic–elastic material (a) eccentrically loaded footing; (b) final adapted mesh and deformed configuration showing displacement discontinuity. *Source:* Based on Zienkiewicz and Huang (1990).

difficulties, we have re-analyzed the problems in each of the previous cases from the start of loading for every new mesh developed. Indeed such a procedure has also been used quite effectively in transient analysis of the Lower San Fernando dam by Zienkiewicz and Xie (1991) and Zienkiewicz et al. (1995a, 1995b) with results shown in Figure 5.16. However, further new procedures of transferring data have been developed and it is now possible to change the mesh at each load increment, thus ensuring a constant degree of accuracy (Zienkiewicz et al. 1999b); see also Section 6.5.

5.4.2 Adaptivity in Time[4]

Adaptivity in time can be achieved by means of the Discontinuous Galerkin Method in the time domain (DGT). This method involves an increased number of degrees of freedom but this effect can be offset by taking advantage of adaptivity.

The Discontinuous Galerkin (DG) method has become increasingly popular in recent years as a method for solving partial differential equations. For mathematical aspects, the reader is referred to Johnson et al. (1984) and to Johnson and Pitkäranta (1986). Hughes and Hulbert (1988) investigated the DG time integration scheme and introduced the Galerkin least-square term to impose the inter-element consistency. Li and Wiberg (1998) applied an adaptive space-time DG method to structural dynamics problems and made use of energetic arguments to enforce continuity between the elements. Palaniappan et al. (2004) introduced the flux treatment into space-time finite element methods for a scalar conservation law. Chen et al. (2006) used a time-discontinuous Galerkin method (DGT) for the dynamic analysis of fully saturated porous media where the continuity across the time interval is weakly enforced by a flux function. We use here such a method in connection with time adaptivity for dynamic and consolidation problems, following Secchi et al. (2007). The DGT method lends itself quite naturally for time adaptivity in a simple and automatic

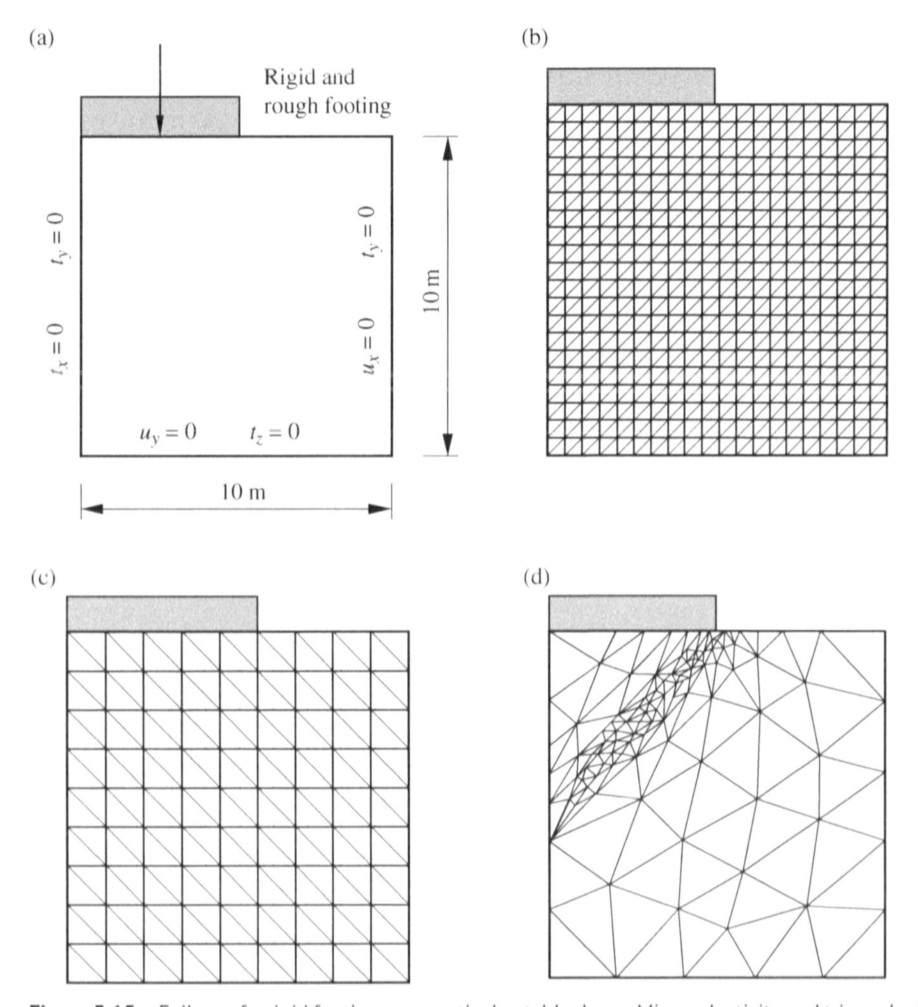

Figure 5.15 Failure of a rigid footing on a vertical cut. Ideal, von Mises, plasticity and triangular T6C/3C element (quadratic, continuous displacements, and linear continuous pressure) (a) geometry data; (b) Mesh 2 (fine "lucky"); (c) Mesh 3 (coarse, "bad"); (d) Mesh 6 (adaptive solution obtained from Mesh 3); (e) displacement vectors; (f) effective strain contour for Mesh 1 (fine, bad) (g) load, displacement results for various meshes. *Source:* Based on Zienkiewicz and Huang (1990).

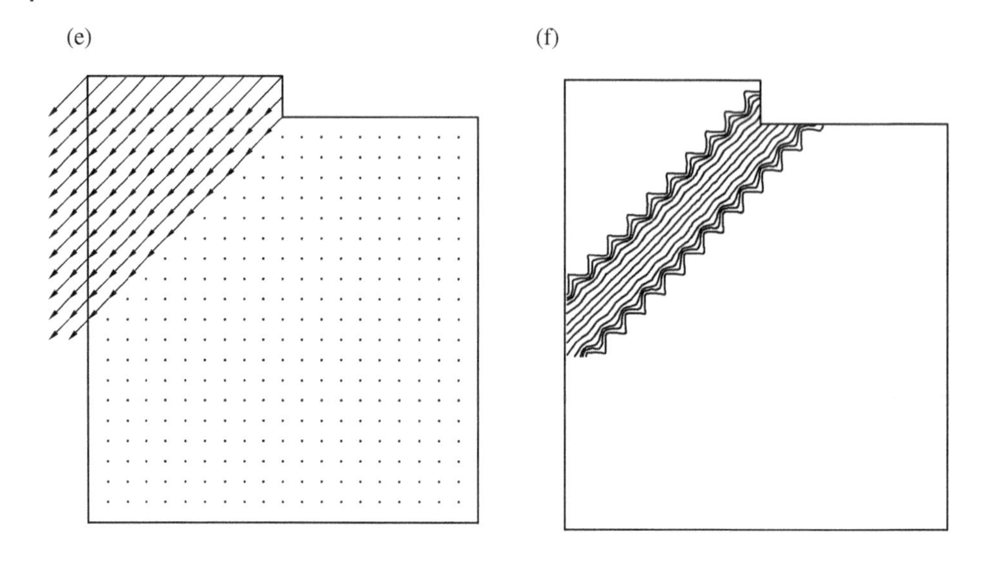

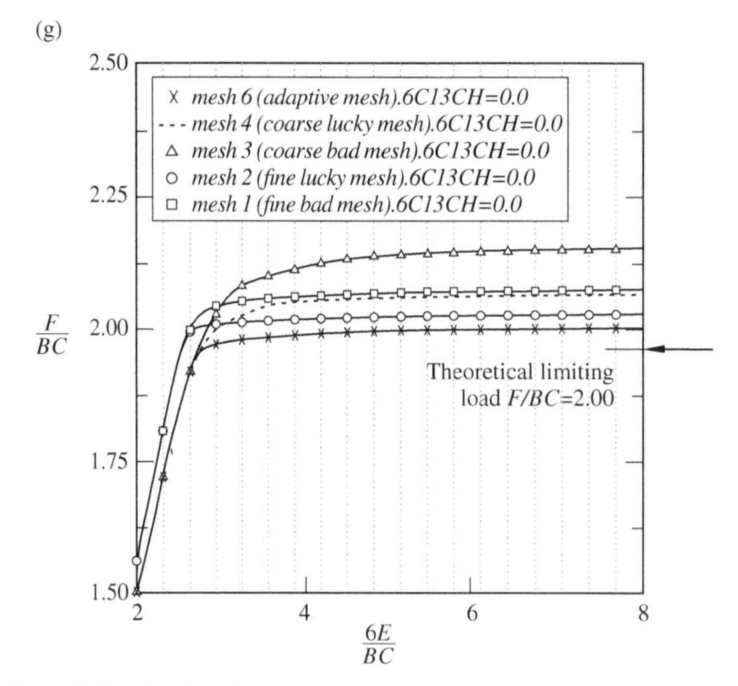

Figure 5.15 (Continued)

way where the time step length is adapted based on energy error measures of the jump of the solution at each time station, see Figure 5.17.

The governing equations are discretized in space as in Section 3.2.2. This results in the following system (dot represents time derivative) at the element level, where the overbar and the superscript w have been omitted for sake of simplicity

$$\begin{aligned}\mathbf{M}\ddot{\mathbf{u}} + \mathbf{K}\mathbf{u} - \mathbf{Q}\mathbf{p} &= \mathbf{f}^{(1)}\\ \mathbf{Q}\dot{\mathbf{u}} + \mathbf{H}\mathbf{p} + \mathbf{S}\dot{\mathbf{p}} &= \mathbf{f}^{(2)}\end{aligned} \tag{5.35}$$

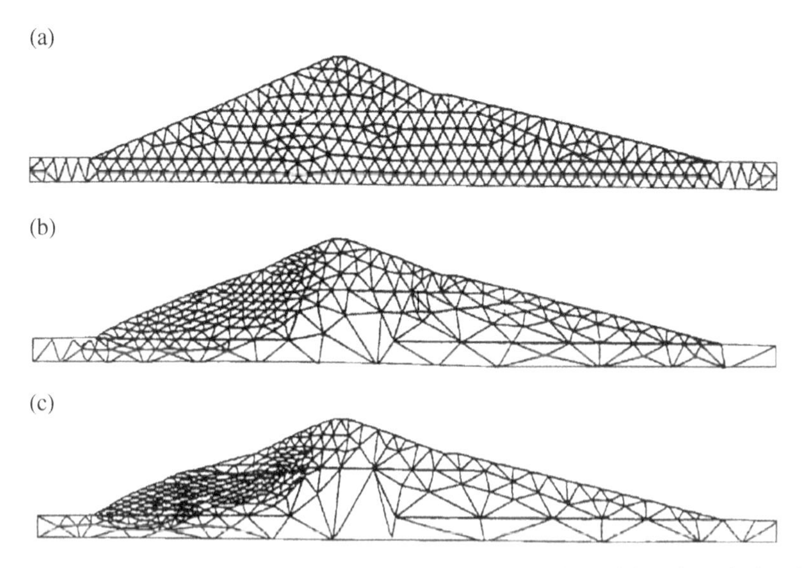

Figure 5.16 Earthquake analysis of lower San Fernando Dam (a) initial mesh; (b) adaptive refinement at t-75 seconds; (c) adaptive refinement at t = 30 seconds. *Source:* Based on Zienkiewicz and Huang (1990).

Figure 5.17 Discontinuous discretization in time with linear elements.

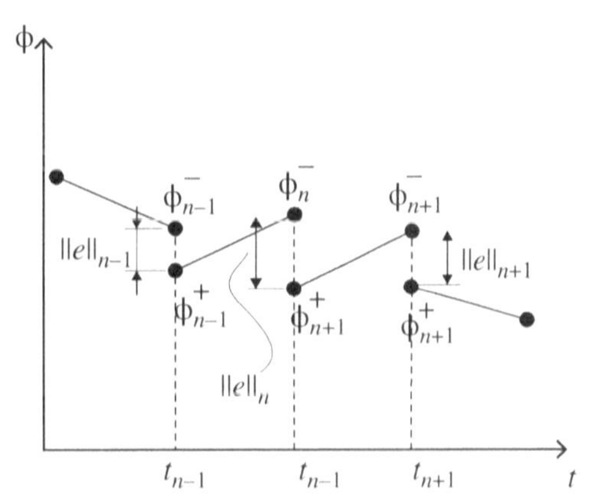

- unknown values ϕ are field variables at the beginning and at the end of the time step (displacements and pressure or derivative)
- $\|e\|_n$ is the 'jump'

Discretization in time is then performed with time discontinuous Galerkin approximation following Li and Wiberg (1996). Denoting with $I_n = \left(t_n^-, t_{n+1}^+\right)$ a typical incremental time step of size $\Delta t = t_{n+1} - t_n$, the weighted residual forms are:

$$\int_{In} \delta\mathbf{v}^{\mathrm{T}}\left(\mathbf{M}\dot{\mathbf{v}} + \mathbf{K}\mathbf{u} - \mathbf{Q}\mathbf{p} - \mathbf{f}^{(1)}\right)dt + \int_{In} \delta\mathbf{u}^{\mathrm{T}}\mathbf{K}(\dot{\mathbf{u}} - \mathbf{v})dt + \\ + \delta\mathbf{u}^{\mathrm{T}}\big|_{t_n}\mathbf{K}\left(\mathbf{u}_n^+ - \mathbf{u}_n^-\right)dt + \delta\mathbf{v}^{\mathrm{T}}\big|_{t_n}\mathbf{M}\left(\mathbf{v}_n^+ - \mathbf{v}_n^-\right) = \mathbf{0} \tag{5.36}$$

$$\int_{In} \delta\mathbf{p}^{\mathrm{T}}\left(\mathbf{Q}^{\mathrm{T}}\mathbf{v} + \mathbf{S}\mathbf{s} + \mathbf{H}\mathbf{p} - \mathbf{f}^{(2)}\right)dt + \int_{In} \delta\mathbf{p}^{\mathrm{T}}\mathbf{S}(\dot{\mathbf{p}} - \mathbf{s})dt + \\ + \delta\mathbf{p}^{\mathrm{T}}\big|_{t_n}\mathbf{S}(\mathbf{p}_n^{+} - \mathbf{p}_n^{+})dt = \mathbf{0} \tag{5.37}$$

with the constraint conditions

$$\dot{\mathbf{u}} - \mathbf{v} = \mathbf{0} \\ \dot{\mathbf{p}} - \mathbf{s} = \mathbf{0} \tag{5.38}$$

Superscripts −/+ indicate quantities immediately before and after the generic time station. Field variables and their first-time derivatives at the time $t \in [\, t_n, t_{n+1}\,]$ are interpolated by linear time shape functions:

$$\mathbf{u} = N_1(t)\mathbf{u}_n + N_2(t)\mathbf{u}_{n+1} \\ \mathbf{p} = N_1(t)\mathbf{p}_n + N_2(t)\mathbf{p}_{n+1} \\ \mathbf{v} = N_1(t)\mathbf{v}_n + N_2(t)\mathbf{v}_{n+1} \\ \dot{\mathbf{v}} = N_1(t)\dot{\mathbf{v}}_n + N_2(t)\dot{\mathbf{v}}_{n+1} \\ \mathbf{s} = N_1(t)\mathbf{s}_n + N_2(t)\mathbf{s}_{n+1} \tag{5.39}$$

Substituting equation (5.39) into (5.35), after simple manipulations, the following discretized equations are obtained:

$$\mathbf{u}_n = \mathbf{u}_n^{-} + \frac{\Delta t}{2}(\mathbf{v}_n - \mathbf{v}_{n+1}) \\ \mathbf{u}_{n+1} = \mathbf{u}_n^{-} + \frac{\Delta t}{2}(\mathbf{v}_n + \mathbf{v}_{n+1}) \\ \mathbf{s}_n = \frac{1}{\Delta t}(\mathbf{p}_{n+1} + 3\mathbf{p}_n - 4\mathbf{p}_n^{-}) \\ \mathbf{s}_{n+1} = \frac{1}{\Delta t}(\mathbf{p}_{n+1} + 3\mathbf{p}_n + 2\mathbf{p}_n^{-}) \tag{5.40}$$

$$\left(\frac{1}{2}\mathbf{M} - \frac{5}{36}\Delta t^2\mathbf{K}\right)\mathbf{v}_n + \left(\frac{1}{2}\mathbf{M} + \frac{1}{36}\Delta t^2\mathbf{K}\right)\mathbf{v}_{n+1} + \frac{\Delta t}{3}\mathbf{Q}\mathbf{p}_n + \\ + \frac{\Delta t}{6}\mathbf{Q}\mathbf{p}_{n+1} = -\frac{\Delta t}{2}\mathbf{K}\mathbf{u}_n^{-} + \mathbf{M}\mathbf{v}_n^{-} + \int_{I_n} N_1(t)\mathbf{f}_n^{(1)}dt \\ \left(-\frac{1}{2}\mathbf{M} - \frac{7}{36}\Delta t^2\mathbf{K}\right)\mathbf{v}_n + \left(\frac{1}{2}\mathbf{M} + \frac{5}{36}\Delta t^2\mathbf{K}\right)\mathbf{v}_{n+1} + \frac{\Delta t}{6}\mathbf{Q}\mathbf{p}_n + \\ + \frac{\Delta t}{3}\mathbf{Q}\mathbf{p}_{n+1} = -\frac{\Delta t}{2}\mathbf{K}\mathbf{u}_n^{-} + \int_{I_n} N_2(t)\mathbf{f}^{(1)}dt \\ \frac{\Delta t}{3}\mathbf{Q}^{\mathrm{T}}\mathbf{v}_n + \frac{\Delta t}{6}\mathbf{Q}^{\mathrm{T}}\mathbf{v}_{n+1} + \left(\frac{1}{2}\mathbf{S} + \frac{\Delta t}{3}\mathbf{H}\right)\mathbf{p}_n + \left(\frac{1}{2}\mathbf{S} + \frac{\Delta t}{6}\mathbf{H}\right)\mathbf{p}_{n+1} = \\ = \mathbf{S}\mathbf{p}_n^{-} + \int_{I_n} N_1(t)\mathbf{f}^{(2)}dt \\ \frac{\Delta t}{6}\mathbf{Q}^{\mathrm{T}}\mathbf{v}_n + \frac{\Delta t}{3}\mathbf{Q}^{\mathrm{T}}\mathbf{v}_{n+1} + \left(-\frac{1}{2}\mathbf{S} + \Delta t\mathbf{H}\right)\mathbf{p}_n + \left(\frac{1}{2}\mathbf{S} + \frac{\Delta t}{3}\mathbf{H}\right)\mathbf{p}_{n+1} = \\ = \int_{I_n} N_2(t)\mathbf{f}^{(2)}dt \tag{5.41}$$

The nodal displacement, velocity, and pressure $\mathbf{u}_n^-, \mathbf{v}_n^-, \mathbf{p}_n^-$, for the current step coincide with the unknowns at the end of the previous one, hence are known in the time-marching scheme and coincide with the initial condition for the first time step. The first remark is that the number of unknowns is doubled with respect to the one required by the traditional trapezoidal method.

The error of the time-integration procedure can be defined through the jump of the solution

$$
\begin{aligned}
&[\mathbf{u}_n] = \mathbf{u}_n - \mathbf{u}_n^-;\\
&[\mathbf{v}_n] = \mathbf{v}_n - \mathbf{v}_n^-\\
&[\mathbf{p}_n] = \mathbf{p}_n - \mathbf{p}_n^-\\
&\mathbf{u}_n = \mathbf{u}_n - \mathbf{u}_n^-\\
&\mathbf{v}_n - \mathbf{v}_n - \mathbf{v}_n^-\\
&\mathbf{p}_n = \mathbf{p}_n - \mathbf{p}_n^-
\end{aligned}
\tag{5.42}
$$

at each time station, i.e. the difference between the final point of time step *n–1* and the first point of time step *n*. By adopting the total energy norms as an error measure, we define the following terms:

$$
\begin{aligned}
&\|\mathbf{e}_u\|_n = \sqrt{\mathbf{v}_n^{\mathrm{T}}\mathbf{M}\mathbf{v}_n + \mathbf{u}_n^{\mathrm{T}}\mathbf{K}\mathbf{u}_n}\\
&\|\mathbf{e}_{u,p}\|_n = \sqrt{\mathbf{u}_n^{\mathrm{T}}\mathbf{Q}\mathbf{p}_n}\\
&\|\mathbf{e}_p\|_n = \sqrt{\mathbf{p}_n^{\mathrm{T}}\mathbf{Q}^{\mathrm{T}}\mathbf{u}_n + \mathbf{p}_n^{\mathrm{T}}\mathbf{H}\mathbf{p}_n\Delta t + \mathbf{p}_n^{\mathrm{T}}\mathbf{S}\mathbf{p}_n}\\
&\|\mathbf{e}\|_n = max\left\{\|\mathbf{e}_u\|_n, \|\mathbf{e}_{u,p}\|_n, \|\mathbf{e}_p\|_n\right\}
\end{aligned}
\tag{5.43}
$$

Error measures defined in Equation (5.43) account at the same time for the cross effects among the different fields and the ones between space and time discretizations.

The relative error is defined as in Li and Wiberg (1996)

$$
\eta_n = \frac{\|\mathbf{e}\|_n}{\|\mathbf{e}\|_{\max}}
\tag{5.44}
$$

where $\|\mathbf{e}\|_{\max}$ is the maximum total energy norm:

$$
\|\mathbf{e}\|_{\max} = max\left(\|\mathbf{e}\|_i\right), \qquad 0 < i < n
\tag{5.45}
$$

When $\eta > \eta_{toll}$, the time step Δt_n is modified and a new $\Delta t_n^{'} < \Delta t_n$ is obtained according to the following rule:

$$\Delta t_n^{'} = \left(\frac{\theta \eta_{tol}}{\eta}\right)^{1/3} \Delta t_n \qquad (5.46)$$

where $\theta < 1.0$ is a safety factor. If the error is smaller than a defined value $\eta_{toll,\ min}$ the step is increased using a rule similar to Equation (5.46). An example with a 3-D consolidation is shown in Section 6.4.3. In that case, we assume $\theta = 0.95$ and $\eta_{toll} = 0.05$. It can be seen that the spurious oscillations at the onset of the consolidation process (see also Section 5.5), which appear especially if the incompressible limit is approached can be efficiently eliminated by using DGT.

5.4.3 Localization and Strain Softening: Possible Nonuniqueness of Numerical Solutions

Strain-softening behavior is a phenomenon frequently encountered in soils and invariably it leads to a very localized, sliding surface type of deformation. This is well exhibited in the so-called "slickensides," frequently observed in clays.

The analysis of plasticity problems with a negative hardening (or softening) modulus, H, is in itself a complex task, but the basic difficulties have been overcome many years ago and are described in Zienkiewicz et al. (2005a). However, the reason for localization only becomes clear if some specific cases are examined. Consider, for example, the behavior of a one-dimensional, bar-type, problem illustrated in Figure 5.18.

In this example, we assume plastic, softening, behavior, and consider the analysis of the bar divided into a number of equal elements of length "h". Further, statistically, we perturb the yield stress in an arbitrary element so that only that element yields when the load is applied and thus gives the peak yield stress. If the extension imposed on the specimen continues beyond that peak, then only that one element shows the plastic deformation, all others unloading elastically. Depending on the ratio of the element size h to that of the total length ℓ, a progressively steeper unloading branch of the load-deformation plot will occur. This can reach negative slope values and allows only elastic unloading which, obviously, is not correct. But, certainly, the steepening of this slope will increase to infinity for a finite value of h and will imply a displacement discontinuity or full localization.

In the example quoted, the localization was caused by a small weakness due to the statistical nature of the material strength behavior. However, in other geometrically more complex problems, the stress concentration, etc., will act in precisely the same manner, always causing a localization with softening material behavior.

However, the example discussed shows up another feature of the problem, i.e. that of numerical nonuniqueness as the slope of the unloading portion of the displacement load curve depends largely on the size of the element used.

This nonuniqueness of the problem becomes most serious in multidimensional behavior in many structural problems. In Figure 5.19a, we illustrate the fairly large discrepancies which occur in the estimate of the maximum load for the problem illustrated in

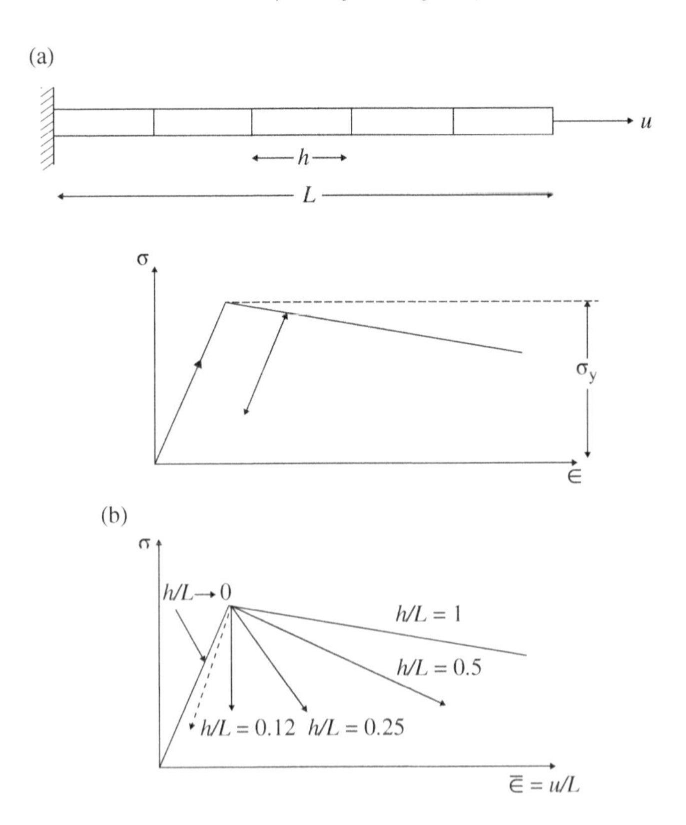

Figure 5.18 Nonuniqueness – mesh size dependence in the extension of a homogeneous bar with a strain-softening material (peak value of yield stress σ, perturbed in a single element). (a) stress σ versus strain ε for material; (b) stress $\bar{\sigma}$ versus average strain $\bar{\varepsilon} = u/L$ assuming yielding in a single element of length h. *Source:* Based on Zienkiewicz and Huang (1990).

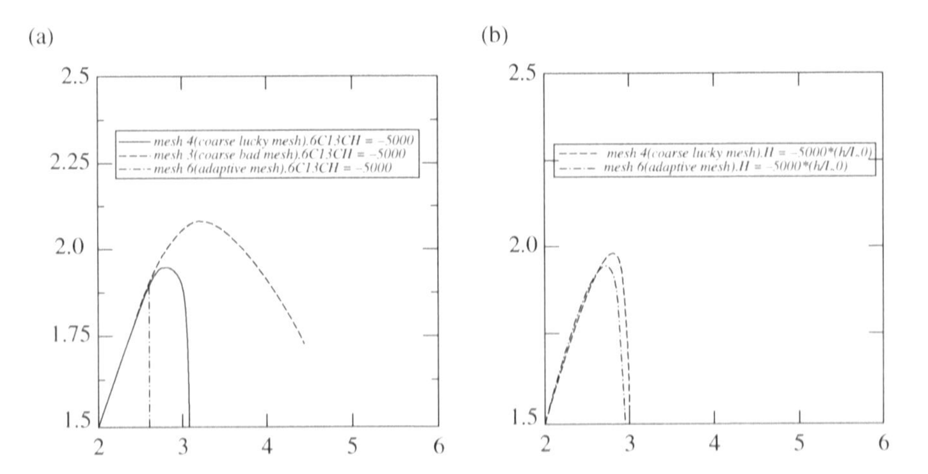

Figure 5.19 Strain softening (H = −5000): comparison of reaction vs. prescribed displacement (horizontal axis) for various meshes using T6C/3C element. (a) constant plastic modulus; (b) mesh-dependent plastic modulus. *Source:* Based on Zienkiewicz and Huang (1990).

Figure 5.15 for which now a softening modulus has been assumed and different mesh subdivisions used in the solution.

Even discounting the results obtained by the use of the coarse, bad, mesh as being very unreliable, we note a difference of about 20% in the estimate of the maximum load capacity when the simulation is achieved by meshes which for ideal plasticity give almost identical answers. While the reason for this has been hinted at it in the simple example of Figure 5.18, the manner in which the problem can be overcome has supplied many researchers with material for exercising their ingenuity. De Borst et al. (1993), Ortiz et al. (1987), Bazant and Lin (1988), Belytschko et al. (1988), and Belytschko and Tabarrok (1993) describe some of the possible procedures which range from the consideration of material as a Cosserat medium, through the so-called gradient plasticity (see Section 5.4.4), to a simple failure energy consideration introduced in the last of these references. We shall only refer here to that last procedure which, in the opinion of the authors, deals adequately and in a simple manner with the difficulties encountered.

The procedure considers, in the manner common to that of early theories of fracture, namely Griffiths (1921), the constancy of work required for failing the material and requires the energy to be independent of the discretization used and therefore to be a pure material property.

In Figure 5.20, we show a typical stress–strain relation with strain-softening in which failure is reached. The area under the full triangle is the work required to cause this failure and, in a unit volume of material, becomes

$$\frac{1}{2}\varepsilon_u \sigma_y \equiv \frac{1}{2}\frac{\sigma_y^2}{H} \tag{5.47}$$

where H is the softening modulus. If an element of size h in the direction of maximum straining is to model failure correctly, the work requirement to fail a unit width of the element which must be kept constant is

$$\frac{1}{2}\frac{\sigma_y^2 h}{H} = \text{constant} \tag{5.48}$$

This would be invariant only if

$$\frac{h}{H} = C \tag{5.49}$$

where C is a constant.

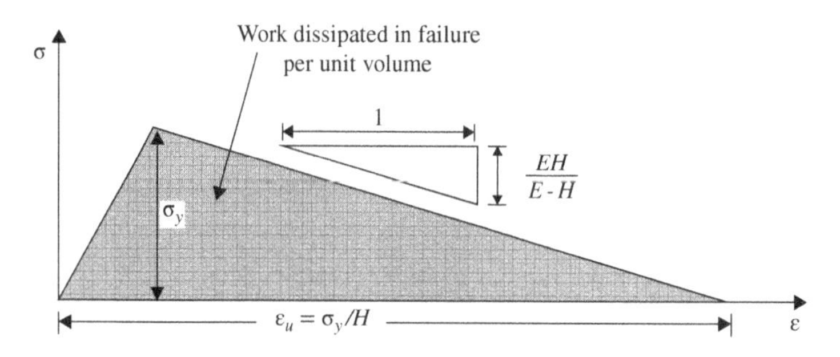

Figure 5.20 Work dissipation in failure of the material. *Source:* Based on Zienkiewicz and Huang (1990).

It appears therefore necessary to reduce the softening modulus in the manner of (5.49) as the size of the elements modeling the localization decreases. This indeed was done in the case of a problem illustrated in Figure 5.15a and the results are shown in Figure 5.19a and b, which gives an almost identical failure load obtained by two very different meshes.

It will be observed that the above discussion leads to two conclusions:

i) that with strain softening, localization will always occur in the failure zone and this will show a continuous decrease in size with the element size h, and
ii) that the softening modulus cannot remain a material constant but must tend to zero (i.e. giving no softening) as the size of the element also tends to zero to present a consistent work estimation. This idea can be incorporated in a material model with concentrated localization singularity and has been introduced by Simo et al. (1993) and Oliver (1995). It is clear that such a model will, in the limit, tend to give identical results to the adaptive refinement if equation (5.49) is used.

The adaptive refinements of the type here discussed have been introduced by Zienkiewicz et al. (1995a, 1995b) from which the examples and previous figures have been quoted. This and other papers in the field indicate that adaptive refinement is a feature that can improve the results of analysis significantly, although, with experience, reasonable engineering estimates can be obtained without this feature.

5.4.4 Regularization Through Gradient-Dependent Plasticity[5]

In the previous section, strain localization was addressed in general and the example shown refers to a single-phase material. Mesh dependency of the solution in case of material softening was pointed out and some solutions to overcome this problem were hinted at. We focus now our attention on the case of porous media. Rice (1975) and Rudnicki (1984) investigated the stability of saturated inelastic porous media in the quasi-static situation. Strain localization in porous media is however most appropriately studied as a dynamic problem, i.e. a problem of wave propagation. This allows the clarification of the reason for the occurrence of mesh dependence which is due to the local change of the character of the governing equations resulting in a loss of their well-posedness when instability occurs. This means that the wave speed becomes imaginary and, consequently, a solution without physical significance is obtained in dynamic localization problems. Further, the model does not contain an internal length scale and the corresponding numerical finite element solutions show pathological mesh dependence. This has been elucidated by Read and Hegemier (1984), Lasry and Belytschko (1988), and de Borst et al. (1993).

Dynamic strain localization in fully saturated porous media was studied by Loret and Prevost (1991) in a large mass of saturated soil in a high-frequency situation using a u-U model. The numerical solution was regularized by means of a rate-dependent model (viscoplasticity).

Gradient dependence of the constitutive model was recognized to regularize the solution in single-phase materials by Muhlhaus and Aifantis (1991), de Borst and Muhlhaus (1992), and Sluys et al. (1993). In fact, the use of a higher-order gradient model results in a well-posed set of partial differential equations. From a dispersion analysis, it is clear that the governing equations remain hyperbolic under strain-softening states and the continuum is capable of transforming a traveling wave into a stationary localization wave. An internal length scale exists in the model and the numerical results do not suffer from pathological mesh dependence.

A gradient-dependent constitutive model was adopted by Zhang and Schrefler (2000, 2004) for fully and partially saturated porous media using a u-p approach and an internal length scale l_g was derived. The **u**-p approach has the advantage to evidence directly the onset of cavitation in the case of dense geomaterials [see Schrefler et al. (1995)]. In multiphase porous media, the situation is complicated with respect to single-phase media because it was found that in the former, the Laplacian existing in the mass balance equation when Darcy's law is used, provides already regularization for axial waves and appropriate wave numbers, with a proper length scale l_w which for small permeability reads (Zhang et al. 1999a)

$$l_w = \frac{2c_m\eta}{kK^2\alpha^2{S_w}^2Q} \tag{5.50}$$

where

$$c_m = \sqrt{\frac{E_2}{\rho}}, \quad \eta = \frac{E_2}{Q}, \quad E_2 = \alpha^2{S_w}^2Q + h$$

$$Q^{-1} = \left[\frac{n\partial S_w}{\partial p} + \frac{nS_w}{K_w} + \frac{S_w(\alpha - n)}{K_s}\left(S_w + \frac{\partial S_w}{\partial p}p\right)\right]$$

and K is the wave number; for an estimate, put $K = 1$.

The two length scales are competing ones and may interact during shear band development as studied extensively by Zhang and Schrefler (2006). This aspect has been confirmed by Mroginski and Etse (2014).

However, to achieve regularization in all situations, including quasi-static ones, gradient-dependent plasticity is the best choice according to the authors. A link with experimental observation was shown by Vardoulakis and Sulem (1995), where the material length introduced directly in a gradient-dependent model was estimated by analyzing some experimental results from a biaxial test on fine-grained sand. The material length was related to the mean grain diameter $d_{50\%}$. Otherwise, an inverse analysis could be done to relate the gradient parameter to the bandwidth.

The above-mentioned aspects will be clarified on a one-dimensional bar of gradient-dependent water-saturated porous medium under dynamic compressive loading. This analysis illustrates the dispersive character of the gradient-dependent multiphase model and the influence of permeability.

The geometrical, material, and loading data for the bar are given in Figure 5.21. von Mises plasticity with gradient-dependence is used for simplicity, which reduces for the one-dimensional problem to

$$f = |\sigma_{xx}| - \sigma_s - h\chi + c^* \nabla^2\chi \tag{5.51}$$

where χ is the equivalent plastic strain, h the hardening modulus, and c^* a material constant.

First, we investigate the behavior of the bar with respect to mesh refinement with a dynamic permeability $k = 10^{-12}$ m^3s/kg (small permeability), and $c^* = 0.4$ MN. For the bar, we use 20, 30, 40, 50, and 60 linear elements, respectively. We consider a block wave ($t_0 = 0$) traveling through the bar in a linearly elastic regime until reflection occurs and the stress increase causes the initiation of the localization process. Extra boundary conditions $\partial\chi/\partial x = 0$ at both sides of the bar are used.

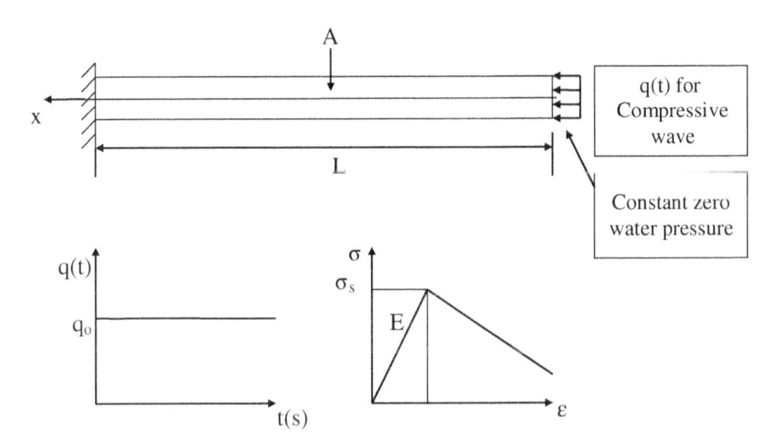

Figure 5.21 One-dimensional soil bar in pure compressive loading

Because a very small permeability is adopted ($k \to 0$), the width of the plastic zone is based on the gradient plasticity regularization and can be obtained from the wavelength with wave number given by K_{up} (Zhang and Schrefler 2000)

$$K_{up} = \sqrt{-\frac{h + h'}{c^*}}, \quad K_{lower} = \sqrt{-\frac{h}{c^*}}, \quad h' = \frac{E\alpha^2 S_w^2 Q}{E + \alpha^2 S_w^2 Q} \tag{5.52}$$

K_{lower} is the wave number limit in case of high permeability ($k \to \infty$).

Using the selected material parameters, the width of the localization zone will be equal to $l_g = 3.176\text{m}$. We use the term "plastic" or "localization zone" and not "shear band" because the involved zone is rather large, as explained below.

The strain profiles plotted for different meshes in Figure 5.22 demonstrate the convergence of the solution to a finite width of the localization zone for compressive waves. This width is half of the length scale l_g predicted above because of the boundary conditions, i.e. only half of a symmetric problem is investigated.

The interaction of the two length scales l_g and l_w in the plastic strain distribution along the bar are shown next. The results obtained with $k = 10^{-6}$ m^{3}s/kg, $k = 10^{-4}$ m^{3}s/kg, and different gradient-dependent parameters are shown in Figures 5.23 and 5.24, respectively. It can be observed that in the case of smaller permeability (Figure 5.23), the width of the plastic zone during the computational process is mainly governed by the seepage regularization because it is independent of the gradient parameter, while in the case of larger permeability, see Figure 5.24, the width of the plastic zone is controlled by gradient-dependent plasticity.

In Figure 5.23, the independence of the width of the plastic zone from the gradient-dependent parameter is due to the fact that the results are not sensitive to relatively small gradient-dependent parameters. For $c^* = 0$, the ratio between the width of the plastic zone of Figures 5.23 and 5.24 is not 100 (ratio of permeabilities) because in Figure 5.24, the smallest width is limited by the element size where we adopted only 40 elements. Once the length is larger than the element size, the width of the plastic zone in Figure 5.24 is consistent with the prediction for K_{lower} of (5.52) (half of the value) due to the use of a larger permeability.

For an intermediate value of the permeability, the problem becomes complicated because the width of the localization zone will be somewhere in the region between the lower and upper bounds of (5.52).

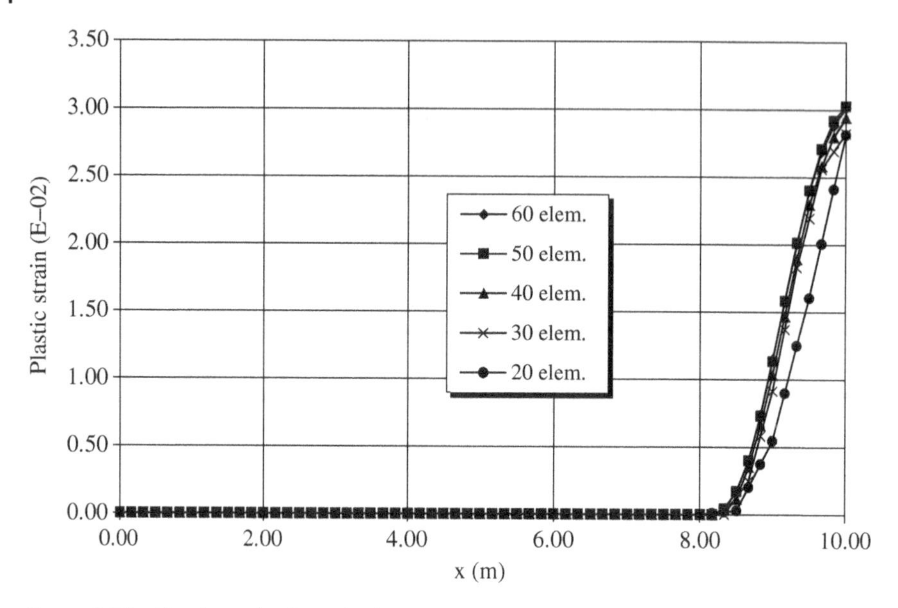

Figure 5.22 Plastic strain along the bar using the gradient-dependent porous media model for different meshes (permeability $k = 10^{-12}$ m^{3}s/kg, $c^* = 0.4$ MN).

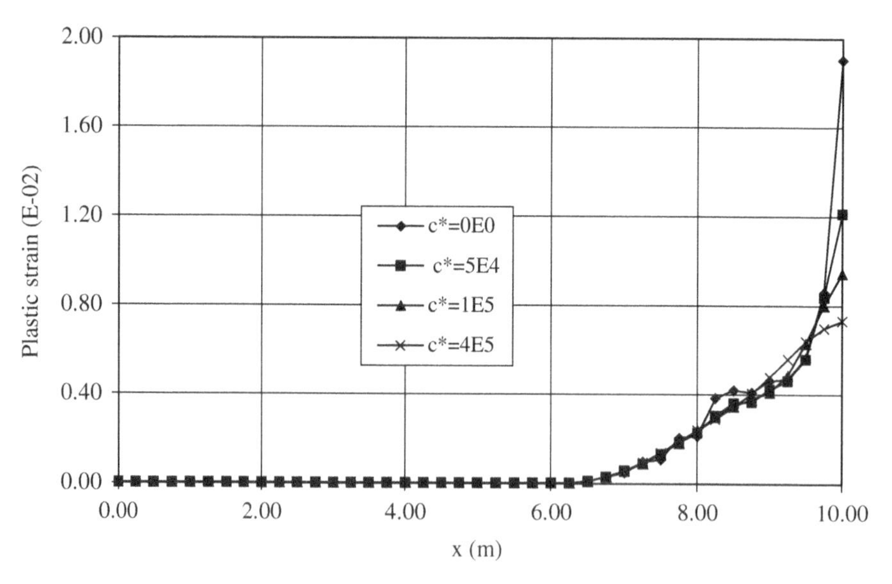

Figure 5.23 Distribution of plastic strain along the bar with different values of c^* [N] and permeability of $k = 10^{-6}$ m^{3}s/kg

The width of the plastic zone in the numerical examples depends on the length scales l_g and/or l_w of the model. It appears that these scales are dependent on the material parameters. In fact, l_g is the square root of the ratio between c^* and h and l_w is, in particular, inversely proportional to the permeability of the medium. A ratio between the thickness of the plastic zone and the length of the bar of (0.1–0.2) has been obtained on several

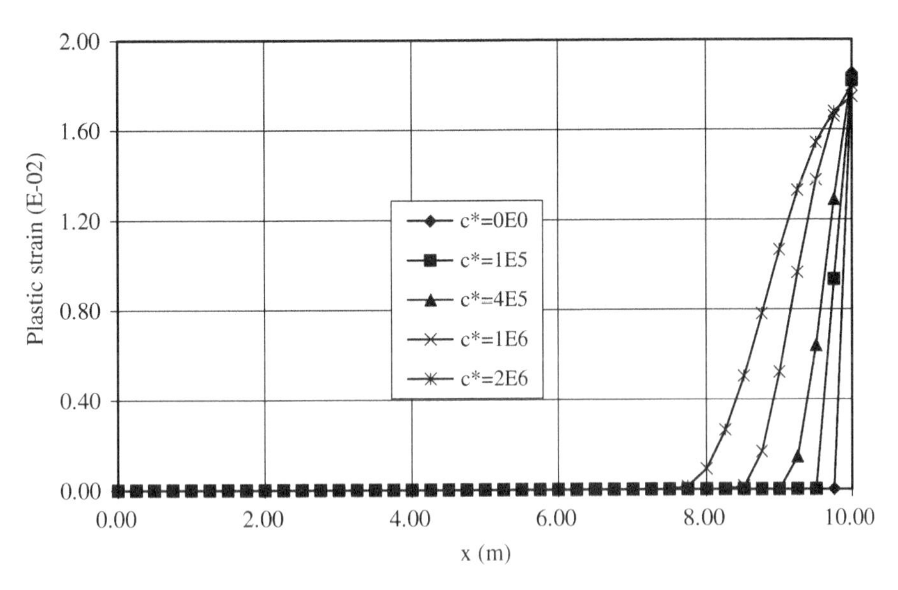

Figure 5.24 Distribution of plastic strain along the bar with different values of c^* [N] and permeability $k = 10^{-4}$ m^{3}s/kg

occasions in the case of single-phase gradient-dependent models and the width we obtained is not in contradiction with what can be observed in nature, where the shear bandwidth in soils also depends on the scale of the problem. For the common laboratory tests, it is usually related to the grain size (about 15–30 times the mean grain diameter $d_{50\%}$) as shown by Vardoulakis and Sulem (1995), while in case of landslides or at geological level, shear bands or faults can usually be of the order of a meter. In fact, for the well-known Vajont landslide in Italy (see Chapter 1, Hendron and Patton (1985)) identified a shear band made of multiple clay layers varying from 1 to 3 m in thickness along much of the surface of sliding.

The examples shown are fully saturated. If partial saturation occurs during strain localization, as in the case of initially water-saturated dense sands under globally undrained conditions, a strong reduction of the seepage process is expected and, hence, the partially saturated zones behave similarly to single-phase materials. The fluid regularization then does not play a role anymore and an internal length scale should be introduced in the multiphase model for the description of its behavior during strain localization. A 2-D example with large strains of an initially fully saturated sample undergoing desaturation upon localization has been presented by Sanavia et al. (2002). Cavitation itself linked to localization has been modeled by Gawin et al. (1998). Finally, a different procedure is shown in Section 5.5.7.

5.5 Stabilization of Computation for Nearly Incompressible Behavior with Mixed Interpolation[6]

5.5.1 The Problem of Incompressible Behavior Under Undrained Conditions

In Chapters 2 and 3, we have already mentioned the difficulties which can be encountered when the standard finite element approximation is used to model incompressible, or nearly incompressible, behavior. Such behavior will be attained when the permeability is very small and when the compressibility ($1/Q$) decreases. In other words, this happens when the elastic bulk modulus of the pore water is very high. However, in the $\mathbf{u}$–p approximation, we have already mentioned that a satisfactory behavior of the solution can be obtained under all circumstances when the Babuska–Brezzi condition or the equivalent of the patch test is satisfied. Some admissible interpolations are shown in both Chapters 2 and 3 and in the two examples of Section 5.5.3, we shall show how the unstable behavior of the "illegal" Q4/P4 element can be eliminated by the use of the acceptable Q9/P4 interpolation. In the first element of quadrilateral form, a bilinear interpolation is used for both the $\mathbf{u}$ and p variables, while in the second element, a quadratic approximation is used for the displacement.

While the use of such correct interpolations is desirable and we have based our code on this assumption, much research effort has been devoted to the introduction of stabilizing procedures that would allow arbitrary interpolation (say equal interpolation) of both variables to be used effectively. Such stabilization can, without doubt, lead to more efficient and simple formulations and the paper by Pastor et al. (1999) shows the various approaches suggested in the literature. Here the work of Schneider et al. (1978), Brezzi and Pitkaranta (1984), Hughes et al. (1986), Hafez and Soliman (1991), and De Sampaio (1991) suggests many alternatives. Some of these were shown by Zienkiewicz and Wu (1991) to derive very simply from the same roots of time-stepping analysis.

The motivation for most of this work lies in problems of fluid mechanics and their numerical solution and it was shown by Zienkiewicz and Codina (1995) that an algorithm using the operator split procedure suggested by Chorin in 1967 automatically provides the desired stabilization. The use of such stabilization in the context of geomechanics was first made by Zienkiewicz and Wu (1994) and extended by Pastor et al. (1999).

In addition to this method, it is worth mentioning the contributions by Mira et al. (2003), Salomoni and Schrefler (2005, 2006), Truty and Zimmermann (2006), and White and Borja (2008).

In this chapter, we shall discuss the fractional step method which is valid for both static and dynamic problems, together with two methods proposed by the authors for consolidation phenomena.

5.5.2 The Velocity Correction and Stabilization Process

In this section, we shall outline the semi-explicit time-stepping, operator-split procedure which is effective in dealing with the incompressibility problems arising in geomechanics and which follows the methodology originally suggested by Chorin (1967, 1968) and extended by Zienkiewicz and Codina (1995).

It is convenient to introduce the velocity, $\mathbf{v}$, as the basic variable and to compute the displacement increment by subsequent integration. Thus, we have the definition

$$\mathbf{v} = \frac{\mathrm{d}\mathbf{u}}{\mathrm{d}t} \tag{5.53}$$

and in each time step, it is simple to establish

$$\Delta\mathbf{u} \equiv \mathbf{u}^{n+1} - \mathbf{u}^{n} = \left(\frac{\mathbf{v}^{n+1} + \mathbf{v}^{n}}{2}\right)\Delta t \tag{5.54}$$

once the value of $\mathbf{v}^{n+1}$ has been computed.

The starting points for the development of the algorithm are equations (2.20) and (2.21) of Chapter 2 rewritten in terms of effective stresses with $\alpha = 1$, again neglecting the source term and introducing the new variable $\mathbf{v}$. We can write these governing equations as

$$\rho\dot{\mathbf{v}} = \mathbf{S}^{\mathrm{T}}\boldsymbol{\sigma}' - \nabla p + \rho\mathbf{b} \tag{5.55a}$$

$$\frac{1}{Q}\dot{p} = \nabla^{\mathrm{T}}k\nabla p - \nabla^{\mathrm{T}}\mathbf{v} \tag{5.55b}$$

noting that

$$\boldsymbol{\sigma} = \boldsymbol{\sigma}' - \mathbf{m}p \tag{5.56}$$

and with

$$\mathrm{d}\boldsymbol{\sigma}' = \mathbf{D}\mathrm{d}\boldsymbol{\varepsilon} \tag{5.57}$$

as constitutive relation and that (3.9) needs to be used for strain calculation.

The operator split algorithm solves (5.55a) in two steps. In the first part, the quantity $\mathbf{v}^*$ is calculated explicitly from

$$\rho\frac{\mathbf{v}^* - \mathbf{v}^n}{\Delta t} = \mathbf{S}^{\mathrm{T}}\boldsymbol{\sigma}' + \rho\mathbf{b} \tag{5.58}$$

where the RHS is computed at $t = t_n$. In the second part, the velocity $\mathbf{v}^*$ is corrected implicitly in terms of known pressures using

$$\rho\frac{\mathbf{v}^{n+1} - \mathbf{v}^*}{\Delta t} = -\nabla p^{n+\theta_2} = -\nabla(p^n + \theta_2\Delta p) \tag{5.59}$$

The above can be evaluated only after Δp is established if $\theta_2 \neq 0$. In what follows, we shall use $\theta_2 = ½$ for good accuracy but any values of it in the range

$$0 < \theta_2 \leq 1 \tag{5.60}$$

can be chosen, provided that Δt satisfies certain stability limits.

Equations (5.58) and (5.59) must be discretized in space before proceeding with numerical calculations. Following the standard procedures of Chapter 3 with

$$\mathbf{u} = \mathbf{N}\overline{\mathbf{u}} \quad \mathbf{v} = \mathbf{N}\overline{\mathbf{v}} \quad p = \mathbf{N}^{\mathbf{p}}\overline{\mathbf{p}} \tag{5.61}$$

we obtain the following after application of the Galerkin process.

$$\mathbf{M}(\overline{\mathbf{v}}^* - \overline{\mathbf{v}}^n) = \Delta t\left(\left(\int_\Omega \mathbf{B}^{\mathrm{T}}\boldsymbol{\sigma}'\mathrm{d}\Omega - \mathbf{f}'\right)\right)^n \tag{5.62}$$

and

$$\mathbf{M}\left(\overline{\mathbf{v}}^{n+1} - \overline{\mathbf{v}}^*\right) = \Delta t\mathbf{Q}(\overline{\mathbf{p}}^n + \theta_2\Delta\overline{\mathbf{p}}) \tag{5.63}$$

All these matrices are defined in Chapter 3 in (3.24–3.26) and need not be repeated here. We must, however, mention that the evaluation of both $\overline{\mathbf{v}}^*$ and $\overline{\mathbf{v}}^{n+1}$ is fully explicit if the mass matrix $\mathbf{M}$ is diagonalized. This can be done in a variety of ways by well-known procedures discussed in finite element texts (see, for example, Zienkiewicz et al. (2005b)).

The determination of the pressure increment Δp and hence of p^{n+1} requires the solution of (5.55b). We now write the implicit time approximation as

$$\frac{1}{Q}\frac{\Delta p}{\Delta t} = \nabla^{\mathrm{T}}(k\nabla p)^n - \nabla^{\mathrm{T}}\mathbf{v}^{n+\theta_1} \tag{5.64}$$

Here various values of θ_1 can be used but

$$\theta_1 = 1 \tag{5.65}$$

is particularly convenient and accurate. We must remark that with $\theta_1 \leq \frac{1}{2}$ no stable solution is possible.

Using the computed values of $\mathbf{u}^*$ and (5.59), we can rewrite (5.64) as

$$\frac{1}{Q}\frac{\Delta p}{\Delta t} = \nabla^{\mathrm{T}}(k\nabla p)^n - \nabla^{\mathrm{T}}\left(\mathbf{v}^* + \frac{\Delta t}{\rho}\nabla(p^n + \theta_2\Delta p)\right) \tag{5.66}$$

from which Δp can be established after discretization. This again proceeds in the manner previously described and we now have

$$\left(\tilde{\mathbf{S}} + \frac{\Delta t^2\theta_2}{\rho}\mathbf{H}^*\right)\frac{\Delta\bar{\mathbf{p}}}{\Delta t} = -\mathbf{H}\bar{\mathbf{p}}^n - \mathbf{Q}^{\mathrm{T}}\bar{\mathbf{v}}^* - \frac{\Delta t\theta_2}{\rho}\mathbf{H}^*\bar{\mathbf{p}}^n - \mathbf{f}^{(2)} \tag{5.67}$$

In the above, the matrices $\tilde{\mathbf{S}}$, $\mathbf{H}$ and $\mathbf{Q}$ are defined in Chapter 3 by (3.29–3.32). The only new matrix occurring now is $\mathbf{H}^*$ which is the approximation to the Laplacian operator.

$$\nabla^{\mathrm{T}}\nabla = \nabla^2 \tag{5.68}$$

By the usual procedures, we find

$$\mathbf{H}^* \equiv \int_\Omega (\nabla\mathbf{N}^p)^{\mathrm{T}}\nabla\mathbf{N}^p \mathrm{d}\Omega \tag{5.69}$$

We shall delay the explanation of the reasons why the split operator procedure permits the use of arbitrary interpolations for $\mathbf{u}$ and p ($\mathbf{N}$ and $\mathbf{N}^p$ respectively) and shall first illustrate its effectiveness in examples.

5.5.3 Examples Illustrating the Effectiveness of the Operator Split Procedure

Two examples are quoted here. The first of these is the soil layer subject to a periodic surface load. Indeed this problem is identical to the one used in Section 2.2.3 where the limits of applicability of various formulations are tested and for which exact solutions are readily available. Here we shall only use the $\mathbf{u}$–p formulation and shall demonstrate how the very oscillatory results obtained by an equal interpolation can be improved by the use of the stabilization just described.

In Figure 5.25, we show the details of the problem and in Figure 5.26, we show solutions obtained by the use of 2D elements. The first (a) uses 20 Q4P4 elements and shows oscillations which are very pronounced. The second one (b) shows the solution obtained with the classical Q8P4 element. Finally, the third (c) shows the very close approximation and suppression of oscillation obtained using the Q4P4 element as well as the new stabilizing algorithm.

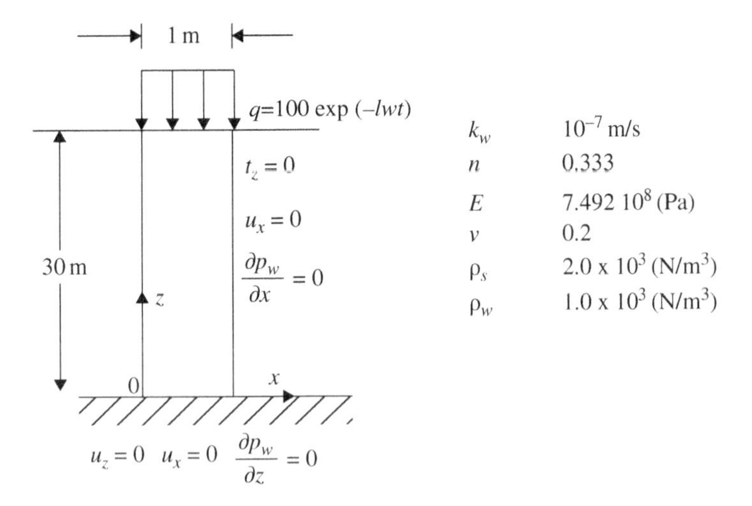

The height of the column has been taken as $L = 30$ m and the excitation frequency chosen is $\omega = 3.379$ rad/s

Figure 5.25 Example 1. A saturated soil layer under a periodic load.

In Figures 5.27 and 5.28, a fully two-dimensional problem of a foundation load is solved again showing similar results.

5.5.4 The Reason for the Success of the Stabilizing Algorithm

In Chapter 3, we have indicated the main reasons for the difficulties encountered in solving the problem where incompressibility is approached. We first made a comment on these difficulties when discussing the Jacobian matrix used in the solution of an iterative step by the Newton–Raphson procedure where the matrices of (3.31) and (3.30) tend to zero, i.e. when

$$\tilde{\mathbf{S}} \equiv \int_{\Omega} \mathbf{N}^{p^{\mathrm{T}}} \frac{1}{Q} \mathbf{N}^{p} \mathrm{d}\Omega \to 0 \tag{5.70a}$$

and

$$\mathbf{H} \equiv \int_{\Omega} (\nabla \mathbf{N}^{p})^{\mathrm{T}} k \nabla \mathbf{N}^{p} \mathrm{d}\Omega \to 0 \tag{5.70b}$$

which occurs when the compressibility and the permeability of both tend to zero.

This zero limit leads to a zero diagonal which appears also in steady-state equations of Section 3.2.5 giving a linear form

$$\begin{bmatrix} \mathbf{K} & -\tilde{\mathbf{Q}} \\ -\tilde{\mathbf{Q}}^{\mathrm{T}} & \mathbf{0} \end{bmatrix} \begin{Bmatrix} \bar{\mathbf{u}} \\ \bar{\mathbf{p}} \end{Bmatrix} = \begin{Bmatrix} \mathbf{f}^{(1)} \\ 0 \end{Bmatrix} \tag{5.71}$$

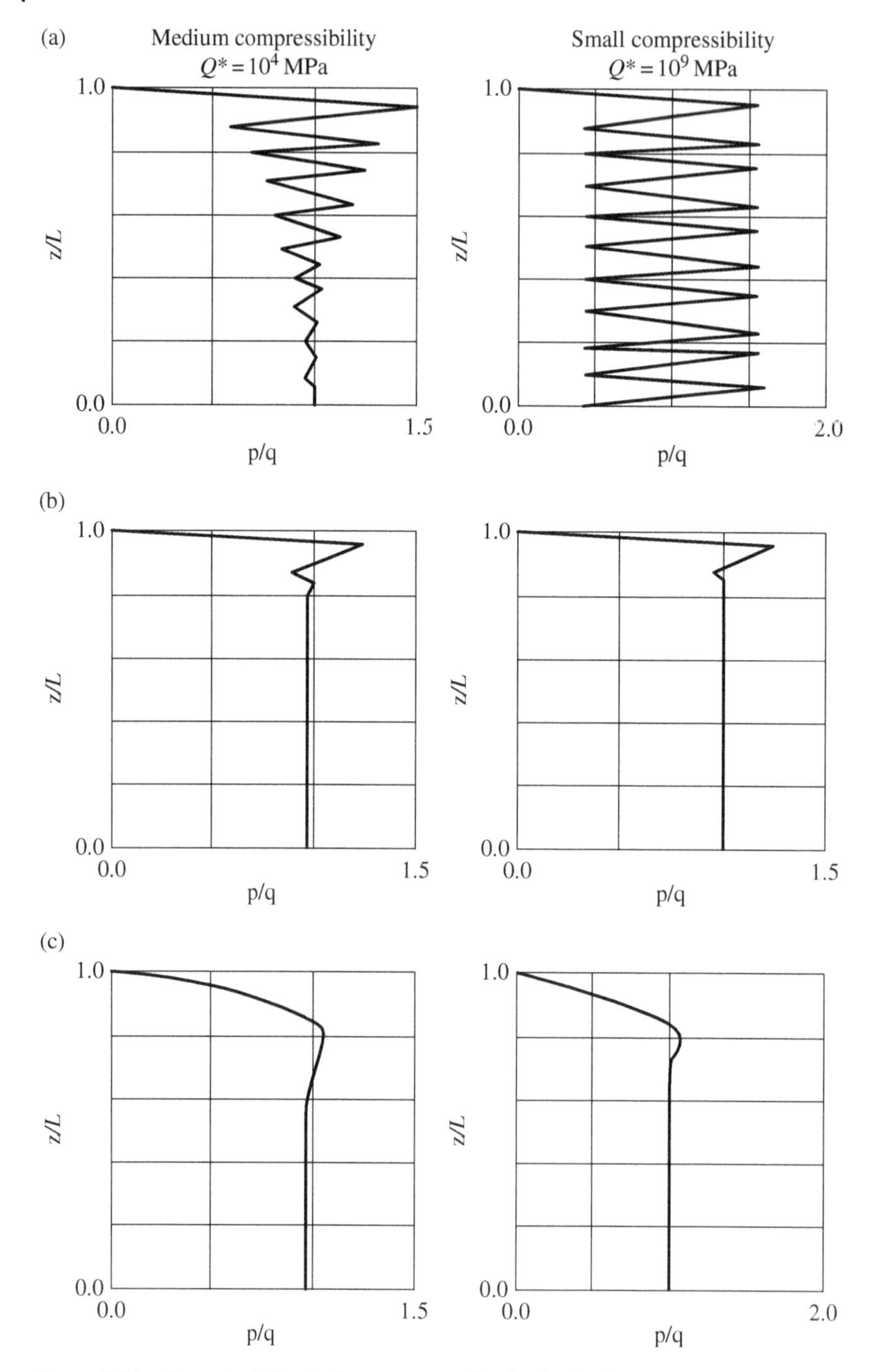

Figure 5.26 Example 1. Vertical pressure amplitude distribution. Note: Exact solution is very close to the stabilized solution. (a) Solution with standard column with 20 Q4P4 elements. (b) Solution with standard column with 20 Q8P4 elements. (c) Stabilized procedure with 20 Q4P4 elements.

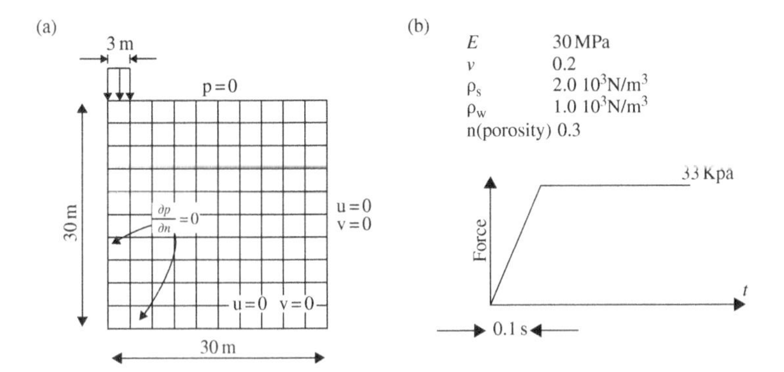

Figure 5.27 Example 2. A saturated soil foundation under transient load; (a) the problem domain; (b) applied data of transient load. *Source:* Based on Pastor et al. (1991).

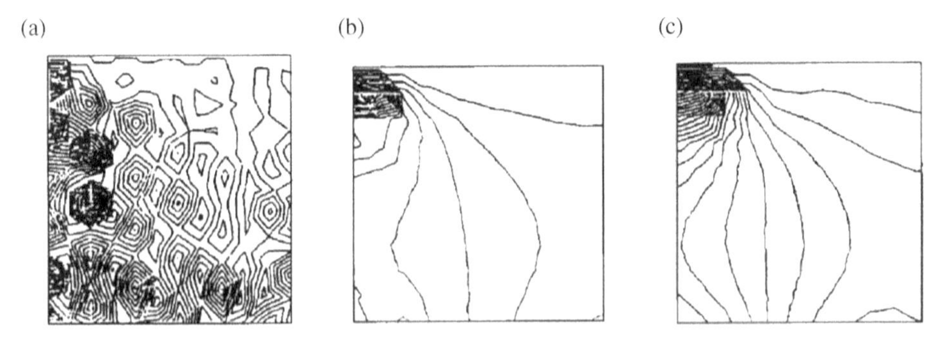

Figure 5.28 Example 2. Two-dimensional foundation pressure contours computed for small permeability and compressibility $Q^* = 10^9$ MPa, $k = 10^{-7}$ m/s; (a) direct use of implicit algorithm with Q4P4 elements; (b) direct use of implicit algorithm with Q8P4 elements; (c) Q4P4 stabilized elements. *Source:* Based on Pastor et al. (1991).

As we mentioned there, satisfactory solutions can still be obtained but these require that the number of parameters describing the variable $\bar{\mathbf{u}}$ must be greater than these describing the variable $\bar{\mathbf{p}}$, i.e.

$$n^u \geq n^p \tag{5.72}$$

This is a necessary condition for avoiding singularities and can be readily achieved with certain interpolations. However, if the problem is recast in the manner given in Section 5.5.2, we shall find that even in the limiting case (i.e. with zero compressibility and permeability), a nonzero diagonal will be obtained and stability can always be achieved.

As we have recast the problem in terms of velocities, we shall linearize using these variables and write

$$\int_{\Omega} \mathbf{B}^{\mathrm{T}} \boldsymbol{\sigma}' \mathrm{d}\Omega = \widetilde{\mathbf{K}} \bar{\mathbf{v}} \tag{5.73}$$

where $\widetilde{\mathbf{K}}$ includes a time integration operator.

In the steady state

$$\overline{\mathbf{v}}^n = \overline{\mathbf{v}}^{n+1} = \overline{\mathbf{v}}$$
$$\overline{\mathbf{p}}^n = \overline{\mathbf{p}}^{n+1} = \overline{\mathbf{p}} \tag{5.74}$$
$$\Delta\overline{\mathbf{p}} = 0$$

and we can write the sum of (5.62) and (5.63) as

$$\widetilde{\mathbf{K}}\overline{\mathbf{v}} - \mathbf{Q}\overline{\mathbf{p}} - \mathbf{f}^{(1)} = \mathbf{0} \tag{5.75}$$

Eliminating $\overline{\mathbf{v}}^*$ in (5.67) by using (5.63), we arrive at:

$$-\mathbf{Q}^{\mathrm{T}}\overline{\mathbf{v}} + \Delta t(\mathbf{H}^* - \mathbf{Q}^{\mathrm{T}}\mathbf{M}^{-1}\mathbf{Q})\overline{\mathbf{p}} = \mathbf{0} \tag{5.76}$$

The two equations (5.75) and (5.76) can be written as

$$\begin{bmatrix} \mathbf{K} & -\mathbf{Q} \\ -\mathbf{Q}^{\mathrm{T}} & \Delta t(\mathbf{H}^* - \mathbf{Q}^{\mathrm{T}}\mathbf{M}^{-1}\mathbf{Q}) \end{bmatrix} \begin{Bmatrix} \overline{\mathbf{u}} \\ \overline{\mathbf{p}} \end{Bmatrix} = \begin{Bmatrix} \mathbf{f}^{(1)} \\ \mathbf{0} \end{Bmatrix} \tag{5.77}$$

and a nonzero diagonal is found to exist in its finite time steps. This seems to achieve complete stabilization and any interpolation of the $\overline{\mathbf{v}}/\overline{\mathbf{u}}$ and $\overline{p}$ variables can be used with equal interpolation, of course, being the obvious choice.

The procedure outlined unfortunately results only in conditional stability, although the time-step length is now given by the speed of the shear wave and hence is not too restrictive.

$$\Delta t_{\mathrm{crit}} \leq \frac{h}{c_{\mathrm{s}}} \tag{5.78}$$

We find that the integration of the new stabilization procedure into the computer code is reasonably economic and can well be made use of in many programs, especially those in which a nearly explicit solution is going to be used.

5.5.5 An Operator Split Stabilizing Algorithm for the Consolidation of Saturated Porous Media

A similar procedure as in Section 5.5.2 has been obtained by Salomoni and Schrefler (2005) for slow phenomena such as consolidation. The starting point for the procedure is the linear momentum balance equation of the fluid (2.76), specialized for a single fluid phase, where the fluid acceleration relative to the solid has been expressed by means of (2.59)

$$\mathbf{a}^{ws} = \frac{\partial \mathbf{v}^{ws}}{\partial t} + grad\mathbf{v}^{ws}.\mathbf{v}^{ws} \tag{5.79}$$

Equation (2.76) hence reads after neglecting the solid velocity with respect to the fluid one

$$\mathbf{w} = n\mathbf{v}^{ws} = \frac{\mathbf{k}}{\mu}\left[-grad p^w + \rho^w\left(\mathbf{g} - \frac{\partial \mathbf{v}^{ws}}{\partial t} - grad\mathbf{v}^{ws}.\mathbf{v}^{ws}\right)\right] \tag{5.80}$$

where $\mathbf{w} = n\mathbf{v}^{ws}$ is the Darcy velocity (2.22). With $\mathbf{v} = \mathbf{v}^w = \mathbf{v}^{ws}$ which are true velocities, (5.80) can be written as

$$\frac{\mathbf{k}}{\mu}\rho^w\frac{\partial \mathbf{v}^w}{\partial t} = -\frac{\mathbf{k}}{\mu}\rho^w(grad\mathbf{v}^w \cdot \mathbf{v}^w) - n\mathbf{v}^w + \frac{\mathbf{k}}{\mu}(-grad p^w + \rho^w \mathbf{g}) \tag{5.81}$$

We now use the following form of temporal (Characteristic-Galerkin type) discretization for this equation as in Massarotti et al. (2001)

$$\rho^w\frac{k_i}{\mu}\frac{v_i^{n+1} - v_i^n}{\Delta t} = \rho^w\frac{k_i}{\mu}\left[-v_j\frac{\partial v_i}{\partial x_j} + \frac{\Delta t}{2}v_k\frac{\partial}{\partial x_k}\left(v_j\frac{\partial v_i}{\partial x_j}\right)\right]^n$$
$$+ \vartheta_1[nv_i]^{n+1} - (1-\vartheta_1)[nv_i]^n - \vartheta_2\frac{k_i}{\mu}\left(\frac{\partial p}{\partial x_i}\right)^{n+1} +$$
$$-(1-\vartheta_2)\frac{k_i}{\mu}\left[\frac{\partial p}{\partial x_i} - \frac{\Delta t}{2}v_k\frac{\partial}{\partial x_k}\left(\frac{\partial p}{\partial x_i}\right)\right]^n + \rho^w\frac{k_i}{\mu}g_i \tag{5.82}$$

where $p = p^w$ and k_i represents the k_{ii} component of the permeability matrix and n the time step: A semi-implicit procedure of the Newmark-type is used where $\vartheta_1 = 1$, $\vartheta_2 = 1$. Taking all terms linked to the velocity at the $(n+1)^{th}$ step to the left-hand side of the equation and multiplying by Δt, we obtain

$$\left(\rho^w\frac{k_i}{\mu} - n\,\Delta t\right)v_i^{n+1} = \rho^w\frac{k_i}{\mu}v_i^n + \rho^w\frac{k_i}{\mu}\Delta t\left[-v_j\frac{\partial v_i}{\partial x_j} + \frac{\Delta t}{2}v_k\frac{\partial}{\partial x_k}\left(v_j\frac{\partial v_i}{\partial x_j}\right)\right]^n$$
$$+ \rho^w\frac{k_i}{\mu}\Delta t g_i - \frac{k_i}{\mu}\Delta t\left(\frac{\partial p}{\partial x_i}\right)^{n+1} \tag{5.83}$$

The method is implicit in the pressure variable and explicit in the second RHS term, i.e. the pressure gradient is calculated at the end of the step but the second RHS term is calculated at the beginning of the step. Only in this semi-implicit form or in an explicit one, once some form of artificial speed of sound is assumed – can incompressible problems (in which the sound velocity is infinite) be solved in fluid mechanics environments. Hence, with the above choice for ϑ_1 and ϑ_2, the algorithm is conditionally stable. The major restrictions on the time step in implicit methods arise from accuracy requirements and the decreasing robustness of the Newton procedure as the time step increases. The Newmark method is second-order accurate, i.e. the truncation error is of the order of Δt^2, the same order as for the central difference method. Therefore, for large time steps, truncation error becomes a concern. However, for slightly compressible or incompressible problems, the semi-implicit form has been found to be efficient (Zienkiewicz et al. (1999b)).

Additionally, it is important to notice that the second RHS term in Eq. (5.83) is a diffusion-like term, proportional to Δt; this is precisely the *anisotropic balancing diffusion* needed to stabilize the Galerkin spatial discretization of convection, Kelly et al. (1980).

The first step of this operator split algorithm consists of the calculation of the intermediate velocity $\tilde{\mathbf{v}}_i$ (auxiliary variable) from (5.83) without incorporating the pressure terms $-\frac{k_i}{\mu}\Delta t\left(\frac{\partial p}{\partial x_i}\right)^{n+1}$ which is known as the *Eulerian velocity correction procedure*. Hence,

$$\left(\rho^w\frac{k_i}{\mu} - n\,\Delta t\right)\tilde{v}_i^{n+1} = \rho^w\frac{k_i}{\mu}v_i^n + \rho^w\frac{k_i}{\mu}\Delta t\left[-v_j\frac{\partial v_i}{\partial x_j} + \frac{\Delta t}{2}v_k\frac{\partial}{\partial x_k}\left(v_j\frac{\partial v_i}{\partial x_j}\right)\right]^n + \rho^w\frac{k_i}{\mu}\Delta t g_i \tag{5.84}$$

Once the pressure is obtained, at the (n+1) step, the value of $\widetilde{v}_i$ is corrected

$$\left(\rho^w \frac{k_i}{\mu} - n\,\Delta t\right) v_i^{n+1} = \left(\rho^w \frac{k_i}{\mu} - n\,\Delta t\right) \widetilde{v}_i^{n+1} - \frac{k_i}{\mu} \Delta t \left(\frac{\partial p}{\partial x_i}\right)^{n+1} \tag{5.85}$$

With the differential operator **S** (2.10), the divergence of **w**, div **w** = $\mathbf{m}^{\mathrm{T}}\mathbf{S}(n\mathbf{v})$, and the volumetric strain $\varepsilon_{ii} = \mathbf{m}^{\mathrm{T}}\mathbf{S}(\delta u_i/\delta t)$, the mass balance equation for water (2.16b) with incompressible fluid and solid grains and with the underlined terms neglected can now be written as

$$\mathbf{m}^{\mathrm{T}}\mathbf{S}\frac{\partial \mathbf{u}}{\partial t} + \mathbf{m}^{\mathrm{T}}\mathbf{S}n\mathbf{v}^{ws} = 0 \tag{5.86}$$

Following the previous scheme, a temporal semi-discretized expression of the mass balance equation for water is obtained,

$$\mathbf{m}^{\mathrm{T}}\mathbf{S}\frac{\mathbf{u}^{n+1} - \mathbf{u}^n}{\Delta t} = -\vartheta_3\left(\mathbf{m}^{\mathrm{T}}\mathbf{S}n\mathbf{v}\right)^{n+1} + (1-\vartheta_3)\left(\mathbf{m}^{\mathrm{T}}\mathbf{S}n\mathbf{v}\right)^n \tag{5.87}$$

Substitution of $\mathbf{v}^{n+1}$ from (5.57) multiplied by $\frac{\mu\mathbf{k}^{-1}}{\rho^w}$ and putting $\vartheta_3 = 1$ yields

$$\mathbf{m}^{\mathrm{T}}\mathbf{S}\frac{\mathbf{u}^{n+1} - \mathbf{u}^n}{\Delta t} = -\mathbf{m}^{\mathrm{T}}\mathbf{S}n\widetilde{\mathbf{v}}^{n+1} + \Delta t\mathbf{m}^{\mathrm{T}}\mathbf{S}\frac{n}{\rho^w}\left[\widetilde{\mathbf{H}}(\nabla p)\right]^{n+1} \tag{5.88}$$

where

$$\left(\mathbf{I} - \frac{\mu n\,\Delta t}{\rho^w}\mathbf{k}^{-1}\right) = \widetilde{\mathbf{H}}^{-1} \tag{5.89}$$

The linear momentum balance equation for the multiphase medium (2.20) with inertial terms neglected, after the introduction of the effective stress (2.1) and of (2.8), becomes at time station n+1 in case of incompressible grains

$$-\mathbf{S}^{\mathrm{T}}\mathbf{D}\mathbf{S}\mathbf{u}^{n+1} + \mathbf{S}^{\mathrm{T}}\mathbf{m}p^{n+1} = \rho\mathbf{b}^{n+1} \tag{5.90}$$

The proposed algorithm works in the following way: solve first (5.84) giving $\widetilde{\mathbf{v}}^{n+1}$; then solve the coupled equations (5.88) and (5.90), giving $\mathbf{u}^{n+1}$, p^{n+1}

$$\begin{cases} -\mathbf{S}^{\mathrm{T}}\mathbf{D}\mathbf{S}\mathbf{u}^{n+1} + \mathbf{S}^{\mathrm{T}}\mathbf{m}p^{n+1} = \rho\mathbf{b}^{n+1} \\ \mathbf{m}^{\mathrm{T}}\mathbf{S}\dfrac{\mathbf{u}^{n+1} - \mathbf{u}^n}{\Delta t} = -\mathbf{m}^{\mathrm{T}}\mathbf{S}n\widetilde{\mathbf{v}}^{n+1} + \Delta t\,\mathbf{m}^{\mathrm{T}}\mathbf{L}\dfrac{n}{\rho^w}\left[\widetilde{\mathbf{H}}(\nabla p)\right]^{n+1} \end{cases} \tag{5.91}$$

and, finally, solve (5.85) for $\mathbf{v}^{n+1}$ which is needed for the next time step.

The system of equations (5.91) is implicit due to the choice of ϑ_3 and to the definition of the equilibrium equation at time t_{n+1}: in this case, the results are the same as for the backward difference method in the FD context. This system of equations is very similar in its form to the corresponding one used in standard one-phase flow consolidation problems; in addition it ensures, through the splitting procedure presented here, the disappearance of pressure oscillations which are present in the standard approach.

The above equations must now be discretized in space following the standard procedures of Chapter 3 with

$$\mathbf{v} = \mathbf{N}\overline{\mathbf{v}}, \mathbf{u} = \mathbf{N}\overline{\mathbf{u}}, p = \mathbf{N}^p\overline{\mathbf{p}} \tag{5.92}$$

After application of the standard Galerkin approximation and assuming for simplicity constant material properties appearing in $\widetilde{\mathbf{H}}$, (5.84) becomes

$$\widetilde{\mathbf{H}}^{-1}\mathbf{M}\overline{\overline{\mathbf{v}}}^{n+1} = \mathbf{M}\overline{\mathbf{v}}^{n} + \Delta t(-\mathbf{C}\overline{\mathbf{v}}^{n} + \Delta t\,\mathbf{K}_{\mathbf{v}}\overline{\mathbf{v}}^{n}) + \Delta t\,\mathbf{f}_{e} \tag{5.93}$$

where

$$\mathbf{M} = \int_{\Omega} \mathbf{N}^{\mathrm{T}}\mathbf{N}d\Omega \tag{5.93a}$$

$$\mathbf{K}_{\mathbf{v}} = -\frac{1}{2}\int_{\Omega} \frac{\partial(v_k\mathbf{N}^{\mathrm{T}})}{\partial x_k}\frac{\partial(v_k\mathbf{N})}{\partial x_k}d\Omega \tag{5.93b}$$

$$\mathbf{C} = \int_{\Omega} \mathbf{N}^{\mathrm{T}}v_i\frac{\partial \mathbf{N}}{\partial x_i}d\Omega \tag{5.93c}$$

$$\mathbf{f}_{\mathbf{e}} = \int_{\Omega} \mathbf{N}_{v}^{\mathrm{T}}\mathbf{g}d\Omega + b.t. \tag{5.93d}$$

and where the matrix $\mathbf{K}_{\mathrm{v}}$ of the stabilization term has been obtained after integration by parts and b.t. in eq. (5.93d) stands for integrals along the boundaries. The residuals on the boundaries coming from the integration by part of the stabilizing term are usually negligible. See Zienkiewicz et al. (2005c).

The system (5.91) becomes

$$\begin{cases} -\mathbf{K}\overline{\mathbf{u}}^{n+1} + \mathbf{Q}\overline{\mathbf{p}}^{n+1} = \mathbf{f}^{n+1} \\ \mathbf{Q}^{\mathrm{T}}\overline{\mathbf{u}}^{n+1} - (\Delta t)^2\mathbf{C}_p\overline{\mathbf{p}}^{n+1} = \mathbf{Q}^{\mathrm{T}}\overline{\mathbf{u}}^{n} - \Delta t\mathbf{P}_v\overline{\overline{\mathbf{v}}}^{n+1} \end{cases} \tag{5.94}$$

where $\mathbf{K}$, $\mathbf{Q}$, and $\mathbf{f}$ are defined in Chapter 3 (3.22, 3.29, and 3.47) and

$$\mathbf{C}_p = \frac{n}{\rho^w}\widetilde{\mathbf{H}}\int_{\Omega}\left(\frac{\partial \mathbf{N}^p}{\partial x_i}\right)^{\mathrm{T}}\frac{\partial \mathbf{N}^p}{\partial x_i}d\Omega \tag{5.95a}$$

$$\mathbf{P}_v = n\int_{\Omega}\mathbf{N}^{\mathrm{T}}\frac{\partial \mathbf{N}^p}{\partial x_i}d\Omega \tag{5.95b}$$

The system (5.94) can be written in the following way:

$$\begin{bmatrix} -\mathbf{K} & \mathbf{Q} \\ \mathbf{Q}^{\mathrm{T}} & -(\Delta t)^2\mathbf{C}_p \end{bmatrix}\begin{bmatrix} \overline{\mathbf{u}} \\ \overline{\mathbf{p}} \end{bmatrix}^{n+1} = \begin{bmatrix} 0 & 0 \\ \mathbf{Q}^{\mathrm{T}} & 0 \end{bmatrix}\begin{bmatrix} \overline{\mathbf{u}} \\ \overline{\mathbf{p}} \end{bmatrix}^{n} - \Delta t\begin{bmatrix} 0 & 0 \\ \mathbf{P}_v & 0 \end{bmatrix}\begin{bmatrix} \overline{\overline{\mathbf{v}}} \\ 0 \end{bmatrix}^{n+1} + \begin{bmatrix} \mathbf{f} \\ 0 \end{bmatrix}^{n+1} \tag{5.96}$$

Finally, the discretized form of (5.85) is

$$\mathbf{M}\overline{\mathbf{v}}^{n+1} = \mathbf{M}\overline{\overline{\mathbf{v}}}^{n+1} - \frac{\Delta t}{\rho^w}\mathbf{H}\mathbf{P}_v^{\mathrm{T}}\overline{\mathbf{p}}^{n+1} \tag{5.97}$$

The resulting scheme is now a u-v-p implementation that yields fluid velocities as a by-product. The loop formed by steps (5.93), (5.96), and (5.97) is iterated within one time step until a suitable convergence criterion is met. For the examples shown in the next section, we found it sufficient to pass only once through this loop in each time step. The same was experienced by Markert et al. (2010) in the case of a splitting scheme for a dynamic u-v-p implementation.

5.5.6 Examples Illustrating the Effectiveness of the Operator Split Stabilizing Algorithm for the Consolidation of Saturated Porous Media

The procedure has been validated in Salomoni and Schrefler (2005) with respect to the one-dimensional consolidation problem of Siriwardane and Desai (1981). It has been shown that the CBS-type procedure allows for balancing the pressure oscillations generally occurring at the onset of FEM analyses of consolidation problems. This is confirmed by the 2D analysis shown next. The pressure balance is strictly connected with Step 3 of the CBS-type algorithm: at low permeability, the effect of pressure is mainly controlled by matrix **H** and at high permeability by its gradient.

The details of the problem are depicted in Figure 5.29. The analyzed domain is a square of size 10a × 10a with a = 0.8 m, with a uniform mesh of quadrilateral elements.

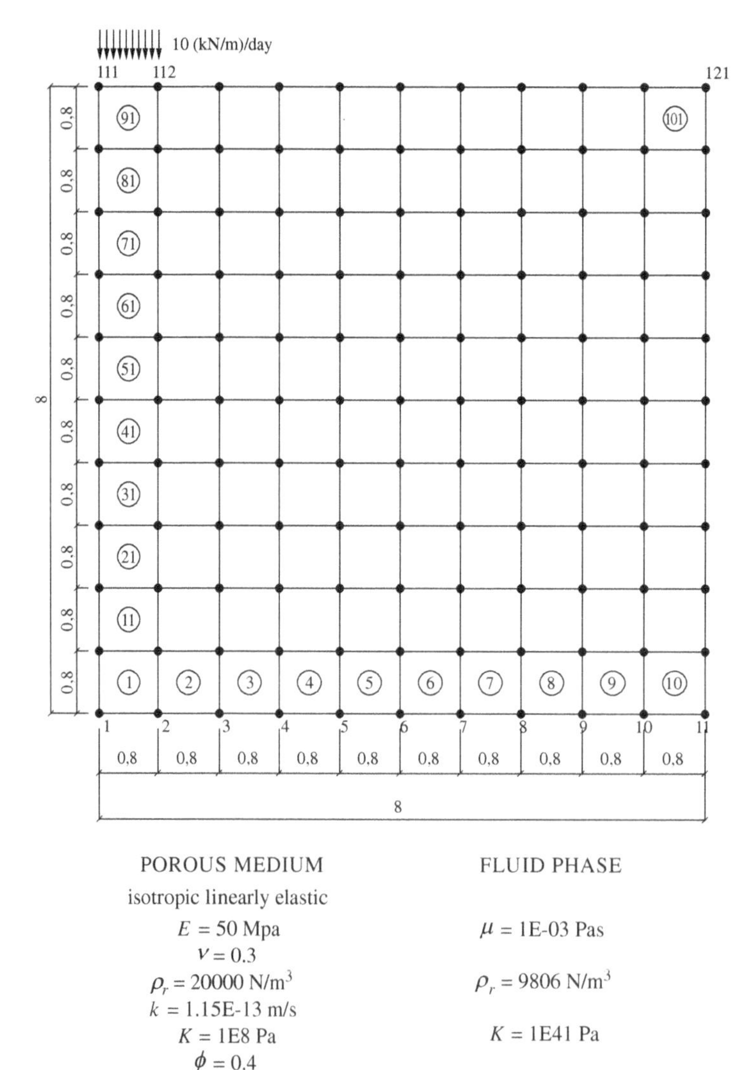

Figure 5.29 Test sample for validating the stabilizing characteristic of the algorithm. *Source:* From Salomoni and Schrefler (2005).

The computation has been carried out assuming a fully saturated medium and incompressible pore fluid. The adopted time step size is 345 seconds.

We show in Figures 5.30 and 5.31 pore overpressures and displacement amplitudes after 10 hours, 1 day, and 4 days of the onset of the phenomenon where usually oscillations both in space and in time occur.

The properties of robustness and stability of the formulation are demonstrated by the absence of spurious pressure oscillations, as shown by Figure 5.32 for different time-stations. Truty and Zimmermann (2003) solved a similar problem with elastoplastic behavior using direct stabilization methods (Truty and Zimmermann 2006).

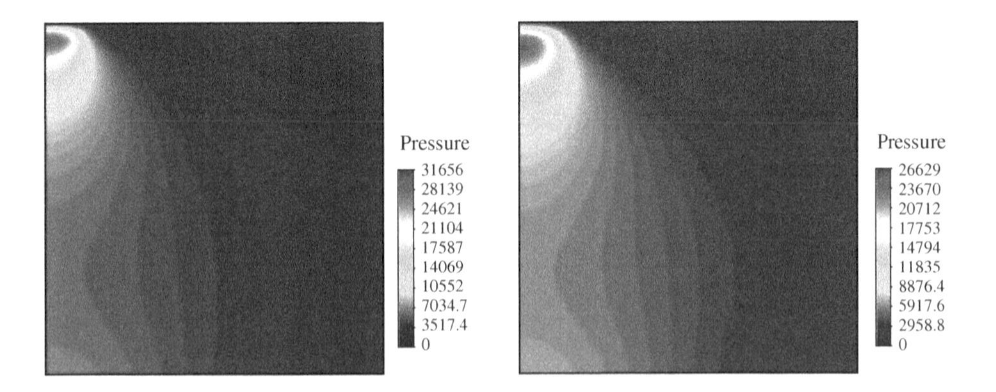

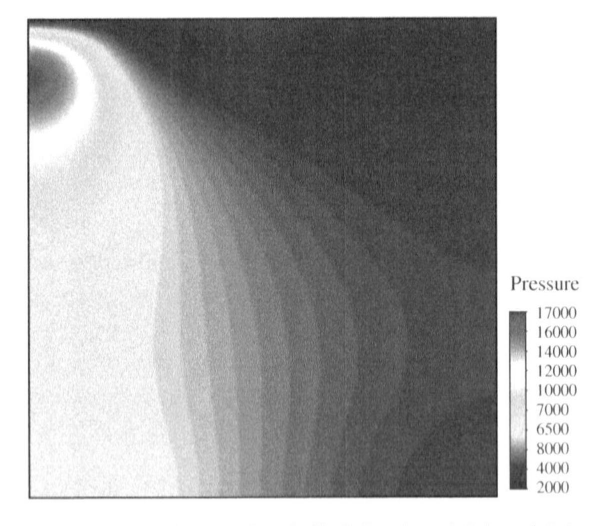

Figure 5.30 Pore pressures at t = 10 hours (top left), 1 day (top right), and 4 days (bottom)

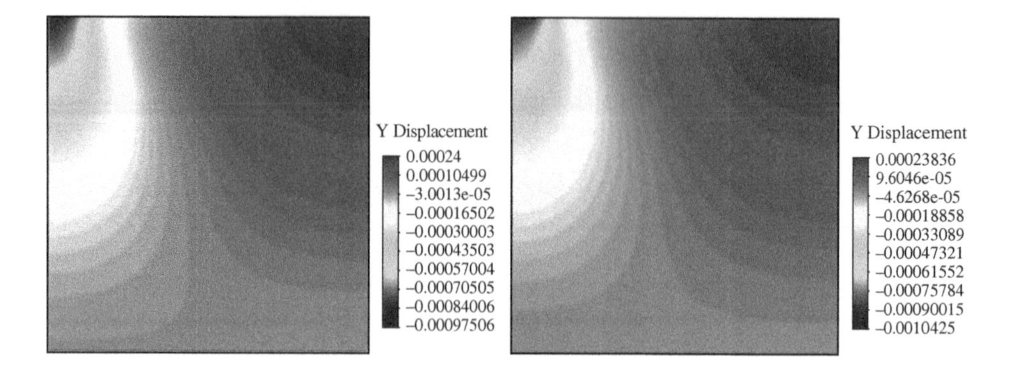

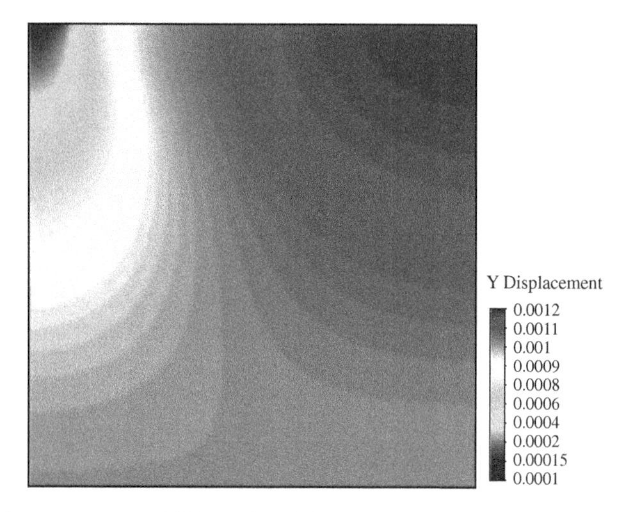

Figure 5.31 Displacements at t = 10 hours (top left), 1 day (top right), and 4 days (bottom)

5.5.7 Further Improvements[7]

So far, we have considered the problem of developing stabilized formulations circumventing the limitations imposed by Babuska-Brezzi conditions or, alternatively, the patch test. Therefore, simple and computationally fast elements such as triangles and tetrahedra can be used in coupled problems of geomechanics. It is important to notice that, even after stabilization, those simple elements present important limitations, among which we can mention:

i) numerical dispersion, which causes oscillations to appear close to fronts,
ii) numerical damping, which decreases the amplitude of propagating shocks,
iii) dependence on mesh alignment which results in spurious mechanisms of failure, and
iv) volumetric locking, which results in failure loads higher than real.

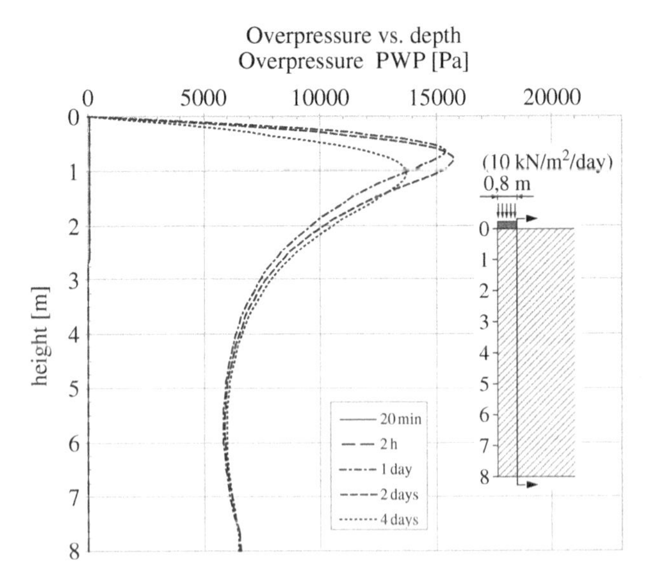

Figure 5.32 Pore pressures at different time-stations. *Source:* From Salomoni and Schrefler (2005).

Several alternative ways exist to improve these simple elements, of which it is worth mentioning (i) the stabilized enhanced strain quadrilateral proposed by Mira et al. (2003), and the mixed stress-velocity-pressure triangles and tetrahedra stabilized with the fractional step technique described above (Mabssout et al. 2006). This technique has recently been extended to SPH by Blanc and Pastor (2011), see also Chapter 10.

In the former work, Mira et al. (2003) applied the inf-sup condition to justify the achieved stabilization. It is important to notice that the patch test is a sufficient condition for stability, while the inf-sup is both necessary and sufficient. The interested reader can find more details in Chapelle and Bathe (1993). As a result, enhanced strain quadrilateral elements in 2D, bilinear in displacements and pore pressures can be safely applied. This element has an excellent performance in localized failure and bending dominated situations.

Concerning the latter, we will outline the method which consists of the following steps:

1) The first step **FS1** of the algorithm is:

$$\mathbf{v}^{*} = \mathbf{v}^{n} + \frac{\Delta t}{\rho}(\rho \mathbf{b} + div\boldsymbol{\sigma}'^{n}) \tag{5.98}$$

$$\boldsymbol{\sigma}'^{*} = \boldsymbol{\sigma}'^{n} + \mathbf{D}^{e} : (\mathbf{d}^{n} - \mathbf{d}^{vp\,n}) - \boldsymbol{\sigma}'^{n}.\dot{\boldsymbol{\omega}}^{n} + \dot{\boldsymbol{\omega}}^{n}.\boldsymbol{\sigma}'^{n}$$

which can be discretized using the Taylor Galerkin algorithm described by Mabssout et al. (2006) or the Taylor Smoothed Particle Hydrodynamics (SPH) method proposed by Blanc and Pastor (2011, 2012, 2013). See also Section 10.4.2.

2) The second step, **FS2**, where the pore pressure at time $n+1$ is obtained from:

$$\left(\frac{1}{Q}\frac{1}{\Delta t} - \mathrm{div}\,(k_w \mathrm{grad}\,) - \frac{\Delta t}{\rho}\mathrm{div}\,\mathrm{grad}\,\right)p_w^{n+1} = -\,\mathrm{div}\,v^{*} + \frac{1}{Q}\frac{1}{\Delta t}p_w^{n} \tag{5.99}$$

3) The third and final step **FS3** consists of obtaining the velocities and stresses at time $n+1$ as:

$$\mathbf{v}^{n+1} = \mathbf{v}^{*} - \frac{\Delta t}{\rho}\,\mathrm{grad}\, p_w^{n+1} \tag{5.100}$$
$$\boldsymbol{\sigma}^{n+1} = \boldsymbol{\sigma}'^{*} - p_w^{n+1} I$$

It is important to notice that the stress is obtained by adding two increments which consist precisely of the increment of effective stress and that of the pore pressure.

As an example, we will consider the case of the 2D soil sample presented in Figure 5.33, under plane strain conditions. The sample is 2 m wide and 1 m high. The soil is subjected to an imposed velocity on its upper face. The material behavior is described by the viscoplastic Perzyna law with a modified Cam-Clay yield surface. The sample is symmetric. Therefore, only its right half part will be considered in the calculation to decrease the computational time.

The sample is subjected to compression. The applied boundary conditions are the following:

- On the left-hand side Γ_1, the nodes can only move along the vertical axis, thus $v_x = 0$ and $\sigma_{xy} = 0$. On this boundary, the condition on the pore pressure is $\frac{\partial p_w}{\partial n} = 0$.
- On the bottom Γ_2, $\sigma_{xy} = 0$ and $v_y = 0$ and the condition on the pore pressure is $\frac{\partial p_w}{\partial n} = 0$.
- On the right-hand side Γ_3, the nodes are free to move, $\sigma_{xx} = 0$ and $\sigma_{xy} = 0$ and the pore pressure is imposed equal to 0.

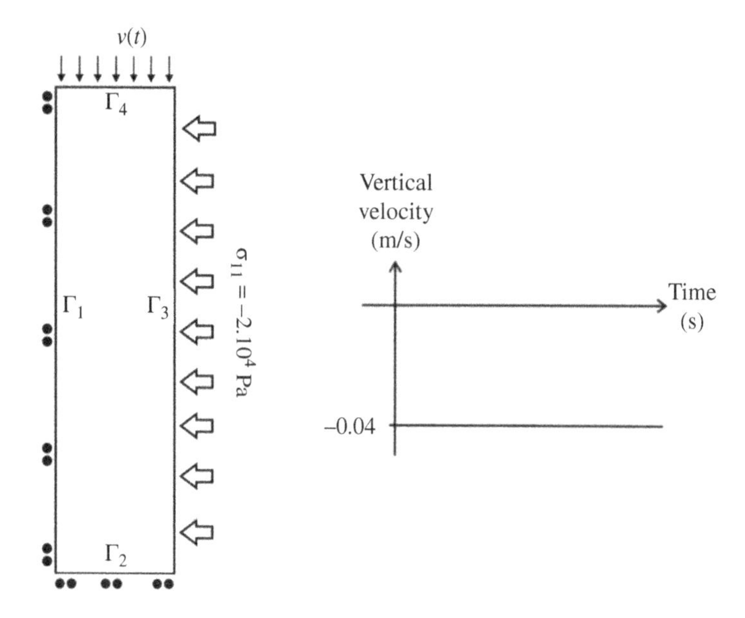

Figure 5.33 Sketch of the soil sample. *Source:* From Blanc and Pastor (2012).

- On the top Γ_4, the vertical component of the velocity of the nodes is imposed, $v_y = v(t)$ and $\sigma_{xy} = 0$. The pore pressure is imposed equal to 0.

The elastic parameters of the material are the elastic modulus $E = 8 \cdot 10^7$ Pa, the Poisson's coefficient $\nu = 0.2$ and the density of the mixture, $\rho_m = 2000\ kg/m^3$. The parameters of the Perzyna's model are the fluidity parameter $\gamma = 20\ \text{s}^{-1}$ and the model parameter, $N = 1$. The parameters describing the coupling properties of the soil are the combined compressibility of the fluid and the soil particles $Q^* = 1 \cdot 10^{10}$ MPa, the kinetic permeability $k_w = 1 \cdot 10^{-6}$ ms^{-1} and the soil is saturated. Thus, $S_w = 1$. The parameters of the modified Cam-Clay model are the initial size of the yield surface $p_c = 2 \cdot 10^6$ Pa, the void ratio, $e = 1$ and $M = 1.2$; $\lambda = 0.1$; $\kappa = 8 \cdot 10^{-2}$.

The problem is solved using the Smooth Particle Hydrodynamics method, which will be described in Chapter 10. The soil sample is discretized in an SPH grid where the horizontal and vertical distances between the SPH nodes are $\Delta x = \Delta y = 0.025$ m. In total, there are 861 SPH nodes and 800 SPH elements. The time step is $\Delta t = 5 \cdot 10^{-5}$ second.

The value of the imposed velocity has been chosen in such a way that the stresses do not reach the yield surface until that the wave is reflected at the bottom. The reflection occurs at time $t = 0.005$ second. At this time, viscoplastic shear strain appears in the right-bottom corner. The strain localizes in the form of a shear band from the right-bottom corner to the left side of the sample. Once the shear band reaches the left side, it is reflected and a new shear band appears from the left side to the right side (Figure 5.34a)

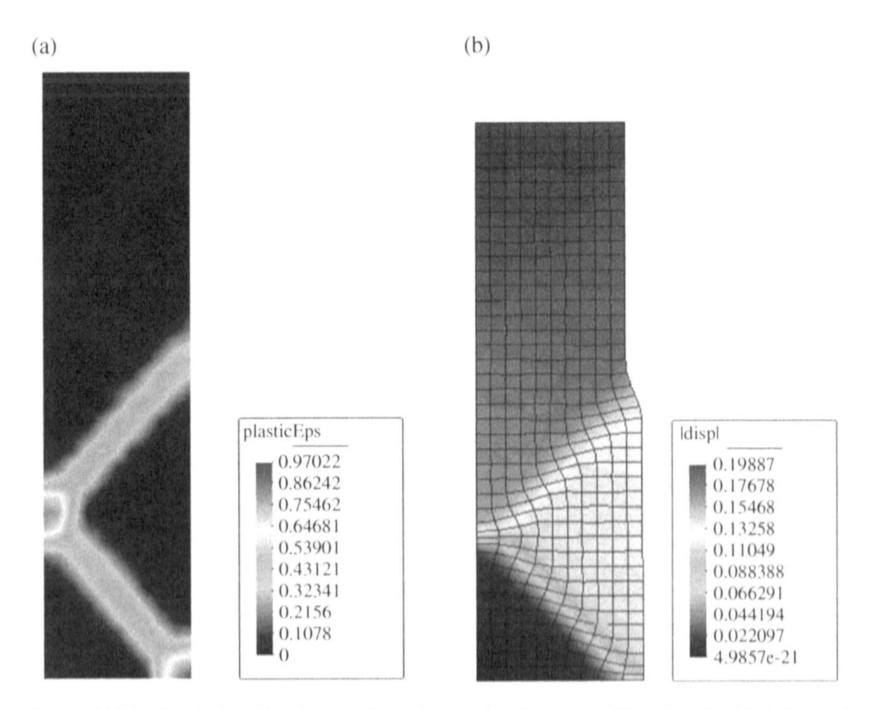

Figure 5.34 Strain localization on the soil sample: a) contour fills of strain; b) deformed mesh (factor 0.8). *Source:* From Blanc and Pastor (2012).

Observing the deformed mesh, we can clearly see that the failure occurs along the shear bands figure 5.34b.

5.6 Conclusion

In this chapter, we have addressed some special aspects regarding the modeling of infinite domains, radiating boundary conditions, adaptivity in space and time, the capture of localized phenomena, and stabilization procedures permitting to use equal order interpolation for displacements and pressures. Several examples have been shown to illustrate some of the concepts put forward. The stage is now set for presenting practical applications of the method and the constitutive relationships.

Notes

1 The authors gratefully acknowledge the collaboration of L. Simoni in the elaboration of this Section.
2 In the USA, such records can be obtained from, e.g. the Pacific Earthquake Engineering Research Center (http://peer.berkeley.edu/) at the University of California at Berkeley and similar sources are available in other countries.
3 The authors are grateful to Ch. Song who contributed to this section.
4 The authors gratefully acknowledge the collaboration of S. Secchi and L. Simoni in the elaboration of this Section.
5 The authors gratefully acknowledge the collaboration of H.W. Zhang in the elaboration of this Section.
6 The authors gratefully acknowledge the collaboration of V. Salomoni, P. Mira, T. Li, X. Liu and T. Blanc in the elaboration of this Section.
7 This Section has been written in collaboration with T. Blanc.

References

Bazant, Z. P. and Lin, F. B. (1988). Non-local yield limit degradation, *Int. J. Num. Methods. Eng.*, **26**, 1805–1823.

Bazyar, M. H. and Song, C. (2008). A continued-fraction based high-order transmitting boundary for wave propagation in unbounded domains of arbitrary geometry, *Int. J. Numer. Methods Eng.*, **74**, 209–237.

Belytschko, T. and Tabarrok, M. (1993). H-adaptive finite element methods for dynamic problems with emphasis on localization, *Int. J. Num. Methods. Eng.*, **36**, 4245–4265.

Belytschko, T., Fish, J. and Englemann, B. E. (1988). A finite element with embedded localization zones, *Comp. Methods. Appl. Mech. Eng.*, **79**, 59–89.

Bettes, P. (1992). *Infinite Elements*, Penshaw Press, Cleadon.

Blanc, T. and Pastor, M. (2011). Towards SPH modelling of failure problems in geomechanics. A fractional step Taylor-SPH model, *Eur. J. Environ. Civ. Eng.*, **15-SI 2011**, 31–49. doi:10.3166/ejece.15si.31-49

Blanc, T. and Pastor, M. (2012). A stabilized fractional step, Runge Kutta Taylor SPH algorithm for coupled problems in geomechanics, *Comput. Methods Appl. Mech. Eng.*, **221–222**, 41–53 DOI: 10.1016/j.cma.2012.02.006

Blanc, T. and Pastor, M. (2013). A stabilized smoothed particle hydrodynamics, Taylor-Galerkin algorithm for soil dynamics problems, *Int. J. Numer. Anal. Methods Geomech.* **37**, 1, 1–30 DOI: 10.1002/nag.1082

Brezzi, F. and Pitkaranta, J. (1984). On the stabilization of finite element approximations of the Stokes problem in Efficient solutions of elliptic problems, *Notes on Numerical Fluid Mechanics*, Vieweg, Wiesbaden.

Chapelle, D. and Bathe, K. J. (1993). The inf-sup test, *Comp. Struct.*, **47** (4–5),537–545.

Chen, Z., Steeb, H. and Diebels, S. (2006). A time-discontinuous Galerkin method for the dynamical analysis of porous media, *Int. J. Numer. Anal. Methods Geomech.*, **30**, 1113–1134.

Chorin, A. J. (1967). A numerical method for solving incompressible viscous problems, *J. Comput. Phys.*, **2**, 12–26.

Chorin, A. J. (1968). Numerical solution of incompressible flow problems, *Studies in Numer. Anal.*, **2**, 64–71.

Clough, R. W. and Penzien, J. (1993). *Dynamics of Structures* (2nd edn), McGraw-Hill, Inc., New York.

De Borst, R. and Muhlhaus, H. B. (1992). Gradient-dependent plasticity: formulation and algorithmic aspects. *Int. J. Num. Meth. Engng.*, **35**, 521–539.

De Borst, R., Sluys, L. J., Hühlhaus, H. B. and Pamin, J. (1993). Fundamental issues in finite element analysis of localization of deformation, *Eng. Comput.*, **10**, 99–121.

De Sampaio, P. A. B. (1991). A Petrov-Galerkin formulation for the incompressible Navier-Stokes equations using equal order interpolation for velocity and pressure, *Int. J. Num. Meth. Engrg.*, **31** 6, 1135–1149.

Gawin, D., Sanavia, L. and Schrefler, B. A (1998). Cavitation modelling in saturated geomaterials with application to dynamic strain localization, *I.J. Num. Methods Fluids*, **27**, 109–125.

Griffiths, A. A. (1921). Brittle fracture, *Proc. Roy. Soc. (A)*, **221**, 163.

Hafez, M. and Soliman, M. (1991). Numerical solution of the incompressible Navier-Stokes equations in primitive variables on unstaggered grids, *Proceedings AIAA Conferences*, **91**–1561–CP, 368–379.

Heider, Y., Markert, B. and Ehlers, W. (2012). Dynamic wave propagation in infinite saturated porous media half spaces, *Comput. Mech.*, **49**, 319–336.

Hendron, A. J. and Patton, F. D. (1985). *The Vaiont Slide, a Geotechnical Analysis Based on New Geologic Observations of the Failure Surface. Technical Report GL-85-5.* Washington DC, Department of the Army US Corps of Engineers, **I**, pp 20, 21.

Hughes, T. R. J. and Hulbert, G. M. (1988). Space-time finite element methods for elastodynamics: formulation and error estimates, *Comp. Methods. Appl. Mech. Eng.*, **66**, 339–363.

Hughes, T. J. R., Franca, L. P. and Balestra, M. (1986). A new finite element formulation for fluid dynamics. V. Circumventing the babuska-brezzi condition: a stable Petrov-Galerkin formulation of the stokes problem accomodating equal order interpolation, *Comp. Methods. Appl. Mech. Eng.*, **59**, 85–99.

Johnson, C. and Pitkäranta, J., (1986). An analysis of the discontinuous galerkin method for a scalar hyperbolic equation, *Math. Comput.*, **46**, 1–26.

Johnson, C., Nävert, U. and Pitkäranta, J., (1984). Finite element method for linear hyperbolic problems. *Comp. Methods. Appl. Mech. Eng.*, **45**, 285–312.

Kelly, D. W., Nakazawa, S. and Zienkiewicz, O. C. (1980). A note on anisotropic balancing dissipation in the finite element method approximation to convective diffusion problems, *Int. J. Numer. Methods Eng.*, **15**, 1705–1711

Kunar, R. R. and Marti, J. N. (1981). A non-reflecting boundary for explicit calculation in computational models for infinite domain media-structure interaction, *ASME – Eng. Mechanics Division*, **46**, 183–204.

Lasry, D. and Belytschko, T. B. (1988). Localization limiters in transient problems, *Int. J. Solids Struct.*, **24**, 581–597.

Lehmann, L. (2007). *Wave Propagation in Infinite Domains, Lecture Notes in Applied and Computational Mechanics*, **32**, Springer Verlag, Berlin.

Li, X. D. and Wiberg, N. E. (1996). Structural dynamic analysis by a timediscontinous Galerkin finite element method, *Int. J. Numer. Methods Eng.*, **39**, 2131–2152.

Li, X. D. and Wiberg, N. E. (1998). Implementation and adaptivity of a space-time finite element method for structural dynamics, *Comp. Methods. Appl. Mech. Eng.*, **156**, 211–229.

Loret, B. and Prevost, J. H. (1991). Dynamic strain localisation in fluid-saturated porous media, *J. Eng. Mech.*, **11**, 907–922.

Lysmer, J. and Kuhlemeyer, R. L. (1969). Finite dynamic model for infinite media, *ASCE EM*, **95**, EM4, 859–877.

Mabssout, M., Herreros, M. I. and Pastor, M. 2006. Wave propagation and localization problems in saturated viscoplastic geomaterials. *Int. J. Numer. Methods Eng.* **68**:425–447.

Markert, B., Heider, Y. and Ehlers, W. (2010). Comparison of monolithic and splitting solution schemes for dynamic porous media problems, *Int. J. Num. Methods. Eng.*, **82**,1341–1383.

Massarotti, N., Nithiarasu, P. and Zienkiewicz, O. C. (2001) Natural convection in porous medium-fluid interface problems, *Int. J. Num. Meth. Heat Fluid Flow*, **11**, 473–490.

Mira, P., Pastor, M., Li, T. and Liu, X. (2003). A new stabilized enhanced strain element with equal order of interpolation for soil consolidation problems. *Comput. Methods Appl. Mech. Eng.*, **192**, 4257–4277.

Mroginski, J. L. and Etse, G. (2014). Discontinuous bifurcation analysis of thermodynamically consistent gradient poroplastic materials, *Int. J. Solids Struct.*, **51**, 1834–1846.

Muhlhaus, H. B. and Aifantis, E. C. (1991). A variational principle for gradient plasticity, *Int. J. Solids Struct.*, **28**, 845–858.

Oliver, J. (1995). Continuum modelling of strong discontinuities in solid mechanics, *Proceedings COMPLAS IV, CINME, Barcelona*, p. 455–479. (see also *Int. J. Num. Meth. Eng.* 1996)

Ortiz, M. and Quigley, J. J. (1991). Adaptive mesh refinement in strain localization problems, *Comp. Methods. Appl. Mech. Eng.*, **90**, 781–804.

Ortiz, M., Leroy, Y. and Needleman, A. (1987). A finite element method for localized failure analysis, *Comp. Methods. Appl. Mech. Eng.*, **61**, 189–214.

Palaniappan, J., Haber, R. B. and Jerrard, R. L. (2004). A space time discontinuous Galerkin method for scalar conservation laws. *Comp. Methods. Appl. Mech. Eng.*, **193**, 3607–3631.

Pastor, M., Peraire, J. and Zienkiewicz, O. C. (1991). Adaptive remeshing for shear band localization problem, *Archive of Applied Mechanics-Ingenieur Archiv*, **61**, 30–39.

Pastor, M., Rubio, C., Mira, P., Peraire, J., Vilotte, J. P. and Zienkiewicz, O. C. (1992). *Numerical analysis of localization in Numerical Models in Geomechanics*, A. A. Balkema, Rotterdam.

Pastor, M., Quecedo, M. and Zienkiewicz, O. C. (1997). A mixed displacement pressure formulation for numerical analysis of plastic failure, *Compos. Struct.*, **62**, 13–23.
Pastor, M., Zienkiewicz, O. C., Li, T., Li, X. and Huang, M. (1999). Stabilized low-order finite elements for failure and localization problems in undrained soils and foundations, *Comp. Meth. Appl. Mech. Engng.*, **174**, 219–234.
Peraire, J., Vahadati, M., Morgan, K. and Zienkiewicz, O. C. (1987). Adaptive remeshing for compressible flow computations, *J. Comput. Phys.*, **72**, 449–466.
Radmanovic, B. and Katz, C. (2010). A high performance scaled boundary finite element method, *IOP Conf. Ser.: Mater. Sci. Eng.* **10**, 012214.
Read, H. E., Hegemier, G. A. (1984). Strain softening of rock, soil and concrete- a review article, *Mech. Mater.*, **3**, 271–294
Reed, M. B. (1984). An investigation of numerical errors in the analysis of consolidation by finite elements, *Int. J. Numer. Anal. Methods Geomech.*, **8**, 243–257.
Rice, J. R. (1975). On the stability of dilatant hardening for saturated rock masses, *J. Geophys. Res.*, **80**, 1531–1536.
Rudnicki, J. W. (1984). Effect of dilatant hardening on the development of concentrated shear-deformation in fissured rock mass, *J. Geophys. Res.*, **89**, 9259–9270.
Salomoni, V.and Schrefler, B. A. (2005) A CBS-type stabilizing algorithm for the consolidation of saturated porous media, *Int. J. Numer. Methods Eng.*, **63**, 502–527.
Salomoni, V. and Schrefler, B. A. (2006). Stabilized-coupled modeling of creep phenomena for saturated porous media, *Int. J. Numer. Methods Eng.*, **66**, 1587–1617.
Sanavia, L., Schrefler, B. A. and Steinmann, P. (2002). A formulation for an unsaturated porous medium undergoing large inelastic strains *Comput. Mech.*, **28**, 137–151.
Schauer, M. and Lehmann, L. (2009). Large scale simulation with scaled boundary Finite Element Method, *Proc. Appl. Math. Mech.*, **9**, 103–106.
Schneider, G. E., Raithby, G. D. and Yovanovich, M. M. (1978). Numerical methods in laminar and turbulent flow. *Proceedings of the First International Conference*, Swansea, Wales (17–21 July 1978). (A79-29801 11-34) London: Pentech Press.
Schrefler, B. A., Majorana, C. E. and Sanavia, L. (1995). Shear band localization in saturated porous media. *Archives of Mechanics*, **47**, 577–599.
Secchi, S., Simoni, L. and Schrefler, B. A. (2007). Mesh adaptation and transfer schemes for discrete fracture propagation in porous materials, *Int. J. Numer. Anal. Methods Geomech.*, **31**, 331–345.
Simo, J. C., Oliver, J. and Armero, F. (1993). An analysis of strong discontinuities induced by strain softening in rate independent inelastic solids, *Comput. Mech.*, **12**, 277–296.
Simoni, L. and Schrefler, B. A. (1987). Mapped' infinite elements in soil consolidation, *Int. J. Num. Meth. Engng.*, **24**, 513–27.
Simoni, L. and Schrefler, B. A. (1989). Numerical modelling of unbounded domains in coupled field problems, *Meccanica*, **24**, 98–106.
Siriwardane, H. J., Desai, C. S. (1981). Two numerical schemes for nonlinear consolidation, *Int. J. Num. Meth. Engng.*, **17**, 405–26.
Sluys, L. J., De Borst, R. and Muhlhaus, H. B (1993). Wave propagation, localization and dispersion in a gradient-dependent medium, *Int. J. Solids Struct.*, **30**, 1153–1171.
Smith, W. D. (1973). A non-reflecting boundary for wave propagation problems, *J. Comput. Phys.*, **15**, 492–503.

Song, C. and Bazyar, M. H. (2008). Development of a fundamental-solution-less boundary element method for exterior wave problems, *Commun. Numer. Methods Eng.*, **24**, 257–279.

Truty, A. and Zimmermann, T. (2003). Stabilized mixed FE formulations for materially nonlinear two-phase fully saturated media, *Proceedings of CMM-2003*, June 3–6, Gliwice, Poland

Truty, A. and Zimmermann, T (2006) Stabilized mixed finite element formulations for materially nonlinear partially saturated two-phase media, *Comp. Methods. Appl. Mech. Eng.*, **195**, 1517–1546.

Vardoulakis, I. and Sulem, J. (1995). *Bifurcation Analysis in Geomechanics*, Blackie Academic & Professional.

White, J. A and Borja, R. I (2008). Stabilized low-order finite elements for coupled solid-deformation/fluid-diffusion and their application to fault zone transients, *Comput. Methods Appl. Mech. Eng.*, **197** 4353–4366.

White, W., Valliappan, S. and Lee, I. K. (1977). Unified boundary for finite dynamic models, *ASCE EM*, **103**, EM5, 949–965.

Wolf, J. P. (1988). *Soil-Structure-Interaction Analysis in Time Domain*, Prentice-Hall, NJ.

Wolf, J. P. and Song, C. (1996). *Finite Element Modelling of Unbounded Media*, John Wiley & Sons, Chichester.

Wolf, J. P. and Song, C. (1996). *Finite-Element Modelling of Unbounded Media*, John Wiley & Sons Ltd.

Yan, J. Y., Zhang, C. H. and Jin, F. (2004). A coupling procedure of FE and SBFE for soil–structure interaction in the time domain, *Int. J. Num. Methods. Eng.*, **59**, 1453–1471.

Zhang, H. W. and Schrefler, B. A (2000). Gradient dependent plasticity model and dynamic strain localisation analysis of saturated and partially saturated porous media: one dimensional model. *Eur. J. Mech. A Solids.*, **19**, 503–524.

Zhang, H. W., Schrefler, B. A. (2004). Particular aspects of internal length scales in strain localization analysis of multiphase porous materials, *Comput. Methods Appl. Mech. Eng.*, **193**, 2867–2884.

Zhang, M. and Shu, C.-W. (2003). An analysis of three different formulations of the discontinuous Galerkin method for diffusion equations, *Math. Models Methods Appl. Sci.*, **13**, 395–413.

Zhang, H. W, Sanavia, L. and Schrefler, B. A. (1999a). An internal length scale in dynamic strain localisation of multiphase porous media, *Mechanics of Cohesive-Frictional Materials*, **4**, 443–460.

Zhang, X., Wegner, J. L. and Haddow, J. B (1999b). Three dimensional dynamic soil-structure interaction analysis in the time domain, *Earthq. Eng. Struct. Dyn.*, **28**, 1501–1524.

Zienkiewicz, O. C. and Codina, R. (1995). A general algorithm for compressible and incompressible flow. Part I: The split characteristic based scheme, *Int. J. Numer. Methods Fluids*, **20**, 869–885.

Zienkiewicz, O. C. and Huang, G. C. (1990) A note on localization phenomena and adaptive finite element analysis in forming processes, *Comm. Appl. Num. Meth.*, **6**, 71–76.

Zienkiewicz, O. C. and Newton, R. E. (1969). Coupled Vibrations of a Structure Submerged in a Compressible Fluid, *Proceedings of the International Symposium on Finite Element Techniques*, Stuttgart, p1–15, May.

Zienkiewicz, O. C. and Taylor, R. L. (1991). *The Finite Element Method – Volume 2: Solid and Fluid Mechanics, Dynamics and Non-linearity* (4th edn), McGraw-Hill Book Company, London.

Zienkiewicz, O. C. and Taylor, R. L. (2000). *The Finite Element Method – Volume 1: Basic Formulation and Linear Problems* (4th edn), McGraw-Hill Book Company, London.

Zienkiewicz, O. C. and Wu, J. (1991). Incompressibility without tears! How to avoid restrictions of mixed formulations, *Int. J. Num. Methods. Eng.*, **32**, 1184–1203.

Zienkiewicz, O. C. and Wu, J. (1994). Automatic directional refinement in adaptive analysis of compressible flows, *Int. J. Num. Methods. Eng.*, **37**, 13, 2189.

Zienkiewicz, O. C. and Xie, Y. M. (1991). Analysis of the lower san fernando dam failure under earthquake, *Dam Eng.*, **2**, 307–322.

Zienkiewicz, O. C. and Zhu, J. Z. (1992). The superconvergent patch recovery (SPR) and adaptive finite element refinement, *Comp. Methods. Appl. Mech. Eng.*, **101**, 207–224.

Zienkiewicz, O. C., Bicanic, N. and Shen, F. Q. (1987a). Single step averaging generalized Smith boundary – (a) transmitting boundary for computational dynamics, in *Proceedings of the International Conference on Numerical Methods in Engineering: Theory and Applications (NUMETA 87)*, Pande, G. N. and Middleton, J. (Eds), Vol. **II**, paper T49/1, Swansea, 6–10 July 1987, Martinus Nijhoff Publishers, Dordrecht.

Zienkiewicz, O. C., Clough, R. W. and Seed, H. B. (1987b). Earthquake analysis procedures for dams, *CIGB ICOLD Bulletin* 52.

Zienkiewicz, O. C., Bicanic, N. and Shen, F. Q. (1988). Earthquake input definition and the transmitting boundary condition, in *Advances in Computational Nonlinear Mechanics*, I. St. Doltsinis (Ed), Springer-Verlag, 109–138.

Zienkiewicz, O. C., Huang, M., Wu, J. and Wu, S. (1992). A new algorithm for coupled soil-pore fluid problem, *Shock. Vib.*, **1**, 215–233.

Zienkiewicz, O. C., Huang, M., Wu, J. and Wu, S. (1993). A new algorithm for the coupled soil-pore fluid problem, *Shock. Vib.*, **1**, 3–14.

Zienkiewicz, O. C., Huang, M. and Pastor, M. (1995a). Localisation problems in Plasticity using finite elements with adaptive remeshing, *Int. J. Num. Anal. Geomech.*, **19**, 127–148.

Zienkiewicz, O. C., Pastor, M. and Huang, M. (1995b). Softening, localisation and adaptive remeshing. Capture of discontinuous solutions, *Comput. Mech.*, **17**, 98–109.

Zienkiewicz, O. C., Boroomand, B. and Zhu, J. Z. (1999a). Recovery procedures in error estimation and adaptivity Part I: Adaptivity in linear problems, *Comput. Methods Appl. Mech. Eng.*, **176**, 1–4, 6 July 1999, 111–125.

Zienkiewicz, O. C., Nithiarasu, P., Codina, R., Vazquez, M., Ortiz, P. (1999b). The characteristic-based-split procedure: an efficient and accurate algorithm for fluid problems, *Int. J. Numer. Methods Fluids.*, **31**, 359–392.

Zienkiewicz, O. C, Taylor, R. L. and Zhu, J. Z. (2005a). *The Finite Element Method: Its Basis and Fundamentals*. Butterworth-Heinemann

Zienkiewicz, O. C, Taylor, R. L. and Fox, D. D. (2005b). *The Finite Element Method for Solid and Structural Mechanics*. Butterworth-Heinemann

Zienkiewicz, O. C, Taylor, R. L. and Nithiarasu, P. (2005c). *The Finite Element Method – Volume 3 The Finite Element Method for Fluid Dynamics*, Butterworth-Heinemann

6

Examples for Static, Consolidation, and Hydraulic Fracturing Problems

6.1 Introduction

In this chapter, we deal with the solution of static and quasi-static problems in which dynamic (inertial) effects are negligible and with hydraulic fracturing where inertia effects may be included.

We are concerned with the deformation and movement of the soil or of its associated foundation. An excessive amount of the latter is a *measure of failure* generally or loss of serviceability where the stress distribution or its magnitude of pore pressures is indicative of the state of the material stressed. It is the *deformation and displacement* which are observable and must be determined. For these, the knowledge of the constitutive relation discussed in Chapter 4 is of paramount importance, but the simplest constitutive law, which answers the question posed by the engineers and which provides the determination of failure, is to be used at all times. Failure is sometimes associated with continuing displacement without a load increment. This is a definition that is often accepted. However, on occasion, a finite displacement can be specified as a failure by the engineer and knowledge of displacements is important even if these are not excessive. In the first three chapters of this book, we have formulated the dynamic problem and its solution with the time dependence being retained in the final discretized equations.

It is clear (as indicated earlier) (see Equation (3.66)) that the problems of consolidation can be directly solved by the code based on our formulation, as with slow motion, the dynamic effects become automatically negligible. Less computational effort is involved if the acceleration terms are neglected and the GN11 scheme is used for both the skeleton displacement and pore pressure. However, it is not obvious that static problems can be directly dealt with by the general program though it has been customary to create special programs for such analyses. In Section 3.2.4, we have shown how the static problem can be dealt with by the dynamic code without any loss of efficiency when an appropriate time-stepping scheme, i.e. the GN00 scheme, is employed.

In this chapter, we shall introduce first in Section 6.2 some typical static problems using a nonassociative Mohr–Coulomb material model. Particular attention is paid to the effect of the plastic dilatancy effect. Section 6.3 deals with seepage while Section 6.4 is devoted to various problems concerning consolidation, including a 3-D example with adaptivity in time. Lastly, Section 6.5 addresses pressure-driven fracture in fully saturated porous media, both in quasi-static and dynamic situations.

Computational Geomechanics: Theory and Applications, Second Edition. Andrew H. C. Chan, Manuel Pastor, Bernhard A. Schrefler, Tadahiko Shiomi and O. C. Zienkiewicz.

6.2 Static Problems

In this section, we shall present some typical problems of static analysis both of the fully drained and totally undrained kind (the latter usually involving fairly rapid load application during which seepage can be considered negligible).

In this section, we shall endeavor to show the reader that numerical solutions, though costly by comparison with the simple limit methods which are widely and often successfully used in geomechanics, are necessary and provide otherwise unavailable information.

The examples chosen will show that:

a) For a relatively unconstrained situation, such as that involved in the stability of embankments and some foundation problems, the failure loads obtained by numerical computation are not very dependent on the plastic flow rules and again match well the predictions by conventional (limit-based) analysis.
b) For situations in which nonhomogeneous material distribution or geometry provides an appreciable amount of confinement, limit-based methods do not predict the failure loads satisfactorily.
c) Drained and undrained behavior can be studied effectively by a single specification of material properties using the effective-stress concept, and, finally
d) The solution can be quite sensitive to the form of the yield section assumed in the π-plane.

In all the examples, we shall specify the material properties by a yield surface of Mohr–Coulomb type with a straight-line tangent and in which "yield" is assumed to occur when the shear stress τ exceeds a given cohesion c' and a friction angle ϕ':

$$\tau \leq c' - \sigma_n' \tan \phi' \tag{6.1}$$

where σ_n' refers to the normal tensile effective stress. The superscript $'$ implies that these quantities are effective stress parameters. For the dry problem, the effective stress will be the same as the total stress σ. In this section, as effective parameters are always used, there is no confusion if the superscript $'$ is dropped.

6.2.1 Example (a): Unconfined Situation – Small Constraint

In this set of examples, we shall study the importance of associative and nonassociative flow rules and will find that here the choice of the flow rule is of almost negligible importance as far as failure loads are concerned.

6.2.1.1 Embankment

The first example is that of undrained soil in an embankment in which the angle is steeper than that of the internal friction and in which cohesion has been added. The computation was carried out by Zienkiewicz et al. (1975) from which Figure 6.1 is taken. In this problem, the collapse is achieved by progressively reducing the cohesion. The value of the cohesion at the collapse point is identified as that obtained by slip circle analysis (despite the coarse mesh employed). Further, the difference in the collapse load between the assumptions of associative and nonassociative behavior is very small.

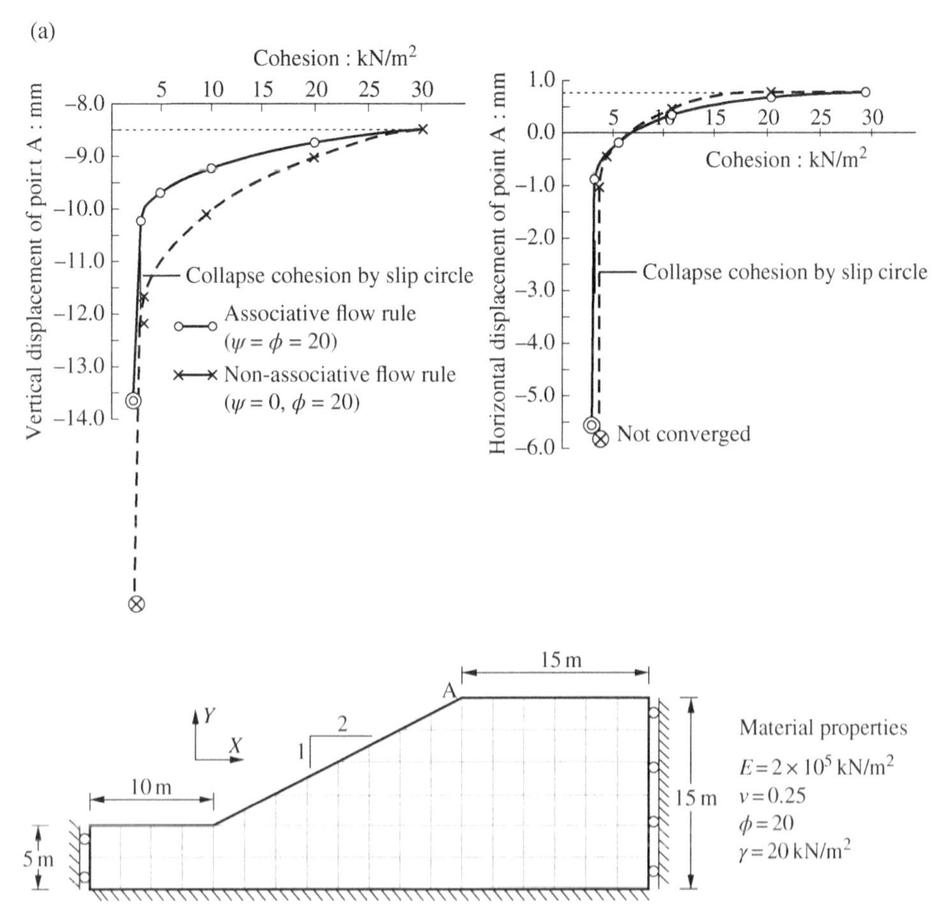

Figure 6.1 Embankment deformation flow patterns and maximum effective shear strain contours (gravity constant; progressive decrease of cohesion). (a) Associative $\theta = \phi = 20°$; (b) Nonassociative $\theta = 0, \phi = 20°$. *Source:* Reproduced from Zienkiewicz et al. (1975) by permission of the Institution of Civil Engineers.

6.2.1.2 Footing

This example, also taken from the same reference as the previous one, shows the collapse of a footing exerting a uniform load on the soil stratum. Here no exact closed-form solution exists, but for collapse, a mechanism suggested by Prandtl (1921) and Terzaghi (1943) gives loads that are compared with the numerical ones in Figure 6.2. Once again very close results were obtained for the limit loads by both associated and nonassociated plasticity models.

The two previous examples suggest that numerical analysis adds little to the solution of the problem from the point of view of practical geomechanics. Indeed, the main addition appears to be only the increased cost of analysis and little additional information has been gained. This, however, is just *not true* as we shall illustrate in the following sections with examples in which some degree of constraint exists.

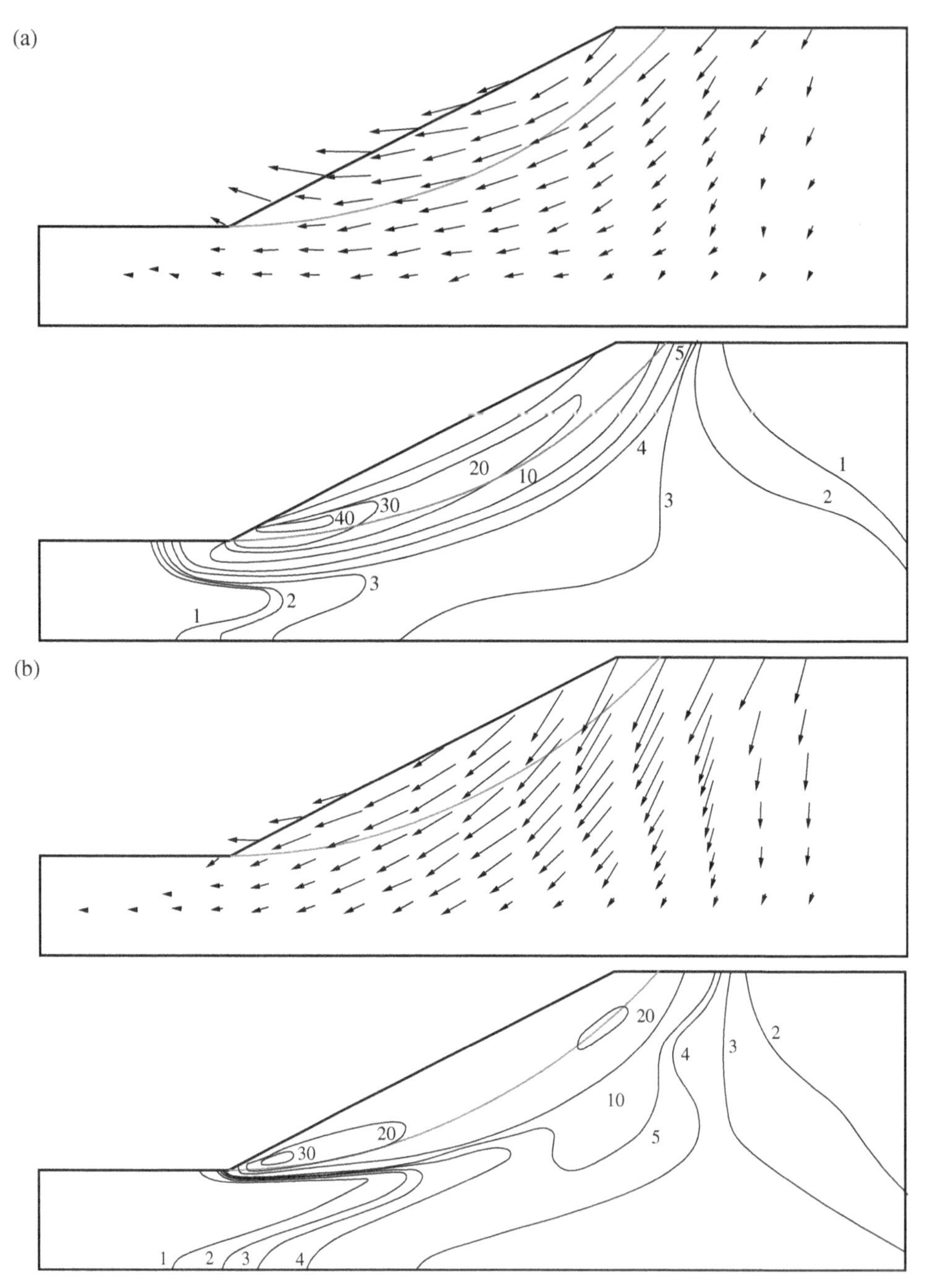

Figure 6.1 (Continued)

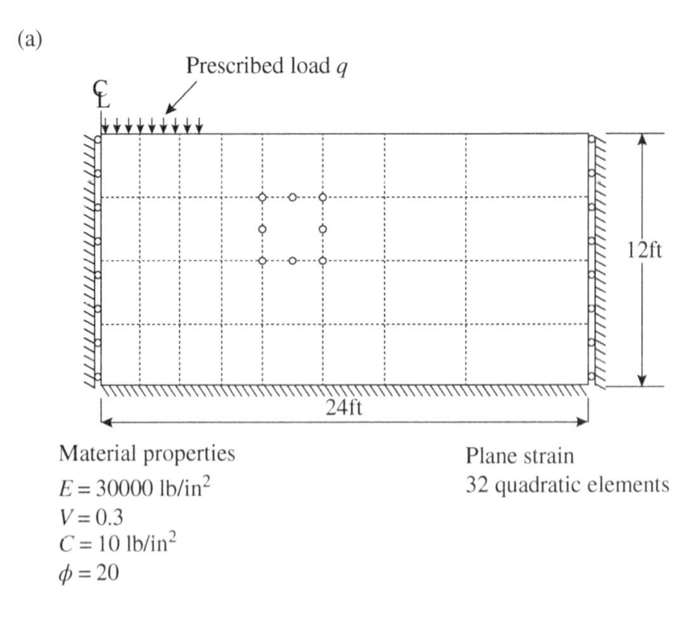

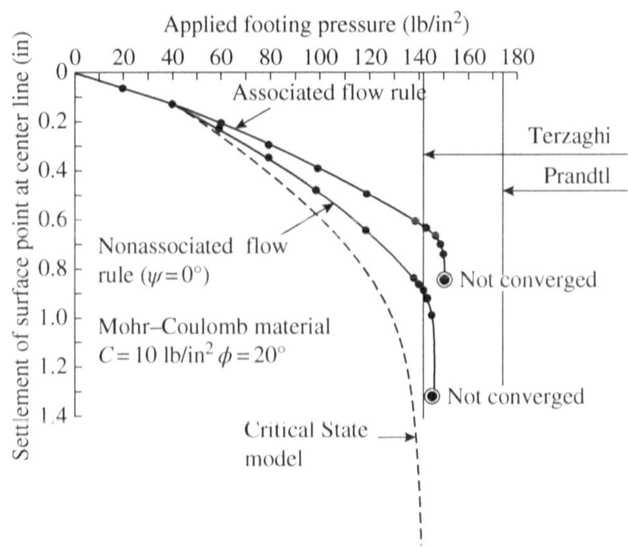

Figure 6.2 (a) Strip load on a foundation of a weightless c–ϕ material; ideal plasticity with associated and nonassociated (non-dilatant) flow rules and strain-hardening plasticity. Mesh and load-deformation behavior. (b) Relative plastic velocities at collapse (drained behavior) (i) associated Mohr–Coulomb (dilatant) (ii) nonassociated Mohr–Coulomb (zero dilatancy) (iii) strain-dependent critical state model. *Source:* Reproduced from Zienkiewicz and Cormeau (1974) by permission of John Wiley & Sons Limited.

(b)

(i)

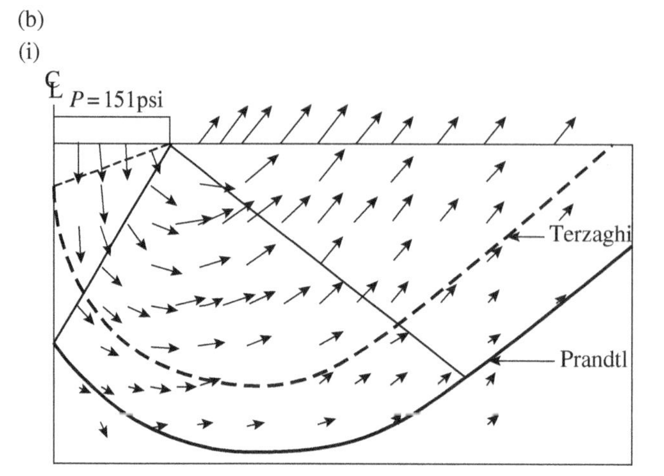

(ii)

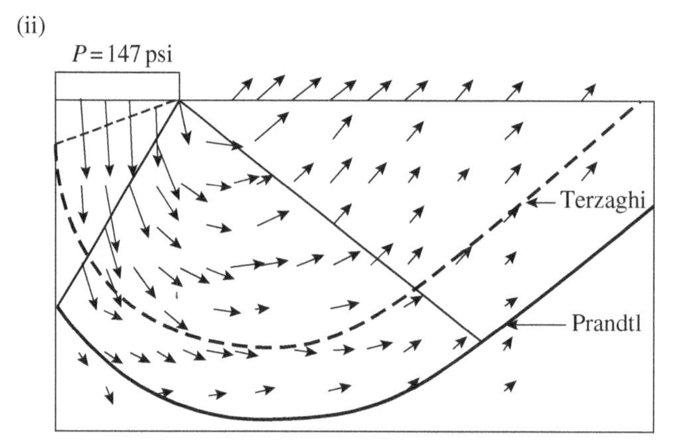

(iii)

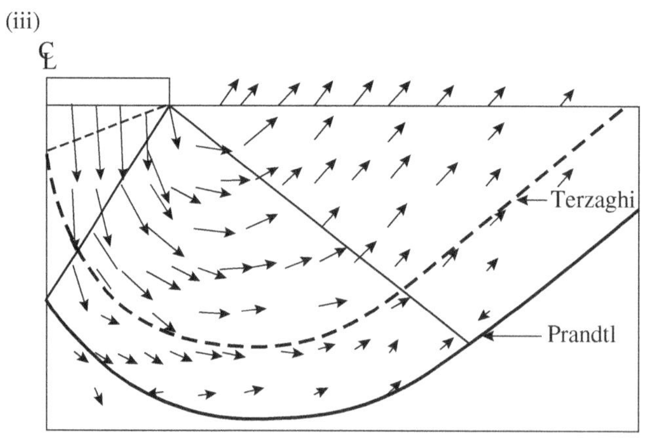

Figure 6.2 (Continued)

6.2.2 Example (b): Problems with Medium (Intermediate) Constraint on Deformation

The first example here is that of a *heterogeneous embankment* illustrated in Figure 6.3. Here the different plastic displacements of each layer are such as to cause a collapsed load for which simple geomechanical methods are not sufficient. In the table attached to the figure, we show that a limit approach gives safety factors ranging from 1.09 to 1.54 while numerical solutions show little influence on the flow rule employed giving 1.165 for both associative and nonassociative flows.

The second example illustrates however a case where differences between associated and nonassociated rules become appreciable. This case is that of an *axisymmetric triaxial sample* loaded between two rough platens as shown in Figure 6.4.

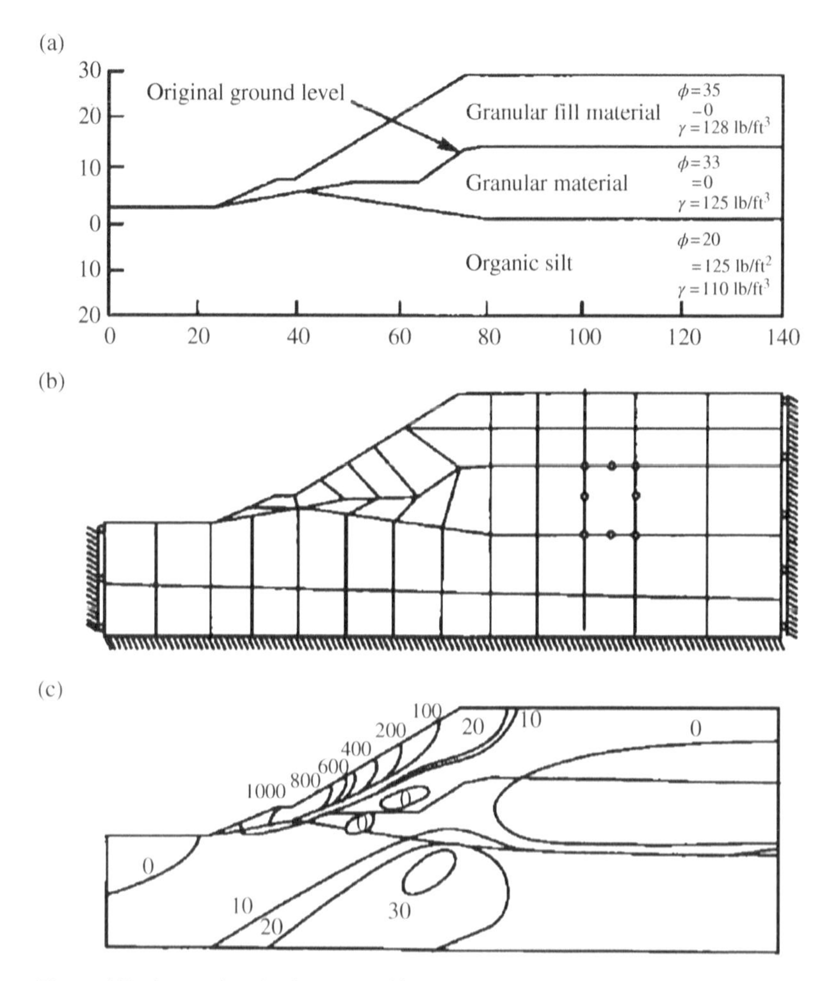

Figure 6.3 Layered embankment problem (a) geometry and material properties; (b) finite element mesh (53 quadratic isoparametric elements); (c) relative shear strain rate contours at collapse; (d) table of safety factors using various methods. *Source:* Reproduced from Zienkiewicz et al. (1975) by permission of the Institution of Civil Engineers.

(d)

Method	Safety factor	References
Whitman–Bailey	1.24–1.26	Whitman–Bailey (1967)
Bishop	1.33	Whitman–Bailey (1967)
Fellenius	1.09	Whitman–Bailey (1967)
Sarma	1.542	Sarma (1973)
Morgenstern–Price	1.557	Sarma (1973)
Bell	1.49	Bell et al. (1969)
Associated Mohr–Coulomb	1.165	
Nonassociated Mohr–Coulomb	1.165	

Figure 6.3 (Continued)

Different degrees of dilatancy now affect the load appreciably.

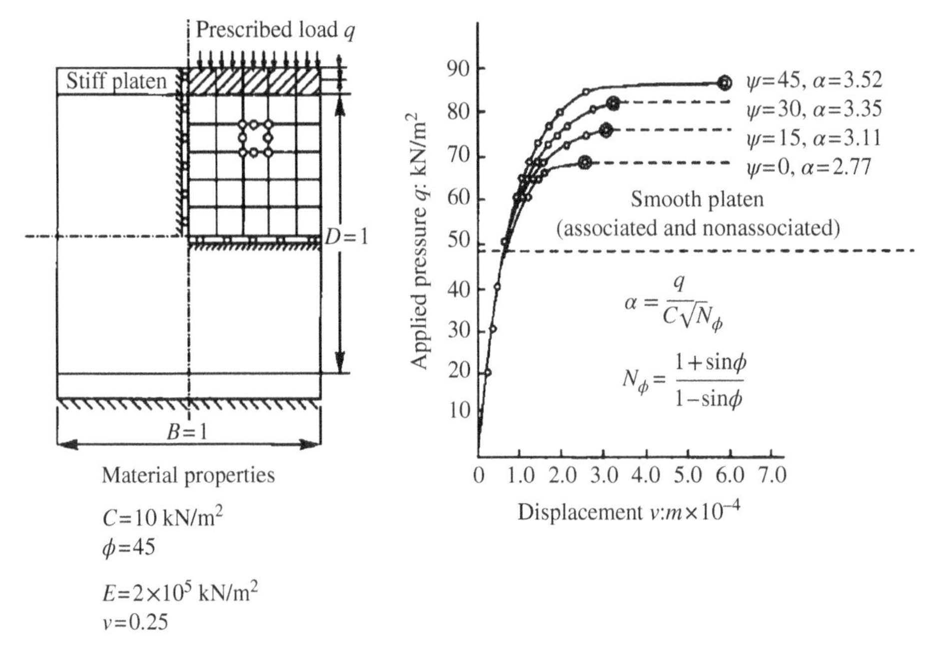

Figure 6.4 Axisymmetric sample between rough platens. Effect of degree of dilatancy $\psi = 0$ (zero dilatancy) and $\psi = \phi = 45°$ (fully associated flow rule). *Source:* Reproduced from Zienkiewicz et al. (1977) by permission of John Wiley & Sons Limited.

6.2.3 Example (c): Strong Constraints – Undrained Behavior

Probably, the highest degree of constraint on the deformation of the soil skeleton is the behavior occurring during undrained deformation. In such conditions, the total volumetric strain is controlled by the compressibility of the pore fluid and this is generally small, resulting in almost no overall change in volume during deformation.

Such behavior has two consequences:

1) If the compressive modulus (bulk modulus) of the skeleton is small compared with that of the fluid, any changes of mean total stress $\Delta\sigma_m$ are compensated by equal and opposite

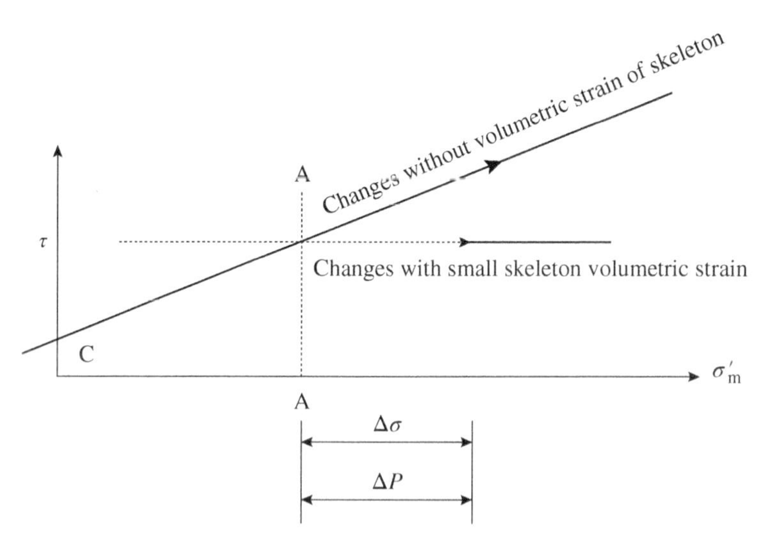

Figure 6.5 The Mohr–Coulomb trace in the mean stress – deviatoric stress plane. Changes of total stress from point A. *Source:* Reproduced from Zienkiewicz and Comeau (1974) by permission of John Wiley & Sons Limited.

changes of pore pressure Δp and the mean *effective* stress remains unchanged, i.e. $\Delta\sigma'_m = 0$.

Thus, if the material is initially at the point of yielding as shown in Point A of Figure 6.5, any deviatoric *stress* changes will occur without changes of the mean stress and the material will behave like a von Mises (or Tresca) solid with respect to total stress changes provided no volumetric stress occurs during such straining.

2) If the plastic yielding is such that deviatoric strain changes of the skeleton must be accompanied by volumetric strain changes (as, for instance, required by associative laws), then the only way to achieve overall straining without volume change is to compress (or expand) the *fluid*. With the volume expansion required by the associated flow rule, the material will develop negative pore pressures during plastic straining and will gain strength continuously.

Figure 6.6 shows the same footing problem as that in Figure 6.2 but now solved (on the same mesh) introducing the undrained assumption. It will be observed that for a nonassociative behavior with no dilatancy in failure, or for a critical state model where failure occurs with no volume changes, the failure loads are almost identical with those of total stress analysis with Tresca assumption. However, the associative, expanding material allows no overall failure.

It must be remarked that in the present case, the failure loads are governed entirely by the cohesion existing in the material as otherwise the strength would be simply zero when gravity was absent giving an entirely different starting point (A) in Figure 6.5 for the material at the different response.

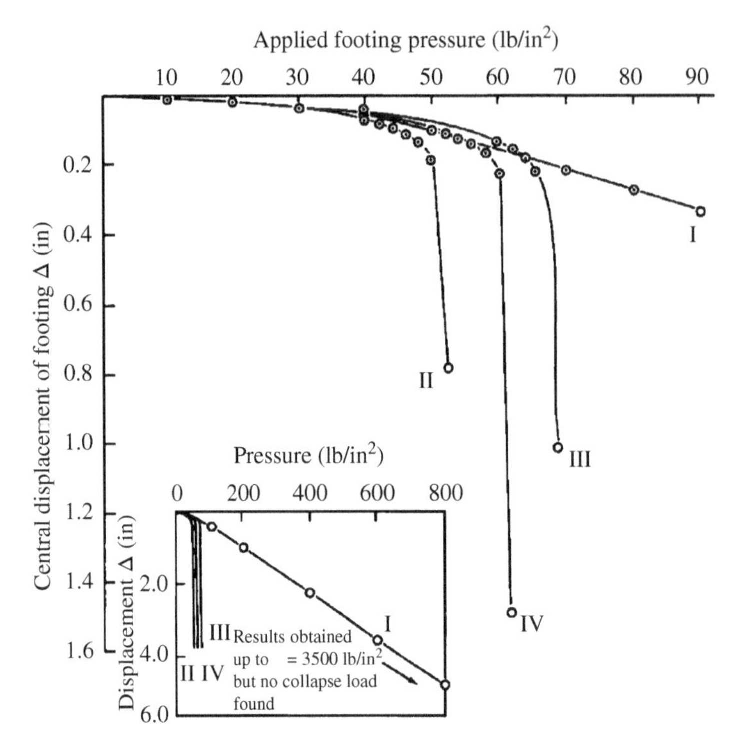

Figure 6.6 Load deformation characteristics (undrained conditions) for plane footing. *Source:* Reproduced from Zienkiewicz and Comeau (1974) by permission of John Wiley & Sons Limited.

6.2.4 Example (d): The Effect of the π Section of the Yield Criterion

In all of the previous examples, we have assumed that the full Mohr–Coulomb relation is employed which requires the yield (and/or flow) surface becoming, in the absence of friction, the Tresca yield surface rather than the better known and simpler von Mises one. However, with this type of definition, difficulties arise at corners that need special treatment (see Zienkiewicz and Taylor 1991) and many simplifications have been suggested in the literature. In Figure 6.7, we show the π plane section of the Mohr–Coulomb surface for $\phi = 20°$ and various approximations of it, which appear feasible. As discussed in Chapter 4, many possibilities arise. One is to use "smoothing" and avoid corners in the manner described by Zienkiewicz and Pande (1977). Another is to replace the surface with a circle as is done in the well-known Drucker–Prager surface (1952).

The effect of such an approximation is surprising as shown in Figure 6.8 where the same foundation problem as that of Figure 6.2 is used as a test bed. It is of interest to note that the full Mohr–Coulomb and the extension cone in the Drucker–Prager give close representation, although the smoothing appears to cause about a 20% difference in the collapse load!!

The simple set of examples presented was computed directly before the advent of the dynamic code. We have, however, recomputed several typical cases using the method suggested in this chapter and have found an identical failure load without extra effort. Some

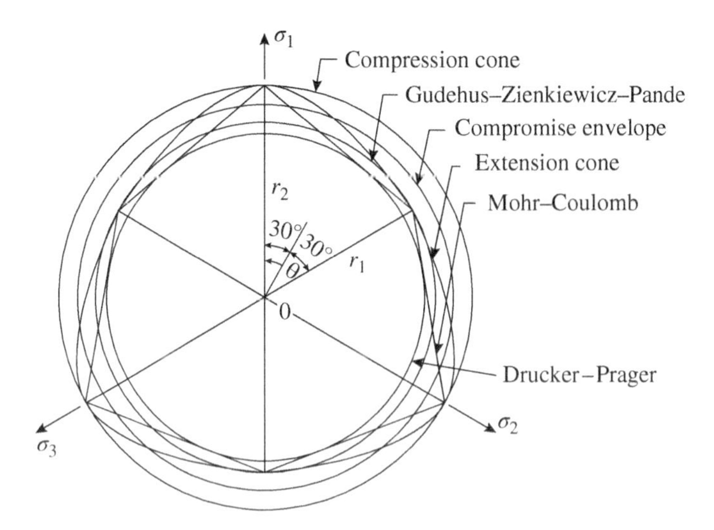

Figure 6.7 The π plane section of the Mohr–Coulomb surface with ϕ = 20° and various smooth approximations. *Source:* Reproduced from Zienkiewicz et al. (1977) by permission of John Wiley & Sons Limited.

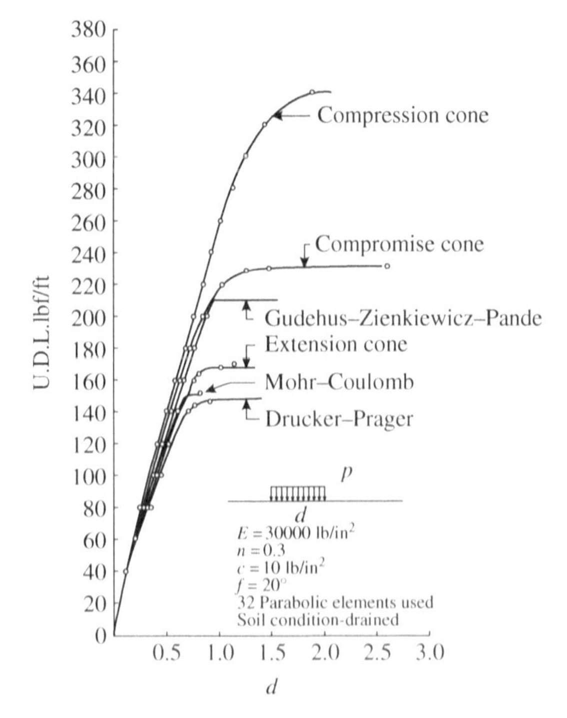

Figure 6.8 Load-deformation curves for ideal associated plasticity for various forms of the Mohr–Coulomb approximation (solution on the same mesh as in Figure 5.2). *Source:* Reproduced from Zienkiewicz (1978) by permission of John Wiley & Sons Limited.

difference in displacement response is obtained due to the difference in plasticity formulation (viscoplasticity formulation had been used in these examples) and a nonlinear iterative scheme has also been used.

6.3 Seepage[1]

In this section, it will be illustrated how the procedures shown in Chapters 2 and 3 can be used to solve seepage problems. Water seepage is encountered, for instance, in a study of earth dams or concrete dams, or in the analysis of foundations or slope stability to mention just a few examples. Seepage problems are either confined when the boundary conditions of the interested domain are known or unconfined when the boundary conditions are not all known, i.e. when the free (Phreatic) surface has to be established.

There are several approaches to deal with seepage analysis. For confined seepage analysis, the flow conservation equation for liquid water (2.15) in steady-state conditions and Darcy's law gives the following equation

$$\nabla^{\mathrm{T}} \boldsymbol{k} \nabla h = 0 \tag{6.2}$$

where $\boldsymbol{k}$ is the permeability tensor and h is the piezometric level or hydraulic head $h = \dfrac{p}{\rho_w g} + z$ with z the elevation above some datum, which is reduced to the Laplace equation $\nabla^2 h = 0$ in case of isotropic permeability.

For unconfined seepage, a large number of seepage problems have been solved since the early 1970s as a free boundary problem following Baiocchi's method (Baiocchi et al. 1973; Bardet and Tobita 2002; Bruch 1991) or more generally based on a Signorini (see, e.g. Chen et al. 2011) type condition. These methods approach the unconfined seepage problem from the viewpoint of the saturated zone and give a black and white picture where the capillary fringe is neglected, i.e. there is only a wet and a dry region.

This approach requires the solution of additional conditions of the type

$$\mathbf{n} \cdot \mathbf{v} = 0 \quad h = z \quad \text{on} \quad \delta\Omega(t) \tag{6.3}$$

which express the fact that the flux $\mathbf{n} \cdot \mathbf{v}$ across the moving free surface $\delta\Omega(t)$ is zero and the pressure too. This is rather cumbersome in 3D situations.

The problem of confined and unconfined seepage can however also be solved with a unique approach with the equations developed in this book. In this case, the problem is seen from the partially saturated point of view, where the free surface is simply a surface with zero capillary pressure and gas pressure at atmospheric value separating the fully and the partially saturated domains. Within this approach, the free surface is given directly by the quasi-static solution of the water and gas mass balance equations (2.41) and (2.42) or simply (2.33) with the static air phase assumption, coupled with the linear momentum balance equation (2.34). This procedure also yields directly the wetting deformation if needed and can also be used for the calculation of the initial hydromechanical conditions for the

1 The authors gratefully acknowledge the collaboration of M. Passarotto and L. Sanavia in the elaboration of this section.

transient analyses. The computation of this numerical solution, which is fully coupled and can be computationally expensive, can be accelerated using fictitiously high permeability values because only a steady-state solution is being looked for.

This approach can be simplified by solving the unsaturated governing equations directly in steady-state conditions. In this case, the transient behavior ceases and thus all the time derivatives in the mentioned governing equations can be neglected, giving the following mass balance equations for water (6.4) and air (6.5) governing the water seepage and water pressure profile independently of the equilibrium equations (6.6):

$$\nabla^{\mathrm{T}}[k_w(\nabla p_w - \rho_w \mathbf{g})] = \mathbf{0} \tag{6.4}$$

$$\nabla^{\mathrm{T}}\left[k_g\left(\nabla p_g - \rho_g \mathbf{g}\right)\right] = \mathbf{0} \tag{6.5}$$

$$\nabla^{\mathrm{T}}\boldsymbol{\sigma} + \rho\mathbf{g} = \mathbf{0} \tag{6.6}$$

with Darcy's equation already incorporated in 6.4 and 6.5.

A few examples using this approach are hereby given.

Confined seepage is considered first. Figure 6.9 represents a submerged foundation, where the capillary pressure profile is defined in the boundaries of the soil involved in the analysis. Figures 6.10 and 6.11 represent respectively an inclined pile and a dam where the water pressures are defined on the boundaries.

Next, we consider examples dealing with unconfined seepage flow. Such analyses are, for instance, necessary to establish the initial hydromechanical conditions and, as a consequence, the water pressure profile, in successive transient analyses.

An example of unconfined seepage is the initial steady-state solution of San Fernando Dam, Figure 2.3.

Another example is related to the study of the seismic behavior of the earth dam of Acciano (Figure 6.12). This dam is located near Perugia in Italy and was stressed by the Umbria–Marche earthquake of September 1997 while the water level in the lake was only 5 m (Briseghella et al. 1999; Schrefler et al. 2011). The simulation was performed with the passive air phase assumption, i.e. with air pressure in the unsaturated region equal to the atmospheric pressure.

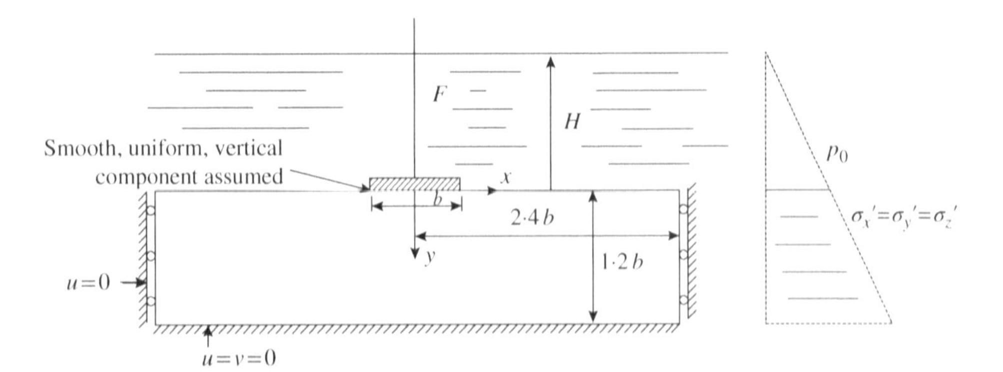

Figure 6.9 A typical example of confined seepage in a submerged structure foundation with plain strain conditions (Zienkiewicz et al. 1977).

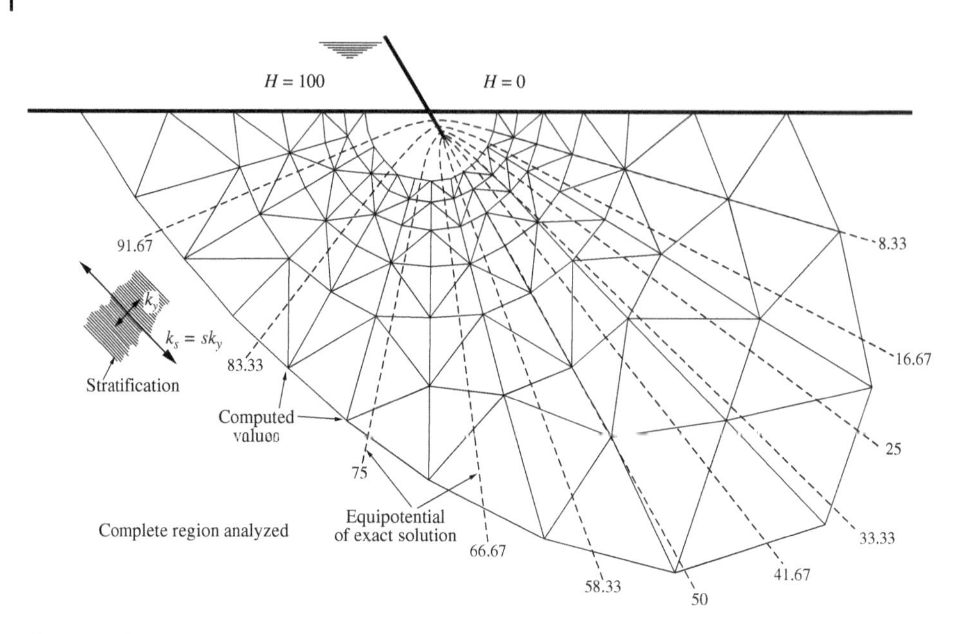

Figure 6.10 Flow under inclined pile wall in a stratified anisotropic foundation. *Source:* Reprinted from Zienkiewicz et al. (1966) with permission from ASCE.

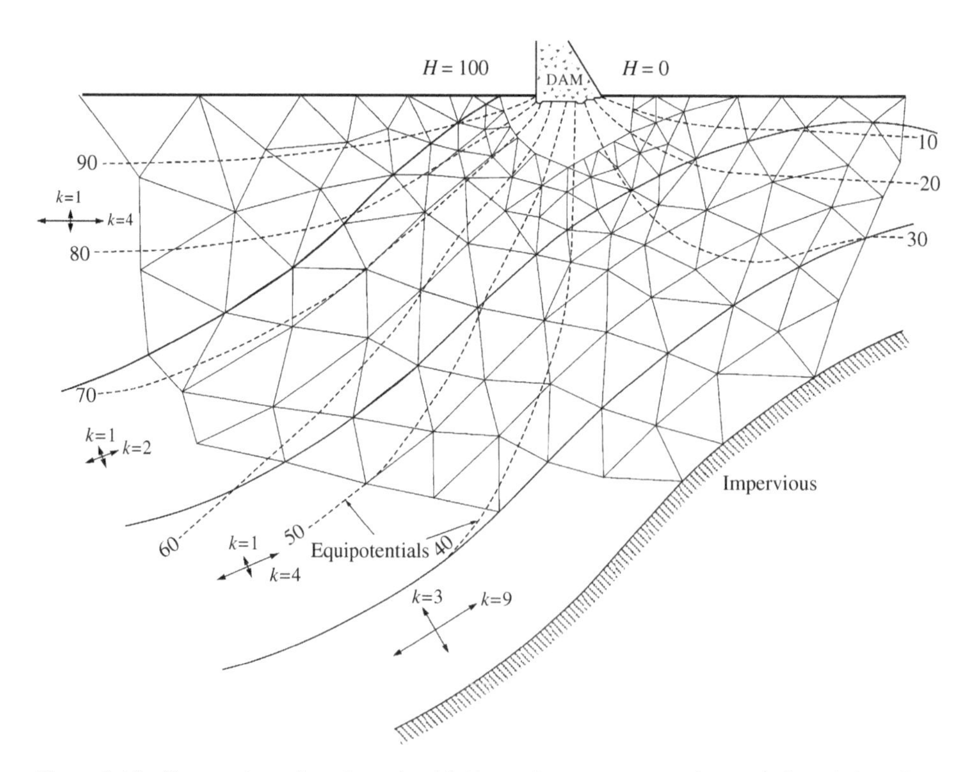

Figure 6.11 Flow under a dam through a highly nonhomogeneous anisotropic foundation. *Source:* Reprinted from Zienkiewicz et al. (1966) with permission from ASCE.

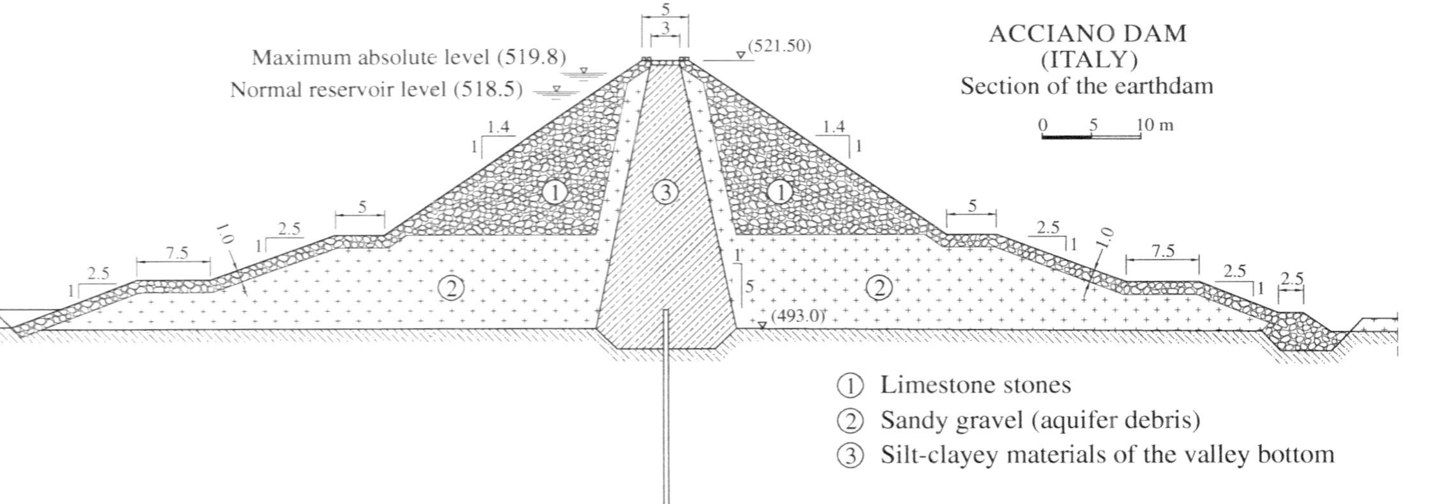

Figure 6.12 Vertical section of Acciano dam (Briseghella et al. 1999; Schrefler et al. 2011).

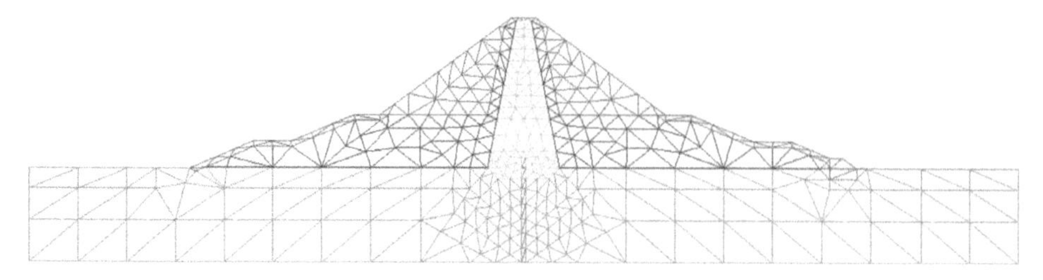

Figure 6.13 Finite element discretization with triangular elements, shades of gray corresponds to different materials (Briseghella et al. 1999; Schrefler et al. 2011).

The Swandyne code (Gawin et al. 1998; Xie 1990; Zhang et al. 2001; Zienkiewicz et al. 1999) was used and a comparison with the in-situ damage has been carried out (Briseghella et al. 1999; Schrefler et al. 2011). The dam and the surrounding soil are discretized with triangular isoparametric finite elements as in Figure 6.13. Small strains, small displacements, and plane strain conditions are assumed. The most important material parameters used in the computation are obtained from the data of the in-situ geotechnical analyses. The others are estimated by direct comparison between the given materials and similar well-known soils. During the seismic analysis, the solid skeleton is considered to be elastoplastic and the Mohr–Coulomb and Pastor–Zienkiewicz (Pastor et al. 1990; Zienkiewicz et al. 1999) laws are used for cohesive and granular materials, respectively (see Chapter 4). The partial saturation conditions are described using Safai–Pinder (1979) law.

The initial hydromechanical conditions for the seismic analysis have been obtained by a seepage analysis solving Equations (6.4)–(6.6) with the water level in the lake of 5 m, considering a linear elastic solid skeleton and assuming the gravity and external water pressure load to be applied without dynamic effects. The phreatic line is hence obtained, as in Figure 6.14, revealing the existence of a zone above it where capillary pressures develop. The cohesion resulting from these capillary pressures is taken into account in the model. The presence of the core dam in clayey silt from the local valley and the vertical wall in concrete (permeability of 1×10^{-12} m/s; for the soil foundation, fractured, a high permeability value 1×10^{-2} m/s has been chosen) do not seem to modify the water pressure profile because the hydraulic load due to the lake is 5 m only.

A second seepage analysis was performed with the maximum reservoir level of 27 m to highlight the influence of different permeabilities. Differently from the previous case, in Figure 6.15, the profile of water pressure is now influenced by the presence of the core dam and the vertical wall in concrete.

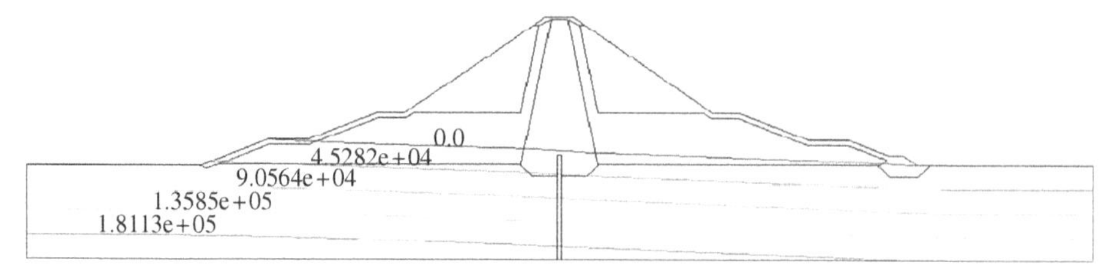

Figure 6.14 Contour lines of water pressure after the initial seepage analysis with a water level in the lake of 5 m (Briseghella et al. 1999; Schrefler et al. 2011).

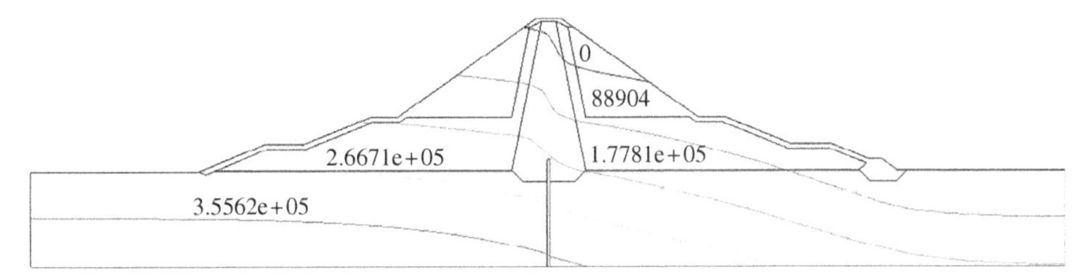

Figure 6.15 Contour lines of water pressure after the initial seepage analysis with a water level in the lake of 27 m (Briseghella et al. 1999).

6.3.1 Concluding Remarks

In this section, some approaches to solve seepage problems have been presented. It is shown that a general procedure to solve both confined and unconfined seepage problems can be obtained by solving the governing equations for the multiphase porous media in quasi-static conditions.

6.4 Consolidation[2]

In this section, we present three examples relating to consolidation in fully saturated soil, i.e. slow transient situations where dynamic effects can be neglected. In the first example, a benchmark problem is solved for which the analytical solution is known. Comparisons are made by controlling the errors of the numerical solution in terms of displacements, pressures, and energy. Since some variables are not given by the analytical solution, these are calculated using a fixed and very fine mesh, for which the overall results have been validated using the analytical solution. The second example deals with the compaction of an aquifer pumped from a fully penetrating well. In this case, infinite elements are used for the far-field solution (see Section 5.2). In the third example, a 3-D consolidation problem is solved by means of a Discontinuous Galerkin procedure in the time domain (DGT) and it is shown that such a procedure allows to effectively eliminate the spurious oscillations which often appear at the onset of the consolidation process when the suddenly applied load produces strong gradients in the involved fields. In such situations, the classical finite element method with continuous weighting and test functions, combined with a finite difference method in time, is not able to reproduce the discontinuous phenomena such as the jump in the pore-pressure field. This jump appears because the sudden applied external load is mainly supported by the pore pressure within the first time steps, the deformation of the solid skeleton being in this interval negligible. This effect leads to a jump in the pore pressure from the initial state value to a value that withstands the external force. Once the fluid is squeezed out through the drained boundary, the external load or part of the load is gradually shifted to the solid skeleton.

2 The authors gratefully acknowledge the collaboration of S. Secchi and L. Simoni in the elaboration of this section.

6.4.1 Benchmark for a Poroelastic Column

This example deals with a poroelastic column presented as a benchmark in Boone and Ingraffea (1990). The geometry and material data are summarized in Figure 6.16. Note that the undrained Poisson's ratio should have been 0.5. The value 0.33 is used for the comparison with Boone and Ingraffea (1990).

External boundary traction is instantaneously applied at the top (1 MPa) together with a fixed pressure (1 MPa). The analytical solution gives an initial vertical displacement at the top of the column

$$u^{0+} = \frac{1-2\nu_u}{1-\nu_u}\frac{qh}{2G} = 0.254 \text{ mm} \tag{6.7}$$

whereas the final vertical displacement is

$$u^{\infty} = (1-\overline{\alpha})\frac{1-2\nu}{1-\nu}\frac{qh}{2G} = 0.079 \text{ mm} \tag{6.8}$$

with a rebound of the top due to inflow. Initial pore pressure (except at the top boundary) is

$$p^{0+} = \frac{B(1+\nu_u)}{3(1-\nu_u)}q = 0.410 \text{ MPa} \tag{6.9}$$

The problem is firstly solved with a fixed fine 2-D mesh where linear interpolation over Delaunay triangles is used for spatial discretization. Results presented in Figure 6.17 show a complete agreement between analytical and numerical approaches.

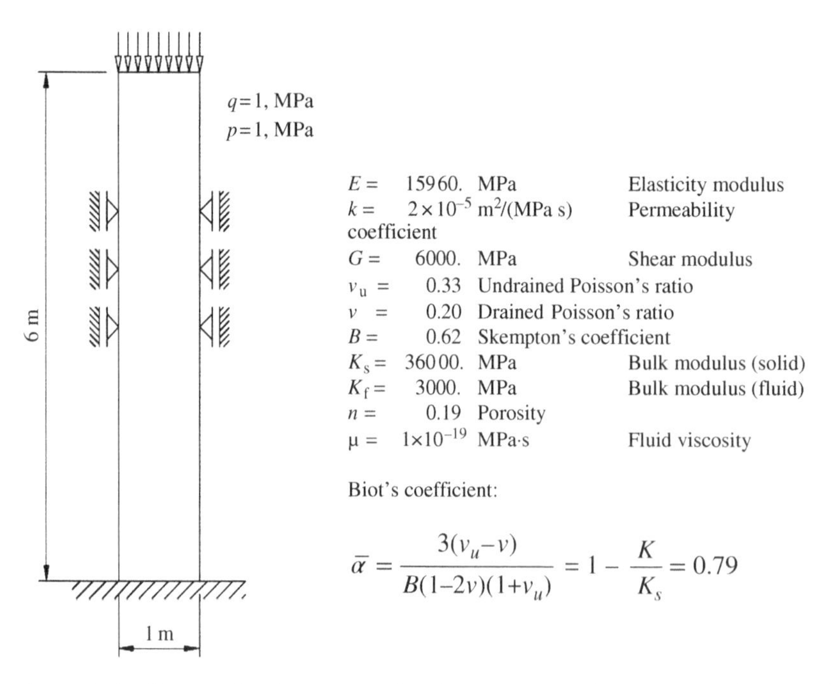

Figure 6.16 The investigated poroelastic column: geometry, boundary conditions, and material properties. *Source:* Reproduced from Simioni et al. (2007) by permission of John Wiley & Sons Limited.

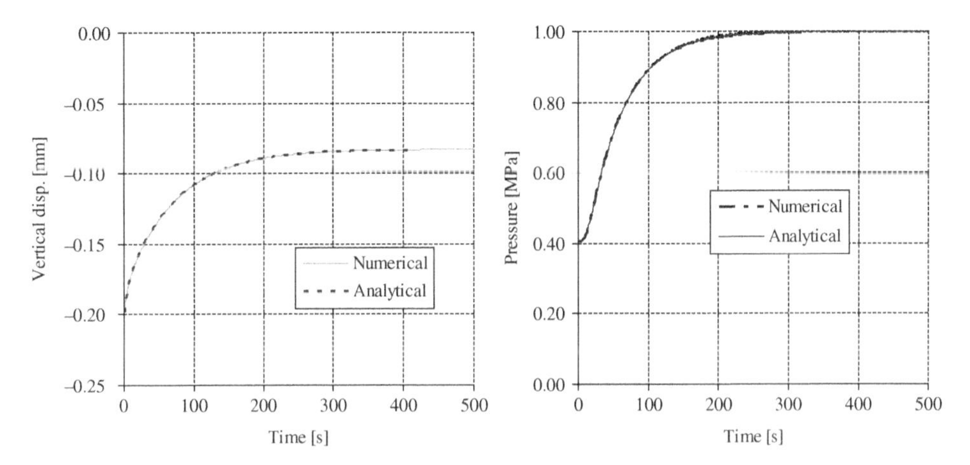

Figure 6.17 Analytical vs. numerical results for displacement at the top and pressure at the bottom of the column. *Source:* Reproduced from Simioni et al. (2007) by permission of John Wiley & Sons Limited.

We also test the nodal force projection algorithm described in Section 6.5 needed in case of remeshing due to adaptivity or fracture. The problem is therefore solved by using a continuously varying mesh. An element source (Secchi and Simoni 2003) is introduced moving from the bottom zone (point A, Figure 6.18) to the top (point B), then back at a fixed velocity, 20 mm/s. This results in a mesh that is artificially changed at each time step, even though the problem does not require this.

Two different calculations are performed, one by repeating the solution at the generic step m with the mesh of step $m + 1$ and the other without this repetition. In both cases, the results coincide with the analytical solution and, for a more convincing comparison, results are presented in terms of percentage errors. Figure 6.19 presents the error time histories $(x_{analytical}-x_{numerical})/x_{analytical}$ for three different variables x (vertical top displacement, strain energy, and base pressure). The agreement is always remarkable from which can be concluded that the errors are very small during the complete transient period. What is interesting to observe is the effect of repeating the solution of the previous step when the mesh is updated at the current one. The benefits of this strategy are evident: errors due to data transfer between the meshes are nearly halved for all parameters. This approach seems hence generally advisable and its computational cost is not very high even in nonlinear problems. Nevertheless, also the results obtained without this repetition are remarkable for quantities depending on the spatial gradient of the field variables, which confirm the efficiency of data transfer.

6.4.2 Single-Aquifer Withdrawal

The compaction of a single aquifer pumped from a fully penetrating well is now discussed. Infinite elements are used for the far-field solution (see Section 5.2). The problem at hand is solved with the full Biot's equations but without inertia terms (see Section 6.5). The following parameters are used for the geometry of Figure 6.20.

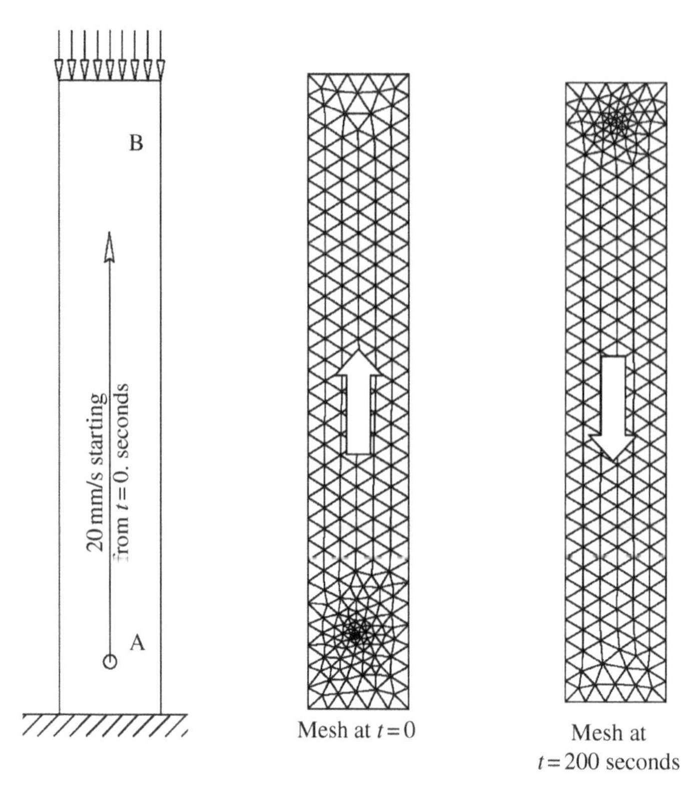

Figure 6.18 Continuously varying meshes: an element source is moving from point A to point B and then in the opposite way at a velocity of 20 mm/s. *Source:* Reproduced from Simioni et al. (2007) by permission of John Wiley & Sons Limited.

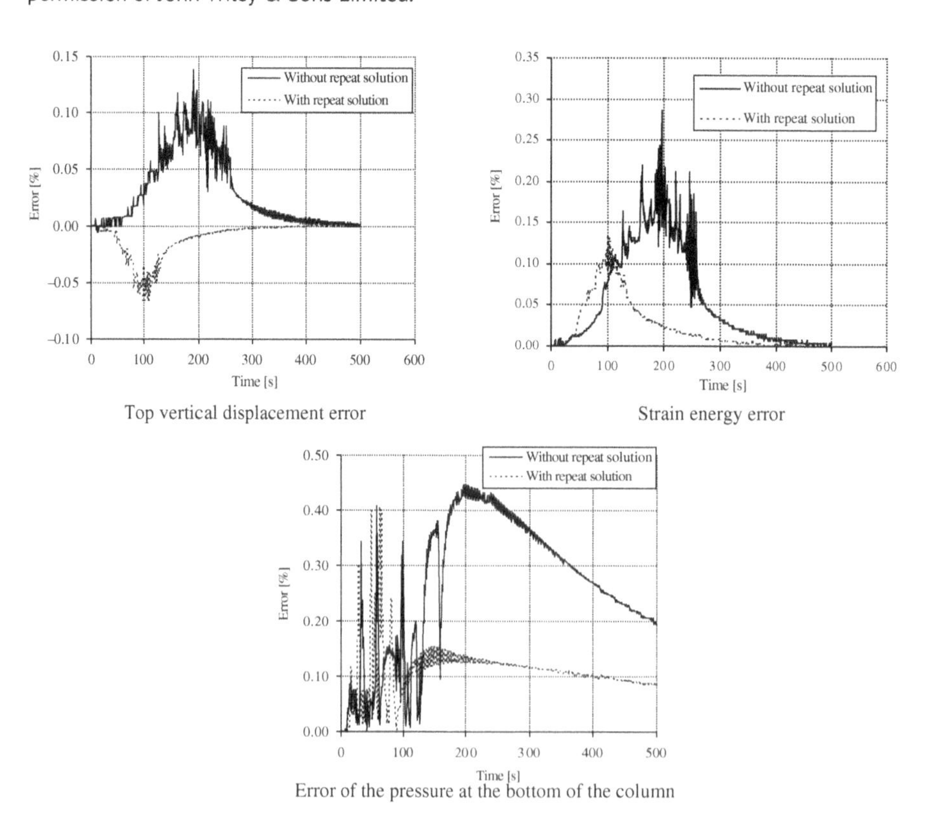

Figure 6.19 Percentage errors of the numerical solutions. *Source:* Reproduced from Simioni et al. (2007) by permission of John Wiley & Sons Limited.

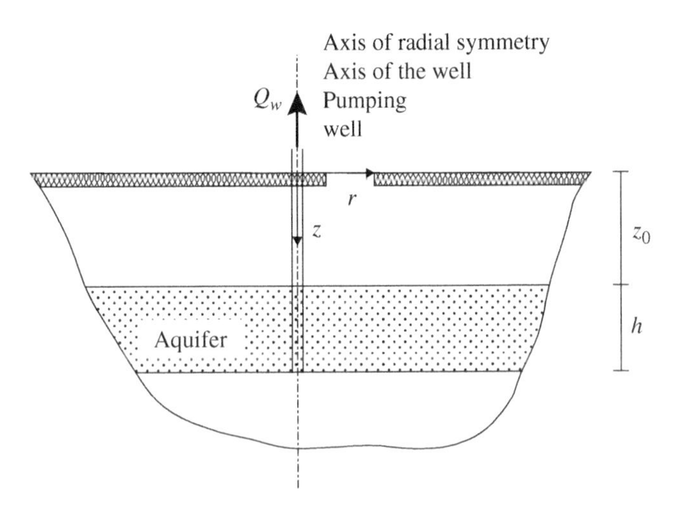

Figure 6.20 Embedded aquifer in a half-space. *Source:* Reproduced from Lewis and Schrefler (1998) by permission of John Wiley & Sons Limited.

Thickness of aquifer $h = 10$ m
Radius of well $r_o = 0.3$ m
Compressibility of water $\beta_w = 5 \times 10^{-10}$ Nm^{-2}
Specific weight of water $\gamma_w = 9800$ Nm^{-3}
Poisson's ratio for solid $\nu = 0.1$
Permeability of medium $k_w = 10^{-4}$ ms^{-1}
Porosity $n = 0.2$
Young's modulus $E = 2.2 \times 10^7$ Nm^{-2}
Pumping rate $Q = 6.81$ m^3/h

The top and bottom surfaces are assumed impervious and vertical displacements at the bottom and horizontal displacements at $r = r_o$ are restrained. The well is assumed to fully penetrate the stratum.

The aquifer is overlain and underlain by sediments with lower permeability values. Its lateral extension is very large so that unbounded conditions and axial symmetry are assumed to exist.

The usual numerical models in the literature consider two different settings for a linear elastic problem: an isolated aquifer with radial flow only and an aquifer embedded in a mechanically homogeneous half-space with zero permeability outside. The ensuing numerical solutions are here discussed and, when possible, compared with analytical ones. These solutions are useful to benchmark problems for testing consolidation programs.

The permeability outside the embedded aquifer is assumed to be 10^{-8} times smaller than in the aquifer to simulate numerically the condition of an impermeable surrounding half-space but Young's modulus is taken to be the same.

Several cases were solved in which the depth z_o of burial of the aquifer ranges from $w = z_o/h = 0$ to $w = z_o/h = 10$. In order to assess the effect of horizontal displacement boundary conditions, a further analysis was performed for the embedded aquifer, which was assumed to lie on smooth bedrock with variable overburden.

The far-field behavior for all the cases was approximated by means of infinite elements (see Section 5.2) since predictions are known to be sensitive to the position of the lateral boundary assumed in the numerical model (Simoni and Schrefler 1989). In this case, the poles are located on the axis of symmetry. The domain below the embedded aquifer is also modeled by finite and infinite elements, with poles located on the lower boundary of the aquifer.

The analytical solution for an isolated aquifer is quite simple under the assumptions of axisymmetric conditions, uniform distribution over the thickness of the aquifer for pore pressure and horizontal displacements, and plane incremental total stress (Bear and Corapcioglu 1981a). An analytical solution for the pressure field was obtained by Theis (1935) under the assumption of uncoupled Terzaghi–Jacob one-dimensional theory. By accounting for the deformability of the aquifer through the storativity coefficient S (Verruijt 1969), the drawdown v is given by

$$v = \frac{Q}{4\pi T} E_1\left(\frac{r^2 S}{4Tt}\right) \tag{6.10}$$

where T is the transmissivity coefficient, t is the time variable, and r is the radial distance. We recall that the storativity and transmissivity coefficients are respectively defined as the product of the aquifer thickness and the averaged values of the capacity term and permeability. Furthermore, E_1 is the exponential integral (*well function*)

$$E_1(x) = \int_x^{\infty} \frac{1}{u} e^{-u} du \tag{6.11}$$

In the coupled analytical solution by Bear and Corapcioglu (1981b), the drawdown has the same expression as (6.10) but the coefficient of consolidation corresponds to actual three-dimensional conditions. The results in terms of excess pore pressures are therefore different (Figure 6.21), which confirms the coupled nature of the problem. In the case analyzed, the differences are approximately 15% for $r = 5$ hours and $t = 12$ days but they increase with increasing distance from the well and decrease with time. Figure 6.21 also shows the good results obtained numerically using infinite elements to model the far-field.

The assumption of a one-dimensional coefficient of consolidation leads to a higher drop in pore water pressure (Figure 6.22). Hence, the use of these pore pressures in a two-step uncoupled model, where pore pressures are given as input (volume forces) to a solid mechanics model, would result in higher compactions with respect to the exact solution.

The isolated aquifer compaction is given as follows (Bear and Corapcioglu 1981a, 1981b)

$$\delta = \frac{Q}{4\pi C_v} E_1\left(\frac{r^2}{4C_v t}\right) \tag{6.12}$$

where C_v is the coefficient of consolidation of the isotropic medium and is defined as

$$C_v = \frac{k_w}{\alpha' + n/K_w} \tag{6.13}$$

The aquifer compressibility α' can be expressed as a function of Lame's constants, λ and μ, and depending on whether one-dimensional consolidation (uncoupled problem) or three-dimensional consolidation (fully coupled problem) is assumed, it takes the form $(\lambda + \mu) - 1$

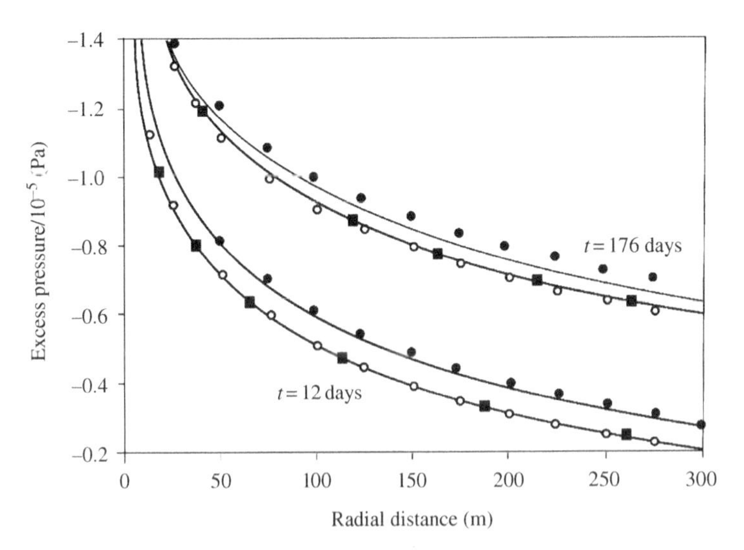

Figure 6.21 Numerical and analytical results for excess pore pressure versus radial distance: (—) embedded aquifers (w = 10), numerical; (-■-) isolated aquifer numerical; (●) Equation 6.10, one-dimensional consolidation; (○) Equation 6.10, three-dimensional consolidation. *Source:* Reproduced from Lewis and Schrefler (1998) by permission of John Wiley & Sons Limited.

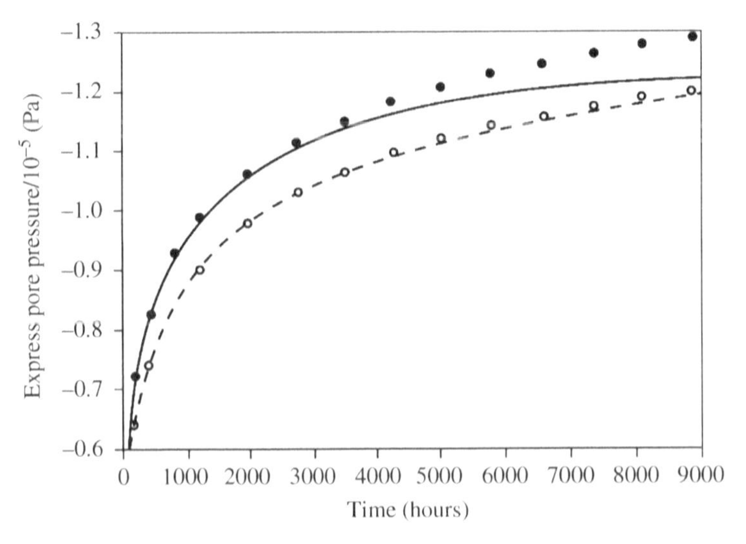

Figure 6.22 Numerical and analytical results for excess pore pressure versus time at r = 50 hours: (—) embedded aquifer, numerical; (- - - -) isolated aquifer, numerical; (•) Equation 6.10, one-dimensional consolidation; (○) Equation 6.10, three-dimensional consolidation. *Source:* Reproduced from Lewis and Schrefler (1998) by permission of John Wiley & Sons Limited.

in a one-dimensional context and $(\lambda + 2\mu) - 1$ in a three-dimensional situation respectively. In the latter case, the horizontal displacement (in the radial direction) is given as

$$u_r = \frac{Q}{16\pi C_v h}\left[E_1(u) - \frac{1-e^{-u}}{u}\right] \tag{6.14}$$

where $u = r^2/(4\pi C_v t)$. By comparing the solution of 6.12 evaluated with the two options for the values of coefficient of compressibility, it follows that for the same pumping rate, the 3-dimensional solution gives approximately $(1 - \nu)$ times the value of compaction given by one-dimensional setting, where ν is Poisson's ratio. This is shown in Figure 6.23 where the two equations are evaluated for $\nu = 0.1$ and $\nu = 0.4$. The differences are relevant; they can be of the order of 70% and they decrease with decreasing Poisson's ratio.

Inspection of Equations (6.12) and (6.14) reveals the necessity of discretizing properly an unbounded domain to avoid errors that appear to be dependent on the distance of truncation of the analyzed area. For instance, when using a mesh truncated at $R = 100$ hours from the axis of symmetry, the pumped aquifer reached a steady-state condition after $t = 200$ days (not shown here) which is not found for the aquifer of infinite radial extent in the analytical solution (see Figure 6.23 for compaction), nor in the numerical solution with infinite elements (see Figure 6.22 for drawdown).

6.4.3 3-D Consolidation with Adaptivity in Time

A 3-D consolidation problem is solved using a Time Discontinuous Galerkin approximation. We investigate in particular the numerical efficiency of the Time Discontinuous Galerkin method (TDG) in coupled problems such as consolidation (Simoni et al. 2008). We adopt a simple test case to assess the convergence properties of the TDG since there exist in the literature applications of the DG method in space that do not exhibit such property (Zhang

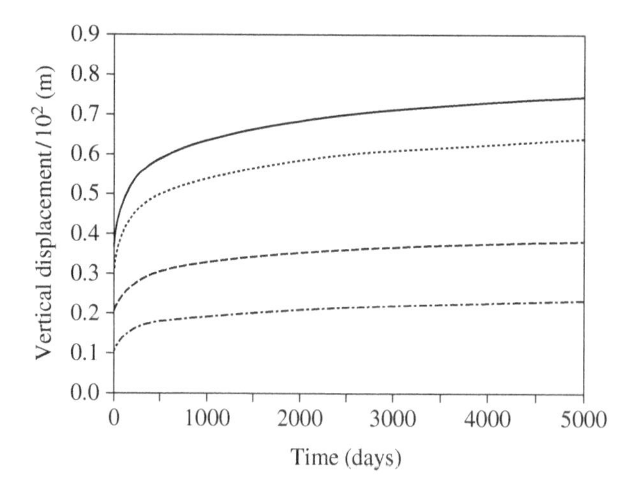

Figure 6.23 Analytical solutions for vertical displacements of an isolated aquifer versus time at $r = 5$ hours from the well (Equation 6.12): (—) 1-D consolidation $\nu = 0.1$; (- - - -) 3-D consolidation $\nu = 0.1$; (- - - -) 1-D consolidation $\nu = 0.4$; (-·-·-·) 3-D consolidation $\nu = 0.4$. *Source:* Reproduced from Lewis and Schrefler (1998) by permission of John Wiley & Sons Limited.

and Shu 2003). To this end, the mechanical behavior of a homogeneous soil specimen compressed by a constant vertical load is analyzed. The specimen is a prismatic cylinder (Terzaghi's column) 650 mm high, with a cross section 200 × 200 mm. The finite element mesh uses standard 3-D isoparametric 8-noded elements as represented in Figure 6.24; material data are summarized in Table 6.1.

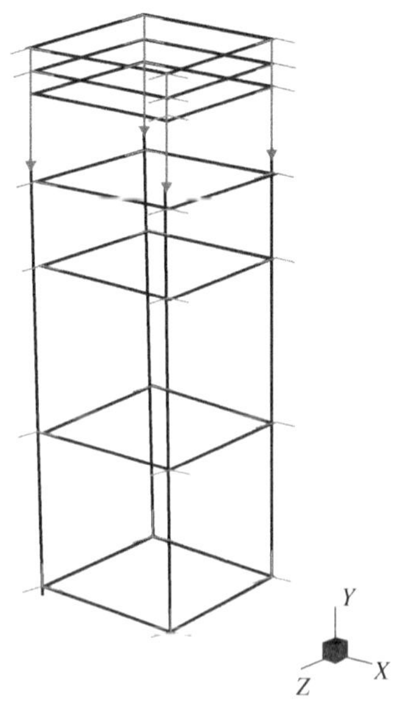

Figure 6.24 Geometry and discretization of the sample. *Source:* Simoni et al. (2008). Reproduced with permission from Springer.

Firstly, we compare the solutions obtained by means of the TDG and the finite difference method. For the latter, we refer to the generalized midpoint rule, in particular, the central (or midpoint) one (TCFD), which is often preferred in coupled problems for its enhanced accuracy. The comparison is made using the same time-discretization shown in Figure 6.25, without time step adaptivity. In Figure 6.26, the time history of the pressure in the first 50 seconds of the transient solution is presented.

In the first case (TCFD), the nearly incompressible fluid causes a numerical oscillation of the solution, whereas with the TDG algorithm, the spurious modes are filtered and the solution is much more accurate. This filtering has been experienced by Li and Wiberg (1998) also in solid dynamics problems and is probably due to the weak enforcement of continuity at time stations as explained above.

Figure 6.27 shows the vertical displacement (at the top) and the pressure (near the top) discontinuities which are the basic information for updating the time step in the time-discontinuous-adaptive-criterion. The total energy norm and the jump energy are shown in Figure 6.28. From the energy variation in time, the role of the different terms and the

Table 6.1 Material data.

Young modulus	60 MPa
Poisson	0.4
ρ	1800 kg/m^3
n (porosity)	0.4
ρ_f	1000 kg/m^3
K_s	10^{10} MPa
K_f	10^{10} MPa
k_x	5.76 mm/s
k_y	5.76 mm/s
k_z	5.76 mm/s

Source: Simoni et al. (2008). Reproduced with permission from Springer.

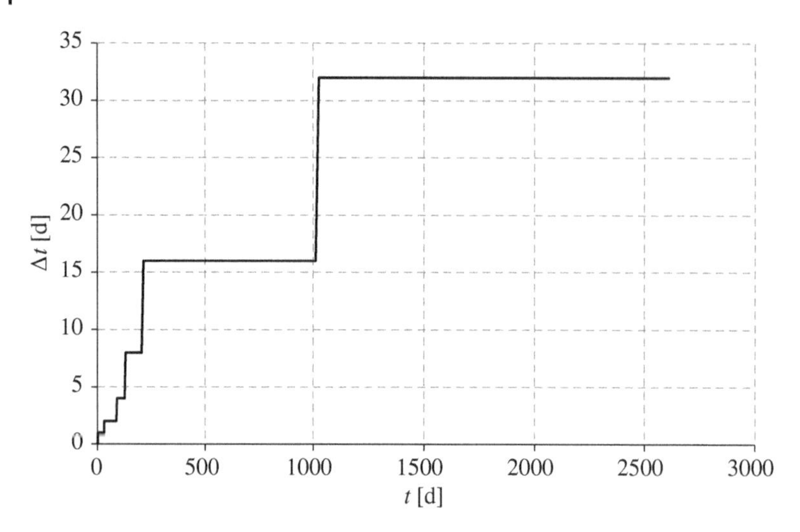

Figure 6.25 Time steps chosen for the analysis.

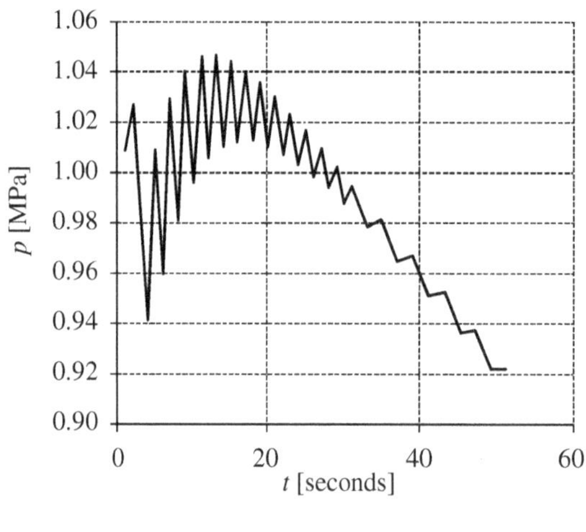

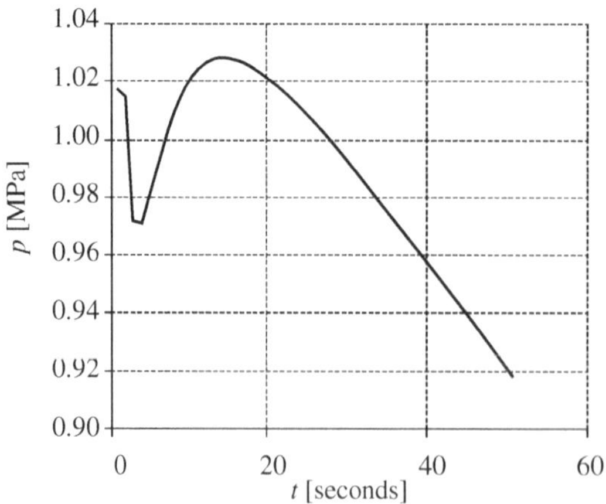

Figure 6.26 Time history of pressure at the top of the sample for the first 50 seconds: midpoint finite-difference solution (top) and discontinuous Galerkin FEM in time (bottom). *Source:* From Simoni et al. (2008). Reproduced with permission from Springer.

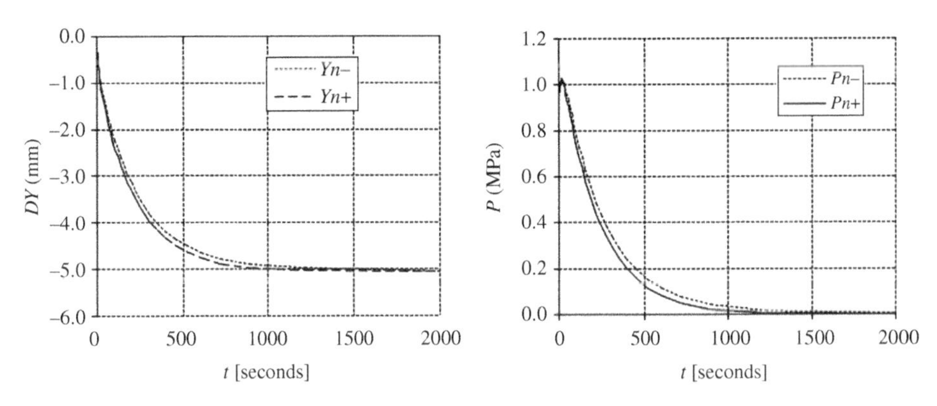

Figure 6.27 Time history of vertical top displacement (left) and top pressure (right). *Source:* From Simoni et al. (2008). Reproduced with permission from Springer.

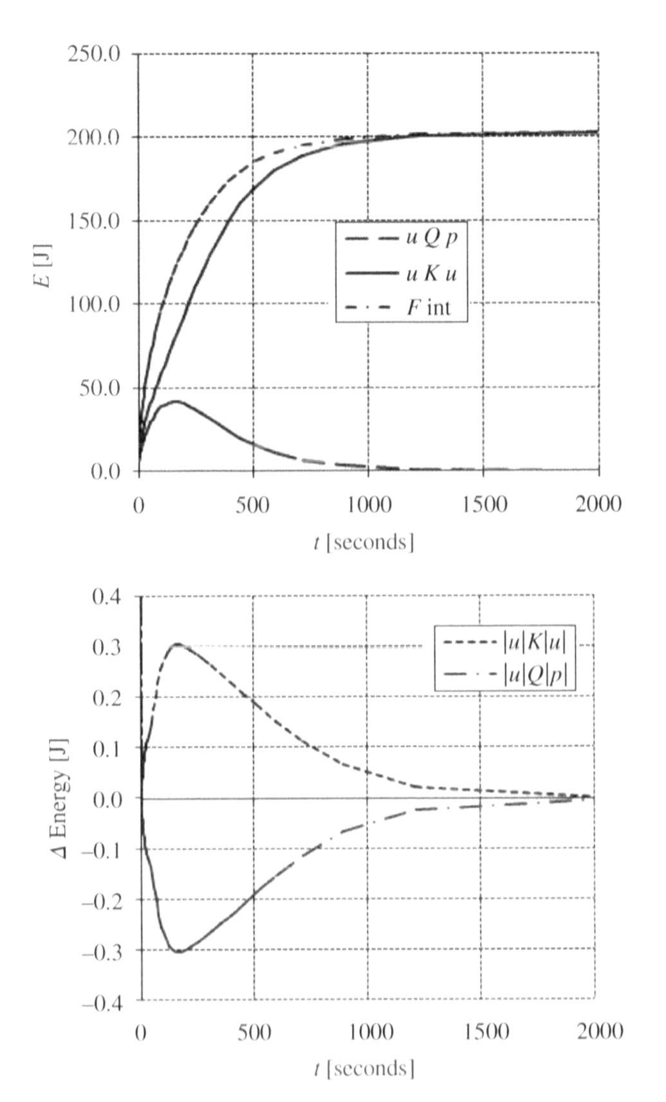

Figure 6.28 Energy norms (internal energy, coupling term energy, and total energy) and jump energy (right). *Source:* From Simoni et al. (2008). Reproduced with permission from Springer.

coupling period are evident. However, the time adaptivity is performed on the basis of energy jumps. As can be seen, the solid jump energy and the one of coupling term are of opposite sign and cancel each one. Hence, their sum, i.e. the jump energy of the complete equilibrium problem, cannot be used for time adaptivity. For correct time adaptivity, all energy jumps, including one of the coupling terms, must be used as formulated in Equation (5.43).

Convergence of the solution is assessed by comparison of the results of TDG with TCFD: the same solution, coincident with the analytical one, for both displacement and pressure field, is obtained as shown in Figure 6.29.

As a second test, we assume that the load applied to the top of the sample suddenly changes in time (no inertia effects are accounted for), oscillating between the limits 1.0 and 0.0. Figure 6.30 shows the time variation of the load and the time step distribution as determined by the automatic time-step adaptive procedure. The adaptivity is based on energy jumps which are shown in Figure 6.31 together with the energy distribution in time.

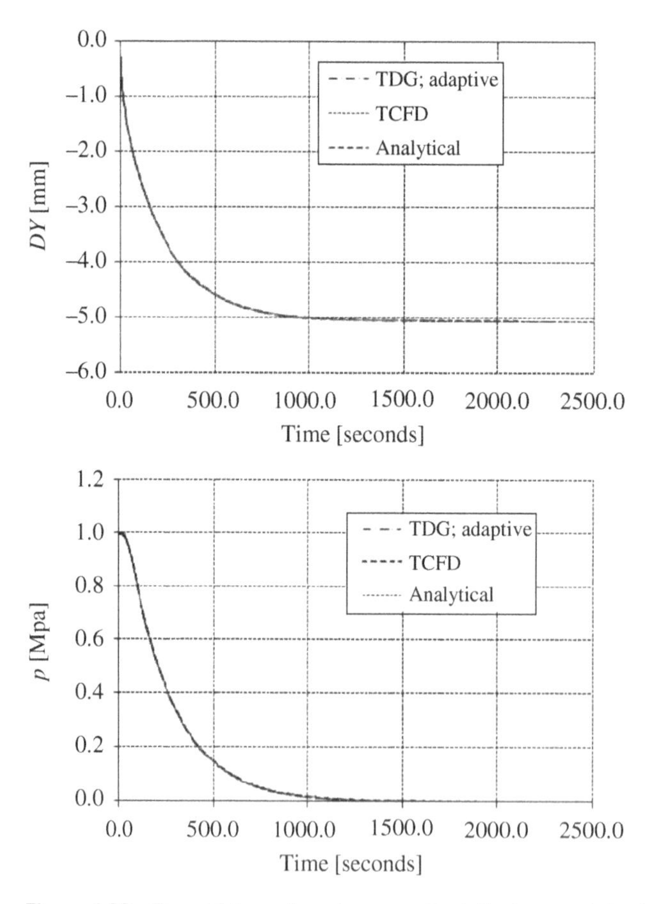

Figure 6.29 Time history of maximum vertical displacement (top) and pore pressure: comparison between TCFD and TDG and analytical solution (bottom). *Source:* From Simoni et al. (2008). Reproduced with permission from Springer.

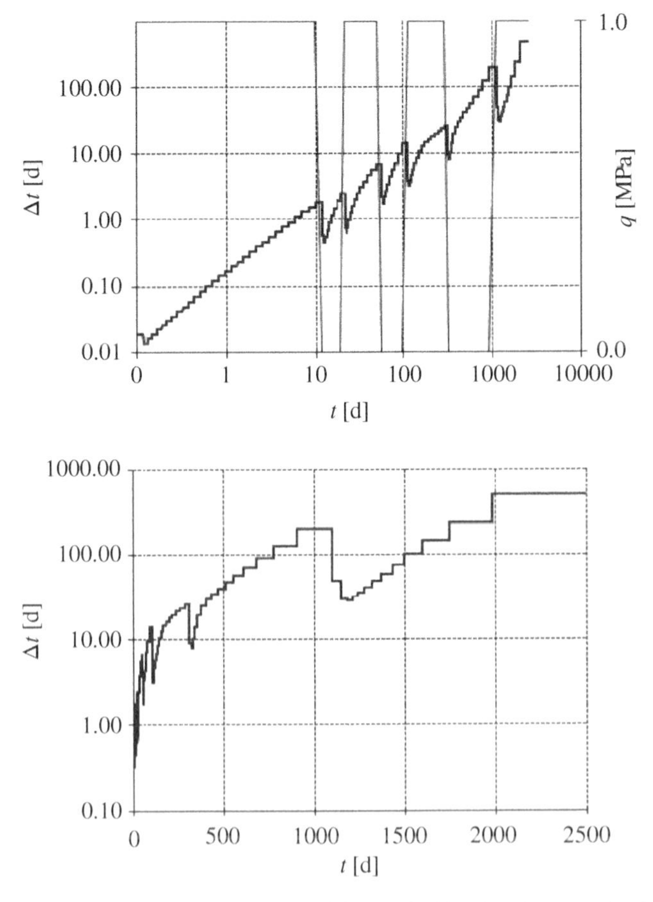

Figure 6.30 Variation in time of the applied load and automatic time stepping (top); final time-step distribution (bottom). *Source:* From Simoni et al. (2008). Reproduced with permission from Springer.

The time span during which coupling takes place is obviously longer than in the previous case and the procedure continues to vary the time step, which is shortened or amplified, depending on the strength of coupling and the calculated energy jumps.

A rough estimate of the costs of TDG in comparison to TCFD can be made on the basis of the number of time steps used in the analysis of the second test case presented. On the one hand, we have a doubling of the number of unknowns for each solution of the time-discretized system of equations, on the other, the procedure for time adaptivity involves a significant reduction of time stations with respect to TCFD. These have been more than halved with respect to a well-balanced time-step distribution in the TCFD analysis, even assuming a relatively small error tolerance η_{toll}. It is obvious that the cost for the solution of the TDG nonsymmetric system is not linearly increasing with the number of unknowns, but TDG has the remarkable additional advantage of producing an optimal distribution of time steps even in the presence of complex variation of the solution in time. Hence, it appears that though TDG is slightly more expensive than TCFD, this is well balanced by the improvement of the results.

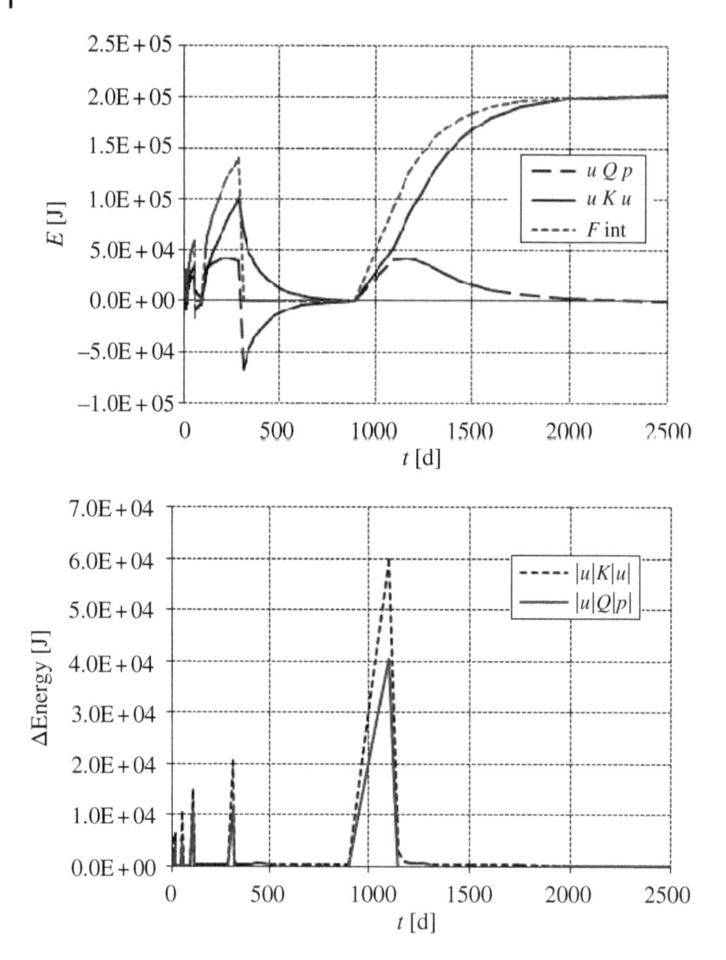

Figure 6.31 Energy norms (internal energy, coupling term energy, and total energy) (top) and jump energy (bottom) for the variable load case. *Source:* From Simoni et al. (2008). Reproduced with permission from Springer.

6.5 Hydraulic Fracturing: Fracture in a Fully Saturated Porous Medium Driven By Increase in Pore Fluid Pressure[3]

Fracture propagation in porous media, such as a fluid-driven fracture propagating in rock to enhance the recovery of hydrocarbons from underground reservoirs, is a common problem in geomechanics. Another application of importance is related to the overtopping stability analysis of dams.

Contributions to the mathematical modeling of fluid-driven fractures have been made continuously since the 1960s, beginning with Perkins and Kern (1961). These authors made various simplifying assumptions, for instance, regarding fluid flow, fracture shape, and

3 The authors gratefully acknowledge the collaboration of T. D. Cao, E. Milanese, C. Peruzzo, S. Secchi, and L. Simoni in the elaboration of this section.

velocity leakage from the fracture. Among other contributions, reference can be found in Rice and Cleary (1976), Cleary (1978), Huang and Russell (1985a, 1985b), and Detournay and Cheng (1991). All these contributions present analytical solutions in the frame of linear fracture mechanics assuming the problem to be stationary. Further, they suffer from the limitations of the analytical approach, in particular, the inability to represent an evolutionary problem in a domain with real complexity. Other papers deal with the analysis of solid and fluid behavior near the crack tip (Advani et al. 1997; Garagash and Detournay 2000). A general numerical model has been presented by Boone and Ingraffea (1990) in the context of linear fracture mechanics. It allows for fluid leakage in the medium surrounding the fracture and assumes a moving crack depending on the applied loads and material properties. Hence, fracture length is a natural product of the solution algorithm, not the result of a priori assumptions. A finite element/difference solution is adopted in this case. In Carter et al. (2000), a fully 3-D hydraulic fracture model is presented. It relies on assumptions similar to Boone and Ingraffea (1990), in particular, assuming the positions where fractures initiate, but completely neglects the fluid continuity equation in the medium surrounding the fracture.

As far as numerical fracture analysis, in general, is concerned, mainly two approaches can be found in literature: smeared crack analysis and discrete crack analysis. We consider here the second one. In this context, embedded discontinuity elements have been proposed by Bolzon and Corigliano (2000), Wells and Sluys (2001), Oliver et al. (2001), Feist and Hofstetter (2006), and Segura and Carol (2008). Another solution for a fracturing solid is that of a moving rosette in a fixed mesh (Wawrzynek and Ingraffea 1989) and a moving fracture in a fixed mesh (Camacho and Ortiz 1996). Extended Finite Elements have been introduced by Moes and Belytschko (2002), and applied to porous media by Al-Khoury and Sluys (2007) and to hydraulic fracturing by Réthoré et al. (2008), Mohammadnejad and Khoei (2013), Remij et al. (2017), and Vahab et al. (2018). Phase-field models for hydraulic fracturing can be found in Mikelic et al. (2015), Miehe and Mauthe (2016), Lee et al. (2017), Ehlers and Luo (2017), Santillan et al. (2018), Zhou et al. (2019), and Ni et al. (2020a). Partition of unity finite elements is used for 2-D mode I crack propagation in saturated ionized porous media by Kraaijeveldt et al. (2013). We present in the following two sections a discrete fracture approach that uses remeshing in an unstructured mesh together with automatic mesh refinement. This method is easily applicable also when the fracture path is a priori unknown and extension to 3-D situations and dynamics is straightforward. At the end of the chapter also, the coupling of FEM for fluid with discrete models for fracturing solids is addressed.

6.5.1 2-D and 3-D Quasi-Static Hydraulic Fracturing

Fluid-driven fracture propagation in a porous medium requires the introduction of an appropriate constitutive relationship for the solid, and for the fluid defining the permeability within the crack and the rest of the domain.

6.5.1.1 Solid Phase: Continuous Medium

In the framework of discrete crack models, the mechanical behavior of the solid phase at a distance from the process zone is usually assumed as simple as possible. The complete domain is decomposed in a set of different homogeneous sub-domains. In the present formulation, a Green-elastic or hyperelastic material is assumed for each component, being the mechanical behavior dependent on effective stress as

$$\sigma'_{ij} = c_{ijrs}\varepsilon_{rs} \tag{6.15}$$

where the elastic coefficients depend on the strain energy function W and can be expressed in terms of Lamé constants as

$$c_{ijrs} = \frac{\partial^2 W}{\partial\varepsilon_{ij}\partial\varepsilon_{rs}} = \mu\left(\delta_{is}\delta_{jr} + \delta_{ir}\delta_{js}\right) + \lambda\delta_{ij}\delta_{rs}\delta_{ij} \quad (i,j,r,s = 1,2,3) \tag{6.16}$$

In the case of plane problems, Equation (6.15) is condensed in the standard way, assuming the pertinent conditions for plane stress or plane strain states.

6.5.1.2 Solid Phase: Cohesive Fracture Model – Mode I Crack Opening

We use the cohesive fracture model shown in Figure 6.32: between the real fracture apex which appears at the macroscopic level and the apex of a fictitious fracture, there is the process zone where cohesive forces act. A constitutive model links the cohesive forces to the crack opening in the process zone.

Within a generic component of the solid skeleton, a fracture can initiate or propagate under the assumption of mode I crack opening, provided that tangential relative displacements of the fracture lips are negligible. The cohesive forces are hence orthogonal to fissure sides. Following the Barenblatt (1959) and Dugdale (1960) model and accounting for Hilleborg et al. (1976), the cohesive law when opening monotonically increases is shown in Figure 6.33.

$$\sigma = \sigma_0\left(1 - \frac{\delta_\sigma}{\delta_{\sigma\mathrm{cr}}}\right) \tag{6.17}$$

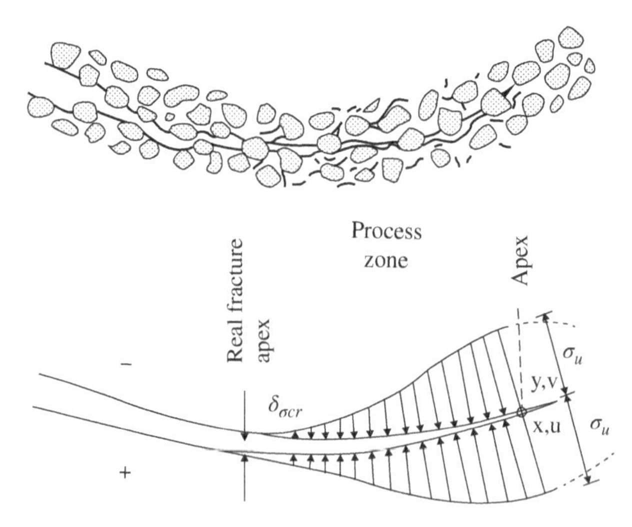

Figure 6.32 Definition of cohesive crack geometry and model parameters. *Source:* From Secchi and Schrefler (2012), reproduced with permission from Springer.

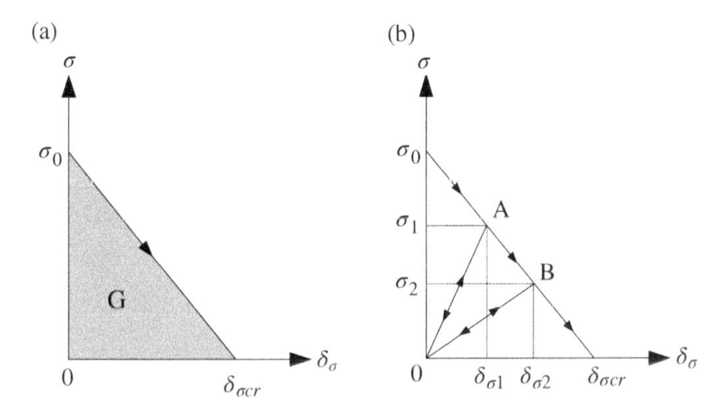

Figure 6.33 Fracture energy (a) and loading/unloading law (b) for each homogeneous component, vertical axis stresses, horizontal axis crack opening.

σ_0 being the maximum cohesive traction (closed crack), δ_σ the current relative displacement normal to the crack, $\delta_{\sigma \mathrm{cr}}$ the maximum opening with the exchange of cohesive tractions and $G = \sigma_0 \times \delta_{\sigma \mathrm{cr}}/2$ the fracture energy. If after some opening $\delta_{\sigma 1}/\delta_{\sigma \mathrm{cr}}$ the crack begins to close, tractions obey linear unloading as

$$\sigma = \sigma_0 \left(1 - \frac{\delta_{\sigma 1}}{\delta_{\sigma \mathrm{cr}}}\right) \frac{\delta_\sigma}{\delta_{\sigma 1}} \tag{6.18}$$

When the crack reopens, Equation (6.18) is reversed until the opening $\delta_{\sigma 1}$ is recovered, then tractions obey Equation (6.17) again. The unloading path is also presented in Figure 6.33. The individual homogeneous components differ only due to different values attributed to the fundamental parameters of Equations (6.17) and (6.18).

The application of this model requires the discretization of the domain to follow the crack evolution.

6.5.1.3 Solid Phase: Cohesive Fracture Model – Mode II and Mixed Mode Crack Opening

When tangential relative displacements of the sides of a fracture in the process zone cannot be disregarded, mixed-mode crack opening takes place. This is usually the case of a crack moving along an interface separating two solid components. In fact, whereas the crack path in a homogeneous medium is governed by the principal stress direction, the interface has an orientation that is usually different from the principal stress direction. A mixed cohesive mechanical model involves the simultaneous activation of normal and tangential displacement discontinuity and corresponding tractions. For pure mode II, the model presented in Figure 6.34 is used, where the relationship between tangential tractions and displacements is

$$\tau = \tau_0 \frac{\delta_\sigma}{\delta_{\sigma \mathrm{cr}}} \frac{\delta_\tau}{|\delta_\tau|} \tag{6.19}$$

τ_0 being the maximum tangential stress (closed crack), δ_τ the relative displacement parallel to the crack and $\delta_{\sigma \mathrm{cr}}$ the limiting value opening for stress transmission. The unloading/loading from/to some opening $\delta_{\sigma 1} < \delta_{\sigma \mathrm{cr}}$ follow the same behavior as for mode I and the traction law is

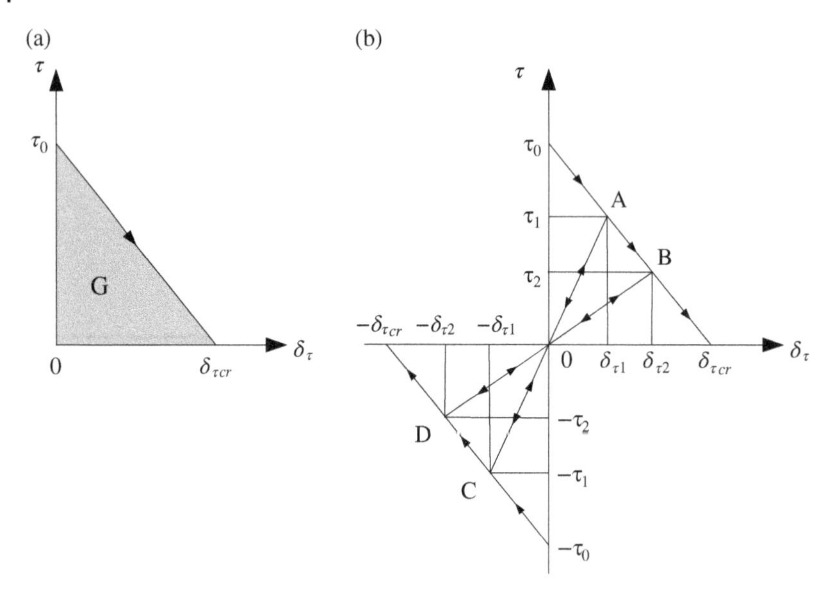

Figure 6.34 Fracture energy (a) and loading/unloading law for the interface and (b) mixed-mode vertical, axis stresses, horizontal axis tangential crack opening.

$$\tau = \tau_0 \left(1 - \frac{\delta_{\sigma 1}}{\delta_{\sigma \mathrm{cr}}}\right) \frac{\delta_\sigma}{\delta_{\sigma 1}} \frac{\delta_\tau}{|\delta_\tau|} \tag{6.20}$$

which is valid for opening in the set $[0 < \delta_{\sigma 1} < \delta_{\sigma \mathrm{cr}}]$, then the original path (Equation 6.19) is followed. Figure 6.34 presents also the unloading/reloading relation.

For the mixed-mode crack propagation, the interaction between the two cohesive mechanisms is treated as in Camacho and Ortiz (1996). By defining an equivalent or effective opening displacement δ and the scalar effective traction t as

$$\delta = \sqrt{\beta^2 \delta_\tau^2 + \delta_\sigma^2}, \quad t = \sqrt{\beta^{-2} \tau^2 + \sigma^2} \tag{6.21}$$

the resulting cohesive law is

$$\mathbf{t} = \frac{t}{\delta} \left(\beta^2 \boldsymbol{\delta}_\tau + \boldsymbol{\delta}_\sigma\right) \tag{6.22}$$

β being a suitable material parameter that defines the ratio between the shear and the normal critical components. The cohesive law takes the same aspect as in Figure 6.33 by replacing displacement and traction parameter with the corresponding effective ones.

6.5.1.4 Linear Momentum Balance for the Mixture Solid + Water

Taking into account the notation of cohesive forces and the symbols in Figure 6.35, the linear momentum balance equation (3.23) without inertia forces is now augmented as follows:

$$\int_\Omega \mathbf{B}^{\mathrm{T}} \boldsymbol{\sigma}'' d\Omega - \mathbf{Q}\overline{\mathbf{p}} - \mathbf{f}^{(1)} - \int_{\Gamma'} (\mathbf{N}^u)^{\mathrm{T}} \mathbf{c} d\Gamma' = \mathbf{0} \tag{6.23}$$

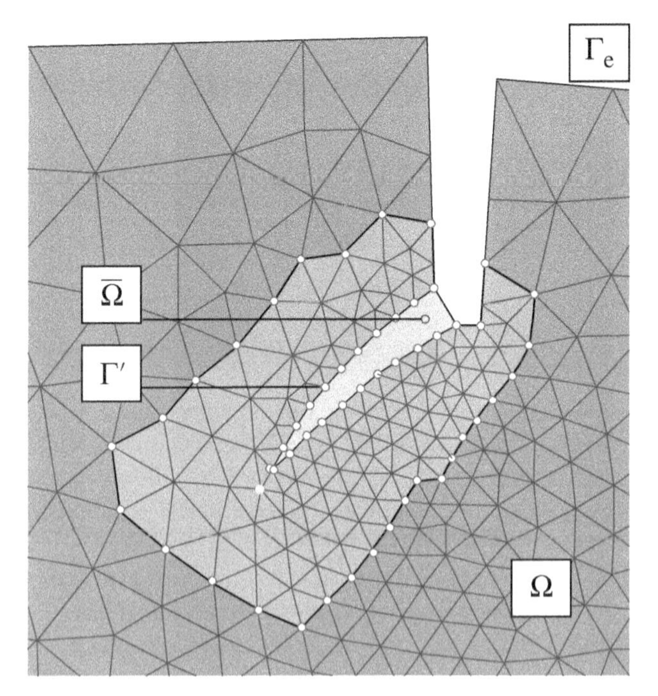

Figure 6.35 Hydraulic crack domain. *Source:* From Simoni et al. (2008). Reproduced with permission from Springer.

where Γ' is the boundary of the fracture and process zone and **c** the cohesive tractions on the process zone as defined above; this term is explicitly given to highlight its importance in the problem being solved.

6.5.1.5 Liquid Phase: Medium and Crack Permeabilities

For simplicity, constant absolute permeability is assumed for the fluid fully saturated medium surrounding the fracture. As far as permeability within the crack is concerned, the validity of Poiseuille or cubic law is assumed. This has been stated in the case of laminar flow through open fractures and its validity has been confirmed in the case of closed fractures where the surfaces are in contact and the aperture decreases under applied stress. The cubic law has been found to be valid when the fracture surfaces are held open or are closed under stress, without significant changes when passing from opening to closing conditions. Permeability is not dependent on the rock type or stress history, but is defined by crack aperture only. Deviation from the ideal parallel surface conditions causes only an apparent reduction in flow and can be incorporated into the cubic law, which can be written as (Witherspoon et al. 1980):

$$k_{ij} = \frac{1}{f}\frac{w^3}{12} \tag{6.24}$$

w being the fracture aperture and f a coefficient in the range 1.04–1.65 depending on the material of the solid phase. In the following, this parameter will be assumed as constant and equal to 1.0. Incorporating the Poiseuille law into the weak form of water mass balance equation within the crack results in

$$\widetilde{\mathbf{H}}\overline{\mathbf{p}} + \widetilde{\mathbf{S}}\dot{\overline{\mathbf{p}}} + \int_{\Gamma'} (\mathbf{N}^p)^{\mathrm{T}} \overline{\mathbf{q}}^w d\Gamma' = \mathbf{0} \tag{6.25a}$$

with

$$\widetilde{\mathbf{H}} = \int_{\overline{\Omega}} (\nabla \mathbf{N}^p)^{\mathrm{T}} \frac{w^2}{12\mu_w} \nabla \mathbf{N}^p d\overline{\Omega} \tag{6.25b}$$

$$\widetilde{\mathbf{S}} = \int_{\overline{\Omega}} (\mathbf{N}^p)^{\mathrm{T}} \frac{1}{Q^*} \mathbf{N}^p d\overline{\Omega} \tag{6.25c}$$

This equation, similar to Equation (3.28) but without solid deformation, represents the fluid flow equation along the fracture. It should be noted that the last term of (6.25a) representing the leakage flux into the surrounding porous medium across the fracture borders, is of paramount importance in hydraulic fracturing techniques. This term can be represented by means of Darcy's law using the medium permeability and pressure gradient generated by the application of water pressure on the fracture lips. No particular simplifying hypotheses are hence necessary for this term. This equation can be directly discretized by finite elements at the same stage as the following Equation (6.26) because both have the same structure. When assembling these equations, only the parameters have to be changed in the appropriate elements depending on whether they belong to the fracture or the surrounding medium. Except for the compressibility term, the above equation is also present in (Boone and Ingraffea 1990), where it is discretized by the finite difference method and integrated separately along a predetermined path by using a staggered approach to obtain the pressure along the crack. Particular relationships for the leakage term have been introduced there such as impermeable boundaries. This staggered procedure resulted in a cumbersome method, requiring several thousand iterations, as the authors noted, due to the strong coupling of displacement and pressure fields, but, more importantly, it needs particular convergence and stability analyses (iteration convergence) to assess the numerical performance of the staggered solution (Turska and Schrefler 1993).

6.5.1.6 Mass Balance Equation for Water (Incorporating Darcy's Law)

The mass balance equation (3.28) for the porous medium surrounding the fracture becomes

$$\widetilde{\mathbf{Q}}\dot{\overline{\mathbf{u}}} + \mathbf{H}\overline{\mathbf{p}} + \widetilde{\mathbf{S}}\dot{\overline{\mathbf{p}}} - \mathbf{f}^{(2)} - \int_{\Gamma'} (\mathbf{N}^p)^{\mathrm{T}} \overline{\mathbf{q}}^w d\Gamma' = \mathbf{0} \tag{6.26}$$

where $\overline{\boldsymbol{q}}_w$ represents the water leakage flux along the fracture toward the surrounding medium of Equation (6.25). This term is defined along the entire fracture, i.e. the open part and the process zone and is given explicitly to highlight its importance in the problem being solved.

It is worth mentioning that the topology of the domain Ω changes with the evolution of the fracture. In particular, the fracture path, the position of the process zone, and the cohesive forces are unknown and must be regarded as products of the mechanical analysis. The discretized governing equations, which are shown next, are solved simultaneously to obtain the displacement and pressure fields together with the fracture path.

6.5.1.7 Discretized Governing Equations and Solution Procedure

Space discretization by means of the Finite Element Method of Equations (6.23), (6.25), and (6.26), incorporating the constitutive equations, results in the following overall system of time differential equations, where the first equation has been written in rate form and multiplied by (−1) as is common in consolidation analysis (but by no means compulsory)

$$\begin{bmatrix} -\mathbf{K}_\mathrm{T} & \tilde{\mathbf{Q}} \\ \mathbf{Q}^\mathrm{T} & \mathbf{S} \end{bmatrix} \begin{bmatrix} \dot{\bar{\mathbf{u}}} \\ \dot{\bar{\mathbf{p}}} \end{bmatrix} + \begin{bmatrix} \mathbf{0} & \mathbf{0} \\ \mathbf{0} & \mathbf{H} \end{bmatrix} \begin{bmatrix} \bar{\mathbf{u}} \\ \bar{\mathbf{p}} \end{bmatrix} = \begin{bmatrix} -\dot{\mathbf{f}}^{(1a)} \\ \mathbf{f}^{(2)} \end{bmatrix} \tag{6.27}$$

For the submatrices of Equation (6.27), see Section 3.2.2 except for

$$\dot{\mathbf{f}}^{(1a)} = \int_\Omega (\mathbf{N}^u)^\mathrm{T} \rho \dot{\mathbf{b}} d\Omega + \int_{\Gamma_t} (\mathbf{N}^u)^\mathrm{T} \dot{\mathbf{t}} d\Gamma + \int_{\Gamma'} (\mathbf{N}^u)^\mathrm{T} \dot{\mathbf{c}} d\Gamma \tag{6.28}$$

This is the only change with respect to the general model, Chapter 3, where $\dot{\mathbf{c}}$ represents the cohesive traction rate and is different from zero only if the element has a side on the lips of the fracture Γ'. Given that the liquid phase is continuous over the whole domain, leakage flux along the opened fracture lips is accounted for through the $\mathbf{H}$ matrix together with the flux along the crack. Finite elements are, in fact, present along the crack, in the domain $\bar{\Omega}$ in Figure 6.35, which account only for the pressure field and have no mechanical stiffness. In the present formulation, nonlinear terms arise through cohesive forces in the process zone and permeability along the fracture.

The overall system of equations is integrated in time by means of the generalized trapezoidal rule. This yields the algebraic system of discretized equations, written for simplicity in a concise form as

$$\mathbf{A}_{n+1}\mathbf{x}_{n+1} = \mathbf{V}_n + \mathbf{Z}_{n+1} \tag{6.29}$$

Being

$$\begin{aligned} \mathbf{x}_{n=1} &= \begin{bmatrix} \bar{\mathbf{u}} \\ \bar{\mathbf{p}} \end{bmatrix}_{n+1} \\ \mathbf{A}_{n+1} &= \begin{bmatrix} -\mathbf{K}_\mathrm{T} & \tilde{\mathbf{Q}} \\ \mathbf{Q}^\mathrm{T} & \mathbf{S} + \alpha \Delta t \mathbf{H} \end{bmatrix} \\ \mathbf{V}_n &= \begin{bmatrix} -\mathbf{K}_\mathrm{T} & \tilde{\mathbf{Q}} \\ \mathbf{Q}^\mathrm{T} & \mathbf{S} - \Delta t(1-\alpha)\mathbf{H} \end{bmatrix}_n \begin{bmatrix} \bar{\mathbf{u}} \\ \bar{\mathbf{p}} \end{bmatrix}_n \\ \mathbf{Z}_{n+1} &= -\begin{bmatrix} \mathbf{f}^{(1a)} \\ \mathbf{0} \end{bmatrix}_{n+1} + \begin{bmatrix} \mathbf{f}^{(1a)} \\ \mathbf{0} \end{bmatrix}_n + \Delta t(1-\alpha) \begin{bmatrix} \mathbf{0} \\ \mathbf{f}^{(2)} \end{bmatrix} - \Delta t \alpha \begin{bmatrix} \mathbf{0} \\ \mathbf{f}^{(2)} \end{bmatrix}_{n+1} \end{aligned} \tag{6.30}$$

As usual, n represents the time station and α the time discretization parameter. Implicit integration is used in the following applications with $\alpha = 0.5$. Hints for a successful implementation of Equations (6.29) and (6.30) can be found in Simoni and Secchi (2003).

Because of the continuous variation of the domain as a consequence of the propagation of the cracks, also the boundary conditions and the related mechanical parameters change. Along the formed crack edges and in the process zone, boundary conditions are the direct result of the field equations while the mechanical parameters have to be updated. The adopted remeshing technique accounts for all these changes (Secchi and Simoni 2003).

In a 2D context, the fracture follows directly the direction normal to the maximum principal stress. In 3D, the fracture follows the face of the element around the fracture tip which is closest to the direction normal to the maximum principal stress; the fracture tip becomes a curve in space (front). If during the advancement, a new node is created at the front, the resulting elements for the fluid in the crack are triangular in 2D and tetrahedral in 3D. At each time t_n, all the necessary spatial refinements are made, i.e. j successive tip (front) advancements are possible within the same time step (see Figure 6.36). Their number in general depends on the chosen time step increment Δt, the adopted crack length increment Δs, and the variation of the applied loads. Note that the possibility of more tip advancements in a time step guarantees consistency because this algorithm does not constrain the crack tip advancement velocity.

The algorithm of Figure 6.36 requires continuous remeshing with a consequent transfer of nodal vector $\mathbf{V}_m$ (Equation 6.29) from the old to the continuously updated mesh (Figure 6.37). Projection of this nodal vector between two consecutive meshes is obtained by using a suitable operator $\mathbf{V}_m(\Omega_{m+1}) = \aleph(\mathbf{V}_m(\Omega_m))$ (Simioni et al. 2007). The solution is then repeated with the quantities of mesh m but recalculated on the new mesh $m + 1$ before advancing the crack tip to preserve as far as possible energy and momentum (see Figure 6.37).

Fluid lag, i.e. negative fluid pressures at the crack tip which may arise if the speed with which the crack tip advances is sufficiently high so that for a given permeability, water cannot flow in fast enough to fill the created space, can be obtained numerically only if an element threshold number is satisfied over the process zone. This number is given by the ratio of elements over the process zone and can be estimated in advance from the problem at hand and the expected process zone length. Hence, a sort of object-oriented refinement is needed locally.

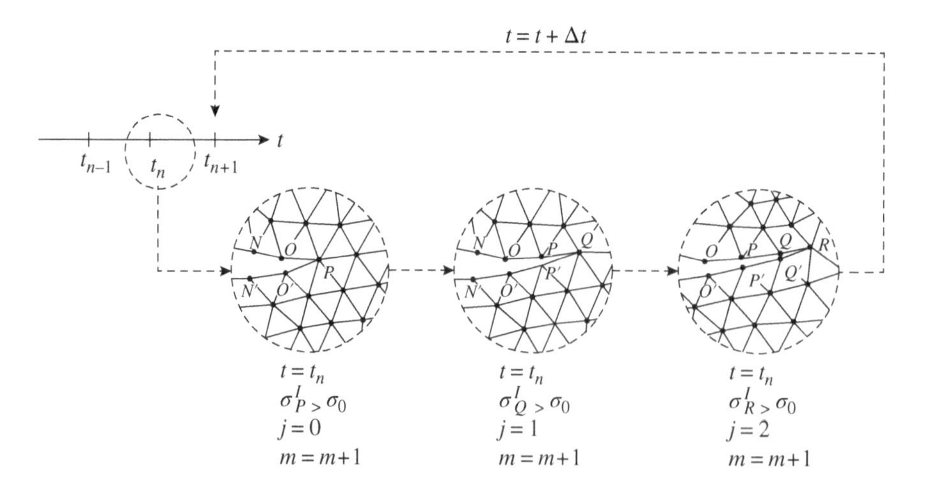

Figure 6.36 Multiple advancing fracture step at the same time station. *Source:* Reprinted from Simioni et al. (2007), Copyright (2007), with permission from Wiley.

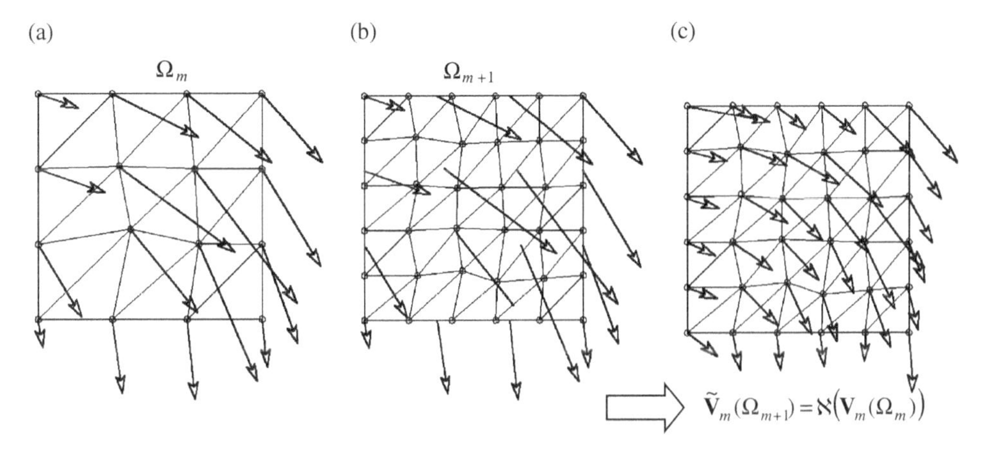

Figure 6.37 Nodal forces projection algorithm. (a) Nodal forces at time station n on mesh m; (b) Nodal forces of time station n on mesh m + 1 before projection; (c) Nodal forces of time n on mesh m + 1 referred to nodes of the latter, $\boldsymbol{\Omega}$ is the domain. *Source:* Reprinted from Simioni et al. (2007), Copyright (2007), with permission from Wiley.

6.5.1.8 Examples

Two numerical applications are presented to demonstrate the capability and efficiency of the adopted adaptive techniques when applied to multi-physics problems.

The first application deals with a hydraulically driven fracture due to a fluid pumped at a constant flow rate. Assuming 2-D plane strain conditions, simplified analytical solutions have been obtained and a more general numerical model has been presented by Boone and Ingraffea (1990). Figure 6.38 presents the geometry of the problem together with the finite element discretization. A notch with a sharp tip is present along the symmetry axis of the analyzed area. Different water inflow rates at the crack mouth give rise to a substantially different behavior.

Material properties are presented in Table 6.2.

The effects of combined spatial/temporal discretization are clearly seen in Figure 6.39, where the crack length is presented versus time for different tip advancement, Δs, and time-step increments, Δt. We can conclude that the more correct time history (case E) is obtained by simultaneously reducing these two parameters, whereas the reduction of only one discretization parameter leads to errors (about $\pm 20\%$) even using a small tip advancement if compared to the crack length. At the same time, the importance of the element threshold number is evident for the choice of Δs (crack length, according to Barenblatt's hypotheses is about 0.9 m). We can also conclude that crack tip velocity is very mesh-sensitive. Hence, the element number threshold must be satisfied to obtain mesh-independent results.

A lower number of elements result in wrong crack tip velocity and the development of fluid lag may be missed. This, in fact, is the result of the interplay between crack velocity and permeability (through continuity equation) and is of great importance because it determines if there is negative pressure in the process zone, hence determines different body forces.

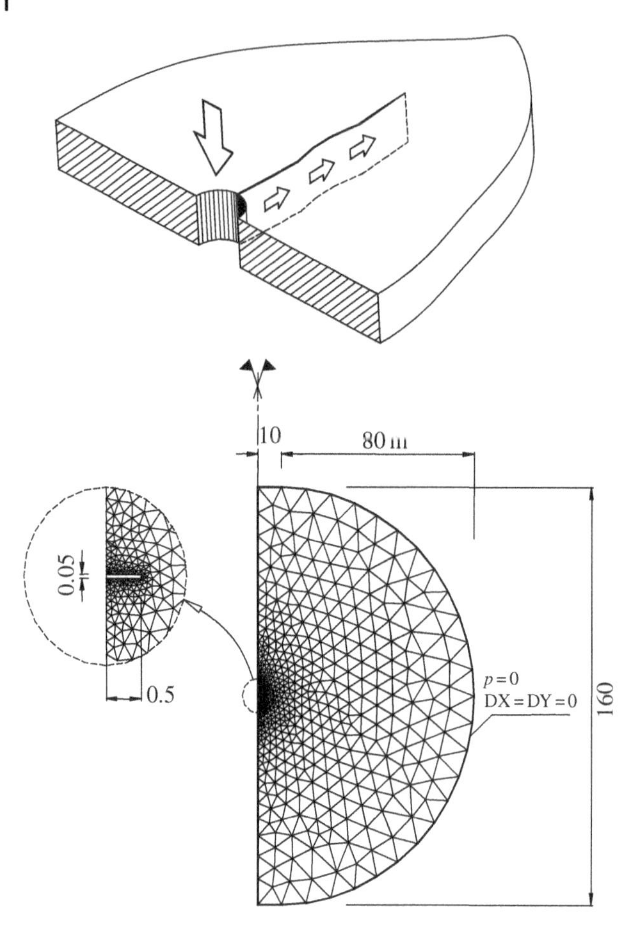

Figure 6.38 Problem geometry for water injection benchmark and overall discretization.
Source: Reprinted from Schrefler et al. (2006), Copyright (2006), with permission from Elsevier.

Table 6.2 Material properties for a water-injected test case.

Permeability coefficient	2×10^{-5} m^2/(MPa · s)
Shear modulus	6000 MPa
Poisson's ratio	0.2
Bulk modulus of solid	36 000 MPa
Bulk modulus of fluid	3000 MPa
Porosity	0.19
Fluid viscosity	10^{-9} MPa s
Max allowable stress (mode I)	1.1 MPa
Critical width	0.26 mm

Source: Reprinted from Schrefler et al. (2006), Copyright (2006), with permission from Elsevier.

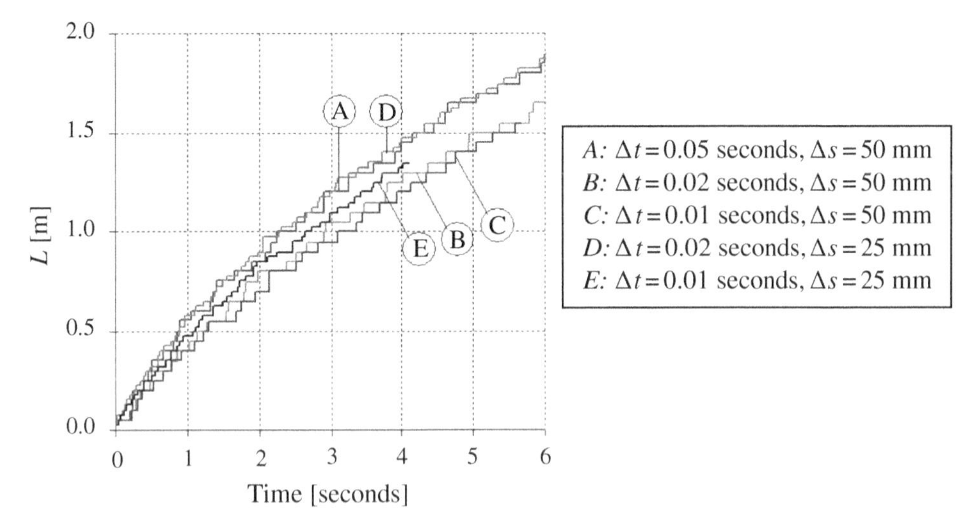

Figure 6.39 Crack length time history. *Source:* Reprinted from Schrefler et al. (2006), Copyright (2006), with permission from Elsevier.

In Figures 6.39 and 6.40, stepwise tip advancement and pressure oscillations can be observed. They do not disappear upon refinement in space and time, including the element number over the process zone. They are not a numerical artefact but reflect the real behavior in case of high injection rate and low fluid viscosity. This fact is known to the oil industry and is even used for steering fracturing operations (Soliman et al. 2014). The observed behavior has also been obtained experimentally, see, e.g. Lhomme et al. (2002) and Okland

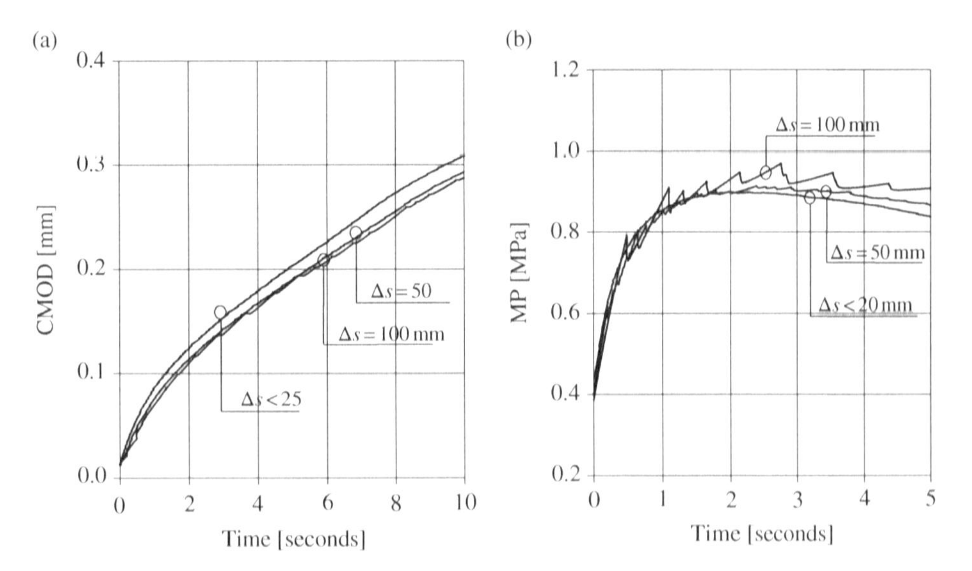

Figure 6.40 Crack mouth-opening displacement (a) and mouth pressure time histories (b). *Source:* Reprinted from Schrefler et al. (2006), Copyright (2006), with permission from Elsevier.

et al. (2002) and confirmed numerically with other models, including lattice models (Milanese et al. 2016a), Peridynamics, see Section 6.5.3, and XFEM (Cao et al. 2017; Milanese et al. 2016b; Vahab et al. 2018). Fewer crack advancement steps in a time interval than the number of time steps used, as in Figure 6.39 is a reliable indicator that the behavior is truly stepwise and not a numerical artefact.

Note that the pressure behavior differs whether the flow is specified or pressure, mechanical load and displacements (Cao et al. 2017; Milanese et al. 2016a). The behavior can be easily explained with Biot's theory: if the flow is specified on the second of Equation (6.27), its effect is transmitted to the solid through the pressure-coupling term in the effective stress. The solid is loaded and, upon rupture, produces a sudden increase of the volumetric strain. This, in turn, causes a drop in pressure (first partial scenario). In this case, stresses and pressures evolve in phase. On the other hand, if a load, pressure, or displacement boundary conditions are applied suddenly (all these conditions acting on the first of Equation (6.27), then the fluid bears initially almost all the induced load because its immediate response is undrained (rigid and non-flowing) causing an overpressure. Only then through the coupling with the mass balance equation of the fluid, the overpressures decrease and the solid gets loaded. Hence, we have a pressure rise upon rupture. Pressure and stresses evolve out of phase (second partial scenario). This explanation applies mainly to mode I fracturing, while in modes II and III, either brittle behavior has been observed or a combination of both scenarios apply. This is so because the sudden changes of displacements would also induce some pressure drop in the second partial scenario due to volumetric strain (effect of the equilibrium of the solid–liquid mixture) and pressure rise in the first partial scenario (effect of the continuity of the fluid) due to the suddenly changing displacements in the crack; therefore the effects of both scenarios have to be superimposed (Milanese et al. 2017; Remij et al. 2017). Finally, stepwise behavior has not been observed in the case of injection of high viscosity fluid at a low injection rate, see Lhomme et al. (2002).

It is interesting to remark that the solution is fully coupled; hence, a change in hydraulic parameters strongly influences the solid behavior as can be seen in Figure (6.41). Here two cases are compared differing only in permeability: this parameter is artificially doubled in the second case. Pressure along the crack and cohesive tractions are substantially different. The consequence of this fact is the difficulty of choosing an optimal static mesh, whereas an adaptive remeshing easily handles such a problem.

The second application deals with the benchmark exercise A2 proposed by ICOLD (1999). The benchmark consists in the evaluation of failure conditions as a consequence of an overtopping wave acting on a concrete gravity dam. The geometry of the dam is shown in Figure 6.42 together with initial, boundary conditions and an intermediate discretization. Differently from the original benchmark, the dam concrete foundation is also considered, which has been assumed homogeneous with the dam body. In such a situation, the crack path is unknown. On the contrary, when a rock foundation is present, the crack naturally develops at the interface between dam and foundation.

As far as initial conditions for water pressure are concerned, it is assumed that during building operations and before filling up the reservoir, pressure can dissipate in the dam body. As a consequence, zero initial pore pressure is assumed in the simulation. A more realistic assumption is the hypothesis of partial saturation of the concrete, which would require a further extension of the present mathematical model.

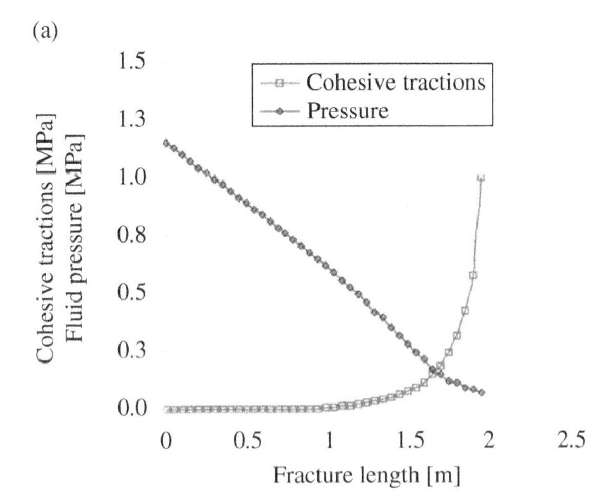

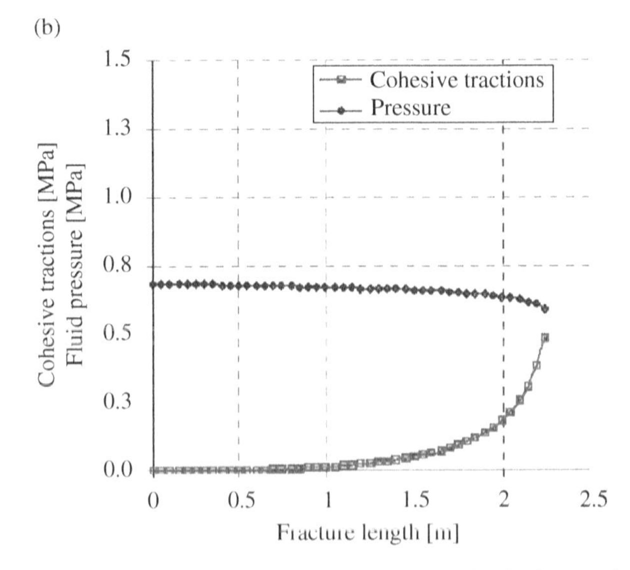

Figure 6.41 Distribution of fluid pressure and cohesive tractions within a fluid-driven fracture: (a) reference case, (b) increased fluid velocity (doubled permeability). *Source:* From Simoni et al. (2008). Reproduced with permission from Springer.

Applied loads are the dam self-weight and the hydrostatic pressure due to water in the reservoir growing from zero to the overtopping level h (which is higher than the dam). The material data for the concrete are those assigned in the benchmark, whereas, for permeability, the value of 10^{-12} cm/s has been assumed. This value could suggest the hypothesis of an impermeable material. This limit case can be analyzed by the present model locating the diffusion phenomenon in restricted areas near the wetted side of the dam and along the crack sides. Such a condition is easily handled by the used mesh generator, but has not been applied in the following. The necessity of adaptive mesh refinement during the solution is demonstrated in Figure 6.42 where two different crack paths are obtained

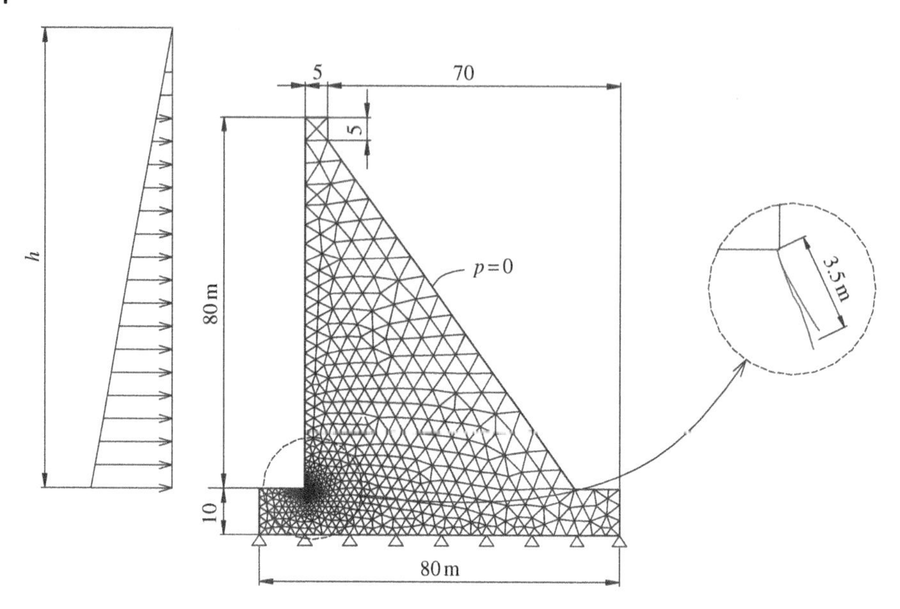

Figure 6.42 Problem geometry for ICOLD benchmark and calculated crack positions (2-D solution). *Source:* Reprinted from Schrefler et al. (2006), Copyright (2006), with permission from Elsevier.

depending on the assumed permeability of the fluid within the fracture. Also, the proper representation of the cohesive forces requires a fine mesh in the area of the process zone. This is shown in Figure 6.43, which represents the process zone when the fracture length is 3.5 m and corresponds to an intermediate step of the analysis when the water level is 78 m. Finally, the formation of the fluid lag is studied. The lag is dependent on the different velocities of propagation of the crack tip and the one of seepage inside the fracture; hence, the simulation of this feature requires a simultaneous correct representation of the solid and fluid field. Zhu–Zienkiewicz's (1988) error estimator for gradient-dependent quantities (stress in the solid and seepage velocity) and goal-oriented refinement for process zone analysis are very useful in this case. Figure 6.44 depicts the fluid lag (water pressure is compression-positive).

For a comparison of the results, see Khoei et al. (2011) where double-node zero-thickness interface elements are used, (Segura and Carol (2008)).

Some important conclusions can be drawn from this application:

- The mechanical behavior of the solid skeleton strongly depends on the characteristic permeability of the fluid within the crack. Crack paths are, in fact, different as a result of the different stress fields.
- The fluid lag is responsible for the differences in the stress field in the process zone; hence, correct modeling of the fluid and solid fields is mandatory.
- The crack path cannot be determined a priori; hence, the traditional use of special/interface elements to simulate fracture propagation in large structures would not be appropriate. An alternative to the successive remeshing is the use of cumbersome discretization of the areas interested by fracture, but this strategy is not viable in the case of large dams.

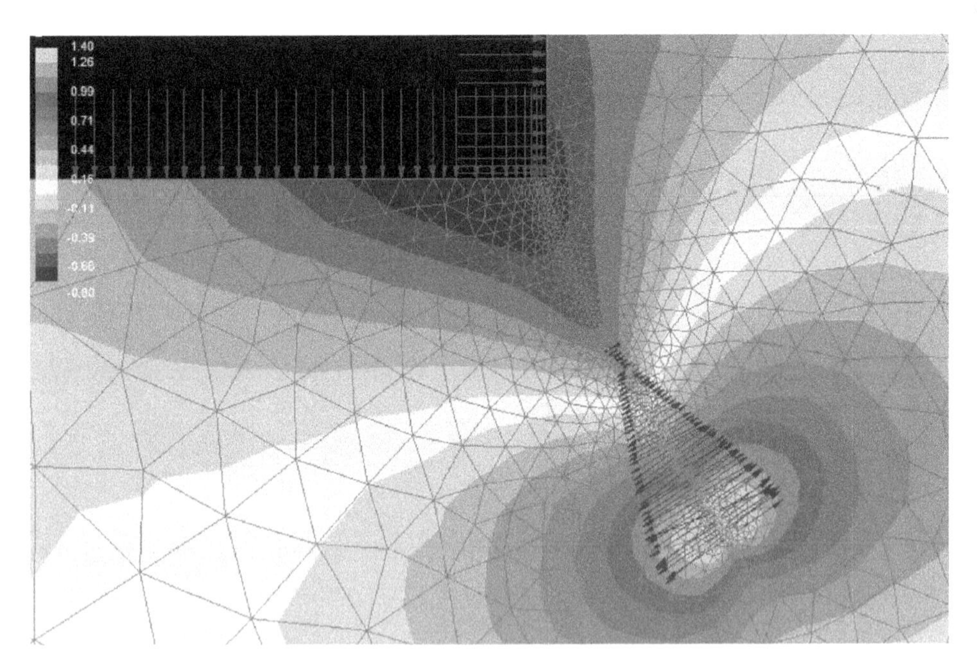

Figure 6.43 Zoom near the fracture for maximum principal stress contour. *Source:* Reprinted from Schrefler et al. (2006), Copyright (2006), with permission from Elsevier.

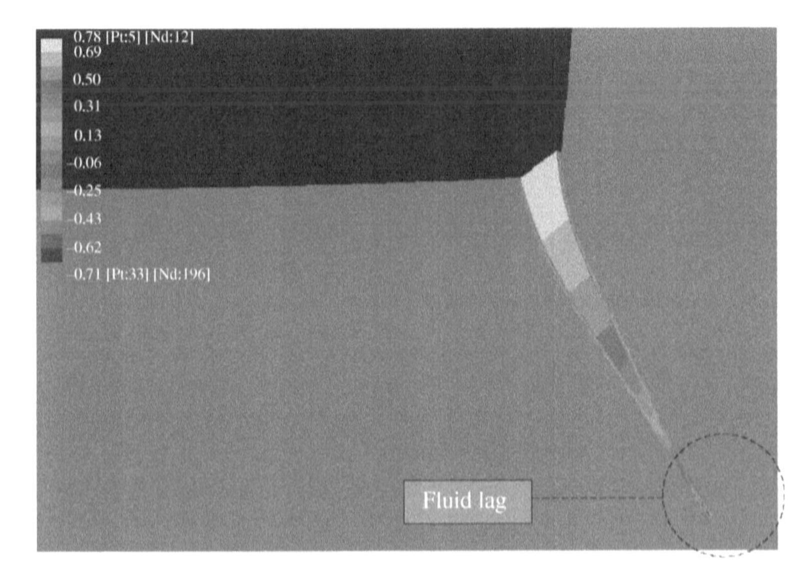

Figure 6.44 Zoom for pressure distribution within the crack and fluid lag. *Source:* Reprinted from Schrefler et al. (2006), Copyright (2006), with permission from Elsevier.

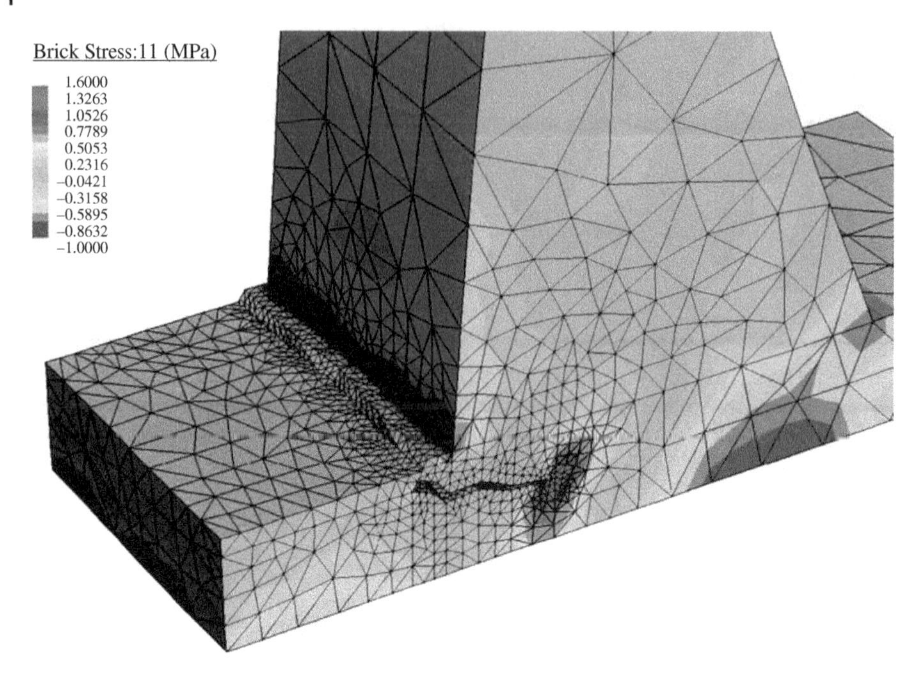

Figure 6.45 Principle stress map contours. *Source:* Reprinted from Secchi and Schrefler (2012), Copyright (2012), with permission from Elsevier.

The technique used here for the analysis of the nucleation of the fracture does not require the presence of an initial notch and requires a very limited amount of information to be initially defined. Furthermore, it can be extended to 3-D analysis where the fracture tip becomes a curve in space (see Secchi and Schrefler (2012)). A 3-D solution for the problem of Figure 6.42 is shown in Figure 6.45.

6.5.2 Dynamic Fracturing in Saturated Porous Media

Repeated hydraulic fracturing tests by Black (1998) reported in Feng et al. (2015) show strong pressure oscillations with high frequency. Experimental evidence for the stepwise tip advancement in the case of mechanical loading has been shown by Pizzocolo et al. (2013). This experiment refers to a single-edge-notch test on a saturated hydrogel tested in air and water. The animation of this experiment clearly evidences that the crack advancement phase is very rapid after each pause. This would suggest that inertia effects may come into play and should not be neglected, see also Hageman and de Borst (2021). Also, applications in geophysics such as simulations of subduction tremor (Burlini et al. (2009)) and volcanic tremor (Burlini and Di Toro (2008)), which likely have a similar triggering mechanism (fluid-induced seismicity) but different seismological signatures (Schwartz and Rokosky 2007) would require inertia effects to be taken into account.

The fracture advancement algorithm shown in Figure 6.36 is applicable also in dynamic situations (see Cao et al. (2018) and Peruzzo et al. (2019)). Equation (6.27) has to be augmented with the inertia terms (see Equation (3.23)). For the time integration of the

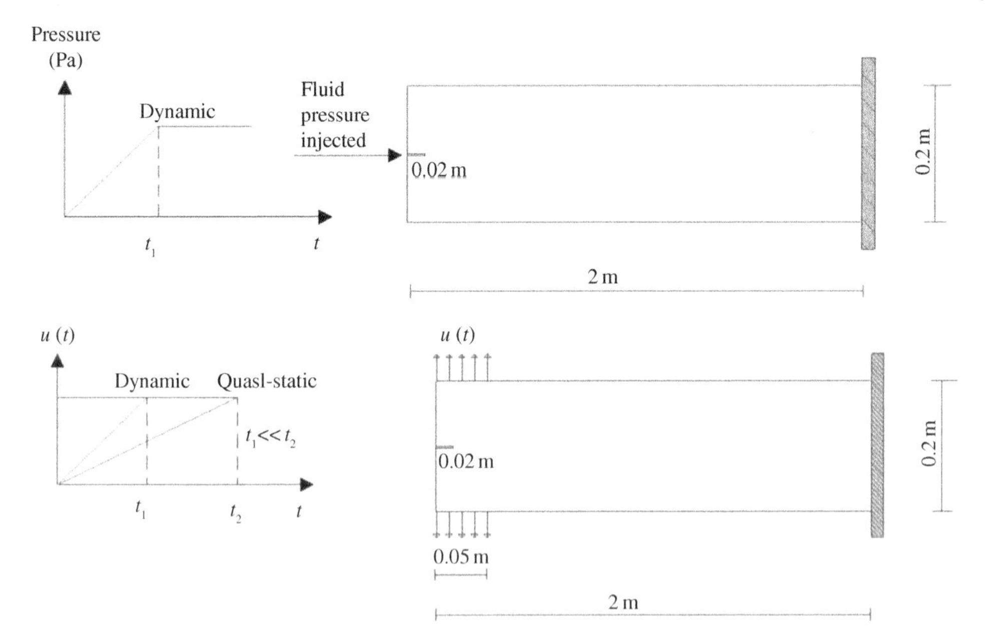

Figure 6.46 Investigated cross section and loading cases: top pressure loading, bottom mechanical loading. *Source:* Reprinted from Peruzzo et al. (2019), Copyright (2019), with permission from Elsevier.

discretized system of equations, the time-stepping scheme GN22 is applied to the nonlinear system of equations (see Section 3.2.3). We show here two applications to fast dynamics, the first one with physical loading and the second one with assigned pressure. The two loading cases are shown in Figure 6.46. The saturated porous medium sample is a rectangle of 0.2 m height and 2 m length. In the case of mechanical loading, the sample is loaded in traction by two uniform vertical velocities with a magnitude 2.35×10^2 m/s applied in opposite directions to the left end of the top and bottom edges. In the case of pressure loading, the same cross section is subjected to a fast pressure increase of 5×10^{11} Pa/s at the crack mouth reaching the maximum value of 100 MPa. The length of the sample has been chosen such that the reflecting wave from the far end does not influence the cracking process in our applications. Vertical and horizontal displacements are constrained at the right edge and the boundary is impervious and adiabatic. The very low intrinsic permeability of the porous medium is 2.78×10^{-21} m^2 as in Réthoré et al. (2008). The material properties of the saturated porous medium are given in Table 6.3. For the fracture analysis, the cohesive fracture parameters of the material are set as follows: the cohesive strength is 2.7 MPa and the cohesive fracture energy is 95 N/m. In the two simulations, square elements of 0.01×0.01 m and a time step size of 1.0×10^{-5} seconds are used. The initial crack at the left side along the symmetry line is 0.02 m. The spatial discretization makes use of four-node isoparametric elements with 2×2 Gauss point integration scheme and the total number of degrees of freedom is 12663. The value of the tolerance for the iterative Newton–Rapson procedure is set to 10^{-5}. During the Newton–Rapson iterations, if convergence is not reached, time-step reduction procedure is automatically applied.

In both cases, stepwise advancement and pressure oscillations are obtained. These are shown in Figure 6.47 for mechanical dynamic loading (second case of Figure 6.46). The time

Table 6.3 Material parameters for the example of Figure 6.46.

i) Porosity: $n = 0.2$;
ii) Intrinsic permeability: $\mathbf{k} = 2.78\times10^{-21}\ \text{m}^2$;
iii) Solid skeleton density: $\rho^s = 2000.0\ \text{kg/m}^3$;
iv) Water density: $\rho^w = 1000.0\ \text{kg/m}^3$;
v) Bulk modulus of the water: $K_w = 0.2$ GPa;
vi) Bulk modulus of the solid phase: $K_s = 13.46$ GPa;
vii) Young's modulus: $E = 25.85$ GPa;
viii) Poisson's coefficient: $\nu = 0.18$;
ix) Biot's constant: $\alpha = 1.0$;
x) Dynamic viscosity of water: $\mu^w = 5.0 \times 10^{-4}\ \text{Pa}\cdot\text{s}$;
xi) Gravity acceleration: $\text{g} = 0.0\ \text{m/s}^2$

Source: Reprinted from Cao et al. (2018), Copyright (2018), with permission from Elsevier.

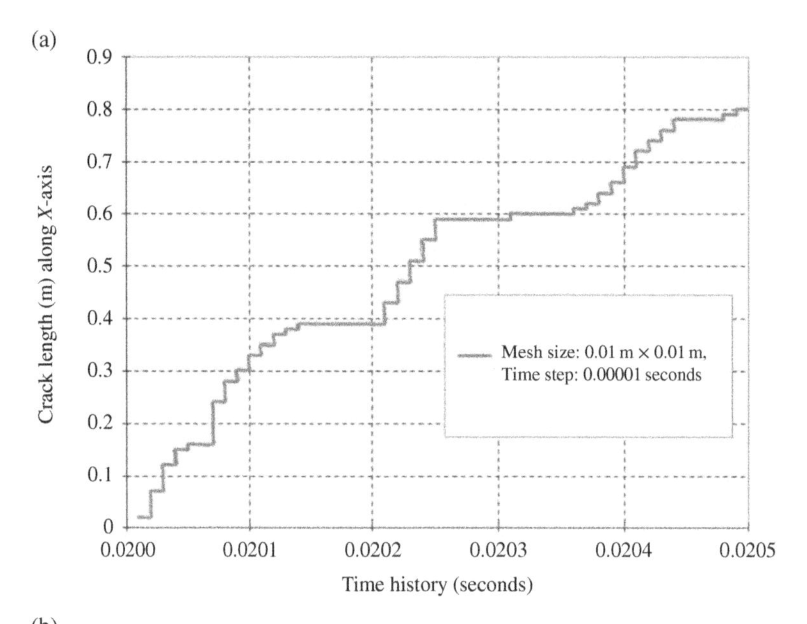

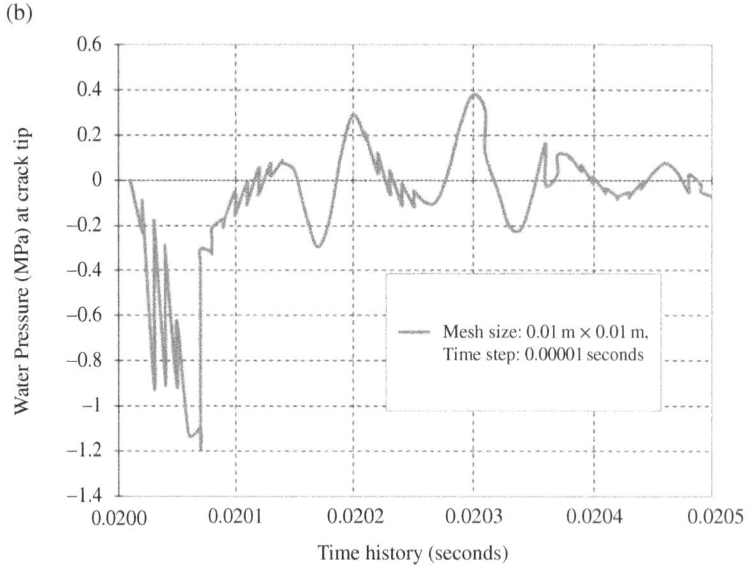

Figure 6.47 Dynamic solutions at the crack tip for fast mechanical loading: (a) crack tip advancement; and (b) pressure oscillation. *Source:* Reprinted from Cao et al. (2018), Copyright (2018), with permission from Elsevier.

to set up the initial conditions takes 0.02 second, while the whole cracking process is over in less than 0.0005 second. The resulting average speed of the fracturing process of 1600 m/s is not far from one-fifth of the speed of sound observed in Phillips (1972) for rocks. In Figure 6.47a, the stepwise crack tip advancement can be clearly seen together with periods of quiescence. The pressure versus time is drawn in Figure 6.47b. Prior to the onset of the crack, the pressure in the sample is declining linearly from zero to roughly −0.2 MPa and afterward jumps appear. As expected, under fast dynamic loading, there is a dominant wave which can also be seen in Figure 6.48 covering the height of the sample. This wave is perturbed by the smaller pressure jumps due to the cracking process which are of the type of quasi-static solution. These smaller jumps are missing in the periods of quiescence of the tip advancement. The coexistence of phenomena taking place with different frequencies can clearly be seen in Figure 6.47b. This was observed by Schwartz and Rokosky (2007) in the case of subduction tremor in Shikoku, Japan, and is believed to be caused by mechanical loading. A fast Fourier transform (FFT) has been used in Cao et al. (2018) for analyzing the spectral properties of the fracturing behavior under fast mechanical loading: a broad spectrum appears with peaks at 2, 12, and 18 kHz. On the contrary, in the case of hydraulic loading, a single peak prevails. This would correspond to what observed by Schwartz and Rokosky (2007) for volcanic tremor in the case of the Arenal volcano where a distinct frequency peak appeared. Note that slugs of magma can rise in a rock up to a speed of 17 m/s (Rosen 2016); hence, a scenario of hydraulic fracturing applies perfectly (see also Burlini et al. (2009)). This example shows that the presented method is also applicable for situations occurring in geophysics (Nolet 2009).

We show in Figure 6.48 for mechanical loading a few significant snapshots for the pressure wave propagation at respectively (a) 0.02025 second; (b) at 0.02035 second; and (c) at 0.0205 second.

For pressure loading, the pressure wave contours are depicted in Figure 6.49 at (a) 0.02005 second; (b) 0.02025 second; and (c) 0.0203 second. Also from these graphs, the different responses can be appreciated. Outgoing waves parallel to the crack can be noticed in this case.

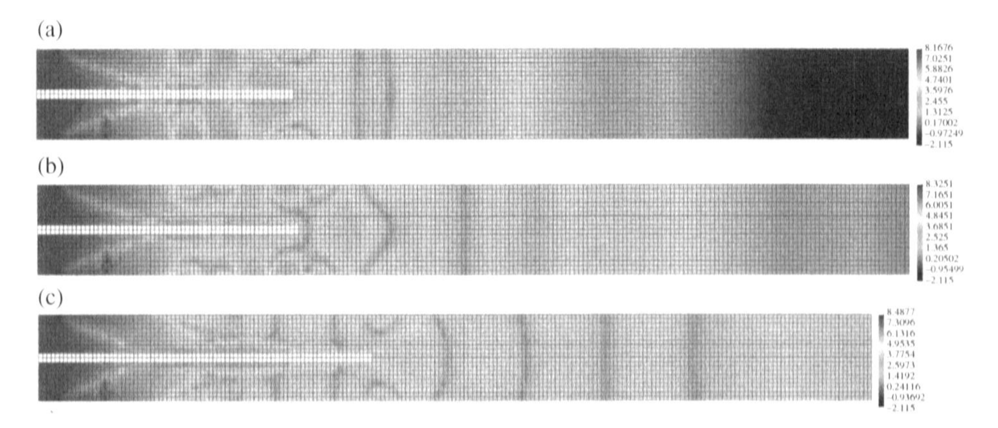

Figure 6.48 Wave propagation of pressure contour for mechanical loading plotted by the logarithmic scale of GID software 12.4 at: (a) 0.02025 second; (b) 0.02035 second; and (c) 0.0205 second. *Source:* Reprinted from Peruzzo et al. (2019), Copyright (2019), with permission from Elsevier.

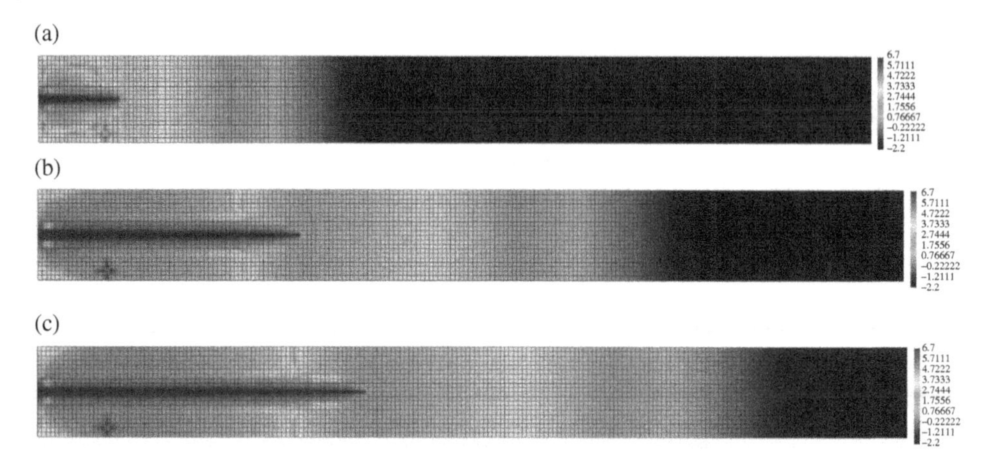

Figure 6.49 Pressure wave contour plots of dynamic solutions for water pressure loading at (a) 0.02005 second; (b) 0.02025 second; and (c) 0.0203 second. *Source:* Reprinted from Peruzzo et al. (2019), Copyright (2019), with permission from Elsevier.

6.5.3 Coupling of FEM for the Fluid with Discrete or Nonlocal Methods for the Fracturing Solid

Discrete methods are commonly employed for the simulation of fracturing solids. The coupling of discrete methods for the solid phase with a FEM solution for the fluid field has also been successfully used for HF problems. Numerical issues of the coupling aspects are discussed in Owen et al. (2009) where a particle method was used for the solid phase. As mentioned in Section 6.5.1, a lattice model for the solid phase was used in Milanese et al. (2016a, 2017) for quasi-static situation. The coupling matrix was obtained within the FEM framework, as well as the matrices for the flow behavior. Lattice models are particularly useful in the case of dry and saturated heterogeneous media. With such media, the avalanche behavior, i.e. the amount of failing elements per loading step of the fracturing process has been investigated. Power-law behavior of the damaging or fracturing events has been recovered under mechanical loading, as happens, for instance, in the case of subduction of tectonic plates (see Section 6.5.2). On the other hand, the power law is destroyed for hydraulic loading, which would also be the case in the lava flow. This method has been extended to a dynamic situation in Peruzzo et al. (2019).

Peridynamics (PD) has been coupled with a FEM model for HF modeling in Ni et al. (2020b) and Sun and Fish (2021). In PD, the Newton equations are solved exactly for the solid within a prescribed horizon radius around a material point; the method is particularly useful in dynamics situations. We show here the case of interaction between an HF crack and a natural crack. An existing crack with an initial length of $l = 0.25$ m is located at the center of a 4 m × 4 m square specimen. The fluid is injected at the center of the initial crack with a constant volume rate of $Q = 1 \times 10^{-3}$ m^3/s. This, together with the assumed dynamic viscosity, is in the range of high flow rate and low viscosity, investigated experimentally in Lhomme et al. (2002), where pressure oscillations have been observed. The whole area is discretized with uniform quadrilateral elements for fluid flow with a grid size of $\Delta x = 1 \times 10^{-2}$ m, and the PD grid shares the same node positions. Figure 6.50 shows the pressure distribution over the current crack length at 0.1, 1.0, and 2.4 second(s).

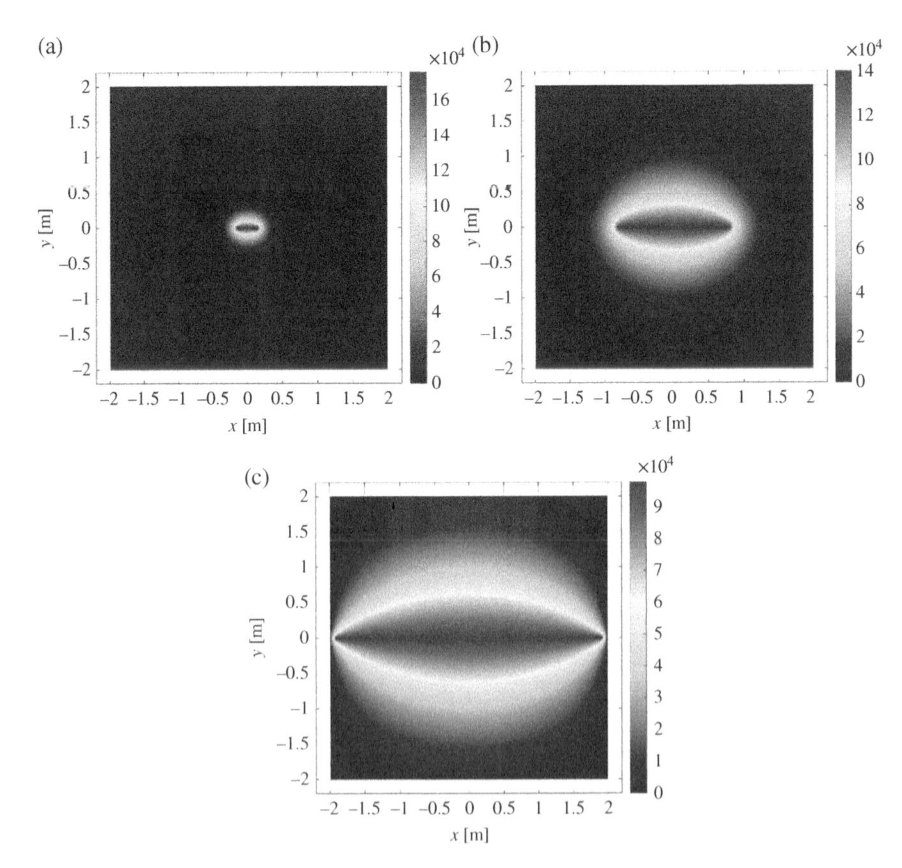

Figure 6.50 Pressure distribution for the current crack pattern at 0.1 second (a); 1.0 second (b); and 2 seconds (c). *Source:* Redrawn from Ni et al. (2020b). Copyright (2020), with permission from Elsevier.

The variations of pressure value at the injection point versus time are drawn in Figure 6.51. At the start of injection, the pressure value at the injection point increases suddenly to a large value, and under its action, the initial central crack opens gradually prior to crack propagation. The permeability in the initial central crack domain increases correspondingly, which leads to a steep drop of the pressure. Due to continuing pumping, the pressure increases again until the crack starts and there is an interplay between crack tip advancement speed and speed of fluid injection. In fact, the magnifying frame of Figure 6.51 evidences that the fluid pressure at the injection point presents a characteristic oscillation, which is consistent with the typical pattern already discussed in the previous two sections. The mean value of the pressure then decreases gradually because the volume of the induced crack increases faster than the injection rate.

6.6 Conclusion

The formulation developed in Chapter 2 and discretized in Chapter 3 is used to analyze various one-dimensional, two-dimensional, and three-dimensional problems with fully saturated soil. The results obtained from static and consolidation analysis are highly

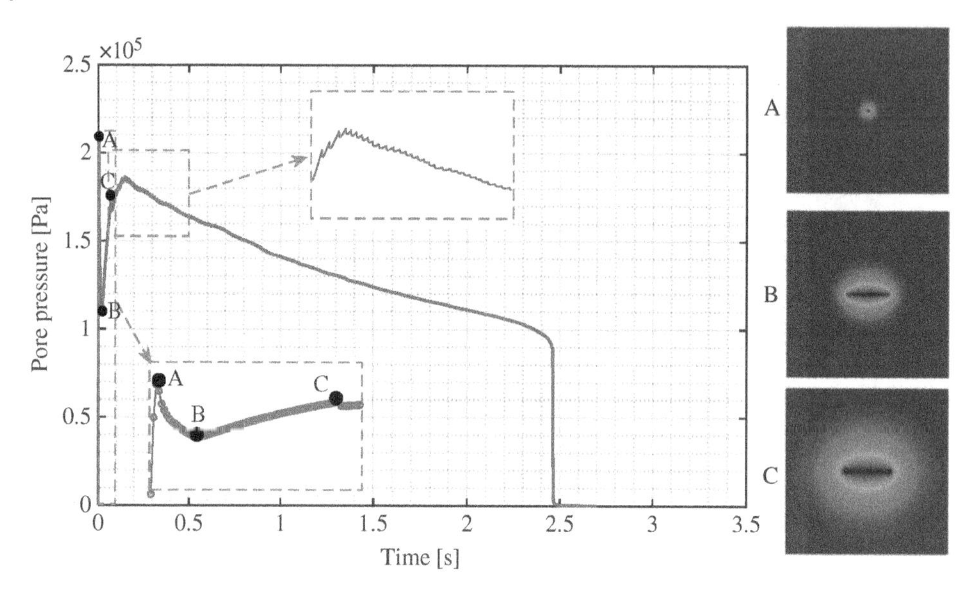

Figure 6.51 Pressure versus time at the injection point. *Source:* Redrawn from Ni et al. (2020b). Copyright (2020), with permission from Elsevier.

satisfactory and compared well with available analytical and experimental solutions. The results for fracturing analysis compare well with experiments and field observations and have been confirmed with other methods such as lattice models and Peridynamics, both coupled with the FEM model for the fluid. However, it would be useful for practical purposes for the formulation to be further validated using model experiments and this will be introduced in Chapter 7.

References

Advani, S. H., Lee, T. S., Dean, R. H., Pak, C. K. and Avasthi, J.M. (1997). Consequences of fluid lag in three-dimensional hydraulic fractures. *Int. J. Numer. Anal. Methods Geomech.*, **21**, 229–240.

Al-Khoury, R. and Sluys, L. J. (2007). A computational model for fracturing porous media, *Int. J. Numer. Methods Eng.*, **70**, 423–444.

Baiocchi, C., Comincioli, V., Magenes, E. and Pozzi, G. A. (1973) Free boundary problems in the theory of fluid flow through porous media: existence and uniqueness theorems, *Annali di Matematica*, **97**, 1–82.

Bardet, J. P.and Tobita, T. (2002). A practical method for solving free-surface seepage problems, *Comput. Geotech.*, **29**, 451–475.

Barenblatt, G. I. (1959) The formation of equilibrium cracks during brittle fracture: general ideas and hypotheses. Axially-symmetric cracks, *J. Appl. Math. Mech.*, **23**, 622–636.

Bear, J. and Corapcioglu, M. Y. (1981a). Mathematical model for regional land subsidence due to pumping. I: Integrated aquifer subsidence equations based on vertical displacement only, *Water Resour. Res.*, **17**, 937–46.

Bear, J. and Corapcioglu, M. Y. (1981b). Mathematical model for regional land subsidence due to pumping. 2. Integrated aquifer subsidence equations for vertical and horizontal displacements, *Water Resour. Res.*, **17**, 947–58.

Bell, J. R., Clarke, J. D. and Johnson, E. L. (1969). Lessons from an embankment failure analysis utilizing vane shear strength data, *Proceedings of the 6th Annual Engineering Geology and Soils Engineering Symposium*, pp 199–210, Boise ID, United States.

Black, A. (1998). DEA 13 (Phase II) Final Report. Investigation of Lost Circulation Problems with Oil-Base Drilling Fluids. Prepared by Drilling Research Laboratory, Inc.

Bolzon, G. and Corigliano, A. (2000). Finite elements with embedded displacement discontinuity: a generalized variable formulation. *Int. J. Numer. Methods Eng.*, **49**, 1227–1266.

Boone, T. J. and Ingraffea, A. R. (1990). A numerical procedure for simulation of hydraulically driven fracture propagation in poroelastic media. *Int. J. Numer. Anal. Methods Geomech.*, **14**, 27–47.

Briseghella, L., Sanavia, L., and Schrefler, B.A. (1999). Seismic analysis of earth dams using a multiphase model. *Proceedings of the IXth Italian National Congress "L'Ingegneria Sismica in Italia"*, Turin, Italy (28–30 September 1999).

Bruch, J. C. Jr (1991). Multisplitting and domain decomposition techniques applied to free surface flow through porous media, in *Computational Modelling of Free and Moving Boundary Problems*, Vol. **1** Fluid Flow, L.C. Wrobel and C.A. Brebbia (Eds), Computational Mechanics Publications, Walter de Gruyter & Co., Berlin.

Burlini, L. and Di Toro, G. (2008). Volcanic symphony in the lab, *Science*, **322**, (5899), 207–208.

Burlini, L., Di Toro, G. and Meredith, P. (2009). Seismic tremor in subduction zones: rock physics evidence, *Geophys. Res. Lett.*, **36**, L08305.

Camacho, G. T. and Ortiz, M. (1996). Computational modelling of impact damage in brittle materials, *Int. J. Solids Struct.*, **33**, 2899–2938.

Cao, T. D., Milanese, E., Remij, E. W., Rizzato, P., Remmers, J. J. C., Simoni, L., Huyghe, J. M., Hussain, F. and Schrefler, B. A. (2017). Interaction between crack tip advancement and fluid flow in saturated porous media, *Mech. Res. Commun.*, **80**, 24–37.

Cao, T. D., Hussain, F. and Schrefler, B. A. (2018). Porous media fracturing dynamics: stepwise crack advancement and fluid pressure oscillations. *J. Mech. Phys. Solid*, **111**, 113–33

Carter, B. J., Desroches, J., Ingraffea, A. R. and Wawrzynek, P. A. (2000). Simulating fully 3-D hydraulic fracturing, in *Modeling in Geomechanics*, M. Zaman, J. R. Booker and G. Gioda (Eds), Wiley, Chichester, 525–567.

Chen, Y., Hu, R., Lu, W., Li, D. and Zhou, C. (2011). Modeling coupled processes of non-steady seepage flow and non-linear deformation for a concrete-faced rockfill dam, *Comput. Struct.*, **89**, 1333–1351.

Cleary, M. P. (1978). Moving singularities in elasto-diffusive solids with applications to fracture propagation, *Int. J. Solids Struct.*, **14**, 81–97.

Detournay, E. and Cheng, A. H. (1991). Plane strain analysis of a stationary hydraulic fracture in a poroelastic medium. *Int. J. Solids Struct.*, **27**, 1645–1662.

Drucker, D. C. and Prager, W. (1952). Soil mechanics and plastic analysis or limit design, *Quart. Appl. Math.*, **10**, 157–165.

Dugdale, D. S. (1960). Yielding of steel sheets containing slits, *J. Mech. Phys. Solids*, **8**, 100–104.

Ehlers, W. and Luo, C. (2017). A phase-field approach embedded in the theory of porous media for the description of dynamic hydraulic fracturing, *Comput. Methods Appl. Mech. Eng.*, **315**, 348–368

Feist, C. and Hofstetter, G. (2006). An embedded strong discontinuity model for cracking of plain concrete, *Comput. Methods Appl. Mech. Eng.*, **52**, 7115–7138.

Feng, Y., Jones, J. F. and Gray, K. E. 2015. Pump-in and flow back test for determination of fracture parameters and in-situ stresses. *AADE National Technical Conference and Exhibition, AADE-15-NTCE-35*, San Antonio, TX (8–9 April 2015).

Garagash, D. and Detournay, E. (2000). The tip region of a fluid-driven fracture in an elastic medium, *J. Appl. Mech.*, **67**, 183–192.

Gawin, D., Sanavia, L. and Schrefler, B. A. (1998). Cavitation modelling in saturated geomaterials with application to dynamic strain localisation, *Int. J. Numer. Methods Fluids*, **27**, 109–125.

Hageman, T., René de Borst, Stick-slip like behavior in shear fracture propagation including the effect of fluid flow, Int J Numer Anal Methods Geomech. 2021;1–25, DOI: 10.1002/nag.3186.

Hilleborg, A., Modeer, M. and Petersson, P. E. (1976). Analysis of crack formation and crack growth in concrete by means of fracture mechanics and finite elements, *Cem. Concr. Res.*, **6**, 773–782.

Huang, N. C. and Russel, S. G. (1985a). Hydraulic fracturing of a saturated porous medium – I: General theory, *Theor. Appl. Fract. Mech.*, **4**, 201–213.

Huang, N. C. and Russel, S. G. (1985b). Hydraulic fracturing of a saturated porous medium – II: Special cases, *Theor. Appl. Fract. Mech.*, **4**, 215–222.

ICOLD (1999). Fifth International Benchmark Workshop on Numerical Analysis of Dams, Theme A2, Denver, Colorado.

Khoei, A. R., Barani, O. R. and Mofid, M. (2011). Modeling of dynamic cohesive fracture propagation in porous saturated media, *Int. J. Numer. Anal. Methods Geomech.*, **35**, 1160–1184.

Kraaijeveldt, F., Huyghe, J. M., Remmers, J. J. C. and de Borst, R. (2013). 2-D mode one crack propagation in saturated ionized porous media using partition of unity finite elements, *J. Appl. Mech.*, **80**, 020907-1-12

Lee, S., Wheeler, M. F. and Wick, T. (2017). Iterative coupling of flow, geomechanics and adaptive phase-field fracture including levelset crack width approaches. *J. Comput. Appl. Math.*, **314**, 40–60.

Lewis, R. W. and Schrefler, B. A. (1998). *The Finite Element Method in the Deformation and Consolidation of Porous Media Second Edition*, Wiley, Chichester.

Lhomme, T. P., de Pater, C. J. and Helferich, P. H. (2002). Experimental study of hydraulic fracture initiation in Colton Sandstone. From *SPE/ISRM 78187, SPE/ISRM Rock Mechanics Conference*, Irving, TX (20–23 October 2002).

Li, X. D. and Wiberg, N.-E., (1998). Implementation and adaptivity of a space-time finite element method for structural dynamics, *Comput. Methods Appl. Mech. Eng.*, **156**, 211–229.

Miehe, C. and Mauthe, S. (2016). Phase field modeling of fracture in multiphysics problems. Part III. Crack driving forces in hydro-poroelasticity and hydraulic fracturing of fluid-saturated porous media, *Comput. Methods Appl. Mech. Eng.*, **304**, 619–655.

Mikelić, A., Wheeler, M. F. and Wick, T. (2015). Phase-field modeling of a fluid-driven fracture in a poroelastic medium, *Comput. Geosci.*, **19**, (6), 1171–1195.

Milanese, E., Yılmaz, O., Molinari, J.-F. and Schrefler, B. (2016a). Avalanches in dry and saturated disordered media at fracture, *Phys. Rev. E*, **93**, (4) 043002

Milanese, E., Rizzato, P., Pesavento, F., Secchi, S. and Schrefler, B. A. (2016b). An explanation for the intermittent crack tip advancement and pressure fluctuations in hydraulic fracturing, *Hydraul. Fract. J.*, **3**, (2), 30–43.

Milanese, E., Yılmaz, O., Molinari, J.-F. and Schrefler, B. (2017). Avalanches in dry and saturated disordered media at fracture in shear and mixed mode scenarios, *Mech. Res. Commun.*, **80**, 58–68.

Moes, N. and Belytschko, T. (2002). Extended finite element method for cohesive crack growth, *Eng. Fract. Mech.*, **69**, (7), 813–833.

Mohammadnejad, T. and Khoei, A. R. (2013). Hydromechanical modelling of cohesive crack propagation in multiphase porous media using extended finite element method, *Int. J. Numer. Anal. Methods Geomech.*, **37**, 1247–1279.

Ni, L., Zhang, X., Zou, L., and Huang, J. (2020a). Phase-field modeling of hydraulic fracture network propagation in poroelastic rocks. *Comput. Geosci.*, 10.1007/s10596-020-09955-4.

Ni, T., Pesavento, F., Zaccariotto, M., Galvanetto, U., Zhu, Q.-Z. and Schrefler, B. A. (2020b). Hybrid FEM and peridynamic simulation of hydraulic fracture propagation in saturated porous media, *Comput. Methods Appl. Mech. Eng.*, **366**, 113101. 10.1016/j.cma.2020.113101.

Nolet, G. (2009). Slabs do not go gently, *Science*, **324**, (5931), 1152–1153.

Okland, D., Gabrielsen, G. K., Gjerde, J., Koen, S. and Williams, E. L. (2002). The importance of extended leak-off test data for combatting lost circulation. *From SPE/ISRM Rock Mechanics Conference*, Irving, TX (20–23 October 2002), SPE-78219-MS.

Oliver, J., Huespe, A. E., Pulido, M. D. G. and Chaves, E. (2001). From continuum mechanics to fracture mechanics: the strong discontinuity approach, *Eng. Fract. Mech.*, **69**, (2), 113–136.

Owen, D. R. J., Feng, Y. T., Labao, M., Han, K., Leonardi, C. R., Yub, J. and Eve, R. (2009). Fluid/structure coupling in fracturing solids and particulate media, *Geomech. Geoeng. Int. J.*, **4**, (1), 27–37.

Pastor, M., Zienkiewicz, O. C. and Chan, A. H. C. (1990). Generalized plasticity and the modelling of soil behaviour, *Int. J. Numer. Anal. Methods Geomech.*, **14**, 151–190.

Perkins, T. K. and Kern, L. R. (1961). Widths of hydraulic fractures, *SPE J.*, **222**, 937–949.

Peruzzo, C., Cao, D. T., Milanese, E., Favia, P., Pesavento, F., Hussain, F. and Chrefler, B. A. (2019). Dynamics of fracturing saturated porous media and selforganization of rupture, *Eur. J Mech./A Solids*, **74**, 471–484.

Phillips, W. J. (1972). Hydraulic fracturing and mineralization, *J. Geol. Soc.*, **128**, 337–359.

Pizzocolo, F., Huyghe, J. M. R. J. and Ito, K. (2013). Mode I crack propagation in hydrogels is stepwise, *Eng. Fract. Mech.*, **97**, 72–79.

Prandtl, L. (1921). Über die Eindringungsfestigkeit plastisher Baustoffe und die Festigkeit von Schneiden, *Z. Angew. Math. Mech.*, **1**, (1), 15–20.

Remij, E. W., Remmers, J. J. C., Huyghe, J. M. and Smeulders, D. M. J. (2017). An investigation of the step-wise propagation of a mode-II fracture in a poroelastic medium, *Mech. Res. Commun.*, **80**, 10–15.

Réthoré, J., de Borst, R. and Abellan, M. A. (2008). A two-scale model for fluid flow in an unsaturated porous medium with cohesive cracks, *Comput. Mech.*, **42**, 227–238.

Rice, J. R. and Cleary, M. P. (1976). Some basic stress diffusion solutions for fluid saturated elastic porous media with compressible constituents, *Rev. Geophys. Space Phys.*, **14**, 227–241.

Rosen, J. (2016). Crystal clocks, *Science*, **354**, (6314), 822–825.

Safai, N. M. and Pinder, G. F. (1979). Vertical and horizontal land deformation in a desaturating porous medium, *Adv. Water Resour.*, **2**, 19–25.

Santillán, D., Juanes, R. and Cueto-Felgueroso, L. (2018). Phase field model of hydraulic fracturing in poroelastic media: fracture propagation, arrest, and branching under fluid injection and extraction, *J. Geophys. Res. Solid Earth*, **123**, (3), 2127–2155.

Sarma, S. K. (1973). Stability analysis of embankments and slopes, *Géotechnique*, **23**, (3), 423–433, https://doi.org/10.1680/geot.1973.23.3.423.

Schrefler, B. A., Secchi, S. and Simoni, L. (2006). On adaptive refinement techniques in multi-field problems including cohesive fracture, *Comput. Methods Appl. Mech. Eng.*, **195**, 444–461.

Schrefler, B. A., Pesavento, F., Sanavia, L., Sciumè, G., Secchi, S. and Simoni, L. (2011). A general framework for modelling long term behaviour of earth and concrete dams, *Front. Archit. Civil Eng. China*, **5**, (1), 41–52. DOI 10.1007/s11709-010-0070-x.

Schwartz, S. Y. and Rokosky, J. M. 2007. Slow slip events and seismic tremor at circum-pacific subduction zones, *Rev. Geophys.*, **45**, RG3004 2006RG000208, 1–32.

Secchi, S. and Schrefler, B. A. (2012). A method for 3-D hydraulic fracturing simulation, *Int. J. Fract.*, **178**, 245–258.

Secchi, S. and Simoni, L. (2003). A improved procedure for 2-D unstructured Delaunay mesh generation, *Adv. Eng. Softw.*, **34**, 217–234.

Secchi, S., Simoni, L. and Schrefler, B. A. (2007). Mesh adaptation and transfer schemes for discrete fracture propagation in porous materials, *Int. J. Numer. Anal. Methods Geomech.*, **31**, 331–345.

Segura, J. M and Carol, I. (2008). Coupled HM analysis using zero-thickness interface elements with double nodes: Part I. Theoretical model, *Int. J. Numer. Anal. Methods Geomech.*, **32**, 2083–2101.

Simoni, L. and Schrefler, B. A. (1989). Numerical modelling of unbounded domains in coupled field problems. *Meccanica*, **24**, 98–106.

Simoni, L. and Secchi, S. (2003). Cohesive fracture mechanics for a multi-phase porous medium, *Eng. Comput.*, **20**, 5/6, 675–698,

Simoni, L. Secchi, S. and Schrefler, B. A. (2008). Numerical difficulties and computational procedures for thermo-hydro-mechanical coupled problems of saturated porous media, *Comput. Mech.*, **43**, 179–189, doi: 10.1007/s00466-008-0302-2.

Soliman, M. Y., Wigwe, M., Alzahabi, A., Pirayesh, E. and Stegent, N. (2014). Analysis of fracturing pressure data in heterogeneous shale formations, *Hydraulic Fract J*,**1**, 8–12.

Terzaghi, K. (1943). *Theoretical Soil Mechanics*, Wiley, New York.

Sun, W., Fish, J., Coupling of non-ordinary state-based peridynamics and finite element method for fracture propagation in saturated porous media, Int J Numer Anal Methods Geomech. 2021;1¨C22. DOI: 10.1002/nag.3200.

Theis, C. V. (1935). The relation between the lowering of the piezometric surface and the rate and duration of discharge of a well using groundwater storage, *Eos Trans. AGU*, **16**, 519–24.

Turska, E. and Schrefler, B. A. (1993). On convergence conditions of partitioned solution procedures for consolidation problems, *Comput. Methods Appl. Mech. Eng.*, **106**, 51–63.

Vahab, M., Akhondzadehb, S., Khoei, A. R. and Khalili, N. (2018). An X-FEM investigation of hydro-fracture evolution in naturally-layered domains. *Eng. Fract. Mech.*, **191**, 187–204.

Verruijt, A. (1969). Elastic storage of aquifers, in *Flow Through Porous Media*, R. J. M. DeWiest (Ed), Academic Press, New York, 331–376.

Wawrzynek, P. A. and Ingraffea, A. R. (1989). An interactive approach to local remeshing around a propagating crack, *Finite Elem. Anal. Des.*, **5**, 87–96.

Wells, G. N. and Sluys, L. J. (2001). Three-dimensional embedded discontinuity model for brittle fracture, *Int. J. Solids Struct.*, **38**, 897–913.

Witherspoon, P. A., Wang, J. S. Y., Iwai, K. and Gale, J. E. (1980). Validity of cubic law for fluid flow in a deformable rock fracture, *Water Resour. Res.*, **16**, 1016–1024.

Whitman, R. V. and Bailey, W. A. (1967). Use of computers for slope stability analysis, *J. Soil Mech. Found. Div.*, **93**, (4), https://doi.org/10.1061/JSFEAQ.0001003.

Xie, Y. M. (1990). Finite element solution and adaptive analysis for static and dynamic problems of saturated-unsaturated porous media. Ph.D. Thesis. C/Ph/136/90, Department of Civil Engineering, Swansea University.

Zhang, M. and Shu, C.-W. (2003). An analysis of three different formulations of the discontinuous Galerkin method for diffusion equations. *Math. Models Methods Appl. Sci.*, **13**, 395–413.

Zhang, H. W., Sanavia, L. and Schrefler, B. A. (2001). Numerical analysis of dynamic strain localisation in initially water saturated dense sand with a modified generalised plasticity model, *Comput. Struct.*, **79**, 441–459.

Zhou, S., Zhuang, X. and Rabczuk, T. (2019). Phase-field modeling of fluid-driven dynamic cracking in porous media, *Comput. Methods Appl. Mech. Eng.*, **350**, 169–198.

Zhu, J. Z. and Zienkiewicz, O. C. (1988). Adaptive techniques in the finite element method, *Commun. Appl. Numeri. Methods*, **4**, 197–204.

Zienkiewicz, O. C. and Cormeau, I. C. (1974). Visco-plasticity-plasticity and creep in elastic solids–a unified numerical solution approach. *Int. J. Num. Meth. Eng.*, **8**, (4), 821–845.

Zienkiewicz, O. C. and Pande, G. N. (1977). Time-dependent multi-laminate model of rocks – a numerical study of deformation and failure of rock masses, *Int. J. Num. Anal. Geomech.*, **1**, 219–247.

Zienkiewicz, O. C. and Taylor, R. L. (1991). *The Finite Element Method-Volume 2 Solid and Fluid Mechanics, Dynamics and Non-Linearity*, (4th edn), McGrew-Hill Book Company, London.

Zienkiewicz, O. C., Mayer, P. and Cheung, Y. K. (1966). Solution of anisotropic seepage by finite elements, *J. Eng. Mech. Div.*, **92**, 111–120.

Zienkiewicz, O. C., Humpheson, C. and Lewis, R. W. (1975). Associated and non-associated viscoplasticity and plasticity in soil mechanics, *Géotechnique*, **25**, 671–689.

Zienkiewicz, O. C., Humpheson, C. and Lewis, R. W. (1977). Chapter 4 – A unified approach to soil mechanics problems (including plasticity and viscoplasticity), in *Finite Elements in Geomechanics*, G. Gudehus (Ed) 151–178, Wiley.

Zienkiewicz, O. C., Chan, A., Pastor, M., Schrefler, B. A. and Shiomi, T (1999). *Computational Soil Dynamics with Special Reference to Earthquake Engineering*, Chichester, Wiley.

7

Validation of Prediction by Centrifuge

7.1 Introduction

In the previous (Chapter 6), we presented several examples of application of the full formulation for various static and consolidation problems. This tested effectively the limit behaviour of various constitutive models and also the interaction of slow drainage with deformation during consolidation process. The problem did not however stretch the predictive capacity: in the case of limit behaviour giving answers which were quite well known in general and in the case of consolidation, the departure from linear, elastic, behaviour was small. To test fully the possibilities offered by the formulation and the models presented we should seek examples where

1) the repeated loading generates substantial pore pressures and possibly liquefaction
2) the problem is such that nonlinear, near failure, stresses are present and at least partial inelastic permanent deformation results.

Clearly the study of earthquake response presents most challenge here but it is hardly possible to instrument it on a site where both conditions 1 and 2 above will occur within a reasonable time span or indeed ever.

Further it would be almost impossible to achieve the so-called "Class A" prediction, so beloved by soil mechanicians, in which the computation precedes the actual event. Even if an earthquake of the desired magnitude with the desired effect did happen, its precise detail of input would not be available before it happened.

What is often possible is to reconstruct catatrophic events, particularly if some idea of the input motion is available through measurements within reasonable proximity. In Chapter 9 we shall show such reconstructions known frequently as back-analysis. These in the soil mechanics problems are of course Class C predictions and therefore mistrusted by some as of course the soil parameters could be adjusted to achieve the already known measured results.

For this reason it is desirable to attempt scale model tests of earthquake events for which both Class A and Class B predictions are possible (The last one being computed

Computational Geomechanics: Theory and Applications, Second Edition. Andrew H. C. Chan, Manuel Pastor, Bernhard A. Schrefler, Tadahiko Shiomi and O. C. Zienkiewicz.

simultaneously with the measurement on the model). Two possible scale models of environmental conditions exist, a. the shaking table and b. the centrifuge.

The shaking table has been used with great success in modelling the dynamic effect of structures but unfortunately it is less successful for soil mechanics problems. The reason here is that the behaviour of soil especially sand depends strongly on the mean effective stress (see Chapter 4). Therefore, for typical soil problems, gravity is the most important external force and this is obviously not modelled correctly in a scaled model in which densities are kept constant and the linear dimensions reduced. For this reason, the centrifuge was invented and this device permits a very considerable increase of the gravitational accelerations. The scale model in the centrifuge is usually rather small as the whole frame of the test has to be rotated at high speed producing a fairly uniform field of up to 100 g in the model area. For this reason, we shall draw our comparisons in the following sections entirely from the centrifuge and here, as we mentioned, both Class A and B type predictions will be possible.

In Section 7.2 the basic theory of centrifuge modelling in geotechnical applications is introduced with particular attention to the use of alternative fluid in order to achieve dynamic compatibility of diffusion and inertial behaviour. It is precisely this substitution of pore pressure rendered centrifuge testing unsuitable for the modelling of dynamic events under partially saturated conditions and here lies one of the limitations of the procedure. For the interest of the reader, we recommend a study of various publications of Prof. Schofield and others concerning the physics of centrifuge modelling.

In the first section describing centrifuge test, we shall concentrate on a model of dyke (Venter 1985, 1987) performed on the Cambridge geotechnical centrifuge. Here comparison of computation are done simultaneously with models and perhaps this section should be classified under Class B verifications. In this Section 7.3 and indeed in the later Section 7.5 where we describe a somewhat similar test done on an embankment wall at Colorado by Dewoolkar, same remarks apply. In both cases, we did not use a scaling as the centrifuge itself and the artifical earthquake itself were considered to be the "prototype". Section 7.4 represents results of a very major study understaken in USA under the name of VELACS (VERification of Liquefaction Analysis by Centrifuge Studies – Arulanandan and Scott 1993). This study was funded by the National Science Foundation, USA and involved some twenty laboratories in various parts of the world performing numerical predictions which later were to be compared to several centrifuge studies in USA and Cambridge Geotechnical centrifuge, UK. In this example, all the predictions are of Class A type as results had to be presented to the organisers before the centrifuge tests were attempted. Only one or two cases were the results obtained later and these are specially marked as Class B. During this study, very many alternative situations were investigated and it will be seen later that excellent comparisons were obtained.

It is necessary however to remark that to date few cases has it been possible to perform centrifuge models with a free fluid surface and such structures as dams, retaining embankments with different level of water at different sides etc cannot be modelled because of the restriction about the partially saturated conditions. For this particular case, the only results which are available will be presented by back-analysis in Chapter 9.

7.2 Scaling Laws of Centrifuge Modelling

In this section, a brief description and derivation of the centrifuge scaling laws for models are described. This is included so that it would aid the readers in the interpretation of the centrifuge results and their comparisons with the numerical results. Furthermore, it is the purpose of this section to explain the concept of dynamic compatibility in centrifuge modelling which led to the use of a different pore fluid from water in saturated centrifuge model tests. As we consider in the numerical analysis that the centrifuge experiment is a prototype itself, issues concerning far-field boundary conditions and particle size will not be dealt with. For readers interested in the centrifuge modelling of dynamic events, the following publications can be referred (Schofield 1980, 1981; Lee and Schofield 1988; Schofield and Zeng 1992; Taylor 1994; Steedman and Zeng, 1995).

Assuming a N (typically between 50 to 200) scale model is introduced in the centrifuge so the linear length is scaled by:

$$x_i^M = x_i^P/N \tag{7.1}$$

where the superscript M denotes the model scale and P denotes the prototype it intends to model. We can write the mixture equilibrium equation (viz. Equation 2.11 neglecting only convective terms) for model and prototype respectively.

$$\frac{\partial \sigma_{ij}^M}{\partial x_j^M} + \rho^M b_i^M - \rho^M \ddot{u}_i^M - \rho_f^M \dot{w}_i^M = 0 \tag{7.2}$$

$$\frac{\partial \sigma_{ij}^P}{\partial x_j^P} + \rho^P b_i^P - \rho^P \ddot{u}_i^P - \rho_f^P \dot{w}_i^P = 0 \tag{7.3}$$

If we assume that the density of the mixture and fluid together with the stress state[1] are maintained the same in model and prototype, this would require the acceleration to be scaled by $1/N$ times with:

$$\ddot{u}_i^M = N\ddot{u}_i^P \tag{7.4a}$$

and

$$b_i^M = Nb_i^P \tag{7.4b}$$

$$\dot{w}_i^M = N\dot{w}_i^P \tag{7.5}$$

Comparing with the linear dimension in Equation (7.1), one would conclude that time is also scaled by N times:

$$t^M = t^P/N \tag{7.6}$$

Therefore, in the model, a dynamic event will happen N times as fast as its corresponding prototype thus pushing up the frequency of the dynamic events by N times. However, for

1 The maintanence of same stress level is important for soil behaviour as the stress:strain behaviour of soils are highly stress level and strain history dependant

most practical earthquake events, the frequency and wavelengths of the wave within the pores should still be within the range of laminar Darcy's flow.

The scaling relationship is then applied for the fluid conservation equation (viz Equation 2.16):

$$w_{i,i} + \alpha\dot{\varepsilon}_{ii} + \frac{\dot{p}}{Q} + n\frac{\dot{\rho}_f}{\rho_f} + \dot{s}_0 = 0 \tag{7.7}$$

which implies that pore pressure is the same in model and prototype if the compressibilities and the void ratio remain the same. Also for any source term its rate must also be the same. If any substitute fluid is used, its compressibility should not be too different from that of water though the bulk modulus of water does vary because of the amount of air dissolved in it.

So far, there is no problem in the scaling, however, a problem arises when the fluid flow equation (viz. Equatons (2.13) and (2.14), again neglecting only the convective terms) is considered:

$$w_i = k_{ij}\left(-p_{,j} - \rho_f\ddot{u}_j + \rho_f b_j - \rho_f\frac{\dot{w}_j}{n}\right) \tag{7.8}$$

Considering the equation for model and prototype separately:

$$w_i^M = k_{ij}^M\left(-\frac{\partial p^M}{\partial x_j^M} - \rho_f^M\ddot{u}_j^M + \rho_f^M b_j^M - \rho_f\frac{\dot{w}_j^M}{n}\right) \tag{7.9}$$

and

$$w_i^P = k_{ij}^P\left(-\frac{\partial p^P}{\partial x_j^P} - \rho_f^P\ddot{u}_j^P + \rho_f^P b_j^P - \rho_f\frac{\dot{w}_j^P}{n}\right) \tag{7.10}$$

As velocity is required to be same in the model and prototype if the scaling of the displacement, acceleration and time as described in Equations (7.1), (7.4), 7.5 and 7.6 we require the permeability to be scaled by $1/N$:

$$k_{ij}^N = k_{ij}^P/N \tag{7.11}$$

This cannot be readily achieved if the same solid material and pore fluid with the same porosity are used. One of the solutions to this problem is therefore to use a different fluid. For instance, in the Camrbidge test of the dyke, silicon oil with the same density as water is used. The viscosity is chosen to be N centi-stokes or cs (Dow Corning Limited 1985) because water has a viscosity of 1 cs. This reduces the permeability by N times and the above relation is retained however there is now a possibility that the bulk modulus of the fluid and its damping characteristics are different from those of water. In the VELACS exercise, in order to avoid problems in the interpretation of the centrifuge tests, water was chosen as the pore fluid and consolidation was therefore found to be applied more rapidly than the corresponding prototype. Other substitute fluid used include Metolose (Dewooklar et al. 2009).

However, for semi-saturated material, the use of another liquid will in all probability lead to a different value for surface tension as well as different drying and wetting characteristic.

Although the use of the same fluid in experimental modelling of pollutant transport in semi-saturated environment has been reported (Cooke, 1991, 1993; Cooke and Mitchell 1991a, 1991b) and the scaling of the capillary was reported to be modelled correctly (Hellawell 1994; Taylor 1994; Culligan-Hensley and Savvidou 1995), it would be difficult at this stage to extend such tests to dynamic events.

7.3 Centrifuge Test of a Dyke Similar to a Prototype Retaining Dyke in Venezuela

The test represents a fully dynamic analysis with transient behaviour. The physical model is a centrifuge experiment performed by Venter (1985) on the Cambridge Geotechnical Centrifuge. The principle of the centrifuge has been briefly explained in Section 7.2. Simply, the centrifuge reproduces similar stresses and strain history as experienced by the prototype on the scaled model. If the behaviour of soil is controlled mainly by the stress state and its strain history then the centrifuge model is able to predict the generic behaviour of the prototype under earthquake condition.

The layout of the experiment is given in Figure 7.1. The model is built in a strong box with a dyke lying on a sand bed. An oil reservoir is created behind the dyke to provide seepage condition. Silicon oil is used so that the diffusion equation and the dynamic equation can have the same time scale under the centrifuge condition while using the same soil at the same density. This is done by using silicon oil (Dow Corning Limited 1985) of viscosity of 80 stokes (80 times the viscosity of the water). Also shown in Figure 7.1 are the measurement devices, which includes 11 PPTs – pore pressure transducers, 1 LVDT – Linear Voltage Displacement Transducer, 7 ACCs – accelerometers and 3 TSTs – total stress transducers.

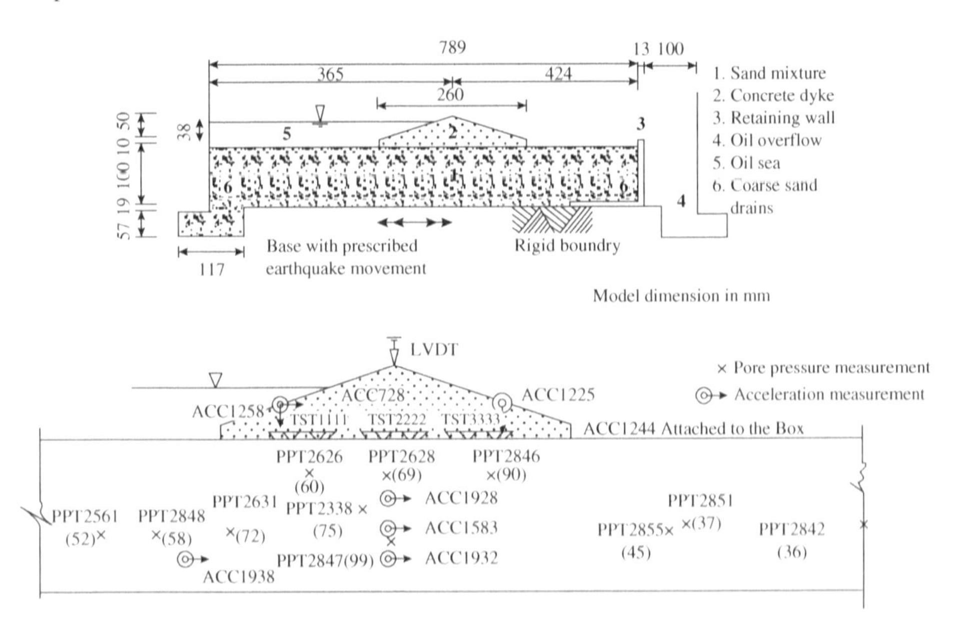

Figure 7.1 Sections through model KVV03 showing dimensions and transducer locations.

With the triaxial test results (Venter 1986) on the Leighton-Buzzard sand used in the centrifuge experiment, soil parameters for the Pastor-Zienkiewicz Mark-III are identified. They are listed in Table 7.1. The Finite Element idealization and the boundary conditions are given in Figure 7.2. Also shown in the figure are the positions of 10 pore pressure transducers, 4 accelerometers and the LVDT presented for displacement comparison purpose. The comparison is done for as many number of points as possible so that an overall picture together with appropriate mechanisms can be obtained. The test was done at 78 g and the material data for the Finite Element analysis are given in Table 7.2. The first study is performed with 4-4 element, i.e. 4-noded for soil displacement (u) and 4-noded for pore pressure (p). The computer code used was SWANDYNE-I (Chan 1988) using SSpj time stepping scheme. For each pair of graphs presented in this section, the left hand side one is measured value from experiments, while the right hand side is the computed value.

The result are given in Figure 7.3a–7.3h. The input motion was taken from the accelerometer ACC1244 attached to the box. Also shown in Figure 7.3 is the vertical displacement at the crest of the dyke. The comparison is excellent. Besides predicting correctly the final displacement, the rising time and the shape of the rise are also predicted. The reader is reminded that this was the first set of soil parameters obtained directly from the soil model tester, no parametric study has been performed i.e. a prediction is stated as one of Class B-1.

Table 7.1 Soil Model Data

	Test	13	14	Predicted
	Adjusted p_o (kPa)	55.0	64.5	55.0
	K_{evo}	20000	23000	20000
(1)	K_{evo}/p_o	360	360	360
	K_{eso}	30000	35000	30000
(2)	K_{eso}/p_o	540	540	540
(3)	M_g	1.15	1.26	1.26
(4)	M_f	0.86	0.55	0.84
	$M_g D_R$	0.51	0.51	0.756
(5)	α_f, α_g*	0.45	0.45	0.45
(6)	β_o*	4.2	4.2	4.2
(7)	β_1*	0.2	0.2	0.2
(8)	H_o	500	600	700
(9)	H_{uo}(kPa)	-	-	60000
(10)	γ_u*	-	-	2
(11)	γ_{DM}*	-	-	-
	e	0.875	0.873	0.80
	D_R**	40%	40.5%	60%

* As suggested in the original paper (Pastor and Zienkiewicz 1986)

** $e_{min} = 0.65$ ($D_R = 100\%$) $e_{max} = 1.025$ ($D_R = 0\%$) $D_R = \frac{e_{max} - e}{e_{max} - e_{min}} x100\%$

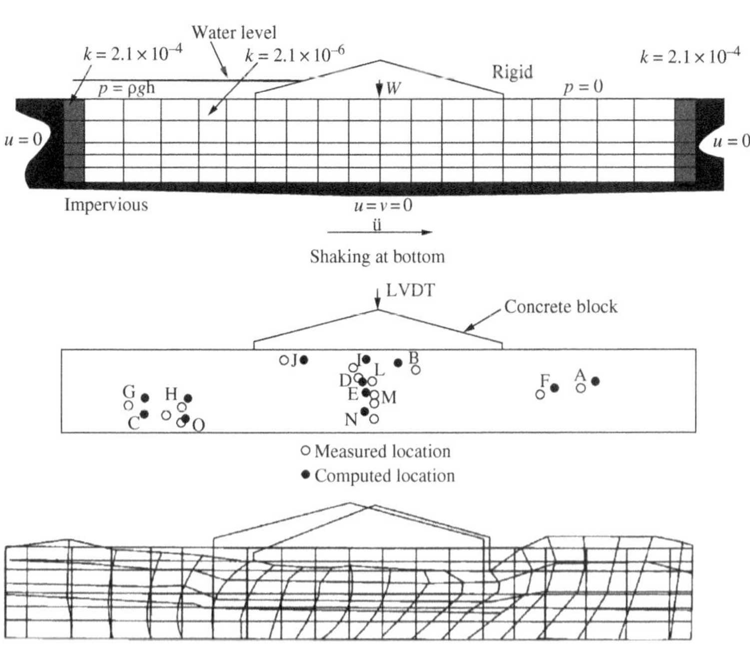

Figure 7.2 Finite element idealisation.

Table 7.2 Material data for Finite Element analysis

Bulk density (average soil-pore fluid)	$= 1908\text{kgm}^{-3}$
Density of the pore fluid	$= 980\text{kgm}^{-3}$
Bulk modulus of the pore fluid	= 1.092GPa
Biot-alpha	= 1.0000
Porosity (n)	= 0.444444
Initial void ratio (e)	= 0.80
Permeability of the sand bulk	$= 2.1\text{x}10^{-6}\text{ms}^{-1}$
Permeability of the sand drain	$= 2.1\text{x}10^{-4}\text{ms}^{-1}$
g-acceleration at 1g	$= 9.81\text{ms}^{-2}$
g-acceleration for the centrifuge test (78g)	$= 765.18\ \text{ms}^{-2}$
Gauss Point for all cases (Gauss-Legendre)	= 2x2
No incremental strain subdivision is performed	
Initial stress method is used for non-linear iterations	
Convergence criteria for non-linear iterations (for all phases):	$\dfrac{\textit{Residual force norm}}{\textit{current external force norm}} \leq 0.1\%$
Properties for the Rigid Block:	
Density	$= 2000\text{kgm}^{-3}$
Mass per unit width	= 18.2kg per metre
Moment of inertia per unit width	$= 0.069503\ \text{kgm}^2$ per metre
Size of time step	= 0.00015 sec (data point spacing of the measurements)
Total number of time steps	= 1024
θ's for the SSpj scheme	= 0.5, 0.5
Bilinear interpolation for both u and p	

The pore pressure transducer A gives good agreement too as shown in Figure 7.3b, though the value is slightly lower. More oscillation is seen in B, however, the predicted trend is still correct. The agreement of C is remarkable though the size of oscillation is larger, nevertheless, the result is very good. The other graph in Figure 7.3c gives the comparison of pore pressure transducer D. The mean value is almost the same as the measured value, though oscillation is more pronounced. E predicts lower pore pressure rise and F is quite satifactory,

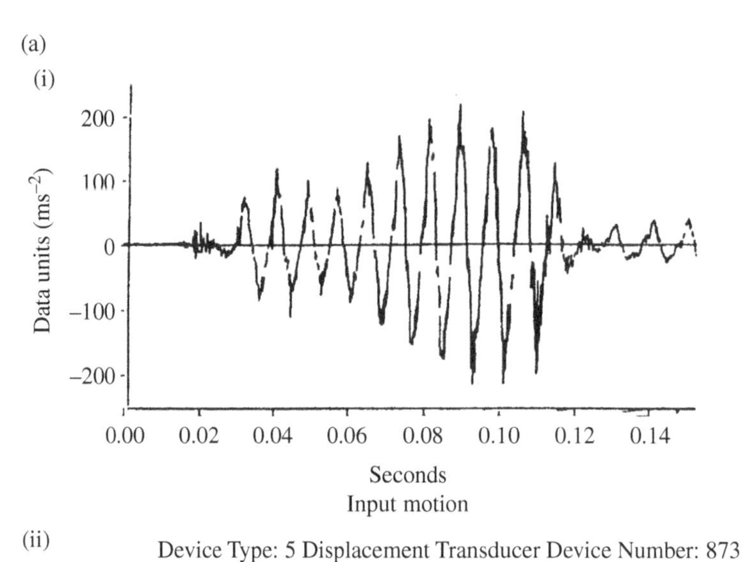

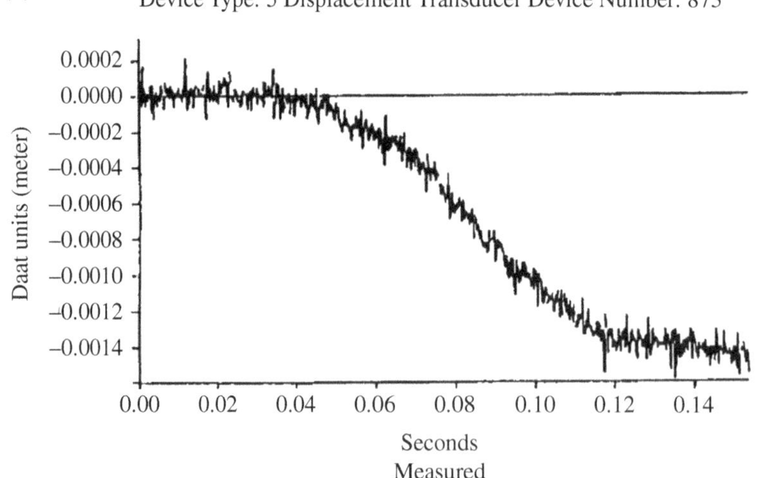

Figure 7.3 (a) Comparison with centrifuge results (top) input motion (bottom) vertical displacement of the dyke (b) comparison with centrifuge results (top) excess pore pressure at point A (bottom) excess pore pressure at point B (c) comparison with centrifuge results (top) excess pore pressure at point C (bottom) excess pore pressure at point D (d) comparison with centrifuge results (top) excess pore pressure at point E (bottom) excess pore pressure at point F (e) comparison with centrifuge results (top) excess pore pressure at point G (bottom) excess pore pressure at point H (f) comparison with centrifuge results (top) excess pore pressure at point I (bottom) excess pore pressure at point J (g) comparison with centrifuge results (top) acceleration at point L (bottom) acceleration at point M (h) comparison with centrifuge results (top) acceleration at point N (bottom) acceleration at point O.

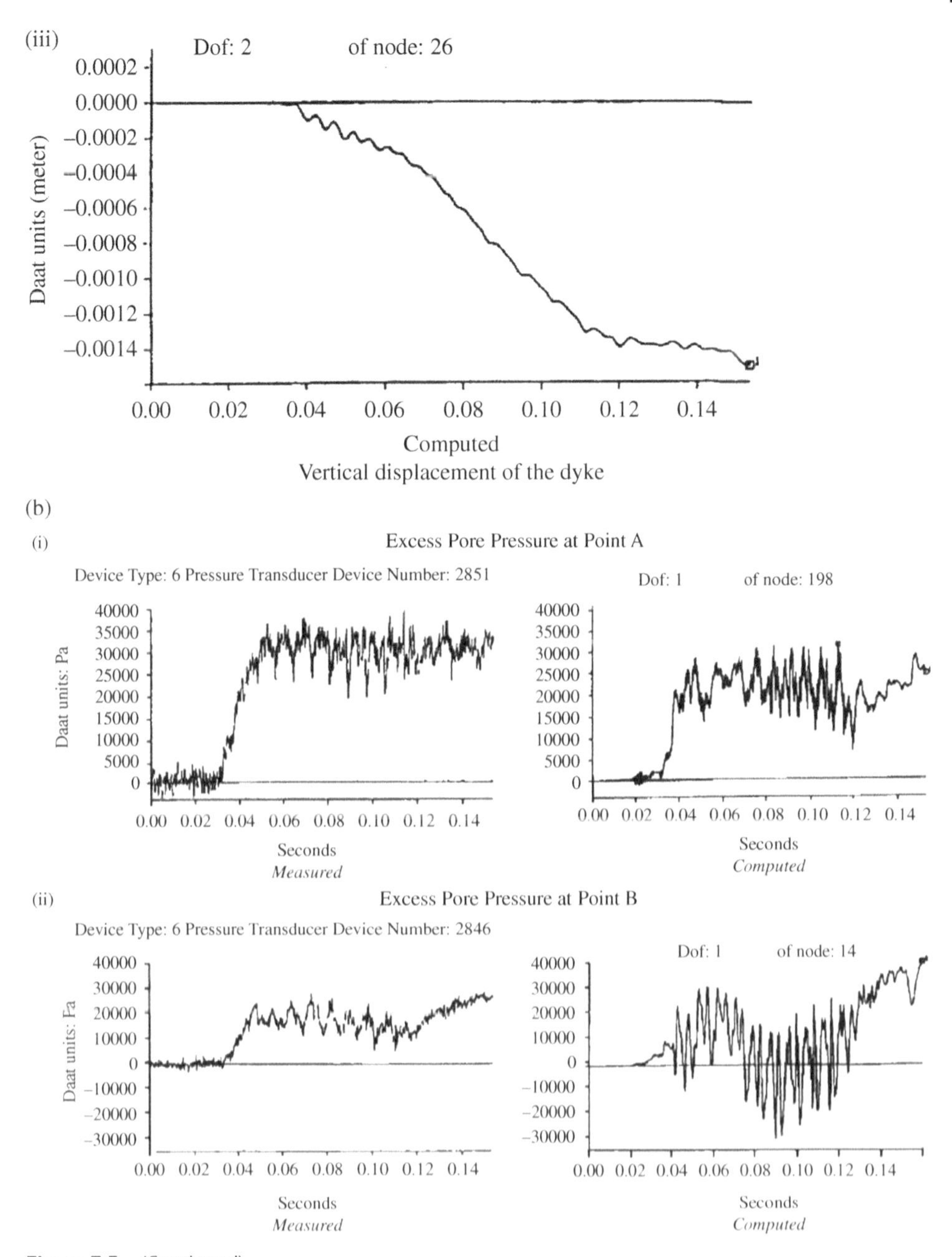

Figure 7.3 (Continued)

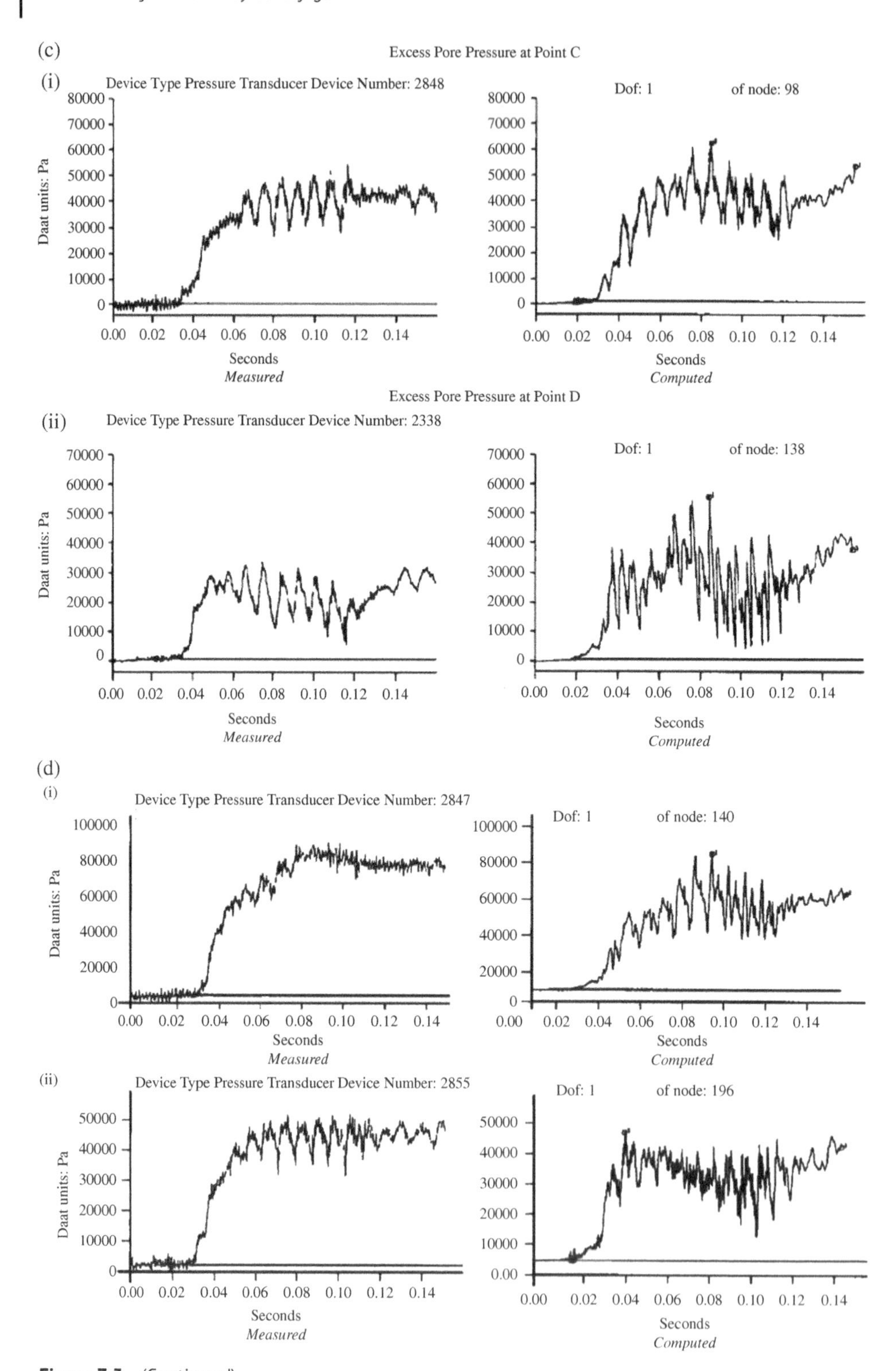

Figure 7.3 (Continued)

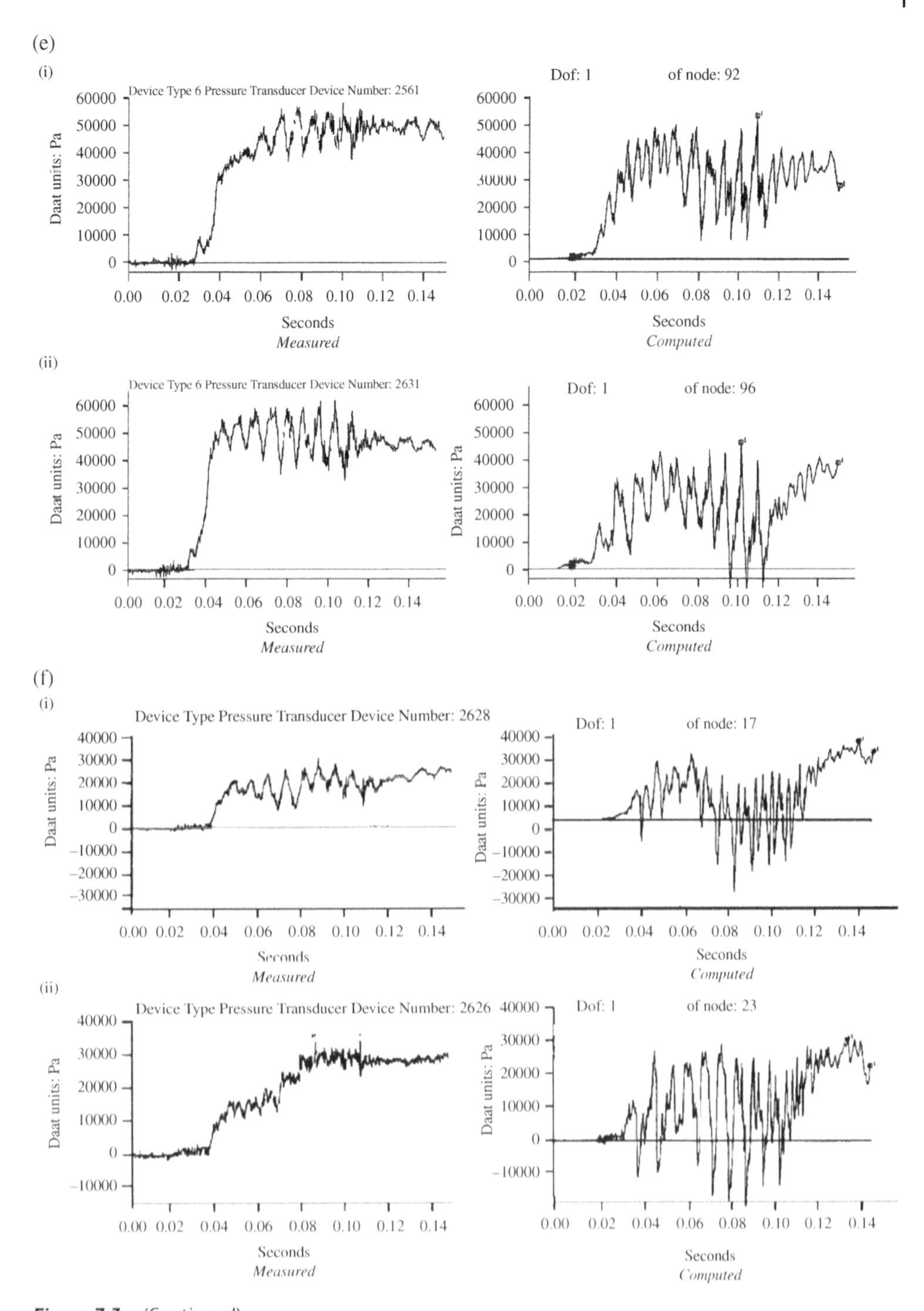

Figure 7.3 (Continued)

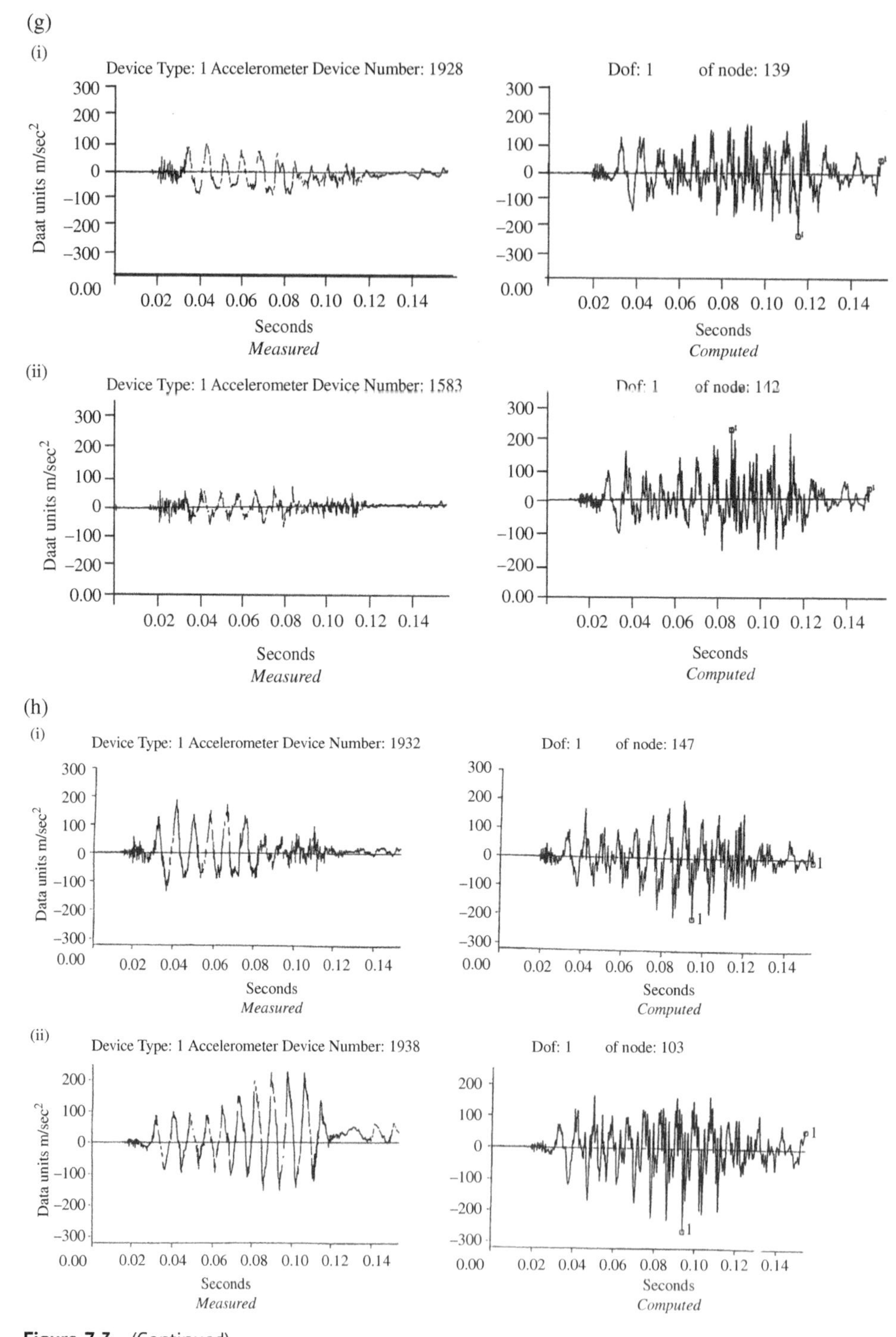

Figure 7.3 (Continued)

Table 7.3 VELACS Project – Summary of centrifuge tests and class A/B predictions (Arulanandan and Scott, 1983)

Models	1		2		3		4a		4b		6		7		11		12			
primary Experimenters	Dobry (RPI)		Dobry (RPI)		Scott (CIT)		Arul. (UCD)		Arul. (UCD)		Arul. (UCD)		Ko (UCB)		Schofield (CAM)		Prevost (PRI)		Prediction conducted	Predicted using achieved input motion
Repeating	Ko (UCB)		Arul (UCD)		Kutter (UCD)		Scott (CIT)		Ko (UCB)		No Repeat		Kutter (UCD)		No Repeat		Kutter (UCD)		Before the test	and Model
Experimenters	Arul (UCD)		Scott (CIT)		Dobry (RPI)		Dobry (RPI)		Elgamal (RPI)				Elgamal (RPI)				Elgamal (RPI)			Configuration
	Class		Class		Class		Class		Class		Class		Class		Class		Class		Class A	Class B
Predictors	A	B	A	B	A	B	A	B	A	B	A	B	A	B	A	B	A	B		
1 Bardet																			3	0
2 Been																			1	0
3 Chan																			7	0
4 Hamada																			2	1
5 Iai																			4	4
6 Ishihara																			1	1
7 Kimura																			4	0
8 Lacy																			5	0
9 Li																			1	0
10 Manzari																			1	0
11 Muraleetharan																			2	0
12 Prevost																			9	0
13 Rollins																			3	0
14 Roth																			3	0
15 Shiomi																			4	4
16 Siddharthan																			4	4
17 Sture																			4	0
18 Towhata																			6	2
19 Yogachandran																			4	0
20 Zienkiewicz																			8	3
21 Anandarajah																			0	3
22 Aubry																			0	3
23 Ito																			0	4
Total	17	8	11	6	8	3	8	2	5	0	5	0	8	2	9	6	5	2	76	29

so does G. H, I and J are slightly worse but the overall prediction on the pore pressure rise is very good.

This oscillation found in the excess pore pressure could be due to the proximity of the rigid boundary condition either at the bottom of the container or near the underside of the dyke. If we consider the transducers in two groups:

1) far-field where the influence of the structure is less at locations A, C, E, F, G and H
2) near-structure where the influence of the structure is more pronouced at locations B, D, I and J

It can be observed that comparatively less oscillations are found at locations which are far-field. Even if there is substantial oscillation such as location H, the dominant frequency is more akin to the input frequency (approximately 120Hz). However for the near-structure locations, the oscillations are much more pronounced and the frequency is more akin to the frequency of the structure and of a higher frequency nature (approximately 240Hz).

One possible reason for this oscillation is the proximity of the impermeable solid boundary and the fact the average relative fluid acceleration has been neglected in the **u**-p formulation. As the average relative fluid acceleration is neglected, any volume change near an impermeable boundary will behave in an undrained manner and lead to a large rate of change in pore water pressure via the fluid continuity equation. Similar behaviour has been found in the proximity of a retaining wall in (Dewooklar et al. 2009). When the same analysis is repeated using the fully explicit **u**-**w** formulation (Chan et al. 1995, 1998), significantly less oscillation is observed in, for instance, point I – PPT2628 (Figure 7.4) when compared with Figure 7.3f.

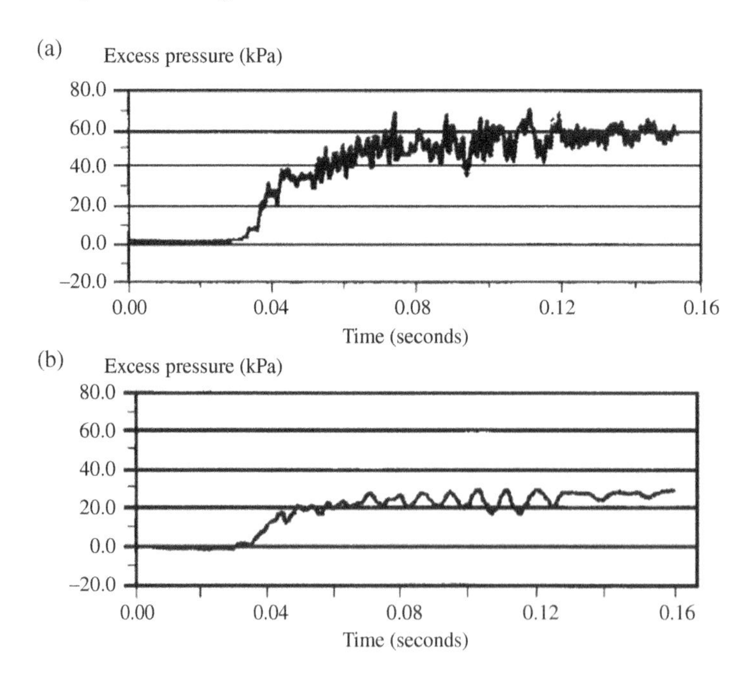

Figure 7.4 Numerical results of excess Pore Pressure at Point I (PPT2628) using fully explicit GLADYS-2E with 3-3 element (a) with 44 elements (b) with 96 elements.

Initially, all the accelerometers show good agreement with the experimental results. As the soil weakens, the value on the higher level (L and M) departs from the experimental value. The predictions of N and O are reasonable. This may due to the amount of shear wave energy being transmitted from the bottom to the top. As the soil weakens, less shear wave should transmitted. However, due to the oscillatory nature of the pore water pressure, the mean effective stress is not reduced as much as in the physical experiment. Therefore the shear modulus is not reduced adequately leading to the excessive transfer of shear wave energy.

Despite a number of shortcomings, the results represent an excellent comparison with the experimental results accounting for possible experimental errors. Nevertheless, these results represent a Class B-1 prediction and a set of more convincing Class A predictions are given in the next and subsequent sections.

7.4 The Velacs Project

Although many verification exercises have been performed by the authors (Chan 1988 and Zienkiewicz et al. 1990) and others using reported centrifuge results – a more systematic study became recently possible through the project VELACS (VErification of Liquefaction Analysis by Centrifuge Studies – Arulanandan and Scott 1993) funded by the National Science Foundation, USA. In this numerical prediction of several postulated tests was requested from "predicting participants" before the experiments are performed and results obtained for "centrifuge experiment participants". The numerical predictions were kept in sealed condition by a third party (Thompson and Lambe 1994) and these were not made available to the "centrifuge experiment participants". This double blind policy was introduced to minimize possible "cheating" thus enhance the credibility of the results. Such "Class A" predictions were submitted by twenty "predicting participants" by 30 September 1992 (Table 6.3) when apparently the centrifuge tests were first commenced by seven universities (University of California, Davis; California Institute of Technology; Cambridge University; University of Colorado, Boulder; Massachusetts Institute of Technology; Princeton Unviersity; and Rensselaer Polytechnic Institute). However, some of the specified centrifuge tests could not be carried out – and additional computations ("Class B") mostly because of the difference in the prescribed and actual input earthquake motion were requested – without however supplying other experimental results. It is instructive for the readers to note that, except for the simpliest model No.1 which represents a level soil layer, all numerical predictors used computer codes based directly or indirectly on the Biot theory and approximation form introduced in this book (Smith 1994).

Nine centrifugal models (see Figure 7.5 taken from Arulanandan and Scott (1993)):

- Model No.1 – Horizontally layered loose sand in laminar box
- Model No.2 – Sloping loose sand layer in laminar box
- Model No.3 – A sand layer one side dense, the other side loose
- Model No.4a – Stratified soil layers in laminar box
- Model No.4b – Stratified soil layers in rigid box
- Model No.6 – A submerged embankment in rigid box

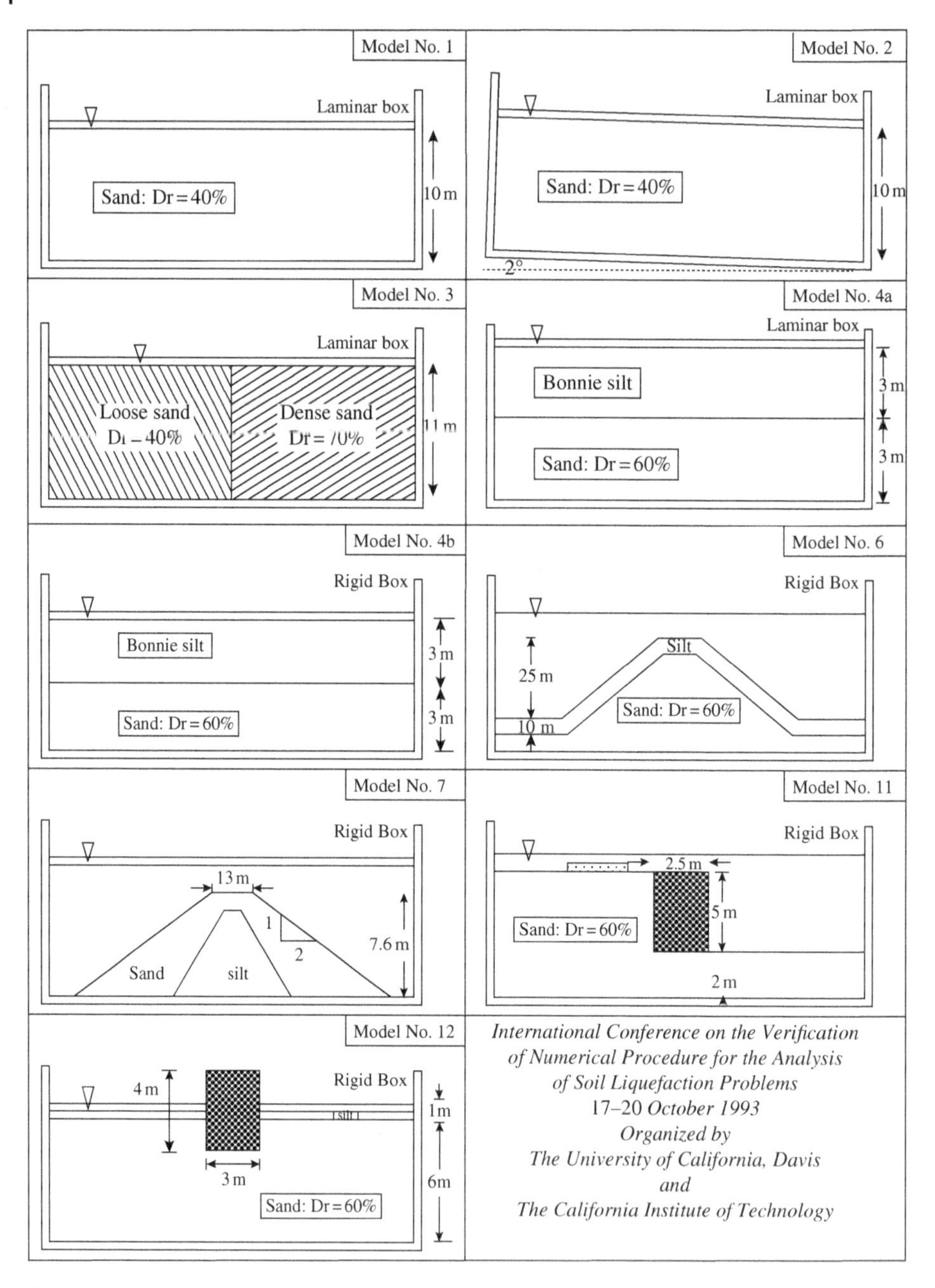

Figure 7.5 Centrifuge model configurations for Class A predictions – VELACS project (Arulanandan and Scott, 1993).

- Model No.7 – A submerged sand embankment with slit core
- Model No.11 – A gravity quaywall with sand backfill
- Model No.12 – A structure embedded in stratified soil layers

were selected for the verification. Most of the tests have been performed at more than one centrifuges for centrifuge validation purpose and considerable scatter of results can be found between centrifuge results. Three of the authors led three different groups of prediction using computer codes SWANDYNE-4 (implicit **u**-p with partial saturation), SWANDYNE-II (implicit **u**-p with full saturation) and MuDIAN (implicit **u**-**U** with full saturation based on their work in University College of Swansea.

The SWANSEA group led by the senior author of the first edition of this book, the late Prof. Zienkiewicz together with one of authors Pastor presented eight predictions for model Nos. 1, 2, 3, 4a, 4b, 6, 7 and 11 (Zienkiewicz et al. 1993a, 1993b, 1993c, 1993d, 1993e, 1993f, 1993g, 1993h). The author Chan presented seven predictions for model Nos. 1, 2, 3, 4a, 4b, 7 and 11 (Chan et al. 1993a, 1993b, 1993c, 1993d, 1993e, 1993f, 1993g) and another of the authors Shiomi together with Prof. Zienkiewicz presented four predictions for model Nos 1, 2, 11 and 12 (Shiomi et al. 1993a, 1993b, 1993c, 1993d). Lastly, Chan (Chan et al. 1994) and Shiomi (Sture et al. 1994) were also involved in the process of overviewing numerical predictions.

Most of the predictive results can be classified as good or excellent. A selection of these will be presented in the subsequent sections in this chapter with brief description of the experimental setup and comparisons with numerical results on the same scale. But before going into the detail of the prediction, the following sub-sections are devoted to the analysing procedure using SWANDYNE-II.

General Analysing Procedure

This analysis procedure is applicable to all the predictions performed by the authors:

i) The information about the centrifuge model is gone through in detail and key data noted.
ii) A finite element mesh is generated using a pre-processor. Time required for the subsequent dynamic and consolidation analysis are also taken into consideration so that a reasonable mesh is chosen.
iii) The appropriate boundary conditions are then applied at the boundaries of the model. Tied nodes are used to model the laminar box behaviour (see Section 7.4.3).
iv) The hydrostatic pressure, which assumed to be constant throughout the analysis, is prescribed at the fluid nodes concerned and the pressure on the solid phase is also applied.
v) The appropriate permeability and gravitational acceleration are then included. The models are modelled at the model scale so the appropriate acceleration level is the centrifugal acceleration imposed.
vi) A static analysis was performed to determine the initial stress state of the model. A Ko value of 0.4 is assumed. In order to avoid tensile stress and high stress ratio, Mohr Coulomb elasto-perfectly plastic model is used for the initial analysis with a reduced frictional angle of 25°.
vii) The output of the static analysis are considered carefully to check if the initial pore pressure distribution is reasonable and also if the stress state is acceptable.
viii) A no-earthquake dynamic run is then performed to check if the initial stress state is in correct equilibrium condition. If it is not, a new static initial analysis is performed with modified parameters to obtain equilibrium.

ix) When the initial stress state is acceptable, a linear elastic analysis is performed to note the basic behaviour of the finite element mesh.

x) Then a nonlinear analysis is performed for the earthquake stage with the supplied horizontal and vertical earthquake with proper scaling. The dynamic analyses were performed using a Generalised Newmark (Katona and Zienkiewicz 1985) scheme with non-linear iterations using initial linear elastic tangential global matrix. The constitutive model used is Pastor Zienkiewicz (1986) mark-III model. The parameters used are described in Section 7.4.4. The time step used is usually equal to at a simple multiple of the earthquake spacing. The choice of the time step depends on the number of the stations in the earthquake input and the frequency of the input. For the analyses, void ratio i.e. permeability and other geometric properties were kept constant during the analysis. Rayleigh damping of (minimum) 5% is applied at 100Hz which is the dominant frequency in the earthquake-like motion input.

xi) The earthquake phase of the analysis is then plotted to check for anomaly.

xii) The consolidation then follows the dynamic analysis. Usually a larger time step is used for the consolidation analysis, a gradual change in time step is used to avoid numerical instability. The full dynamic equation is used for the consolidation stage of the analysis with the appropriate mass matrix.

xiii) The results are first plotted using a simple post-processing program to check its validity. If the result does not seem reasonable, the dynamic analysis is repeated with another set of numerical parameters, iteration schemes etc. until a reasonable and numerically stable result is obtained.

xiv) Various plots are then performed for the final report. Since total quantities e.g. pore pressure and displacement are used in the program, post-processing is required to obtain the excess pore water pressure and relative displacement required by the specification.

xv) Other post-processing e.g. excess pore water pressure ratio, spectral analysis and response spectrum are calculated for the reporting purpose.

7.4.1 Description of the Precise Method of Determination of Each Coefficient in the Numerical Model

The determination of each coefficient of the Pastor-Zienkiewicz mark III model follows the procedure outlined in Section 5.5 of (Chan 1988) and is being reproduced in this sub-section. As drained monotonic, undrained monotonic and undrained cyclic tests are most widely available tests in common engineering applications, they are chosen in the parametric determination process. During earthquake and other rapid loading, the undrained test is more relevant. The tests should be done with samples having relative density around the intended relative density. In this section, the way to identify each of the parameters required by the model is illustrated. One undrained monotonic and cyclic test is taken from each of the loose sand ($D_r = 40\%$), dense sand ($D_r = 60\%$) and silt experimental data sets respectively. The comparison of the constitutive model and the physical undrained triaxial tests has been given in Chan et al. (1992a, 1992b). These results are produced using soil model subroutine for DIANA-SWANDYNE II interfaced with a soil model testing program

SM2D and the experimental results are also plotted on the same graph for the monotonic test. In the following sections, each parameters will be discussed in turn:

i) M_g (dimensionless): can be estimated from the graph plotting stress ratio versus the shear strain or axial strain. M_g is approximately equal to the maximum value of stress ratio the test reaches. It can also be estimated from the q versus p' plot with a tangent drawn from the origin to the resdual stress path in an undrained triaxial test. M_g corresponds to the maximum slope obtained by this method. M_g can also be obtained from drained test using the intercept of dilatancy versus stress ratio plot. In this exercise, the stress ratio plot was used.

ii) M_f (dimensionless): can be determinated by matching the shape of the stress path in the q versus p' in an undrained triaxial test. Alternatively, it can be obtained by matching the critical stress ratio that the soil changes from contractive to dilative behaviour in the case of dense sand. The value of D_R . M_g can serve as a good starting point for the evaluation of its value. In this exercise, the critical stress ratio is used.

iii) α_g (dimensionless): can be obtained from the slope of graph between dilatancy and stress ratio over M_g graph. However, this value is usually taken as 0.45 and it is also used in this exercise.

iv) α_f (dimensionless): is usually taken to be the same as α_g so that the loading locus and plastic potential are having the same shape.

v) *Kev0c* (dimension of stress): represents the value of Bulk modulus at the mean effective stress p'_0. It can be obtained by matching the initial slope of the mean effective stress p' or pore pressure versus axial strain plot in an undrained test. Its value can be adjusted so that a better match of the curve of pore pressure versus axial strain can be obtained. In the VELACS exercise, this is done so that the end point in the predicted curves stayed close to the experimental data.

vi) *Kes0c* (dimension of stress): represents the value of three times of Shear modulus at the mean effective stress p'_0. It can be obtained by matching the initial slope of deviatoric stress q versus axial strain plot in an undrained test. Its value can be adjusted so that a better match of the curve of q versus axial strain can be obtained. In the VELACS exercise, this is done so that the end point in the predicted curves stayed close to the experimental data.

vii) β_0 (dimensionless): is usually taken as 4.2 and this value is taken here.

viii) β_1 (dimensionless): is usually taken as 0.2 and this value is taken here.

ix) H_o (dimensionless): is determined by fitting the curves in p' or q versus axial strain plot. It can be found by matching the shape of the q versus p' plot for undrained tests also.

x) H_{uo} (dimensionless): is determined by matching the initial slope of the first unloading curve.

xi) γ_{Hu} (dimensionless): is determined by matching the rate of change of slope of the first unloading curve or by matching the number of cycles in a series of loading and unloading. The second method is used in this exercise.

xii) γ_{DM} (dimensionless): is determined by matching the slope of the first reloading curve or by matching the number of cycles in a series of loading and unloading. The second method is used in this exercise.

xiii) p'_0 (dimension of stress): is the initial mean effective stress of the undrained triaxial test.

7.4.2 Modelling of the Laminar Box

The way that the boundary conditions were incorporated in the numerical method has been given in the relevant prediction papers. The laminar box (Hushmand et al. 1988) is treated with tied node facility. The horizontal and vertical nodal displacements at the two ends of the soil are restrainted to have the same value. The interface between structure and soil is assumed to be perfectly bonded. No change of boundary condition is made during the analyses. Only rigid block and 4-noded linear isoparametric elements for both the displacement (u) and pore pressure (p) are used in the analyses.

7.4.3 Parameters Identified for Pastor-Zienkiewicz Mark III Model

There are quite a number of parameters in the Pastor-Zinekiewicz mark III model which requires definition. Three CIUC (Isotropically Consolidated followed by Undrained Compression test) experimental results starting from 40kPa were chosen to identify the parameters. The experimental results were taken from (Arulmoli et al. 1992) which provided the standard soil model test results for the numerical predictors. The 40 kPa ones were chosen because it is close to the mean effective stress value at the middle of the centrifuge model. The permeability of silt is also calculated for this level of mean effective stress.

i) Loose sand: ($D_r = 40\%$) Experiment CIUC4051 was used. The parameters obtained are as follow: $M_g = 1.15$ $M_f = 1.03$ $\alpha_f = \alpha g = 0.45$ $Kev0c = 770$ kPa $Kes0c = 1155$ kPa Elastic modulus is proportional to the mean effective stress $\beta_0 = 4.2$ $\beta_1 = 0.2$ $p'_0 = 4$ kPa $H_o = 600$ $H_{uo} = 4000$ kPa $\gamma_{Hu} = 2$ $\gamma_{DM} = 0$.
ii) Dense sand: ($D_r = 60\%$) Experiment CIUC6012 was used. The parameters obtained are as follow: $M_g = 1.32$ $M_f = 1.30$ $\alpha_f = \alpha g = 0.45$ $Kev0c = 2000$ kPa $Kes0c = 2600$ kPa Elastic modulus is proportional to the mean effective stress $\beta_0 = 4.2$ $\beta_1 = 0.2$ $p'_0 = 4$ kPa $H_o = 750$ $H_{uo} = 40000$ kPa $\gamma_{Hu} = 2$ $\gamma_{DM} = 4$
iii) Silt: Experiment CIUCBS13 was used. The parameters obtained are as follow: $M_g = 1.15$ $M_f = 0.50$ $\alpha_f = \alpha g = 0.45$ $Kev0c = 400$ kPa $Kes0c = 1520$ kPa Elastic modulus is proportional to the mean effective stress $\beta_0 = 4.2$ $\beta_1 = 0.2$ $p'_0 = 4$ kPa $H_o = 900$ $H_{uo} = 100000$ kPa $\gamma_{Hu} = 2$ $\gamma_{DM} = 8$

7.5 Comparison with the Velacs Centrifuge Experiment

7.5.1 Description of the Models

Model No. 1

A water-saturated uniform layer of loose sand ($D_r = 40\%$), 10 m (in prototype scale) thick, in a laminar box, was mainly subjected to horizontal base motion. The test was instrumented as shown in Figure 7.6a. Centrifuge test resuts for model No.1 were carried out by three universities. RPI is the primary experimenter and UC Davis and the University of Colorado conducted duplicate tests.

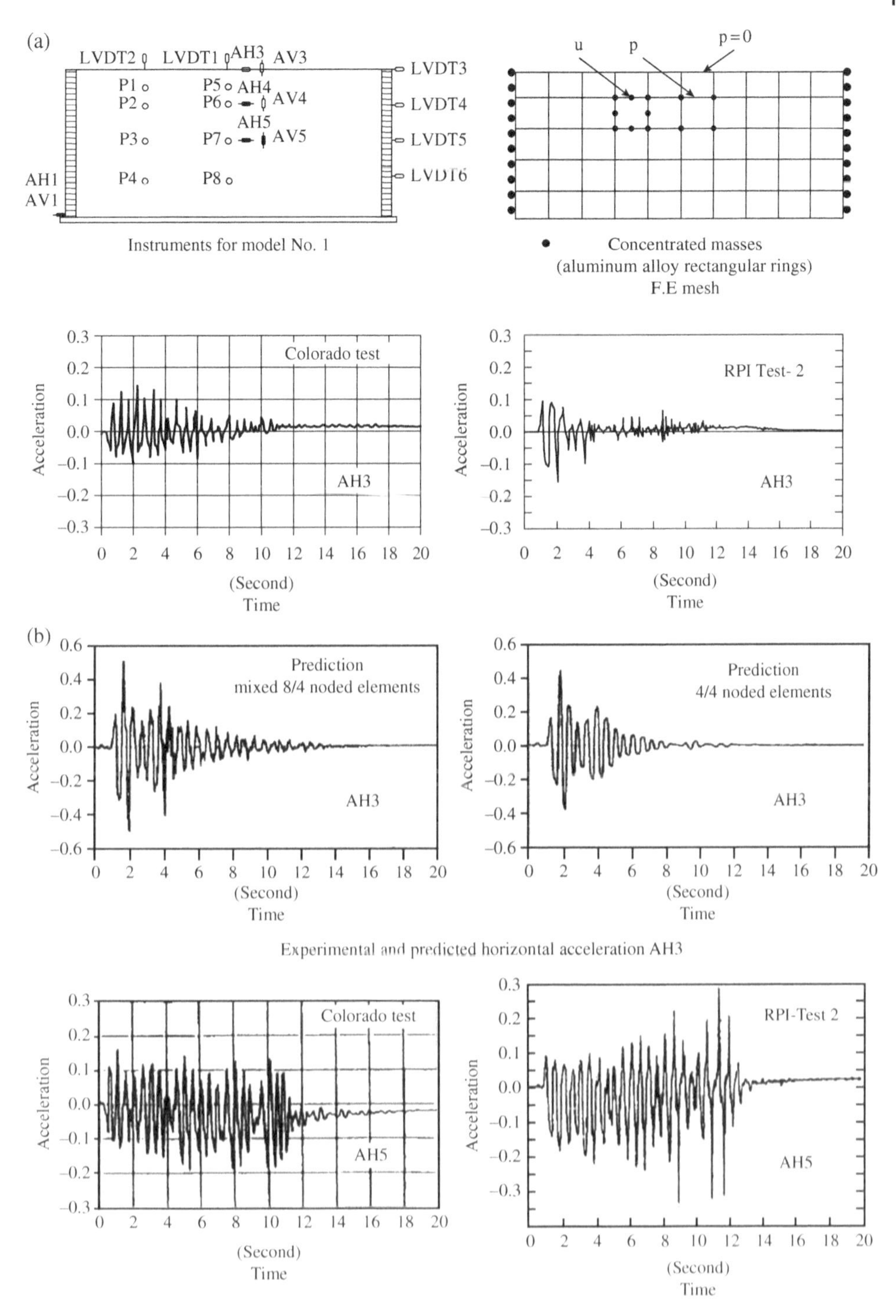

Figure 7.6 VELACS Centrifuge model No. 1: (a) instrumentation, finite element mesh and input acceleration (b) comparison of experimental and predicted acceleration results. (c) comparison of experimental and predicted acceleration and excess pore pressure results. (d) comparison of experimental and predicted excess pore pressure results. (e) comparison of experimental and predicted excess pore pressure results.

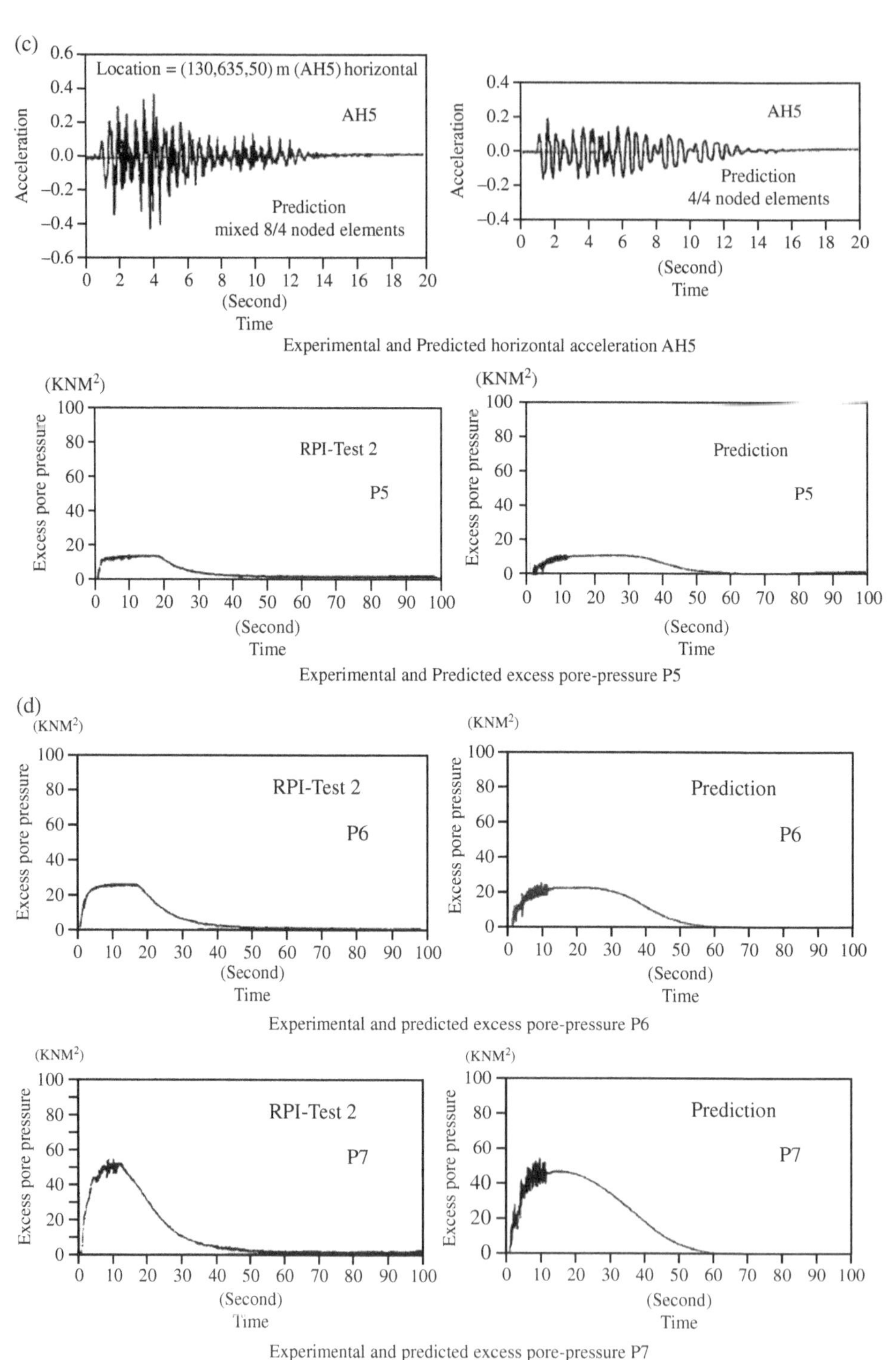

Figure 7.6 (Continued)

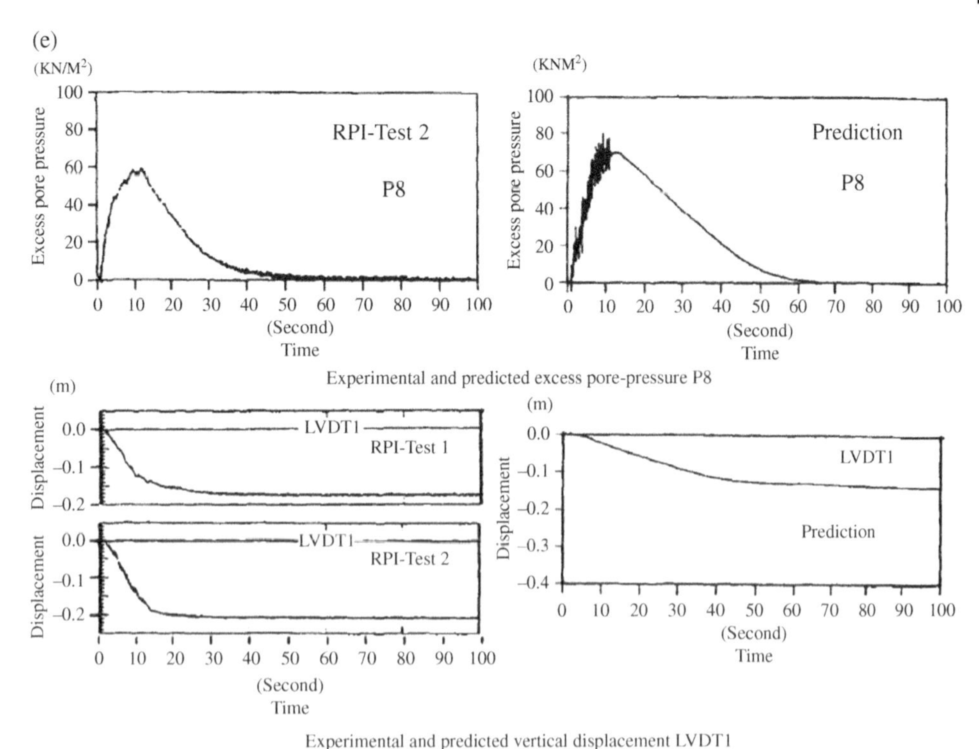

Figure 7.6 (Continued)

Model No. 3

A water-saturated layer of sand deposited at 11 m (in prototype scale) thick, in a laminar box was subjected to base shaking. The test was instrumented as shown in Figure 7.7a. Although the model contained sand deposited at both 40% and 70% relative densities, laboratory data were available only for 40% and 60% relative densities. The predictors were expected to infer properties of sand at 70% relative density based on the properties at 40% and 60% relative densities. In our case, the material parameters for relative density 60% is used. The centrifuge results for this model were carried out by CIT as primary experimenter and UC Davis and RPI as duplicate experimenters.

Model No.11

A soil and water retaining wall was subjected to base shaking. A surcharge was added on the backfill of the retaining wall. The relative density of the sand in this model was approximately 60%. The centrifuge test of this model was carried out by Cambridge University only and was instrumented as shown in Figure 7.8a. No repeated test was conducted.

7.5.2 Comparison of Experiment and Prediction

Overall, predictions of the SWANDYNE program compared well with experiments. The close agreement of the predicted pore pressures and displacements with those measured

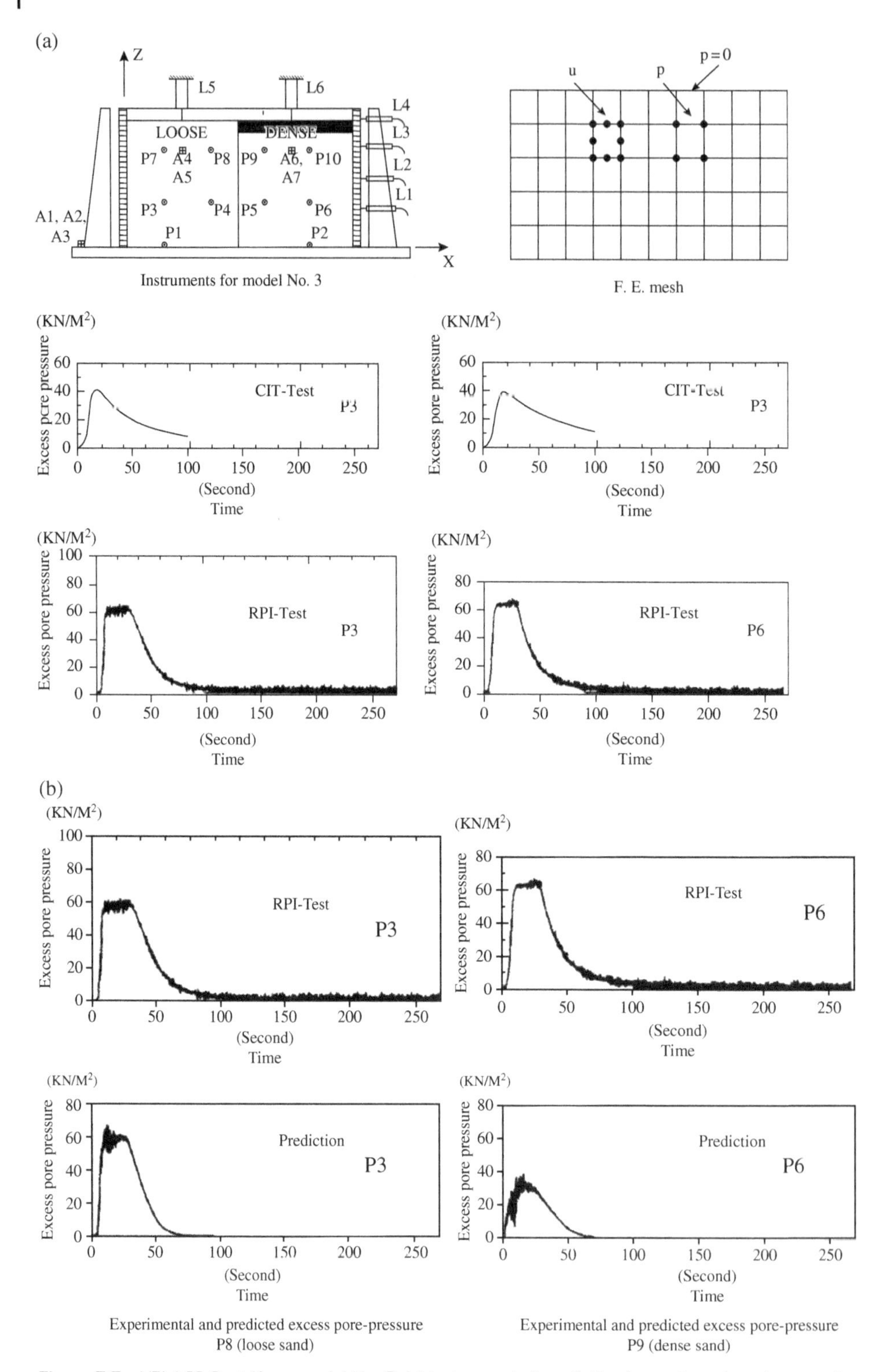

Figure 7.7 VELACS Centrifuge model No. 3: (a) instrumentation, finite element mesh and comparison of experimental and predicted excess pore pressure results. (b) comparison of experimental and predicted excess pore pressure results. (c) comparison of experimental and predicted excess pore pressure results. (d) comparison of experimental and predicted vertical displacement results.

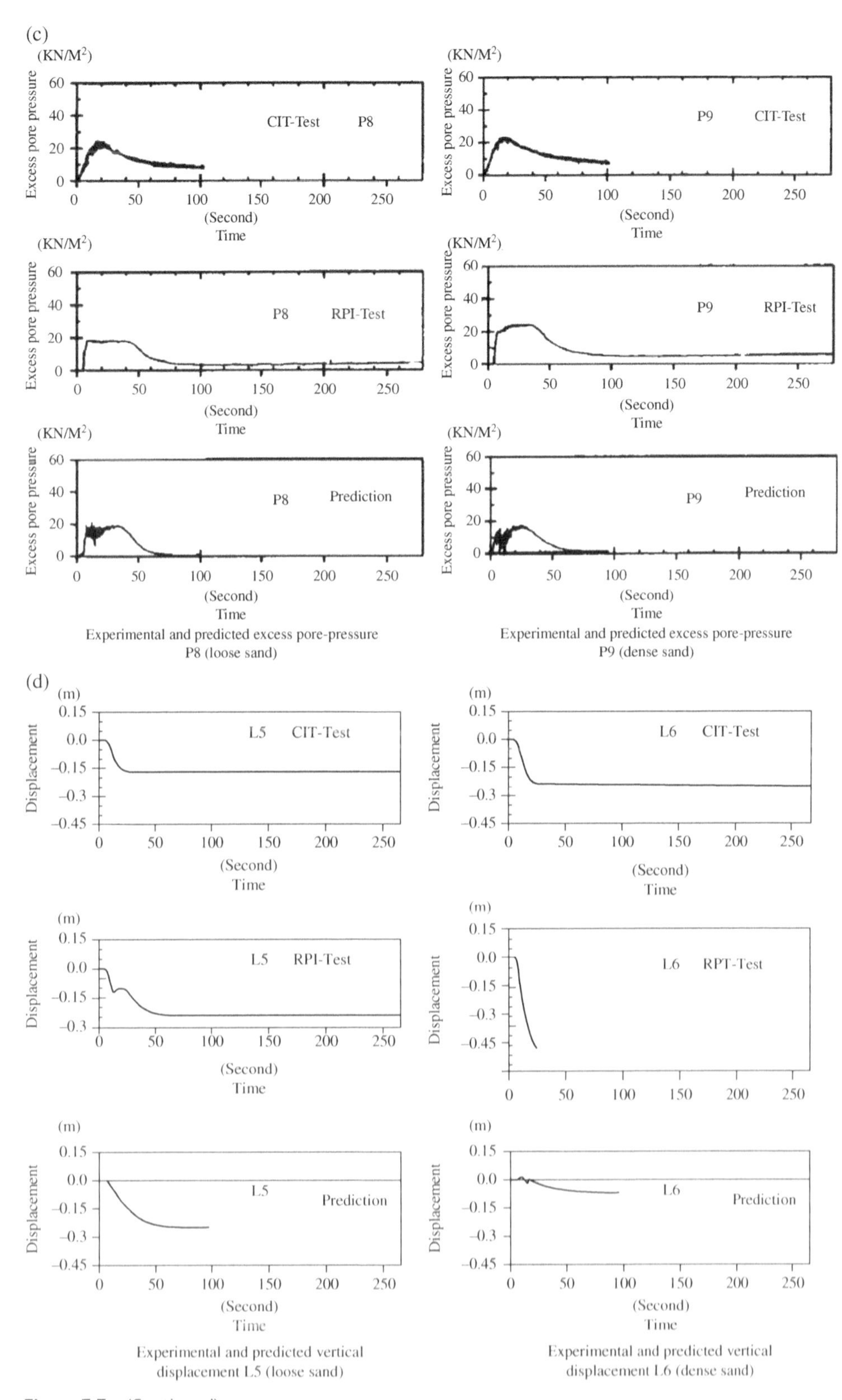

Figure 7.7 (Continued)

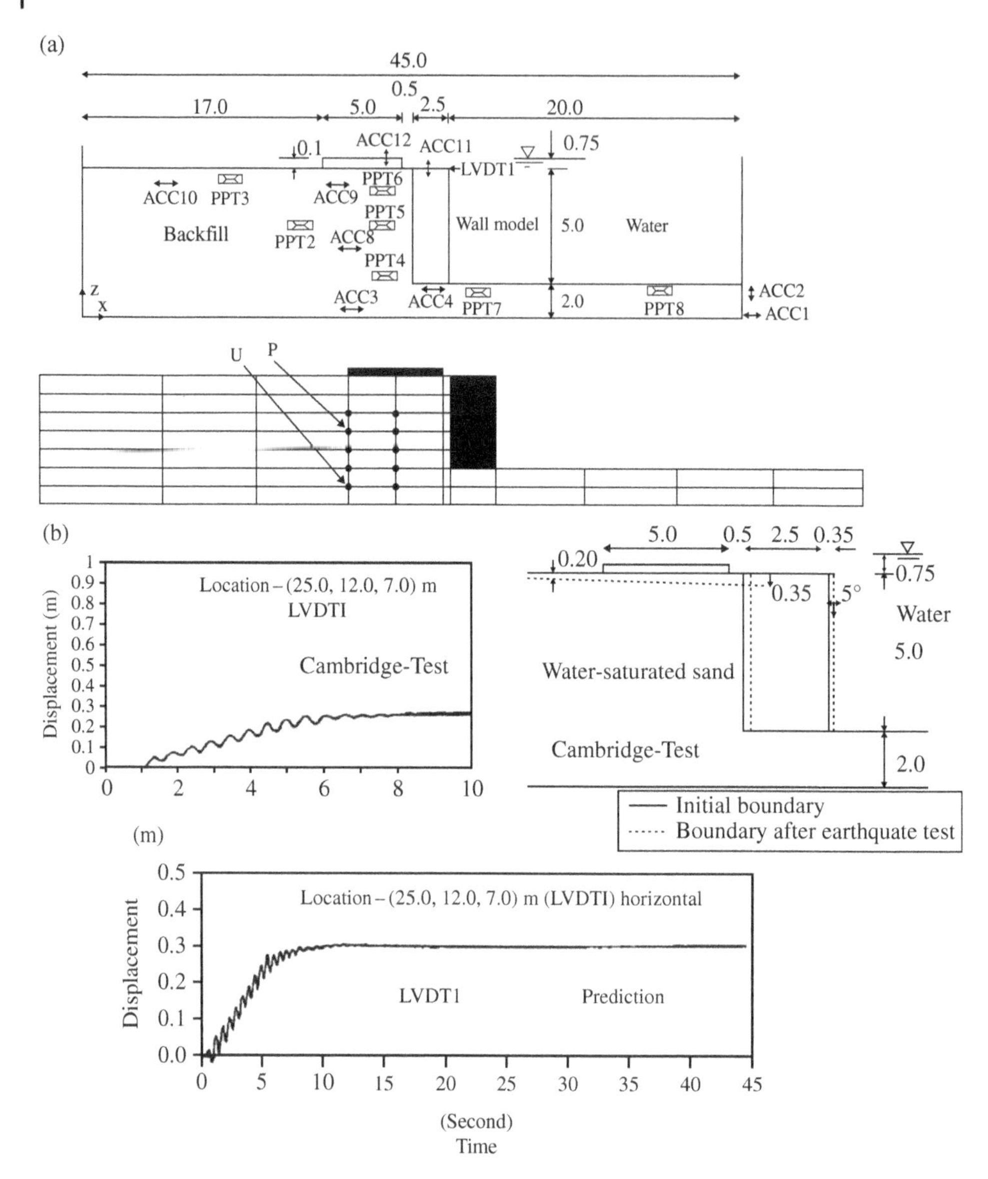

Figure 7.8 VELACS Centrifuge model No. 11: (a) instrumentation, finite element mesh (b) comparison of experimental and predicted horizontal displacement results.

in centrifuge tests affirms the reliability of the computation procedure in SWANDYNE program when used with carefully calibrated material properties and model parameters. The use of fully coupled equations for soil-pore fluid interaction and of a simple soil model based on the generalised plasticity for soil skeleton, both introduced in detail in this volume, form a consistent and powerful prediction procedure. This rigorous analysis procedure promises to be a reliable tool for practical problems in engineering. However, the following problems have to be addressed:

a) Though the predicted horizontal acceleration on the surface of the soil layer is much higher than the experimental results, it is obvious that the experimental results of the surface acceleration is not neceesarily reliable after soil liquefaction. We also have recognised that the numerical result of the surface acceleration obtained by using mixed 8-4 noded elements probably gives much higher peak value of acceleration. This can be improved by using 4-4 noded elements. There is no large difference of the acceleration at the middle points between the prediction and the experiment. The prediction also showed that the peak acceleration on the surface decreased gradually when liquefaction occurred. The similar phenomena can be found in the centrifuge tests and earthquake history.
b) The predicted time histories of pore-pressure agreed closely with those recorded in the centrifuge tests. In model No. 1, the experimental results showed slightly faster pressure dissipation, however the differences were not large. The experimental results of model No. 3 by RPI and CIT gave obviously different peak values of pore-pressure and making comparison very difficult. The predicted post-earthquake consolidation was consistent with the values measured in the centrifuge test.
c) The prediction of displacement showed similarities to those measured during centrifuge experiment. The horizontal displacement of the retaining wall in model No.11 was well predicted. The predicted values of surface settlement in model No.1 were lower than those recorded in the centrifuge test, but the difference was not too marked. In model No. 3, though the prediction of surface settlement in the loose sand agreed closely with that measured in the centrifuge tests, the numerical analysis did not predict more settlement in dense sand than in the loose sand which was recorded in all experiments for this model despite later repeat of the experiments showed that different results could also be obtained (Dobry 1994). Almost all numerical predictors failed to achieve this qualitative difference. The mechanism involves a failed dense sand wedge moving into the liquefied loose sand due to the lack of support. It is of interest to carry out further research into this phenomenon. More reserach is also needed on the modelling and calibration of data for the behaviour of dense and and silt under cyclic loading.

7.6 Centrifuge Test of a Retaining Wall (Dewooklar et al 2009)

The schematic model configuration of test MMD1 in Figure 7.9 has been taken from (**Dewooklar et al 2009**). A layer of slip element can be found behind the retaining wall in Figure 7.9b. Without this layer of slip element, a consistent stress state cannot be found as the initial condition for the dynamic analyses. Good agreement with theoretical and experimental results has been found for the static analysis in Figure 7.10. The experimental

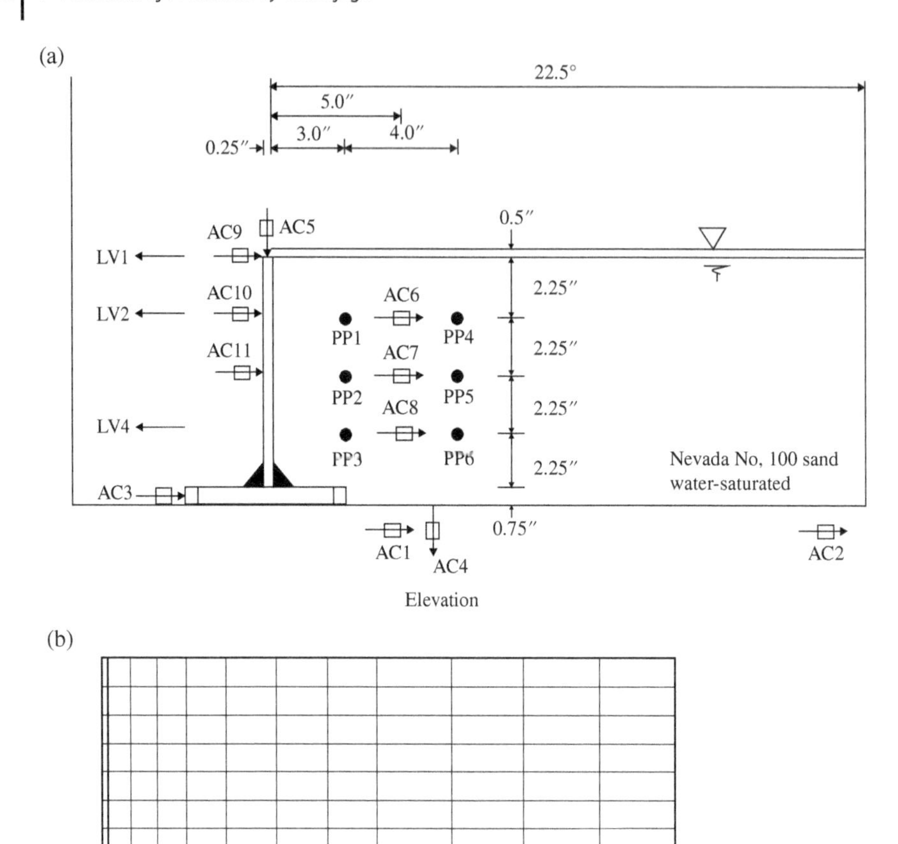

Figure 7.9 (a) Schematic model configuration of test MMD1 (Dewooklar et al 2009) (b) finite element mesh with slip element, in heavy line (Dewooklar et al 2009) *Source:* Based on Dewoolkaret al. (2009).

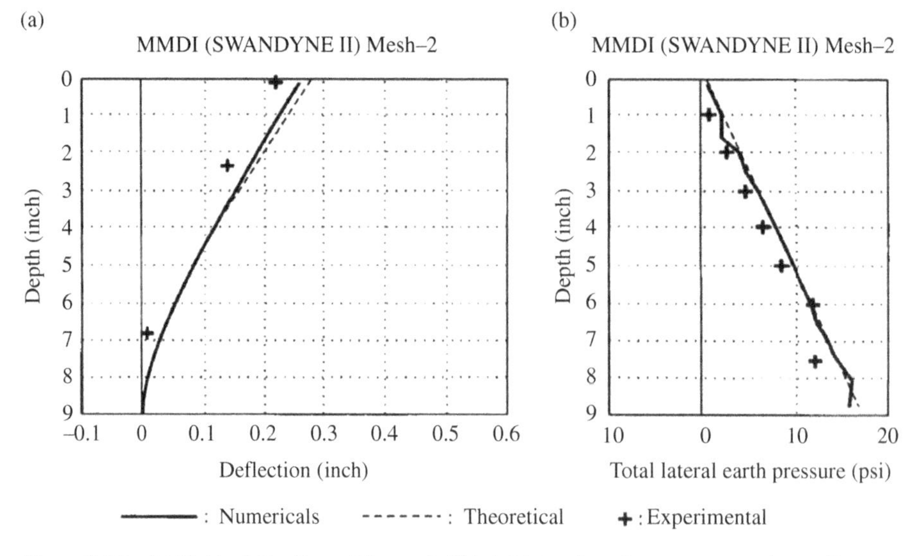

Figure 7.10 (a) Static deflection of the wall (b) total lateral earth pressure on the wall; *Source:* Dewoolkar et al. (2009).

results are stiffer than the theoretical and numerical results because of the stiffening of the retaining wall by the weld at the root of it.

Selected results for the horizontal accelerations at the tip of the wall, horizontal accelerations within the soil and dynamic bending strains in the wall can be found in Figure 7.11. All of them showed good agreement with the experimental results. The same is true for the horizontal displacement at the tip of the wall and long term excess pore pressure trace in Figure 7.12 in all comparisons the program SWANDYNE introduced by the authors was used. Although the numerical modelling were performed at different stages of the centrifuge, all the material parameters used are derived from the VELACS exercise as given in

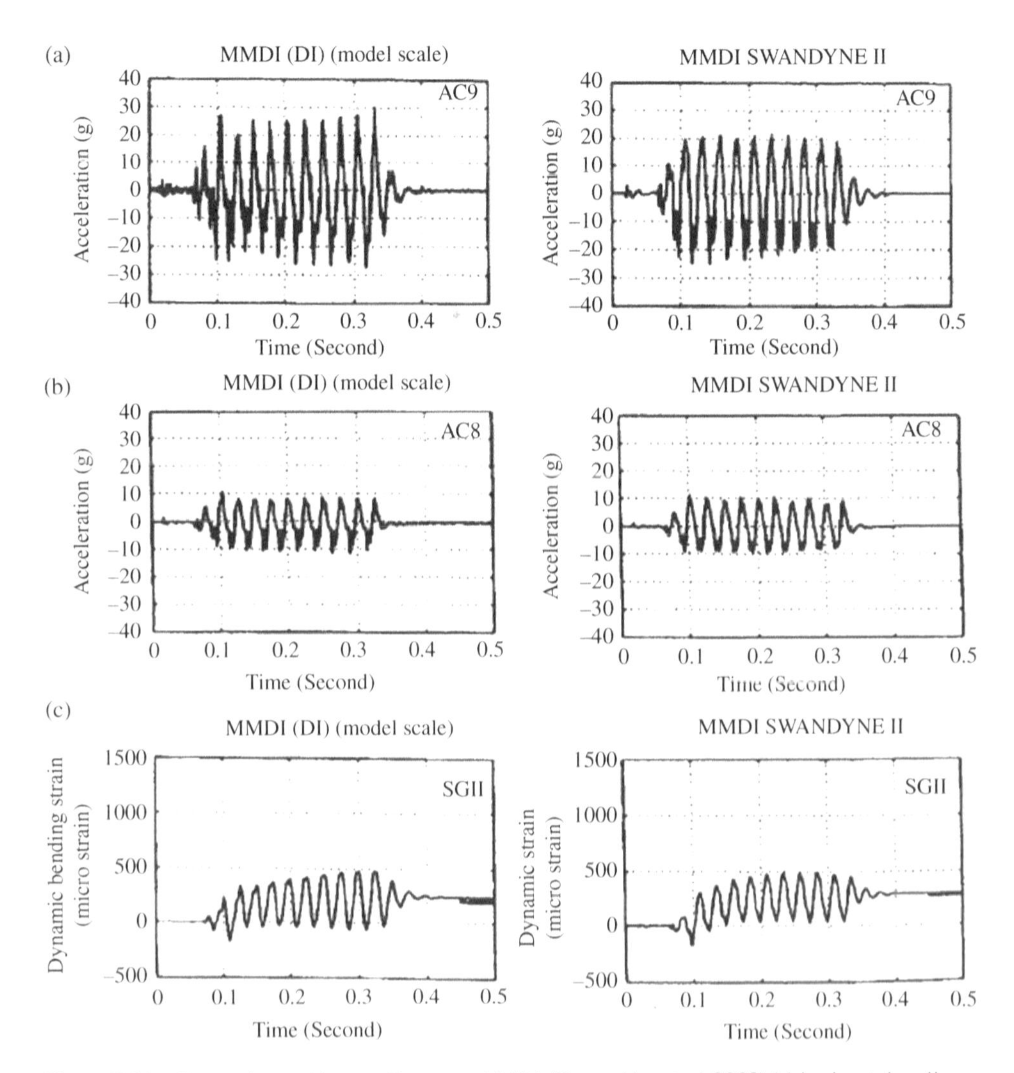

Figure 7.11 Comparison with centrifuge test MMD1 (Dewooklar et al 2009) (a) horizontal wall accelerations at the top (b) horizontal soil accelerations (c) dynamic bending strains of the wall.

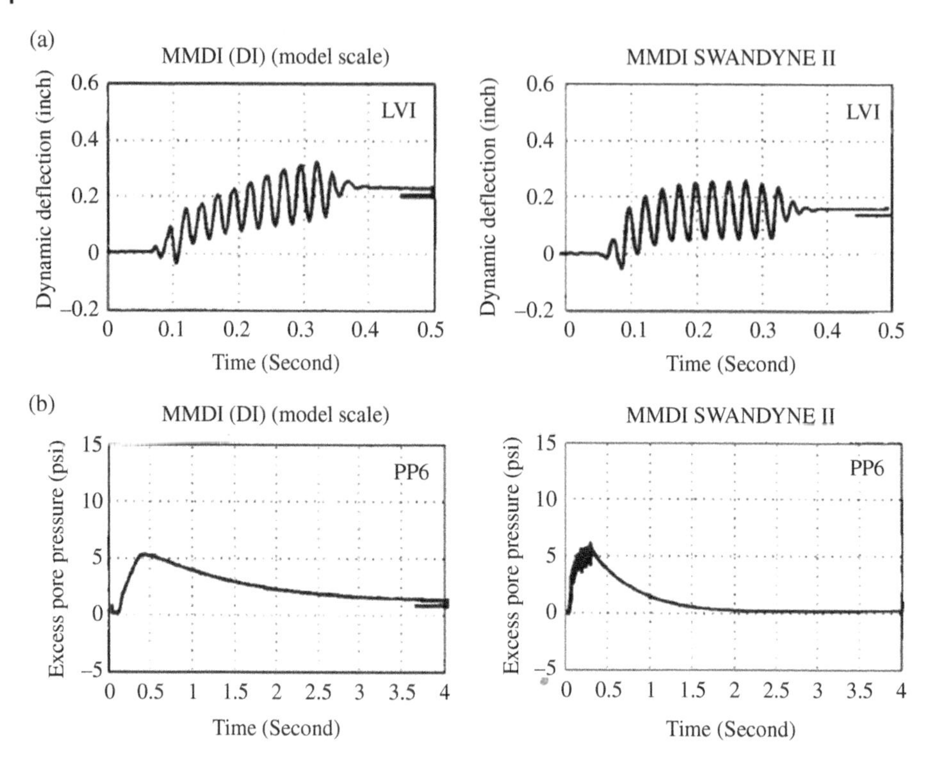

Figure 7.12 Comparison with centrifuge test MMD1 (Dewooklar et al 2009) (a) dynamic wall deflections (b) long term excess pore pressure.

Section 7.4.4. Only the amount of damping has been varied to investigate the effect of different level of damping.

7.7 Conclusions

In this chapter, the **u**-p formulation derived in Chapter 3 has been used to compare with various centrifuge tests performed on the Cambridge Geotechnical Centrifuge and the VELACS project. Very good and excellent agreements have been obtained thus validating the formulation and the computer code for various types of analysis under saturated condition. In the next chapter, we are going to show numerical predictions of practical examples and the use of the numerical procedure in design.

References

Arulanandan, K. and Scott, R. F. (Eds) (1993). *Proceeding of VELACS Symposium*, A. A. Balkema, Rotterdam.

Arulmoli, K., Muraleetharan, M. M. H. and Fruth, L. S. (1992). VELACS Laboratory testing program – Soil data report. Earth Technology Corporation, project No. 90-0562, Irvine, California.

Chan, A. H. C. (1988). A Unified Finite Element Solution to Static and Dynamic Geomechanics Problems. Ph.D. Dissertation, University College of Swansea, Wales.

Chan, A. H. C., Famiyesin, O. O. R. and Muir, W. D. (1992a). Report No. CE-GE92-23-0: Numerical Simulation Report for the VELACS Project – General Description. Department of Civil Engineering, Glasgow University, Glasgow.

Chan, A. H. C., Famiyesin, O. O. R. and Muir, W. D. (1992b). Report No. CE-GE92-23-1: Numerical Simulation Report for the VELACS Project – Class A prediction of RPI model. Department of Civil Engineering, Glasgow University, Glasgow.

Chan, A. H. C., Famiyesin, O. O. and Muir, W. D. (1993b). Numerical prediction for model No. 2, in *Verification of Numerical Procedures for the Analysis of Soil Liquefaction Problems*, K. Arulanandan and R. F. Scott (Eds), UC Davis, (17–20 October 1993), Vol. **1**, A. A. Balkema, Rotterdam, 343–362.

Chan, A. H. C., Famiyesin, O. O. and Muir, W. D. (1993c). Numerical prediction for model No. 3, in *Verification of Numerical Procedures for the Analysis of Soil Liquefaction Problems*, K. Arulanandan and R. F. Scott (Eds), UC Davis, (17–20 Octobrt 1993), Vol. **1**, A. A. Balkema, Rotterdam, 489–510.

Chan, A. H. C., Famiyesin, O. O. and Muir, W. D. (1993d) Numerical prediction for model No. 4a, in *Verification of Numerical Procedures for the Analysis of Soil Liquefaction Problems*, K. Arulanandan and R. F. Scott (Eds), UC Davis, (17–20 October 1993), Vol. **1**, A. A. Balkema, Rotterdam, 623–630.

Chan, A. H. C., Famiyesin, O. O. and Muir, W. D. (1993e). Numerical prediction for model No. 4b, in *Verification of Numerical Procedures for the Analysis of Soil Liquefaction Problems*, K. Arulanandan and R. F. Scott (Eds), UC Davis, (17–20 October 1993), Vol. **1**, A. A. Balkema, Rotterdam, 711–720.

Chan, A. H. C., Famiyesin, O. O. and Muir, W. D. (1993f). Numerical prediction for model No. 7, in *Verification of Numerical Procedures for the Analysis of Soil Liquefaction Problems*, K. Arulanandan and R. F. Scott (Eds), UC Davis, (17–20 October 1993), Vol. **1**, A. A. Balkema, Rotterdam, 835–850.

Chan, A. H. C., Famiyesin, O. O. and Muir, W. D. (1993g). Numerical prediction for model No. 11, in *Verification of Numerical Procedures for the Analysis of Soil Liquefaction Problems*, K. Arulanandan and R. F. Scott (Eds), UC Davis, (17–20 October 1933), Vol. **1**, 909–931.

Chan, A. H. C., Famiyesin, O. O. and Muir, W. D. (1995). *User Manual for GLADYS-2E*, School of Civil Engineering, University of Birmingham, December, Birmingham.

Cooke, B. (1991). Centrifuge Modelling of Flow and Contaiminant Transport Through Partially Saturated Soils. Ph.D. Dissertation. Queen's University, Kingston, Ontario, Canada.

Cooke, B. (1993). *Physical Modelling of Contaminant Transport in the Unsaturated Zone in Geotechnical Management of Waste and Contamination*, Balkema, Rotterdam.

Cooke, B. and Mitchell, R. J. (1991a). *Evaluation of Contaminant Transport in Partially Saturated Soil in Centrifuge 000'91*, Balkema, Rotterdam.

Cooke, B. and Mitchell, R. J. (1991b). Physical modelling of dissolved contaiminant transport in an unsaturated sand, *Can. Geotech. J.*, **28**, 6, 829–833.

Culligan-Hensley, P. J. and Savvidou, C. (1995). Environmental geomechanics and transport process, Chapter 8, in *Geotechnical Centrifuge Technology*, R. N. Taylor (Ed), Blackie Academic & Professional, London, 196–263.

Dobry, R. (1994). Possible lessons from VELACS model No. 2 results. Proceedings of the International Conference on the Verification of Numerical Procedures for the Analysis of Soil Liquefaction Pr 2, 1341–135.

Dewoolkar, M. M., Chan, A. H. C., Ko, H.-Y. and Pak, R. Y. S., (2009). Finite element simulations of seismic effects on retaining walls with liquefiable backfills, *Int. J. Numer. Anal. methods Geomech.*, **33**, 6, 791–816, DOI: 10.1002/nag.748

Dow Corning Limited (1985). Dow Corning 200 Fluid in Information about Silicone Fluids – Bulletin: 22-069D-01, Dow Corning Data Sheet.

Hellawell, E. E. (1994). Modelling transport processes in soil due to hydraulic, density and electrical gradients. Ph.D. Dissertation. University of Cambridge.

Hushmand, B., Scott, R. F. and Crouse, C. B. (1988). Centrifuge liquefaction tests in a laminar box, *Géotechnique*, **38**, 2, 253–262.

Katona, M. G. and Zienkiewicz, O. C. (1985). A unified set of single step algorithms Part 3: The beta-m method, a generalisation of the Newmark scheme, *Int. J. Num. Meth. Engrg.*, **21**, 1345–1359.

Lee, F. H. and Schofield, A. N. (1988). Centrifuge modelling of sand embankments and islands in earthquakes, *Géotechnique*, **38**, 1, 45–58.

Pastor, M. and Zienkiewicz, O. C. (1986). *A Generalised Plasticity Hierarchical Model for Sand Under Monotonic and Cyclic Loading*, NUMOG II, Ghent, April, 131–150.

Schofield, A. N. (1980). Cambridge geotechnical centrifuge operations, 20th Rankine Lecture, *Géotechnique*, **30**, 3, 227–268.

Schofield, A. N. (1981). Dynamic and earthquake geotechnical centrifuge modelling, in *Process International Conferences on Recent Advances in Geotechnical Engineering and Soil Dynamics*, University of Missouri-Rolla, Rolla, MO, USA, Vol. **3**, 1081–1100.

Schofield, A. N. and Zeng, X. (1992). *Design and Performance of an Equivalent-Shear-Beam (ESB) Container for Earthquake Centrifuge Modelling – CUED/D-soils/TR245*, Cambridge University Engineering Department, Cambridge, England.

Shiomi, T., Shigeno, Y. and Zienkiewicz, O. C. (1993a). Numerical prediction for model No. 1, in *Verification of Numerical Procedures for the Analysis of Soil Liquefaction Problems*, K. Arulanandan and R. F. Scott (Eds), UC Davis, (17–20 October), Vol. **1**, 213–219.

Shiomi, T., Shigeno, Y. and Zienkiewicz, O. C. (1993b). Numerical prediction for model No. 2, in *Verification of Numerical Procedures for the Analysis of Soil Liquefaction Problems*, K. Arulanandan and R. F. Scott (Eds), UC Davis, (17–20 October), Vol. **1**, 391–394.

Shiomi, T., Shigeno, Y. and Zienkiewicz, O. C. (1993c). Numerical prediction for model No. 11, in *Verification of Numerical Procedures for the Analysis of Soil Liquefaction Problems*, K. Arulanandan and R. F. Scott (Eds), UC Davis, (17–20 October), Vol. **1**, 977–986.

Shiomi, T., Shigeno, Y. and Zienkiewicz, O. C. (1993d). Numerical prediction for model No. 12, in *Verification of Numerical Procedures for the Analysis of Soil Liquefaction Problems*, K. Arulanandan and R. F. Scott (Eds), UC Davis, (17–20 October), Vol. **1**, 1067–1074.

Smith, I. M. (1994). An overview of numerical procedures used in the VELACS project, in *Verification of Numerical Procedures for the Analysis of Soil Liquefaction Problems*, K. Arulanandan and R. F. Scott (Eds), UC Davis, (17–20 October), Vol. **2**, 1321–1338.

Steedman, R. S. and Zeng, X. (1995). Dynamics, Chapter 7, in *Geotechnical Centrifuge Technology*, R. N. Taylor (Ed), Blackie Academic & Professional, London, 168–195.

Sture, S., Law, H. K., Shiomi, T. and Iai, S. (1994) VELACS: Overview of numerical predictions for model No.12, in *Verification of Numerical Procedures for the Analysis of Soil Liquefaction Problems*, K. Arulanandan and R. F. Scott (Eds), UC Davis, (17–20 October), Vol. **2**, A. A. Balkema, Rotterdam, 1635–1646.

Taylor, R. N. (1994). (Ed) *Geotechnical Centrifuge Technology*, CRC Press, Abingdon.

Venter, K. V. (1985). *KVV03 Data Report: Revised Data Report of a Centrifuge Model Test and Two Triaxial Tests*, N. Andrew (Ed), Schofield and Associates, Cambridge, England.

Venter, K. V. (1986). *Triaxial Data Report: Report on Seven Triaxial Tests*, N. Andrew (Ed), Schofield and Associates, Cambridge, England.

Venter, K. V. (1987). Modelling the response of sand to cyclic loads. Ph.D. Dissertation. Cambridge University Engineering Department.

Zienkiewicz, O. C., Chan, A. H. C., Pastor, M., Paul, D. K. and Shiomi, T. (1990). Static and dynamic behaviour of geomaterials – a rational approach to quantitative solutions, Part I – fully saturated problems, *Proc. Roy. Soc. Lond.*, **A429**, 285–309.

Zienkiewicz, O. C., Huang, M. and Pastor, M. (1993a). Numerical prediction for model No. 1, in *Verification of Numerical Procedures for the Analysis of Soil Liquefaction Problems*, K. Arulanandan and R. F. Scott (Eds), UC Davis, (17–20 October), Vol. **1**, 259–276.

Zienkiewicz, O. C., Huang, M. and Pastor, M. (1993b). Numerical prediction for model No. 2, in *Verification of Numerical Procedures for the Analysis of Soil Liquefaction Problems*, K. Arulanandan and R. F. Scott (Eds), UC Davis, (17–20 October), Vol. **1**, 423–434.

Zienkiewicz, O. C., Huang, M. and Pastor, M. (1993c). Numerical prediction for model No.3, in *Verification of Numerical Procedures for the Analysis of Soil Liquefaction Problems*, K. Arulanandan and R. F. Scott (Eds), UC Davis, (17–20 October, Vol. **1**, 583–591.

Zienkiewicz, O. C., Huang, M. and Pastor, M. (1993d). Numerical prediction for model No. 4a, in *Verification of Numerical Procedures for the Analysis of Soil Liquefaction Problems*, K. Arulanandan and R. F. Scott (Eds), UC Davis, (17–20 October), Vol. **1**, 675–680.

Zienkiewicz, O. C., Huang, M. and Pastor, M. (1993e). Numerical prediction for model No. 4b, in *Verification of numerical procedures for the analysis of soil liquefaction problems*, K. Arulanandan and R. F. Scott (Eds), UC Davis, (17–20 October), Vol. **1**, 731–736.

Zienkiewicz, O. C., Huang, M. and Pastor, M. (1993f). Numerical prediction for model No. 6, in *Verification of Numerical Procedures for the Analysis of Soil Liquefaction Problems*, K. Arulanandan and R. F. Scott (Eds), UC Davis, (17–20 October), Vol. **1**, 777–782.

Zienkiewicz, O. C., Huang, M. and Pastor, M. (1993g). Numerical prediction for model No. 7, in *Verification of Numerical Procedures for the Analysis of Soil Liquefaction Problems*, K. Arulanandan and R. F. Scott (Eds), UC Davis, (17–20 October), Vol. **1**, 873–880.

Zienkiewicz, O. C., Huang, M. and Pastor, M. (1993h). Numerical prediction for model No. 11, in *Verification of Numerical Procedures for the Analysis of Soil Liquefaction Problems*, K. Arulanandan and R. F. Scott (Eds), UC Davis, (17–20 October), Vol. **1**, 997–1006.

8

Applications to Unsaturated Problems

8.1 Introduction

In this chapter, we deal with the solution of quasi-static and dynamic problems for partially saturated soils. Also, finite deformations will be addressed. In many soil mechanics situations, the pressure of the gaseous phase, in general air, can be neglected but not always. The case with zero air pressure is also known as the static air phase assumption. We show examples for both situations. In the first case, equations of Section 2.3 apply, while in the latter, equations of Section 2.4 are relevant.

First, the analysis of isothermal drainage of water from a vertical column of sand is compared with experimental results in Section 8.2. The effect of airflow is considered in Section 8.3 which deals with the numerical modeling of air storage in an aquifer. In Section 8.4, a comparison of consolidation and dynamic results for various fully saturated and partially saturated problems using both small strain and finite deformation formulation is given. The comparison begins with the consolidation of a fully saturated soil column in Section 8.4.1. It is followed by the comparison between the consolidation of a fully saturated and a partially saturated soil column in Section 8.4.2. A two-dimensional consolidation example is introduced in Section 8.4.3 for a soil layer consolidating under fully saturated and partially saturated conditions, respectively. The comparison of the small strain and finite deformation formulation under earthquake loading is given for a one-dimensional soil column example in Section 8.4.4. Lastly, in Section 8.4.5, the elastoplastic large-strain behavior of an initially saturated vertical slope subjected to gravitational load and horizontal earthquake acceleration, followed by a consolidation phase, is given. This part is followed in Section 8.5 by a dynamic analysis with a full two-phase flow solution of a partially saturated soil column subjected to a step load. Compaction of a gas-bearing formation due to its exploitation is addressed in Section 8.6 and, finally, the onset of a landslide triggered by rainfall in a partially saturated slope concludes this chapter.

8.2 Isothermal Drainage of Water from a Vertical Column of Sand

It is very difficult to choose some appropriate tests to validate the two-phase flow partially saturated model because of the lack of any analytical solutions for this type of coupled problem where deformations of the solid skeleton are studied, together with the

Computational Geomechanics: Theory and Applications, Second Edition. Andrew H. C. Chan, Manuel Pastor, Bernhard A. Schrefler, Tadahiko Shiomi and O. C. Zienkiewicz.

saturated–unsaturated flow of mass transfer. There are also very few documented laboratory experiments available. One of these is the experimental work conducted by Liakopoulos (1965), on the isothermal drainage of water from a vertical column of sand. This test was also used by Narasimhan and Witherspoon (1978), Schrefler and Simoni (1988), Zienkiewicz et al. (1990), as well as by Schrefler and Zhan (1993), to check their numerical models. The same example was solved in Gawin et al. (1995), but only for one phase flow. Here (Gawin and Schrefler (1996)) attention is paid to the effects of two-phase flow, as compared to the previous solution (Gawin et al. 1995). In the experiment, a column of one meter high Perspex was packed with Del Monte sand and calibrated to measure the moisture tension at several locations along the column. Before the start of the experiment, i.e. at $t < 0$, water was continuously added from the top and was allowed to drain freely at the bottom through a filter, until uniform flow conditions were established. At the start of the experiment, i.e. $t = 0$, the water supply was stopped and the tensiometer readings were recorded. The porosity $n = 29.75\%$ and the hydraulic properties of Del Monte sand had been measured by Liakopoulos and reported in his PhD dissertation (1965), by a separate set of experiments.

During the numerical simulation, Liakopoulos' saturation-capillary pressure and relative permeability of water-capillary pressure relationships were approximated using the following equations:

$$S = 1 - 1.9722 \times 10^{-11}\left(\frac{p_c}{1\,\text{Pa}}\right)^{2.4279} \tag{8.1}$$

$$k_{rw} = 1 - 2.207(1 - S)^{1.0121} \tag{8.2}$$

where p_c is given in the unit of Pa and k_{rw} is the relative permeability for the water phase.

In this example, isoparametric Lagrangian elements were used with the same interpolation for pressures and displacement fields. Furthermore, linear elastic material behavior was assumed. The column was divided, in turn, into 10 and 20 four-, eight-, and nine-noded finite elements of equal size and different time steps in the time domain with $\Delta t = 10$ seconds, 1 seconds, or 0.5 seconds gave practically the same results. At the beginning, besides uniform flow conditions, i.e. unit vertical gradient of the potential and $p_c = 0$ at the top surface, static equilibrium was also assumed.

The boundary conditions were as follows:

For the lateral surface, the horizontal displacement and the fluid outflow is zero.

For the top surface, $p_a = p_{atm}$ where p_{atm} means atmospheric pressure. For the bottom surface, $p_a = p_{atm}$, $p_c = 0$ for $t > 150$ seconds, while water p_w was assumed to change linearly from the initial value to zero for $t < 150$ seconds and the base is fixed in both displacement directions.

Liakopoulos did not measure the mechanical properties of the soil; so, the Young Modulus of the soil was assumed to be $E = 1.3$ MPa, Poisson's ratio $\nu = 0.4$ and Biot's constant $\alpha = 1$, similar to Gawin et al. (1995), Schrefler and Zhan (1993), and Schrefler and Simoni (1988).

The calculations were performed for one-phase flow with gas pressure fixed at the atmospheric pressure in the partially saturated zone, as well as for two-phase flow. For the latter case, the switching between saturated and unsaturated solution was performed at $p_c = 2$ kPa, i.e. $S = 0.998$, which corresponds to bubbling pressure of the analyzed sand (Liakopoulos

1965), as well as at $p_c = 0$ Pa, in order to analyze the effect of the "switching" value on the solution obtained. For more details about the switching procedure, see Section 8.5.

The relative permeability of the gas phase was assumed to follow the relationship proposed by Brooks and Corey (1966):

$$k_{rg} = (1 - S_e)^2 \left(1 - S_e^{5/3}\right) \tag{8.3}$$

$$S_e = \frac{S - 0.2}{0.8} \tag{8.4}$$

where S_e is the equivalent value of saturation with the additional lower limit of $k_{rg} \geq 0.0001$.

The resulting profiles of water pressure for the two-phase flow with switching at p_c 2 kPa shown in Figure 8.1 as solid curves with different symbols at different time stations and for one-phase flow in dash-dot curves are compared with the experimental results of Liakopoulos (1965) in fine lines. The solution from one-phase flow showed better agreement and this is because in Schrefler and Simoni (1988), Young's modulus value has been "fitted" for the case.

The profiles of vertical displacements, water saturation and capillary pressure for two-phase flow with switching pressure at $p_c = 2$ kPa (solid curves) and $p_c = 0$ Pa (dash-dot curves) are compared with the results for one-phase flow (fine curves) in Figure 8.2.

There are some noticeable differences between one- and two-phase flow solutions for vertical displacements as shown in Figure 8.2, although their final values are similar. The differences between the one- and two-phase flow solutions are more appreciable for saturation of water and capillary pressure as shown in Figures 8.3 and 8.4. In the lower zone where no

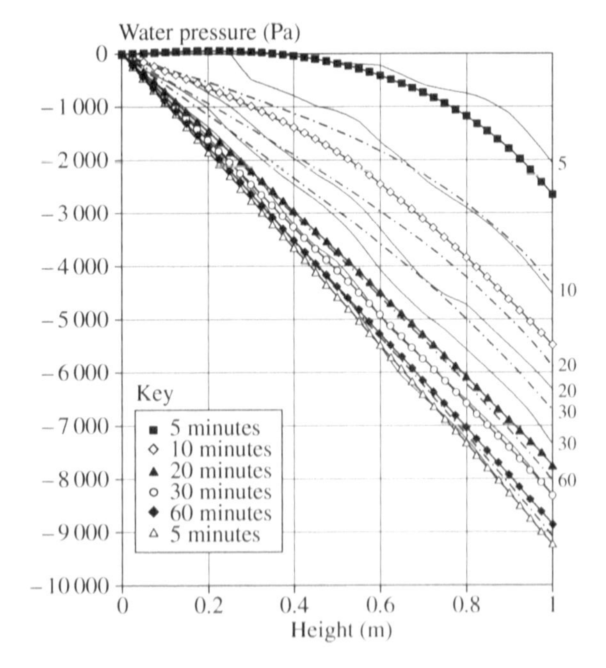

Figure 8.1 Comparison of the numerical solutions for water pressure (two-phase flow with switching at $p_c = 2$ kPa indicated by solid curves and one-phase flow solution indicated by dash-dot curves) with experimental results of Liakopoulos (1965) in fine lines. *Source:* Reproduced from Gawin and Schrefler (1996) by permission of MCB University Press Ltd.

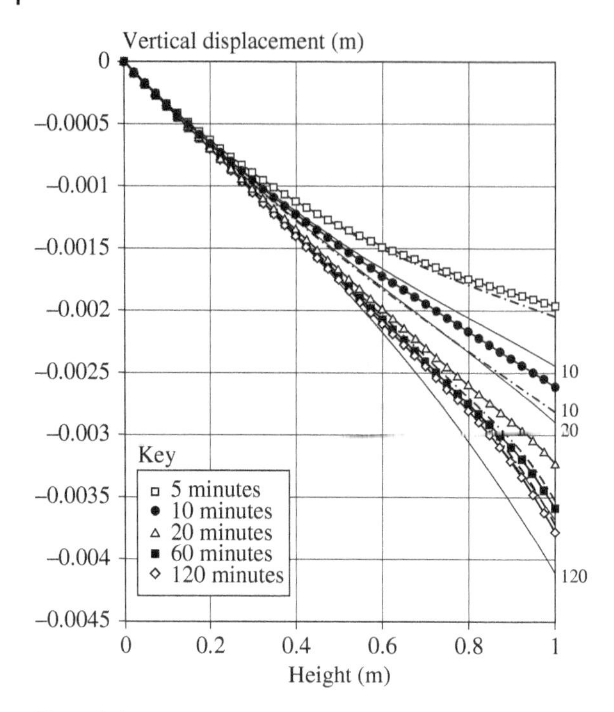

Figure 8.2 Comparison of the two-phase flow solutions with switching at $p_c = 2$ kPa (solid curves) and $p_c = 0$ Pa (dash-dot curves) with one-phase flow solution (fine curves) made for vertical displacements. *Source:* Reproduced from Gawin and Schrefler (1996) by permission of MCB University Press Ltd.

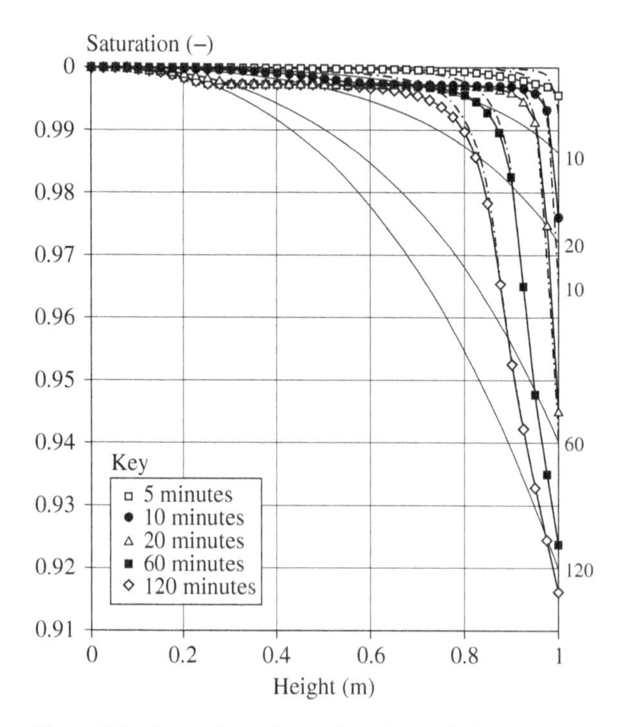

Figure 8.3 Comparison of two-phase flow solutions with switching at $p_c = 2$ kPa (solid curves) and at $p_c = 0$ Pa (dash-dot curves) with one-phase flow solution (fine curves) made for the degree of saturation of water. *Source:* Reproduced from Gawin and Schrefler (1996) by permission of MCB University Press Ltd.

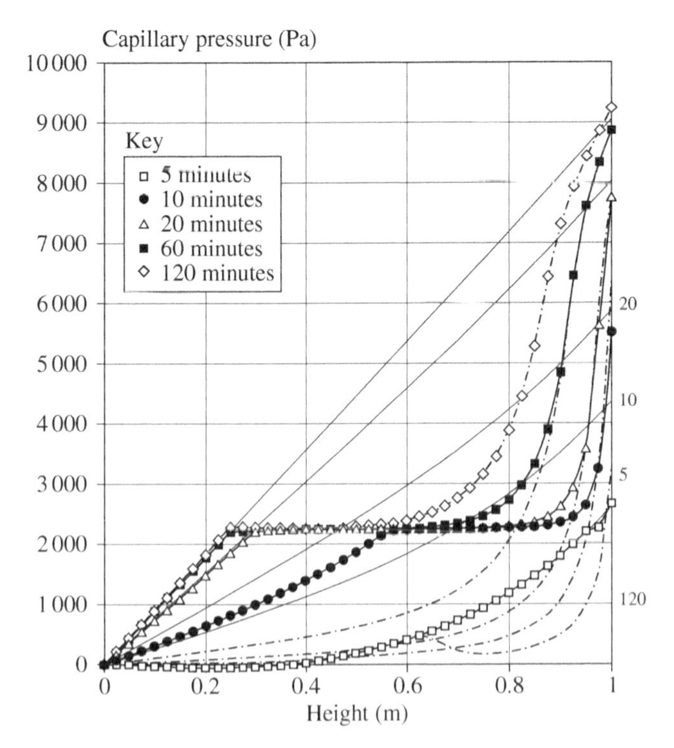

Figure 8.4 Comparison of the two-phase flow solution with switching at p_c = 2 kPa (solid curves) and p_c = 0 Pa (dash-dot curves) with one-phase flow solution (fine curves) made for capillary pressure. *Source:* Reproduced from Gawin and Schrefler (1996) by permission of MCB University Press Ltd.

gas flow occurs, the differences are small, but higher up, there is a qualitative change in the soil behavior caused by the presence of gas under pressure (see also Figure 8.5). The solution for two-phase flow with switching at $p_c = 2$ kPa is similar to the one-phase flow solution at the bottom of the column while in the middle region, it forms some characteristic constant-value zones corresponding to switching values of capillary pressure or saturation. The one-phase flow solution tends to the two-phase flow solution with switching at $p_c = 0$ Pa in the upper part of the sand column. In general, the gradients of saturation and capillary pressure are higher for the two-phase solution in the upper zone where gas flow occurs. This is qualitatively in accordance with the solution obtained by Schrefler and Zhan (1993).

The profiles of gas pressure for two-phase flow with switching at $p_c = 2$ kPa are given as solid curves in Figure 8.5 while at $p_c = 0$ Pa, as dash-dot curves. The discernable differences are caused by the different assumptions about gas flow, i.e. no gas flow for zones where the degree of saturation is greater than 0.998 in the first case. There is a small difference in the pressure amplitudes, nevertheless, the qualitative similarity of the gas pressure profiles in the zone, where gas flow occurs, is obvious.

This example shows that the modeling of the transition from fully saturated to partially saturated condition and vice versa is very sensitive to the procedure adopted.

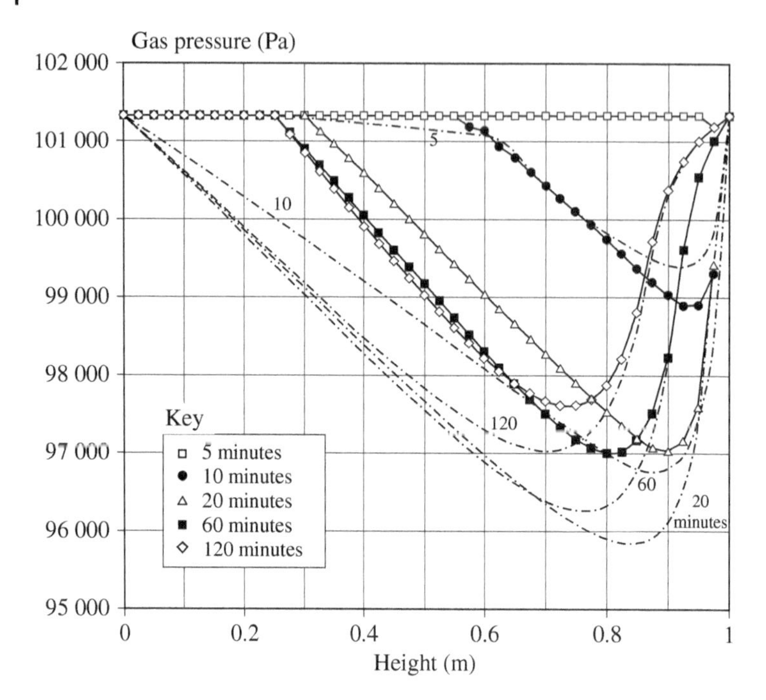

Figure 8.5 Comparison of the gas pressure profiles (the two-phase flow solution) for switching at p_c = 2 kPa (solid curves) and at p_c = 0 Pa (dash-dot curves). *Source:* Reproduced from Gawin and Schrefler (1996) by permission of MCB University Press Ltd.

8.3 Air Storage Modeling in an Aquifer

We simulate air storage in an aquifer of 10 m depth sited on an impervious layer, by a partially saturated model (Schrefler and Zhan (1993)). Meiri and Karadi (1982) simulated this problem by a one-dimensional finite element model, with a rigid soil skeleton. In the present application, the skeleton is considered deformable, with Young's modulus $E = 692$ kN/m^2. The porous medium system is assumed initially fully saturated with a reference permeability of 5×10^{-13} m^2, porosity of 0.2 and initial aquifer pressure of 5.066 MN/m^2 or 50 Patm, under isothermal conditions at 149 °C (or 300 °F). Initial conditions are no displacement and constant water pressure of 5.066 MN/m^2 at all points. The boundary conditions are as follows:

Lateral surface, $q_a = 0$, $q_w = 0$, $u_h = 0$
Bottom surface, $q_a = 0$, $p_w = p_{ref}$, $u_h = u_v = 0$
Top surface, $q_a = 2.44 \times 10^{-4}$kg/s/m^2 (air injection), $q_w = 0$.

The water and air viscosities are selected as 0.3 mNs/m^2 and 24 μNs/m^2, respectively. The water density for standard conditions is 100 kg/m^3 and the air density 1.22 kg/m^3.

The following relationships between the relative permeabilities of water and air, the capillary pressure, and water saturation, proposed by Brooks and Corey (1966), are employed in the simulation:

$$k_{rw} = S_e^{\frac{2+3\lambda}{\lambda}} \tag{8.5}$$

$$k_{ra} = (1 - S_e)^2\left(1 - S_e^{\frac{2+\lambda}{\lambda}}\right) \tag{8.6}$$

$$p_c = \frac{p_b}{S_e^{1/\lambda}} \tag{8.7}$$

where $S_e = (S_w - S_{wc})/(1 - S_{wc})$ is the effective saturation, S_{wc} is the irreducible saturation, λ is the pore size distribution index, and p_b is the bubbling pressure. The values for S_{wc}, λ, and p_b are given as 0.2, 3.0, and 1.68 kN/m^2, respectively, corresponding to sand with an intrinsic permeability of 5×10^{-13} m^2.

The vertical displacement, water pressure, air pressure, and saturation versus depth of the column are shown in Figure 8.6 for different time instances. Water and air pressure have a

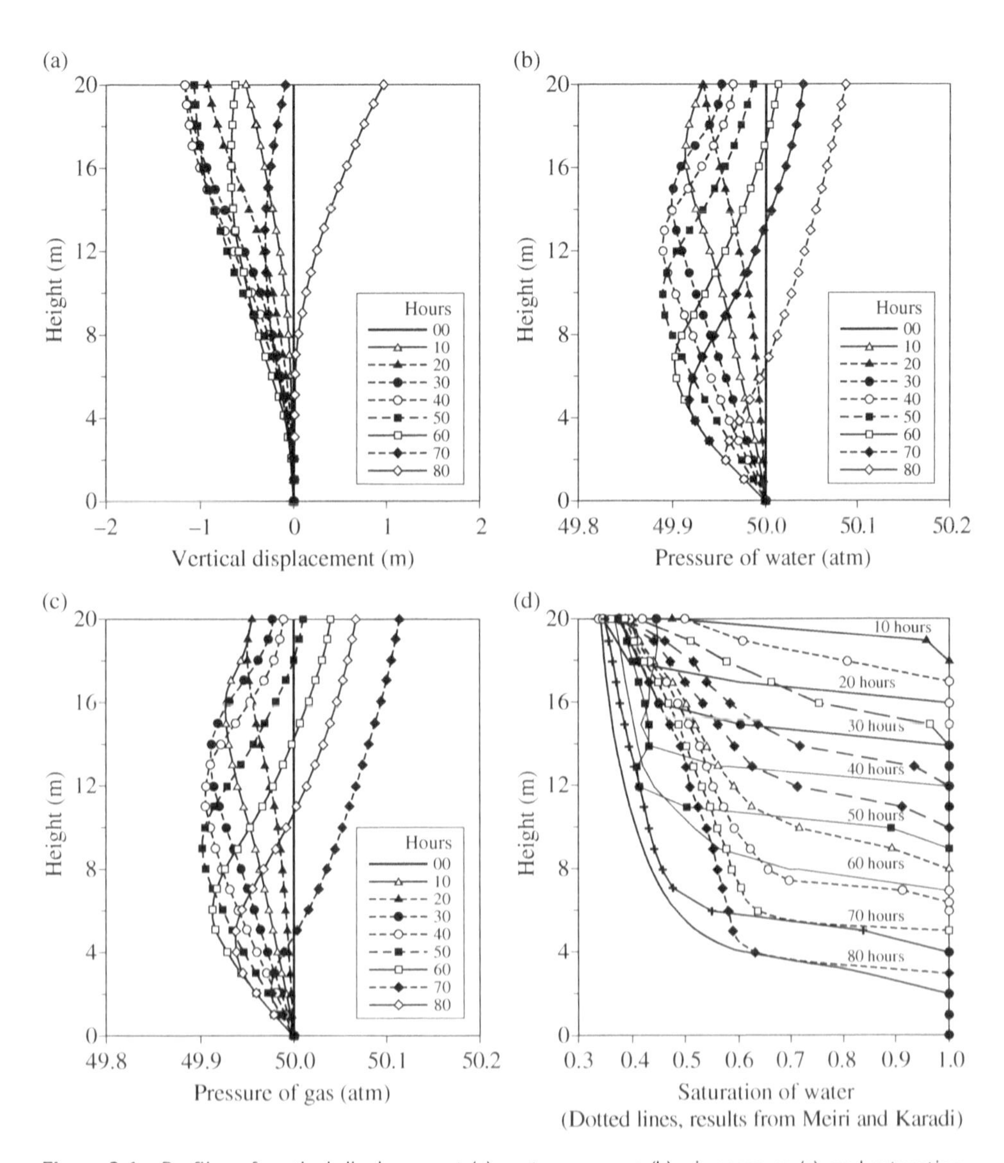

Figure 8.6 Profiles of vertical displacement (a), water pressure (b), air pressure (c), and saturation (d) of air storage modeling in an aquifer. *Source:* Reproduced with permission from Schrefler and Zhan (1993), American Geophysical Union.

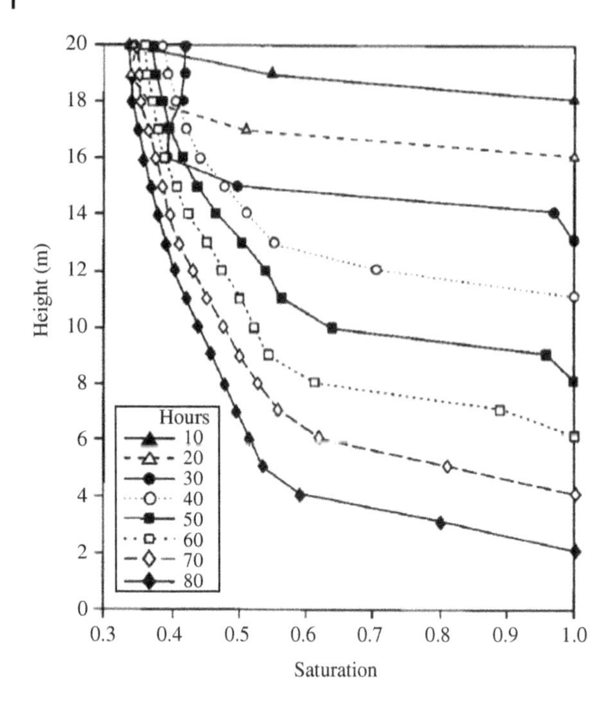

Figure 8.7 Profile of saturation of air storage modeling in an aquifer with doubled Young's modulus for Figure 8.6. *Source:* Reproduced with permission from Schrefler and Zhan (1993), American Geophysical Union.

minimum at the interface of the two fluids. The column is first contracting and then, finally, expanding vertically with continuing air inflow from the top. The saturations are compared with the result reported in Meiri and Karadi (1982), indicated with dashed lines. The effects of a deformable skeleton assumption can be seen clearly. In general, at the onset of airflow, the desaturation is greater in the rigid skeleton model while, later on, the situation is reversed, i.e. the deformable skeleton model has greater desaturations.

A second simulation is performed with a doubled Young's modulus to point out the influence of the solid skeleton deformation. The new saturation profiles are shown in Figure 8.7. While the profile of saturation after 80 hours of pumping is roughly the same with both Young's moduli, the profiles of shorter time spans are different. These simulations confirm that the model proposed can reproduce this quasi-static case and that the main features of the model, i.e. airflow and solid deformation, have their importance.

8.4 Comparison of Consolidation and Dynamic Results Between Small Strain and Finite Deformation Formulation

A large-deformation model with a number of simplifications has been developed by Meroi et al. (1995). Although the finite deformation solution has only been shortly mentioned in this book in Section 2.2.1, it would be of interest to compare the small strain formulation introduced in Chapters 2 and 3 and to investigate the effect of finite deformation. The large-deformation model has been extensively validated with respect to solutions reported in the literature. For the purpose of validation, a number of one-phase static and dynamic problems (Bathe and Ozdemir 1976; Bathe et al. 1975; Heyliger and Reddy 1988; Shantaram et al. 1976) were first solved and then consolidation solutions of fully saturated soils were obtained (Carter et al. 1979; Kim et al. 1993; Lewis and Schrefler 1998; Meijer 1984; Monte and Krizen 1976; Prévost 1981, 1984) showing good agreement with the corresponding results in the literature.

Five examples have been presented in Meroi et al. (1995) to test the new features offered by the large-deformation approach. All the examples contained comparison with small deformation developed in Chapters 2 and 3. All of them are reproduced in this section. For a more advanced approach to finite strain partially saturated elastoplastic consolidation with multiplicative decomposition of the deformation gradient, see Sanavia et al. (2002), and for dynamics at the finite strain in a fully saturated condition, Li et al. (2004) and partially saturated condition, Uzuoka and Borja (2012).

8.4.1 Consolidation of Fully Saturated Soil Column

This example deals with the consolidation of a one-dimensional 10-m deep ground, fully saturated by water, infinitely extended in the horizontal direction, and subjected to a step load applied at the top level, with drainage allowed only through the top surface. The problem, which is one-dimensional since each vertical section can be considered as a plane of symmetry, is modeled as a saturated soil column under plane strain condition. The boundary conditions for the displacement field are such that all nodes are horizontally constrained and the bottom level is fixed with no vertical movement. Atmospheric pressure is assumed at the top level and impermeable boundaries are imposed at the lateral and bottom surfaces. For comparison with Prévost's (1981, 1982, 1984) results of one-dimensional elastic consolidation, different load levels were considered, reaching the maximum intensity for the uniformly distributed load equal to Young's modulus of the ground. An initial porosity of 0.3, a specific permeability of 0.01 m/s, an elastic modulus of 1 GPa and a zero Poisson ratio are adopted, in accordance with Prévost (1982). A backward difference integrator was used because of its superior efficiency in consolidation analyses as reported in Prévost (1982).

With the finite-deformation formulation, the theoretical relationship between the applied load and the final displacement is logarithmic for this example and it is represented in Figure 8.8 by the dot-dash curve. The computational results (curve a) are in very good agreement with the theoretical behavior (curve c). The dashed curve (curve d) represents the results of linear analysis while the dotted curve (curve e) represents the solution obtained by Prévost (1982).

At high load levels, due to the assumption of constant elastic modulus, a near-zero void ratio may occur computationally, and at this stage, the soil behavior should be described by the elastic characteristics of the compacted grain itself: a relationship between the elastic modulus of soil and its void ratio has been given in Monte and Krizen (1976). It is assumed here, for the sake of simplicity, that the soil elastic modulus becomes ten times as large as the initial one when the porosity approaches zero. The results obtained under this

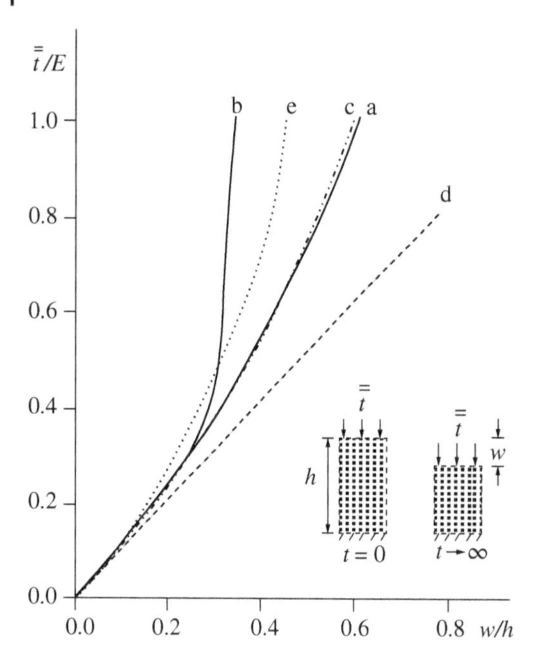

Figure 8.8 Vertical settlement versus load level in one-dimensional elastic consolidation problem with fully saturated conditions: curves a and b, computational results with the finite deformation approach, for constant and stepwise variable elastic behavior, respectively; curve c, theoretical solution for the finite deformation regime, curve d: linear analysis response; curve e, Prevost solution. *Source:* Reproduced from Meroi et al. (1995) by permission of John Wiley & Sons Ltd.

assumption are represented by curve b. It is recalled that the current value of the porosity can be recovered from the mass balance Equation (2.16).

In Figure 8.9, the maximum vertical displacement versus the normalized time T_v ($T_v = c_v t/h^2 = c_t t$), where c_v is the coefficient of consolidation (6.13) and c_t the time factor (Lewis and Schrefler 1998) is shown for different load levels: curves a, b, and c refer to load levels equal to 0.2, 0.4, and 0.6 times the elastic modulus of the soil matrix, respectively.

A further improvement to the finite deformation model consists of the introduction of the dependence of the absolute permeability on the void ratio. In the present analysis, permeability is assumed to be a linear function of the void ratio, varying from the initial value to zero when porosity becomes zero. Figure 8.10, in which time is given in the logarithmic scale, shows the influence on computational results of such a relationship. Consolidation with constant and variable permeability is described, respectively, by curves a and c for the load level equal to 0.3 E and by curves b and d for a load level of 0.5 E. The significant increase in time (in logarithmic scale on the figure) necessary to reach full consolidation can be seen clearly. Other relationships for variable permeability can be found in Monte and Krizen (1976) and Lewis and Schrefler (1998).

8.4.2 Consolidation of Fully and Partially Saturated Soil Column

This example refers to the work by Advani et al. (1993) and Kim et al. (1993), but partial saturation conditions are also considered. A one-dimensional column, 7 m high, is modeled in accordance with Kim et al. (1993) using plane strain elements. The material properties

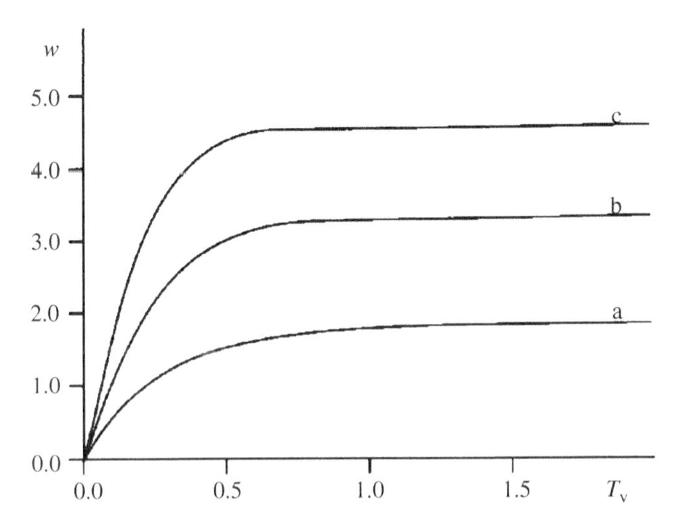

Figure 8.9 Vertical settlement versus normalized time in the one-dimensional consolidation problem with constant permeability and elastic modulus under fully saturated conditions and finite-deformation assumptions. Curves a, b, and c refer to load levels equal to 0.2, 0.4, and 0.6 E. *Source:* Reproduced from Meroi et al. (1995) by permission of John Wiley & Sons Ltd.

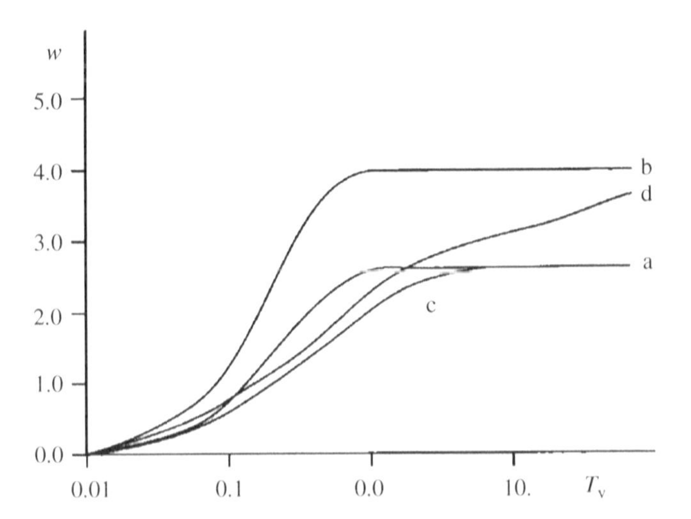

Figure 8.10 Vertical settlement vs. normalized time in the one-dimensional elastic consolidation problem with fully saturated conditions. Curves a and c refer to a load level of 0.3 E while curves b and d to 0.5 E with c and d having variable permeability. *Source:* Reproduced from Meroi et al. (1995) by permission of John Wiley & Sons Ltd.

are: porosity of 0.5, dynamic permeability of 4×10^{-11} m^4/Ns, elastic modulus of 6 GPa and Poisson ratio of 0.4. The resulting time factor $c_t = 1.049 \times 10^{-3}$. The uniformly distributed load applied in the vertical direction reaches a limiting value of 10 GPa. In Figure 8.11, the vertical settlement, normalized with respect to the corresponding values of Terzaghi's theory (1943), is plotted against normalized time for the case of fully saturated conditions for small deformation (curve a) and large deformation (curve b). The results are in substantial agreement with the ones presented in Kim *et al.* (1993). The large deformation analysis is also performed for the case of permeability linearly dependent on the void ratio (curve c).

In the same figure, the large deformation results for three different homogeneous partially saturated initial conditions are also plotted. These partially saturated initial conditions are imposed by assigning the initial capillary pressure distribution corresponding to the degree of saturation via the capillary pressure relationship in Figure 8.12. While the small deformation analysis with air pressure equal to the atmospheric pressure gives the final results in one time step (curve d), in the large deformation analysis, a real consolidation process takes place.

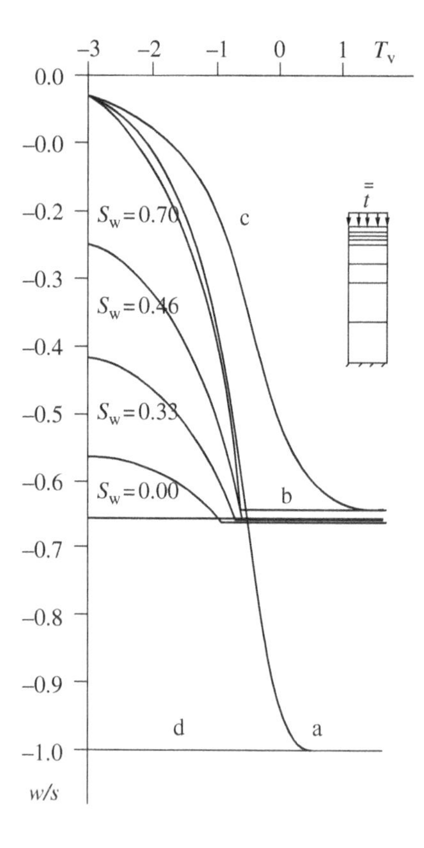

Figure 8.11 For the model drawn, vertical settlement, normalized with respect to the corresponding values of Terzaghi's theory, vs. normalized time: fully (curve a, small-deformation analysis; curves b and c, large-deformation analyses with constant and variable permeability, respectively) and partially saturated conditions (curves d for small deformation analysis). *Source:* Reproduced from Meroi et al. (1995) by permission of John Wiley & Sons Ltd.

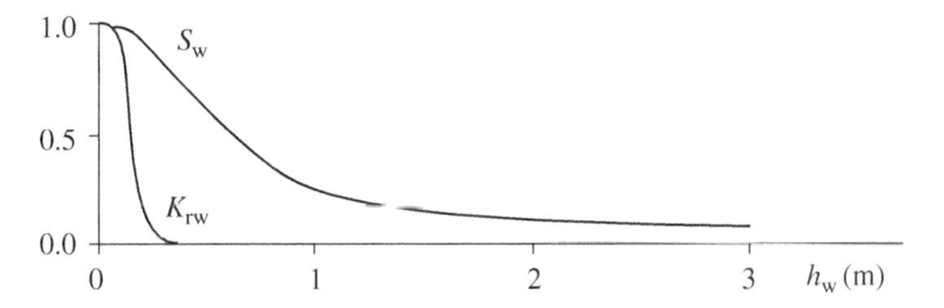

Figure 8.12 Saturation and relative permeability vs. hydraulic head. *Source:* Reproduced from Meroi et al. (1995) by permission of John Wiley & Sons Ltd.

Figure 8.13 Water pressure versus time, normalized with respect to applied load, for the fully saturated case at the top. Curve a, small-deformation approach; curves b and c, large-deformation cases with constant and variable permeability, respectively. *Source:* Reproduced from Meroi et al. (1995) by permission of John Wiley & Sons Ltd.

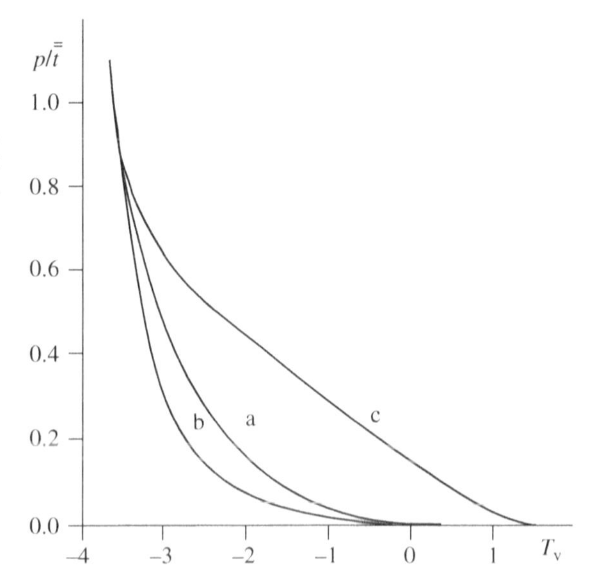

The dissipation of water pressure for the fully saturated case in the top element is shown in Figure 8.13 for small deformation as curve a, and for large deformation with constant and variable permeability as curves b and c.

8.4.3 Consolidation of Two-Dimensional Soil Layer Under Fully and Partially Saturated Conditions

This example is also taken from Kim et al. (1993) and consists of a two-dimensional, plane strain analysis of an 8 m deep soil layer loaded by a uniform pressure = 6 GPa by a 16 m wide foundation (= $2b$). Because of symmetry, only half of the model is considered which extends for 48 m in the horizontal direction.

All the material data are the same as in Section 8.4.2, apart from the Poisson ratio which is assumed to be zero in accordance with Kim et al. (1993), the resulting time factor is $c_t = 0.375 \times 10^{-3}$. The boundary conditions for displacements are taken as: the bottom level is fixed, while laterally, only vertical displacements are allowed. Both rigid and flexible footings are considered. In order to model the rigid footing, a multipoint constraint technique, i.e. tied-nodes technique, is employed.

Pressure is assumed to be zero at the top surface including the underside of the foundation in accordance with Kim et al. (1993). Since a net flow is allowed through the foundation, it is assumed to be permeable. Besides the fully saturated case, a partially saturated model is also investigated with an initial saturation of 92%. The initial condition is obtained by imposing suction in accordance with the relationship given in Figure 8.12. In the case of a rigid foundation under partially saturated conditions, the foundation is now considered impervious.

Displacements of the top node at the center line, normalized according to $w_n = Ew/2bq$, with w displacement, are shown in Figure 8.14a for the rigid foundation and Figure 8.14b, for the flexible one. In both figures, curves refer to small-deformation analysis introduced in Chapters 2 and 3, for the fully saturated case, and are in perfect agreement with results proposed by Kim et al. (1993) and with the analytical solution of Booker (1974).

Curve c refers to large-deformation analysis. In the case of a rigid foundation, the result is also given by Kim et al. (1993) and is in good agreement with the one reported here. Consolidation with the linear dependence of the permeability with the void ratio is indicated by curve d.

Curve e describes the behavior in the case of an initially uniform partial saturation of 92% and large deformations. For permeable flexible foundation, the behavior is almost time-independent, while for a rigid impermeable foundation, some consolidation takes place and the time for transient behavior is longer than in the fully saturated case because the foundation impermeability forced the water not only to follow a longer path, but also one with a smaller relative permeability.

Figure 8.15 represents the deformed mesh in the case of a flexible foundation, giving consolidation patterns at T_v equal to 0.01, 0.1, and 0.55. For the same dimensionless times, Figures 8.16 and 8.17 give the results for a rigid foundation with full and partial saturation, respectively. The comparison of the different deformed shapes allows one to appreciate the influence of the different fluid pressures. In particular, swelling close to the foundation during the first stage of the analysis can be observed under fully saturated conditions.

8.4.4 Fully Saturated Soil Column Under Earthquake Loading

This example of liquefaction, performed by Zienkiewicz et al. (1990) and Xie (1990) makes use of the sand of the Pastor and Zienkiewicz (1986) model as described in Chapter 4. It can be seen that the results are not significantly affected by the use of geometric nonlinearity because the fluid pressure can rise even without the large deformation during a cycle of earthquake loading. The first 10 seconds of the N–S component of the El-Centro 1940

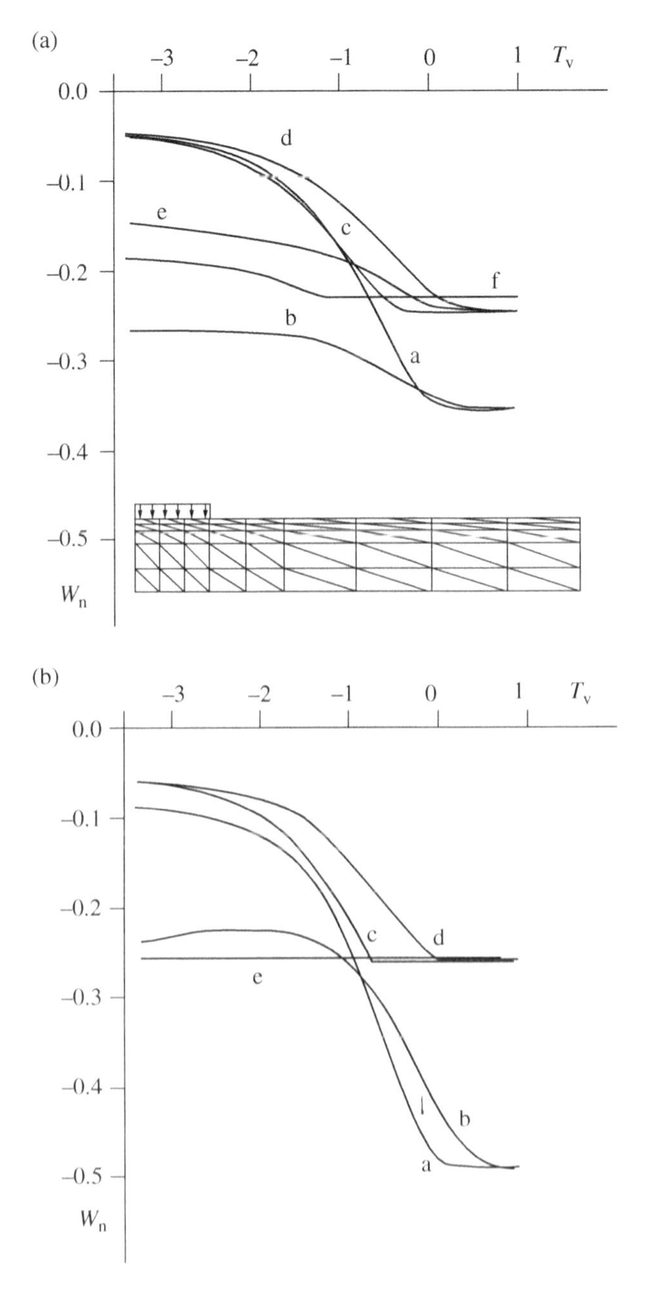

Figure 8.14 (a) Model description and normalized settlement of the top node at the center line versus normalized time for the rigid footing. Curves a and b, small-deformation regime for fully and partially saturated initial conditions, respectively; curves c and d, finite deformation analysis from initial fully saturated conditions with constant and variable permeability, respectively; curve e, finite deformation result from initial partially saturated conditions. (b) Normalized settlement of the top node at the center line for the flexible footing case versus normalized time. Curves a and b, small-deformation analysis for fully and partially saturated initial conditions; curves c and d, finite deformation analysis from initial fully saturated conditions with constant and variable permeability, respectively; curve e, finite deformation results from initial partially saturated conditions. *Source:* Reproduced from Meroi et al. (1995) by permission of John Wiley & Sons Ltd.

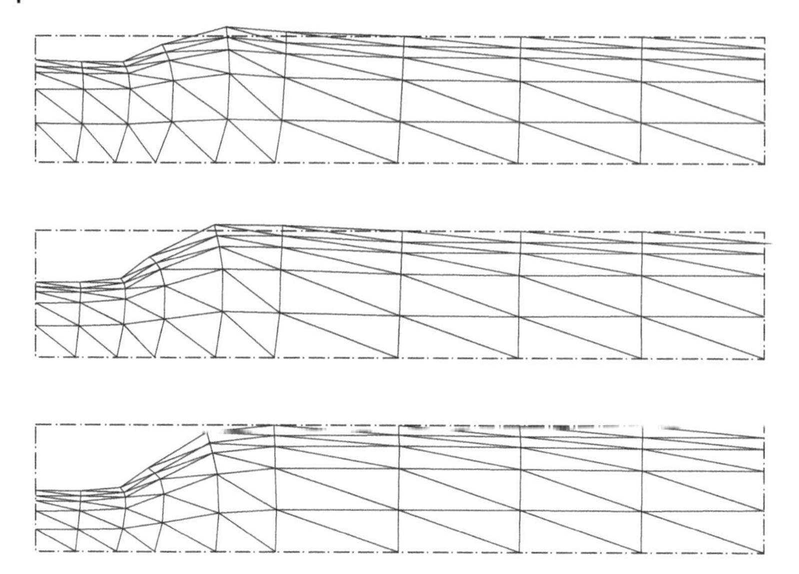

Figure 8.15 Deformed mesh for flexible footing, the fully saturated case: the consolidation pattern is given at the dimensionless time, T_v equal to 0.01, 0.1, and 0.55. *Source:* Reproduced from Meroi *et al.* (1995) by permission of John Wiley & Sons Ltd.

earthquake is taken as the horizontal base acceleration input during the consolidating phase of the sandy soil under a uniformly applied load of 600 kPa. The geometry and mechanical characteristics of the model are the same as those given by Zienkiewicz et al. (1990), and, in particular, an initial elastic modulus of 4.5 MPa is assumed.

Figure 8.18 shows an increase in pore pressure, both for large deformation (upper of the twin curves) and the small-deformation approach during the first 15 seconds for three reference points on the finite-element model. The horizontal displacements obtained using the two different approaches at points A and D are given in Figure 8.19.

Pore pressure and vertical displacements are shown for the whole analyzed period in Figures 8.20 and 8.21, respectively. For any pair of curves, the one with a smaller final value belongs to the large-deformation approach. It can be noted that the amount of vertical displacement induced by the earthquakes is not significant in comparison to the one induced by the surface-applied load.

8.4.5 Elastoplastic Large-Strain Behavior of an Initially Saturated Vertical Slope Under a Gravitational Loading and Horizontal Earthquake Followed by a Partially Saturated Consolidation Phase

This example considers the elastoplastic large-strain behavior of a 9.15 m vertical slope (Chen and Mizuno 1990) subjected to gravitational loading during the first 25 seconds of El-Centro N–S component of horizontal acceleration and during the following

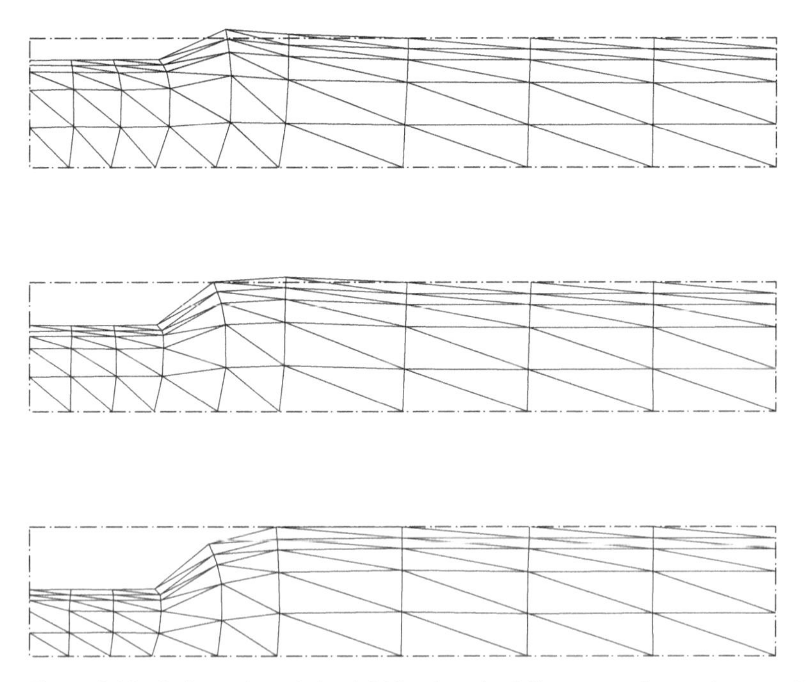

Figure 8.16 Deformed mesh for rigid footing, the fully saturated case: the consolidation pattern is given at the dimensionless time, T_v equal to 0.01, 0.1, and 0.55. *Source:* Reproduced from Meroi et al. (1995) by permission of John Wiley & Sons Ltd.

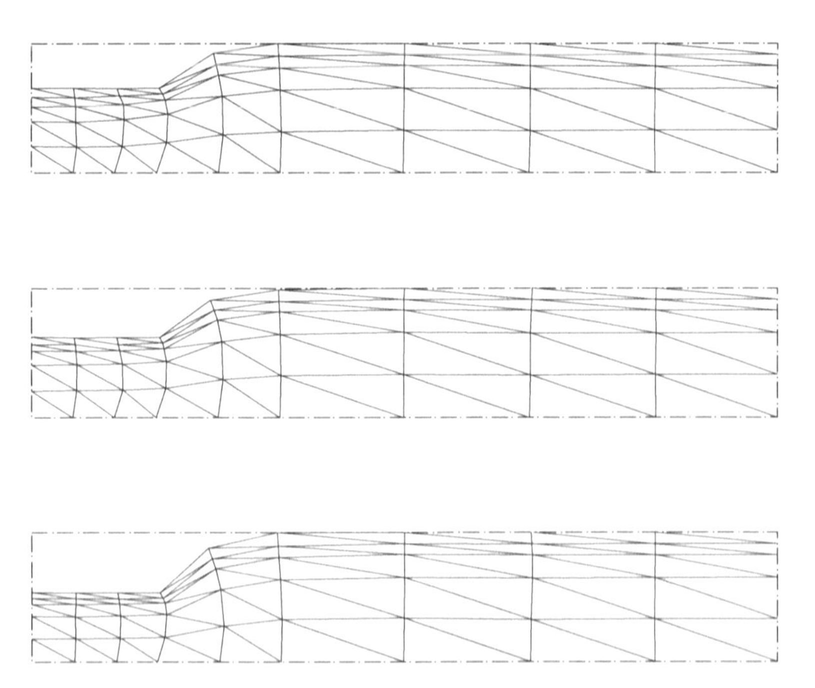

Figure 8.17 Deformed mesh for rigid footing, initial partial saturation of 92%: the consolidation pattern is given at the dimensionless time, T_v equal to 0.01, 0.1, and 0.55. *Source:* Reproduced from Meroi et al. (1995) by permission of John Wiley & Sons Ltd.

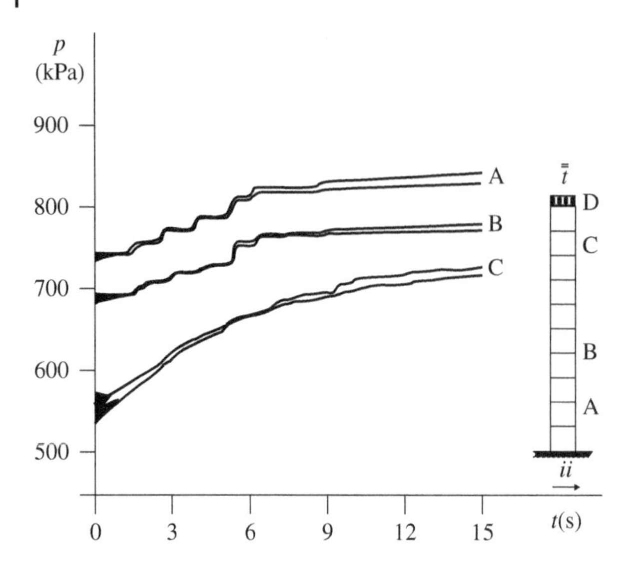

Figure 8.18 Pore pressure versus time in the generation phase of the pore pressure at three reference points of the drawn model, both for large-deformation (upper one of the twin curves) and for the small-deformation approach. *Source:* Reproduced from Meroi et al. (1995) by permission of John Wiley & Sons Ltd.

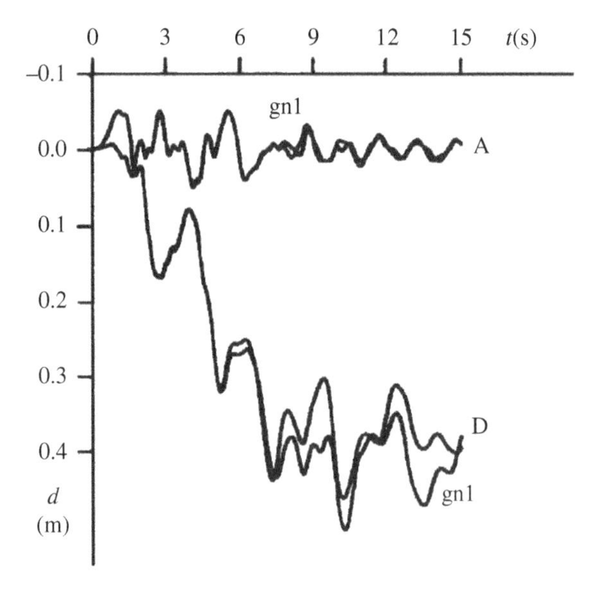

Figure 8.19 Horizontal displacements versus time at points A and D for both large-deformation (upper of the twin curves) and the small-deformation approach. *Source:* Reproduced from Meroi et al. (1995) by permission of John Wiley & Sons Ltd.

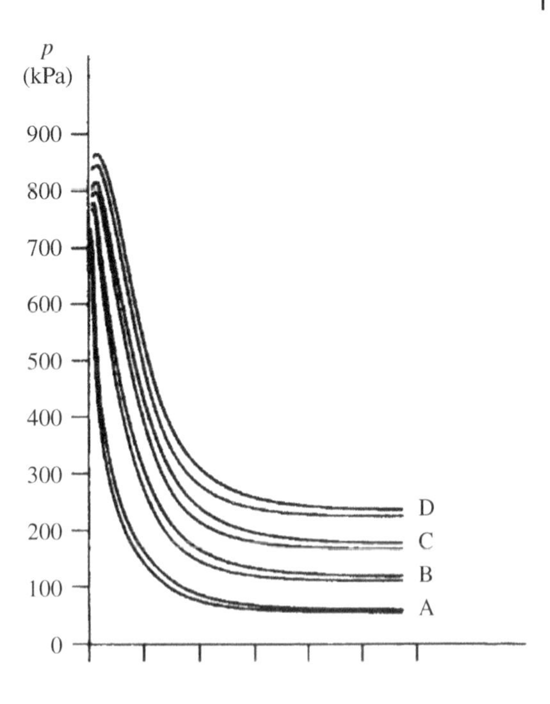

Figure 8.20 Pore pressure versus time in the dissipation phase for the given points, both for the large (the one with the smallest final value of the twin curves) and for the small-deformation approach. *Source:* Reproduced from Meroi et al. (1995) by permission of John Wiley & Sons Ltd.

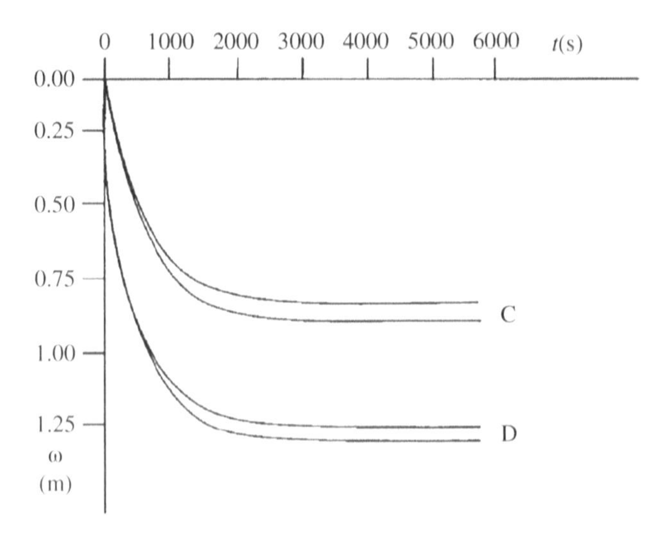

Figure 8.21 Vertical displacement versus time in the consolidation phase at the given points, both for the large (the one with the smallest final value of the twin curves) and for the small-deformation approach. *Source:* Reproduced from Meroi et al. (1995) by permission of John Wiley & Sons Ltd.

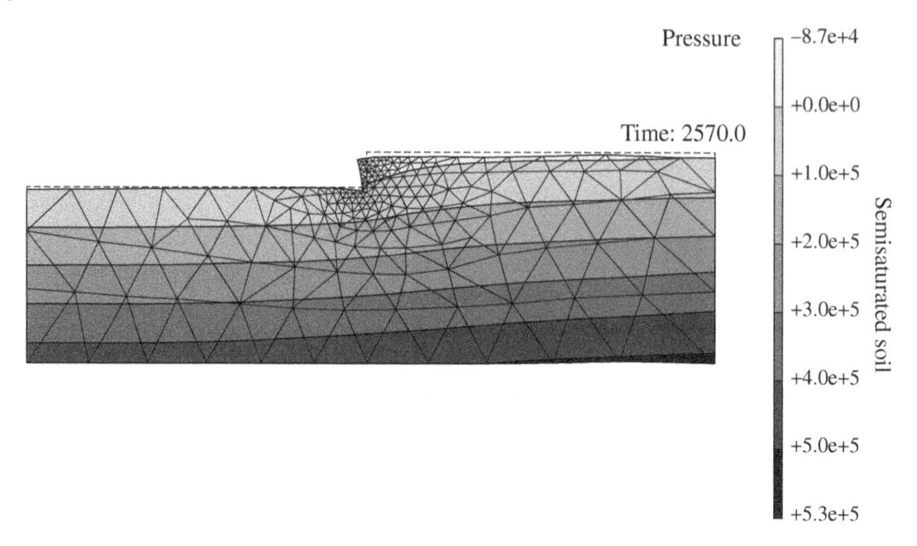

Figure 8.22 Final pressure distribution over the deformed mesh. *Source:* Reproduced from Meroi et al. (1995) by permission of John Wiley & Sons Ltd.

consolidating phase. The fully saturated domain after excavation is modeled by linear triangular finite elements. Figure 8.22 shows the final pressure distribution plotted over the deformed mesh. To impose initial, fully saturated conditions, with the piezometric level at the top nodes of the domain, an undrained analysis is carried out with horizontal constraints at the vertical slope nodes. Then the seismic excitation is applied at the bottom nodes, the above-mentioned horizontal constraints are released and the soil is allowed to desaturate.

At the left and right vertical sides of the domain, horizontal displacements are fixed and hydrostatic pressure distribution with atmospheric value at the two corresponding top nodes is assigned. The material characteristics of soil-saturation relationships included are assumed with reference to the Pastor and Zienkiewicz (1986) model of the clay core of the San Fernando dam, in accordance with Zienkiewicz et al. (1990).

The saturation distributions are represented over the corresponding deformed configurations in the area close to the slope at different times in Figure 8.23, while in Figure 8.24, the time history of vertical and horizontal displacements of the top node of the slope is reported.

8.5 Dynamic Analysis with a Full Two-Phase Flow Solution of a Partially Saturated Soil Column Subjected to a Step Load

This example, solved by Schrefler and Scotta (2001), deals with the dynamic behavior of a saturated/unsaturated sand column to which a step load is applied. The column is 10 m high and 1 m wide. The material characteristics are summarized in Table 8.1. The dependence of

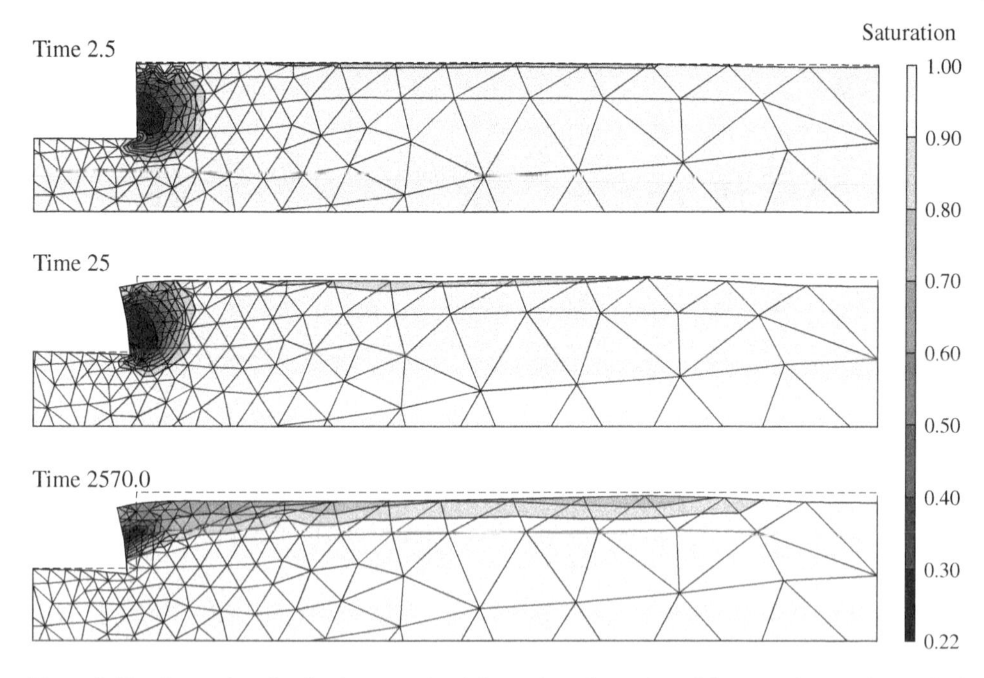

Figure 8.23 Saturation distribution over the deformed configuration of the area close to the vertical slope at three different steps of the analysis. *Source:* Reproduced from Meroi et al. (1995) by permission of John Wiley & Sons Ltd.

water saturation and water permeability on capillary pressure, and the relative permeabilities of gas and water were assumed according to the relationship of Brooks and Corey (1966) with $S_{wc} = 0.2$, $p_{ref} = 0.01$ MPa (equivalent to 1 m of water column), $p_b = 0.001\,68$ MPa and $\lambda = 3$, see Section 8.3.

The following five cases are simulated:

I) Solid skeleton only, with no coupling with the fluid phases ($p_w = p_a = 0$, $S_w = 0$)
II) Single fluid in full saturated condition ($p_a = 0$ and $S_w = 1$)
III) Single fluid in partially saturated condition ($p_a = 0$)
IV) Two fluids in partially saturated condition, sample initially fully saturated ($S_w(t = 0) = 1$)
V) Two fluids in partially saturated condition, sample initially partially saturated ($S_w(t = 0) < 1$)

The comparison of the solutions obtained with the various hypotheses allows appreciating the influence of the initial conditions and the airflow assumptions on the numerical results. In fully coupled two-phase fluid-flow analyses, two strategies are usually adopted to model the transition from fully saturated conditions where only the liquid phase is present to partially saturated conditions where the flow of both liquid and gas takes place (see also Section 8.2). The first strategy consists in a switch from two-phase flow to one-phase flow when saturation goes below a fixed value, in general corresponding to the bubbling

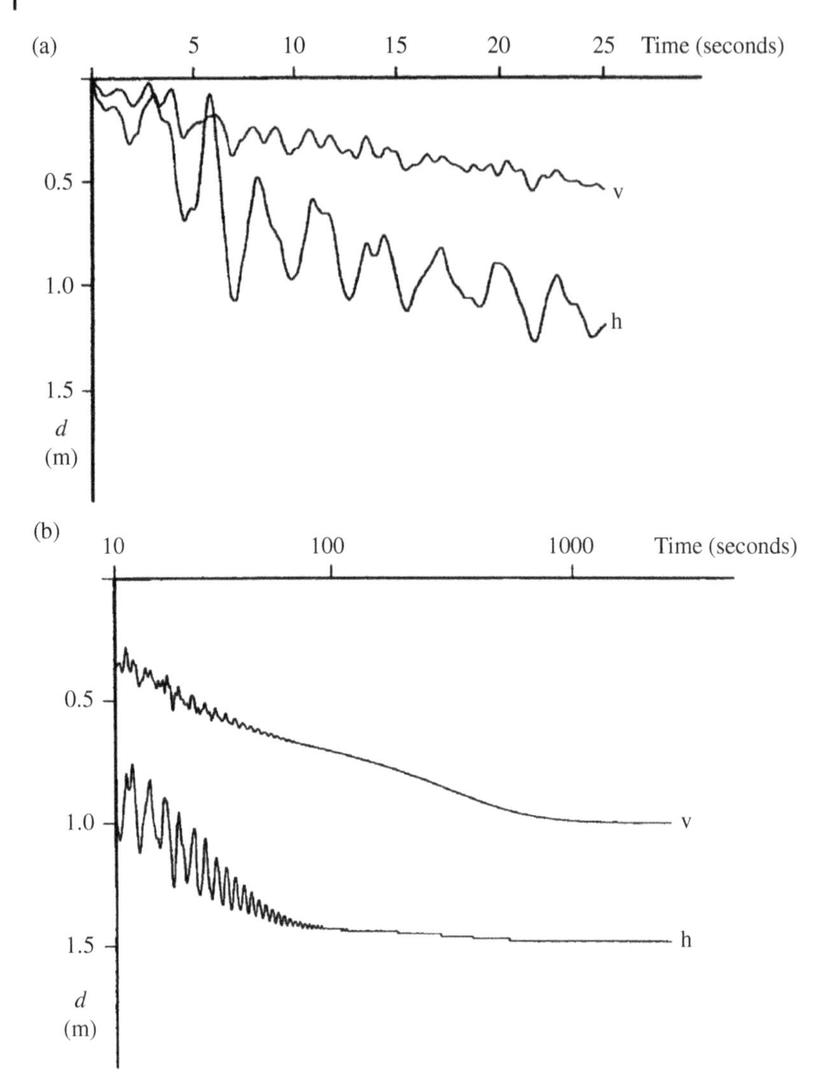

Figure 8.24 Vertical (v) and horizontal (h) displacements versus time during the seismic and consolidation phases of the top point of the slope. (a) Blow up of the time scale. (b) Longer time period. *Source:* Reproduced from Meroi et al. (1995) by permission of John Wiley & Sons Ltd.

pressure. Consequently, in the saturated region, the gas pressure is imposed to be equal to the reference atmospheric value and the gas flow to be null. This procedure, proposed by Gawin et al. (1995) and Gawin and Schrefler (1996), even if the more adherent to the physics of the problem, has the disadvantage of introducing a strong instability in the numerical solution because of the sharp change in the form of the governing equations in the transition zone. In the second strategy, the water saturation has an upper bound close to but always less than 1.0. In this way, a very small, but finite, value of gas-relative permeability

Table 8.1 Material data used in the test of a saturated sand column subjected to a step load.

Young's modulus	$E = 4.5$ MPa
Poisson ratio	$\nu = 0.2$
Solid grain density	$\rho_s = 2000$ kg/m^3
Bulk modulus of solid grains	$K_s = 1.0 \times 10^6$ MPa
Bulk modulus of water	$K_w = 2.0 \times 10^2$ MPa
Bulk modulus of air	$K_a = 0.1$ MPa
Water density	$\rho_w = 1000$ kg/m^3
Air density	$\rho_a = 1.20$ kg/m^3
Initial porosity	$n = 0.30$
Intrinsic permeability	$k = 5 \times 10^{-13}$ m^2
Water viscosity	$\mu_w = 5.0 \times 10^{-4}$ Pa·s
Air viscosity	$\mu_a = 2.4 \times 10^{-5}$ Pa·s
Gravitational acceleration	$g = 9.806$ m/s^2
Atmospheric ref. pressure	$p_{atm} = 0.0$ MPa

results even in a fully saturated state, thus allowing to keep the set of equations unchanged. This approach, used by Schrefler and Zhan (1993), is adopted here because it assures greater numerical stability. A discussion on the differences between these different solution strategies can be found in Gawin et al. (1997).

The initial and boundary conditions assumed are the following:

Initially, the material and the fluids are in static equilibrium under gravitational load. Vertical displacements at the bottom and horizontal displacements at the lateral sides are restricted. All the sides are impermeable to the fluids except the top where air and water pressures are imposed. In all the analyses, at the top surface $p_a = p_w = p_{atm} = 0$ MPa were assumed except in the last case V, where we assumed $p_w = -0.1$ MPa to obtain the initial partially saturated condition.

At time 0, a vertical compressive step surface load of 0.1 MPa is applied at the top of the column and removed at time 1.0 seconds.

The resulting displacement histories are plotted in Figure 8.25 and the free oscillation of the sand column from the first case represents the reference solution: the natural period of the skeleton is of about 0.65 seconds and the oscillation is always around the equilibrium position ($y = 0.2$ m when load is applied, $y = 0$ m without surface load). The velocity of propagation of displacement waves is $v_S = \sqrt{E_S/\rho_S} = 56.7$ m/s as confirmed by the fact that the mid of the column starts to move after about $5.0/56.7 = 0.088$ seconds from the application of the load as shown in Figure 8.25b.

The behavior of the sample during the first second when the applied load induces compression in the column is now examined in detail.

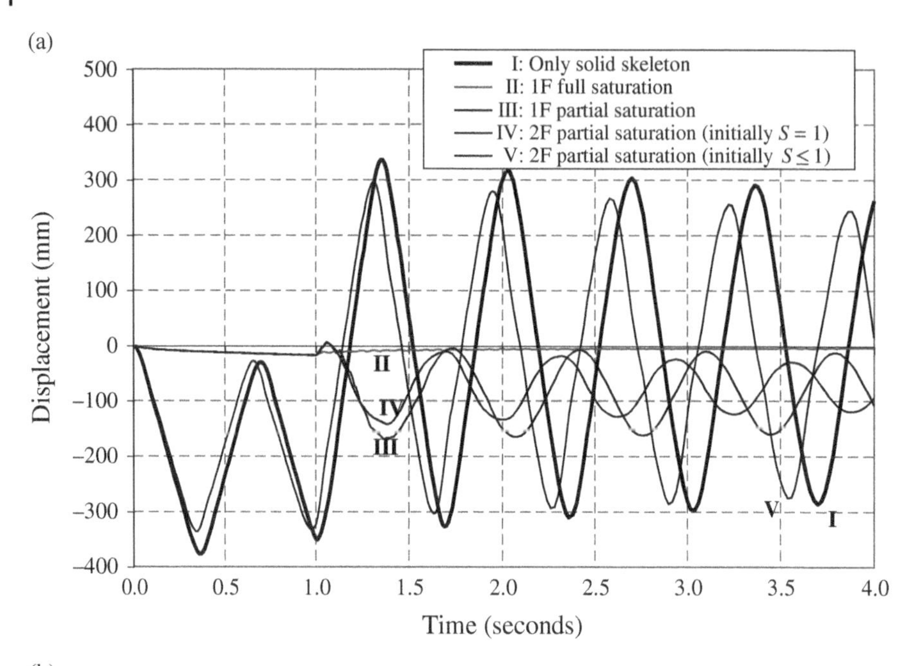

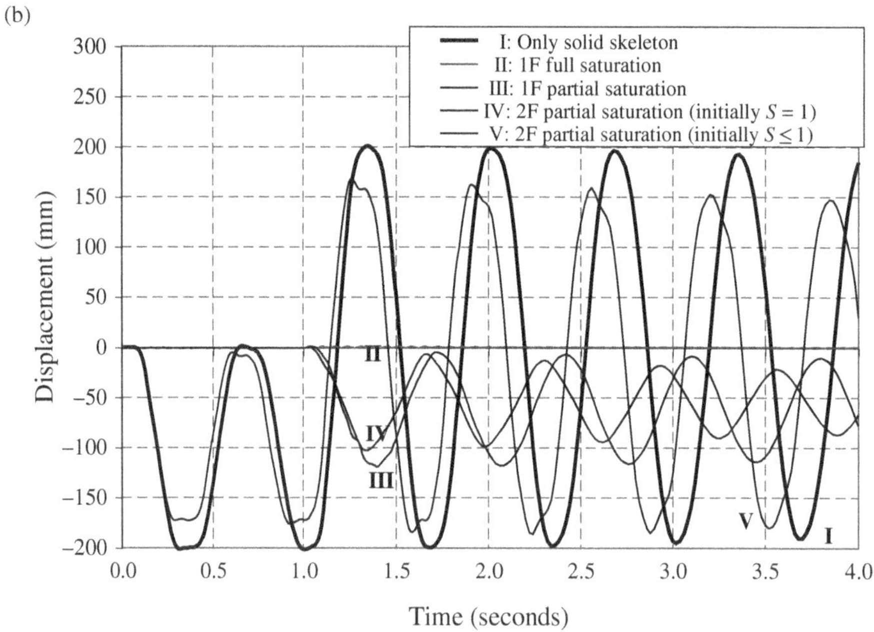

Figure 8.25 Results from the test of a saturated sand column subjected to a step load. Displacement history at (a) top and (b) mid-height of the column. S stands for the degree of saturation S_w.

In analyses II, III, and IV (initial full saturated condition), the behavior is completely different from the previous case: the oscillation amplitude is negligible and the period strongly shortened. The vertical movement of the top surface is due only to the water drainage which produces a slow consolidation of the porous medium, while at the mid-surface, no displacements are registered. In this phase, the compression load is completely absorbed through an increment of the internal pressure of the water (in the cases II and III) or water and air (in case IV). This can be seen in Figure 8.26 where the evolution of water and air pressures at the top, middle, and bottom of the mesh column are shown. Near the top of the column, where, for the imposed boundary conditions, almost free drainage is allowed, the pressures start to decrease already in the first second of analysis, while in the mid and at the bottom of the column, the pressures oscillate around the equilibrium position given by the initial pressure plus the value of the applied surface load. In the first part of the analysis, the sand column, being fully saturated, behaves like a column of incompressible water in which pressure waves propagate at the natural velocity in water.

If initially partially saturated conditions are considered, as in case V, the behavior becomes similar to that of only solid skeleton (I). Only the oscillation frequency is slightly increased and the maximum displacement is reduced because part of the applied total stress is now absorbed by an increase of the internal pressures (see Figure 8.26). The pressure waves propagate from top to bottom with the same delay that has been evidenced for the vertical displacement.

The behavior after 1.0 seconds is now examined, when the applied load is removed. The analyses I (only solid skeleton) and V (two fluids with initially partially saturated condition) give a similar behavior with oscillations around the static equilibrium position $y = 0$ m. Solution II (only water flux with permanent full saturation) coherently still shows a consolidation like behavior with slowly diminishing small displacements. Inversely, solutions III and IV exhibit a peculiar change where oscillations have the same period of the initially unsaturated solutions I and V but have a minor amplitude and the equilibrium position differs from $y = 0$ m. Differences also exist between the two solutions in partially saturated conditions, however they are not too strong.

When the load is suddenly removed ($t = 1$ seconds), the column expands in a very short time span with ensuing negative pore pressures (see Figures 8.25 and 8.26). According to the capillary pressure-saturation relationship adopted, a desaturation follows which is then maintained (Figure 8.27). The sudden decrease of the pore pressure produces a compressive effective stress, hence a shortening of the column, which results then in the observed oscillatory behavior for the displacements. The reference mean value of the displacements about which in cases III and IV, the solid displacements oscillate after removal of the load, is due to the effective stress induced by negative pressures.

In case III, the air is free to flow by assumption, while in case IV, the sudden decrease of saturation produces a sharp increase in air permeability, permitting the necessary inflow of air. This relatively free airflow explains why we do not have strong pressure oscillation ($t > 1$ seconds) even with oscillating solid displacements. The fact that for displacements and pressures solution III (with "passive" air phase) does not differ too much from the solution with the approach adopted here indicates that the assumption of an always present gas phase (with negligible small permeability at fully saturated conditions) is acceptable.

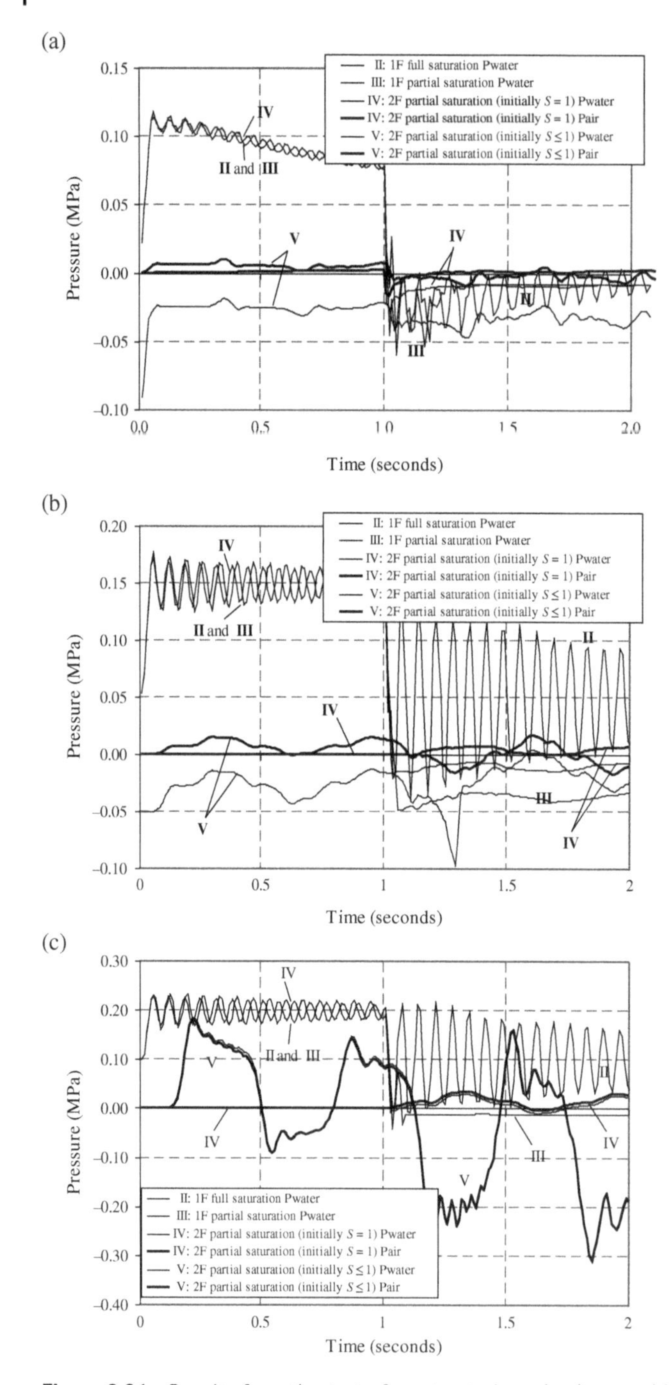

Figure 8.26 Results from the test of a saturated sand column subjected to a step load. Pressures history at (a) near the top, (b) mid-height, and (c) bottom of the column. S stands for the degree of saturation S_w.

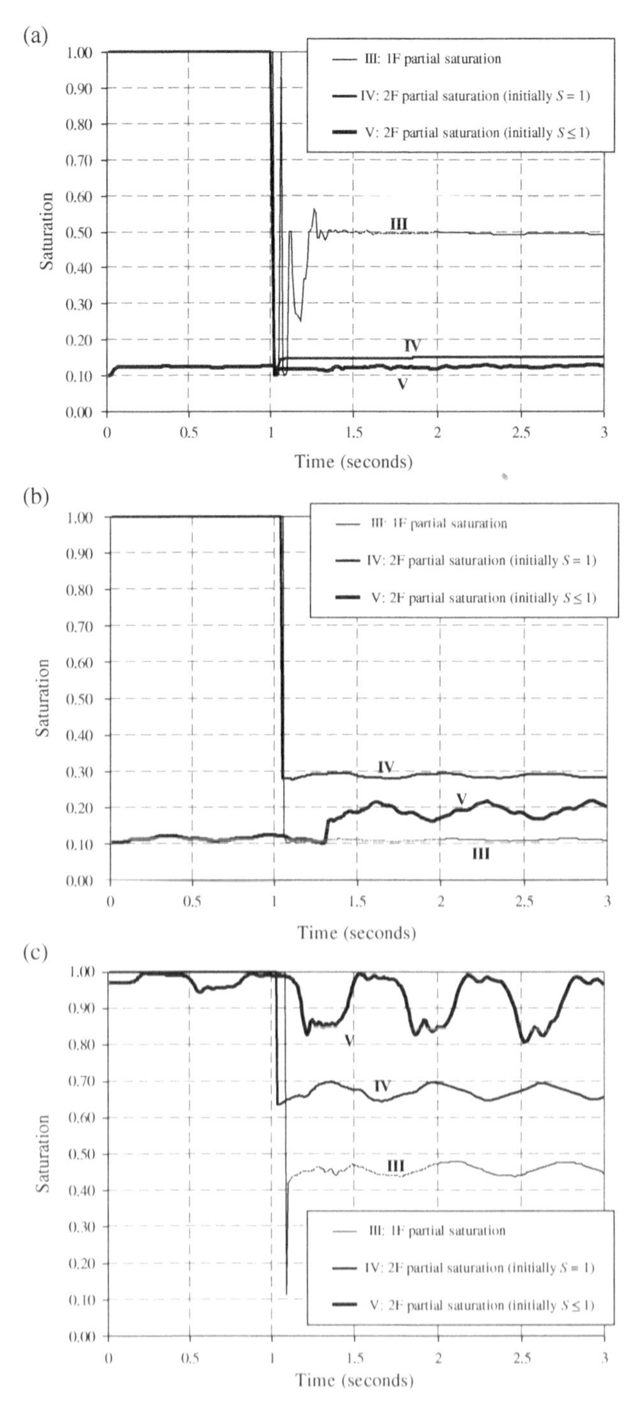

Figure 8.27 Results from the test of a saturated sand column subjected to a step load. Water saturation evolution at (a) near the top, (b) mid-height, and (c) bottom of the column. S stands for the degree of saturation S_w.

8.6 Compaction and Land Subsidence Analysis Related to the Exploitation of Gas Reservoirs[1]

Simulation of compaction and ensuing land subsidence due to the exploitation of gas reservoirs can be found in Lewis and Schrefler (1982), Gambolati et al. (1991), and Schrefler and Delage (2010). During the exploitation of a gas reservoir which generally causes a diminution of the gas pressure, two mechanical effects appear simultaneously:

the decrease of the gas pressure produces compaction in the reservoir formation and acts as a pressure boundary condition on the overburden; the capillary pressure in the reservoir diminishes and, after a small initial elastic expansion, causes in most reservoir rocks irreversible compaction (also called structural collapse). This compaction acts as a displacement boundary condition for the overburden. Both effects produce surface subsidence and can be observed as land lowering which during the production phase has a typical form of a cone, called subsidence bowl. The first effect may be modeled by means of the effective stress concept for two-phase flow (water and gas), Equation (2.69), while the second one requires an appropriate constitutive model such as the generalized plasticity model for unsaturated soil of Section 4.4.7.

In a fully saturated reservoir or aquifer, see Section 6.4.2, compaction stops if the pressure is kept constant and with increasing pressure, even a small rebound may be observed. In case of a gas reservoir which usually is in partially saturated conditions, compaction and ensuing subsidence continues after the end of production even with increasing gas pressure in the formation. This fact may be explained by the capillary pressure decline which continues over long time spans due to encroaching water from the aquifers surrounding the reservoir. This corresponds, for instance, to path CE on Figure 8.30, top, where the yield surface still moves to the right. The observed subsidence after end of production in the case of the reservoir Ravenna Terra, for which data exist (Menin et al. 2008), shows a reversed subsidence bowl, i.e. a hat-like shape as shown in Figure 8.28.

We focus our attention, in particular, at the period after the end of production in an axisymmetric model reservoir with a radius of 2000 m, a thickness of 80 m, at a mean depth of burial of 1840 m. The reservoir is overlaid by a cap rock of 10 m thickness, with the remaining depth comprising sand. An edge aquifer is also present. During the period considered, the reservoir pressure increased by roughly 6 MPa. Hence, a single-phase effective stress-based model, Equation (2.1a), would give a rebound while here subsidence continues (Simoni et al. 1999).

For the simulation, we have used a two-step model. As a first step, by assuming oedometric conditions, we have predicted the compaction at the reservoir center and the reservoir boundary during the production period and the period afterward.

The constitutive model BSZe (Santagiuliana and Schrefler 2006) has been adopted for these simulations. The material properties listed in Table 8.2 have been obtained by

1 The authors gratefully acknowledge the collaboration of A. Menin, V. Salomoni, and R. Santagiuliana in the elaboration of this Section.

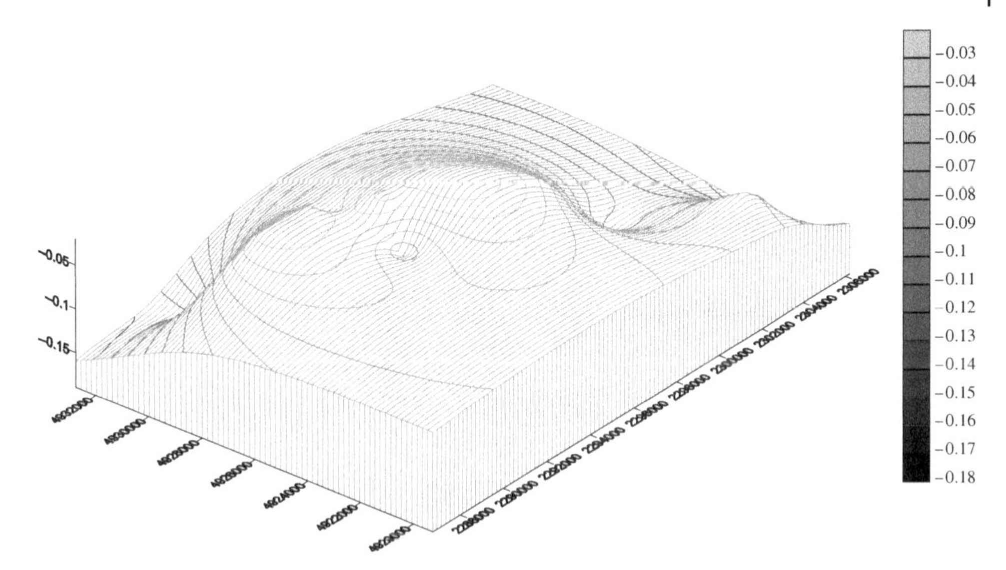

Figure 8.28 3-D rendering of the subsidence above and around the reservoir Ravenna Terra for the period 1982–1998 where reservoir pressure is increasing. The units of all three scales are meters.

Table 8.2 Identified parameters for the constitutive model.

Data	l	m	a	$K_T(0)$	$\lambda(0)$	v_0	i	w
Saturation test Papamichos et al. (1998) and Simoni and Schrefler (2001)	4.816	0.85	2.56	0.0046	0.0296	1.35	1.5	0.1

l, m, and a are material parameters, $\lambda(0)$ and $K_T(0)$ the soil compressibility and the elastic compressibility coefficients in saturated conditions, $v_0(0)$ the initial value of the specific volume, and i and w come from interpolation of experimental data.

parameter identification (Simoni and Schrefler 2001) from experiments carried out by Papamichos et al. (1998) on reservoir rock samples. The initial degree of saturation is $S_w = 35\%$.

The resulting stress path for the reservoir center for the period of gas production with a pressure decline of 8 MPa and constant saturation (Menin et al. 2008) is path AB in Figure 8.29 on the Δv–p' plane, with v the specific volume and p' the mean effective stress. The ensuing volumetric compaction is $\varepsilon_v = (1 - 0.9905)/1.35 = 0.0095/1.35 = 0.70\%$ where 1.35 is the initial value of the specific volume.

For the recovery period which is of interest here, a pressure increase of 6 MPa and a saturation increase of 10% has been assumed for the reservoir center. The LC curve moves from LC_1 to LC_2 due to this increase, with ensuing compaction, path BC and then there is a slight elastic expansion moving from C to D on Figure 8.29, bottom. The ensuing volumetric strain is $\varepsilon_v = (0.0007 - 0.001)/1.35 = -0.02\%$.

At the reservoir boundary, the pressure declines during the production period also by 8 MPa and the saturation increases by 25%, path ABC on Figure 8.30, bottom, with ensuing

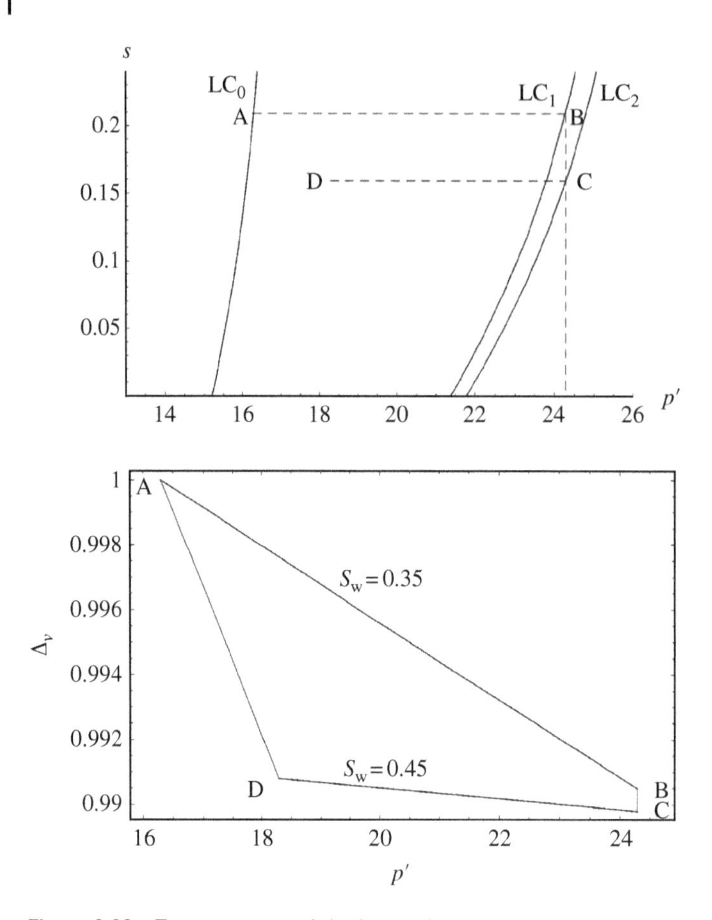

Figure 8.29 Top: – stress path in the suction – mean effective stress (Equation 2.69) plane for the reservoir center; bottom: specific volume versus mean effective stress (Δv, AB = 0.95%, Δv, BC = 0.07%, and Δv, CD = −0.1%).

volumetric compaction of 0.81%. During the recovery period, the pressure is assumed to increase by 6 MPa and the saturation by a further 35%. The resulting volumetric compaction is 0.52% and is shown by path CF in Figure 8.30.

As the second step, we impose the compaction for the recovery period obtained from the volumetric strain at reservoir level on a purely mechanical analysis, with the same geometry and material as above. This corresponds to an equivalent displacement boundary condition also containing the pressure effects in the formation. Linear interpolation has been used for the compaction values between reservoir border and center and a linear decay in the adjacent aquifer, with negligible compaction at a distance of 3 km from the reservoir center. A "hat"-like shape is obtained as shown in Figure 8.31 which is in line with what is experimentally observed (Figure 8.28).

This simplified simulation shows that the used constitutive model and the adopted material parameters can realistically model the observed behavior after exploitation ("hat"-like shape) while conventional subsidence models are not capable of doing so. For the results for the exploitation period, which are also satisfactory, the reader is referred to Menin et al.

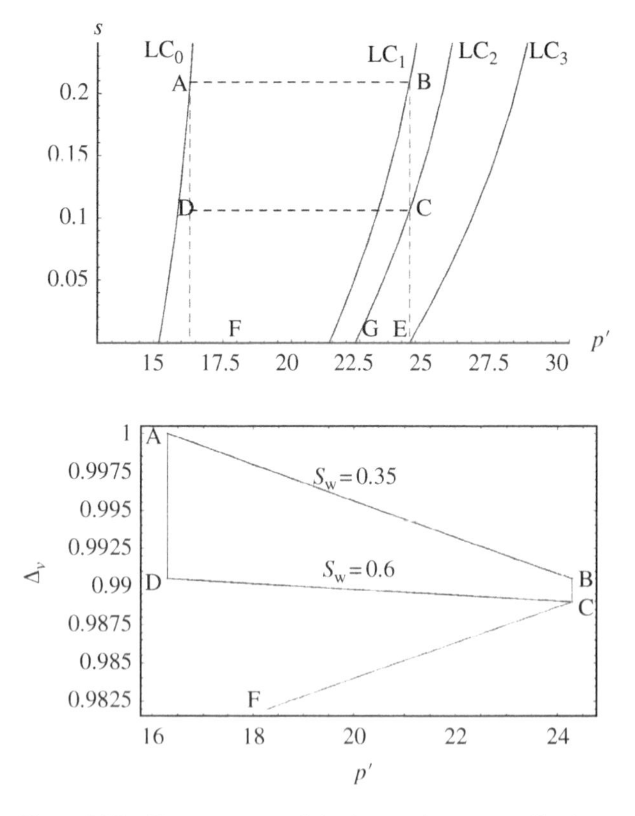

Figure 8.30 Top – stress path in the suction-mean effective stress plane for the reservoir border; bottom – specific volume vs. mean effective stress (Δv, AB = 0.95%, Δv, BC = 0.15%, Δv, CF = 0.7%).

(2008). An alternative to this approach is a more expensive full two-phase flow model for the reservoir and the overburden. This has been adopted in Santagiuliana et al. (2015) on a domain with a radius of 14 km. It yields the whole picture of the analyzed cross section starting from the production data, i.e. the evolution of saturation, pressures, and displacements which are shown in Figure 8.32. As can be seen from a comparison of Figures 8.31 and 8.32, both approaches yield for the recovery phase a similar shape and relative vertical displacement between border and reservoir center.

8.7 Initiation of Landslide in Partially Saturated Soil[2]

In this section, we show that the onset of shallow flow slides can be captured with the procedures presented in Chapters 2 and 3. Since the general problem of landslides will be dealt with more extensively in Chapter 10, we just recall that the onset of flow of the soil–water mixture is triggered by diffuse instabilities while slides are triggered by localized failure

2 The authors gratefully acknowledge the collaboration of R. Ngaradoumbe Nanhorngué, E. Kakogiannou, and L. Sanavia in the elaboration of this section.

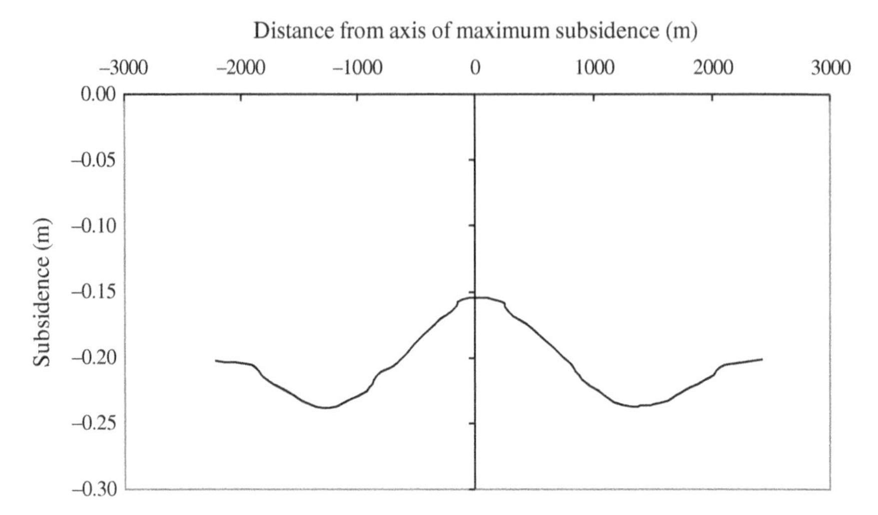

Figure 8.31 Inverse subsidence bowl for the recovery phase obtained by numerical modeling.

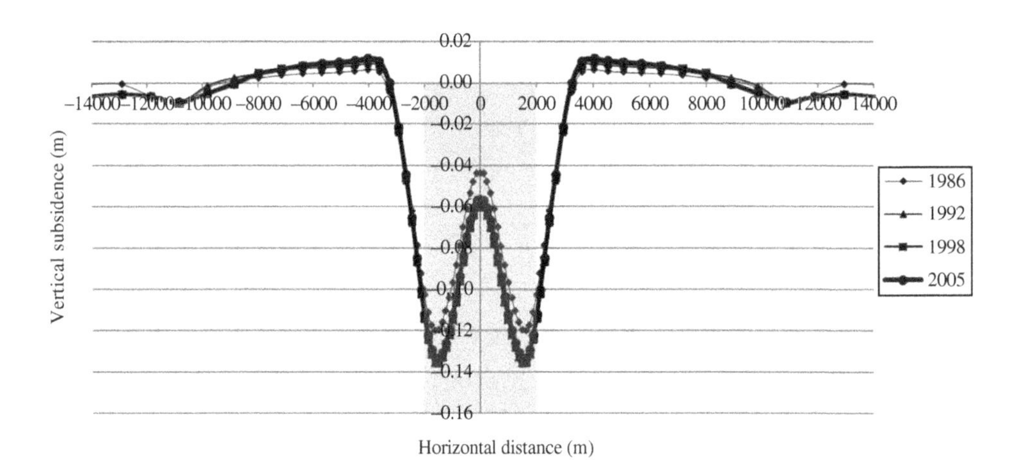

Figure 8.32 Vertical subsidence simulated along the reservoir diameter for different periods always starting from 1982. The grey zone corresponds to the domain of the previous simulation. *Source:* Redrawn with permission from Santagiuliana et al. (2015), Copyright Emerald.

involving failure surfaces and shear bands. This second aspect has been addressed in Section 5.4.1. The post-failure behavior, i.e. the actual flow of the soil mass requires, however, a different approach which will be outlined in Chapter 10. We show here, in particular, the numerical analysis of the initiation of a flow slide in pyroclastic soils which took place in May 1998 in the southern part of Italy and was due to an increase in water pressure induced by heavy and sustained rainfall. Significant examples of flow slides occur periodically in the Campania Region (South Italy) triggered by critical rainfall events. They involve shallow slips on slopes of unsaturated pyroclastic soils. These soils were originated by the explosive phases of the Somma-Vesuvio volcano which covered the limestone and tuffaceous slopes

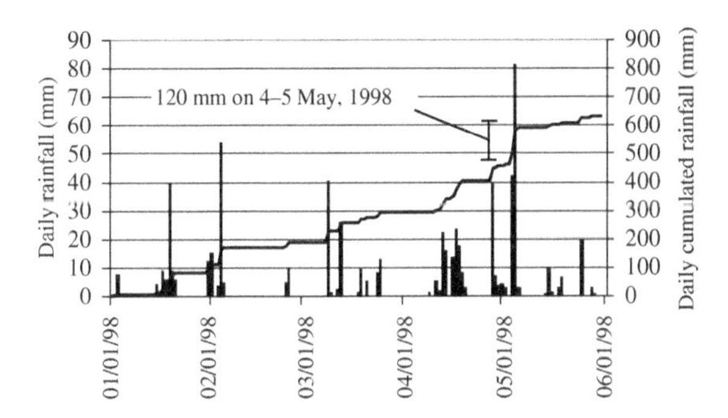

28 April: 40 mm /24 hours
29 April: 8 mm /24 hours
30 April, 1–2 May: 5 mm /24 hours
3 May: **0 mm** /24 hours
4 May: 40 mm /24 hours
5 May: 80 mm /24 hours
6 May: 0 mm /24 hours

Figure 8.33 Rainfall recorded at the toe of Pizzo d'Alvano massif. *Source:* Based on Cascini et al. (2005).

over an area of about 3000 km^2. In this area, one of the most calamitous events occurred on 5–6 May 1998, when 160 casualties and serious damages were recorded in four small towns (Bracigliano, Quindici, Sarno, and Siano) located at the toe of the Pizzo d'Alvano massif. Rain gauges located at the toe of this massif recorded for the period 28 April to 5 May 1998, a cumulative rainfall of 183 mm, of which 120 mm fell in the last two days (Figure 8.33). During and soon after these events, within an interval of about ten hours, more than 100 flow slides occurred in almost all the basins of the massif. A survey conducted following their occurrence revealed that triggering zones were characterized by an average slope angle varying from 30 to 41° and a soil thickness ranging from 0.5 to 5.0 m. As pointed out by Klubertanz et al. (2009), in case of heavy rainfall events, a flow regime parallel to the surface installs itself in the upper partially saturated layers and carries water to the toe region. This increases the degree of saturation and reduces the capillary pressure between soil grains even up to full saturation in the toe zone. A weakening of the slope ensues. Further, due to the flow inside the soil layers, a destabilizing downhill frictional drag may develop. Hence, the key factors involved in the onset of shallow landslides are water flow through the solid matrix, soil water retention behavior, and the effect of matric suction on the mechanical behavior of the skeleton (Eichenberger et al. 2011).

The modeled slope section is shown in Figure 8.34 together with the stratigraphy (Cascini et al. 2005). The pyroclastic soil cover is 400 m in length and is characterized by the presence of several thin layers of pumice and upper and lower ashy layers, for a total thickness varying between 2 and 5 m.

Partially saturated soil behavior, experimentally observed, small strains and quasi-static loading conditions are assumed in the simulation. The solid skeleton is elastoplastic, homogeneous, and isotropic; its mechanical behavior is described within the classical rate-independent elastoplasticity theory for geometrically linear problems. The yield function restricting the effective stress state $\boldsymbol{\sigma}'(\boldsymbol{x},t)$ is of the Drucker and Prager (1952) type (see Section 4.2.3.4) with linear isotropic softening and nonassociated plastic flow to take into account the post-peak and dilatant/contractant behavior of the slope materials. The cohesion c is suction-dependent following Fredlund et al. (1978)

$$c = c_0 + p_c \tan \varphi_b' \tag{8.8}$$

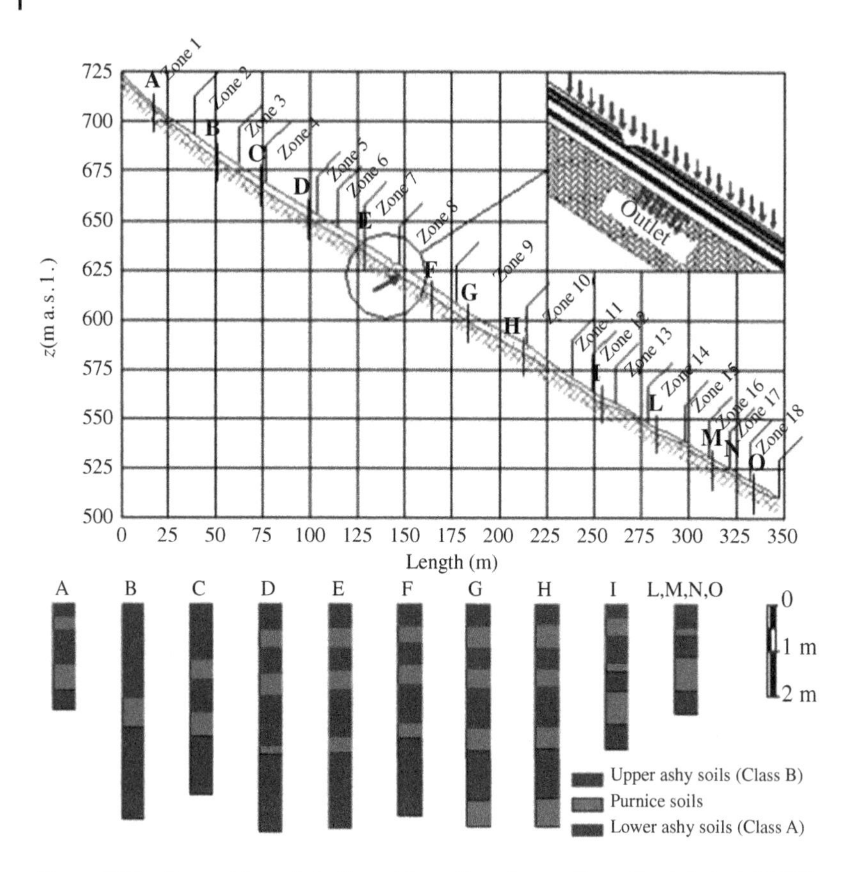

Figure 8.34 Geometric and stratigraphic section. *Source:* From Cascini et al. (2005).

where p_c is the capillary pressure and φ'_b is the shear resistance angle with respect to suction.

The return mapping algorithm and the consistent tangent operator for this model have been developed by Sanavia et al. (2006), solving the singular behavior of the Drucker–Prager yield surface in the apex zone by means of multi-surface plasticity.

The hydromechanical properties of the three types of layers mentioned above have been determined by laboratory tests (Cascini et al. 2005) and are summarized in Table 8.3. The capillary pressure–water saturation relationship and the water-relative permeability used in the numerical simulation are obtained by interpolating the experimental values of Figure 8.35 (case 1).

We have analyzed the time period from 28 April until 6 May, loading the upper surface of the slope with a double rainfall compared to that measured at the toe of Pizzo d'Alvano massif (Figure 8.33) because the analyzed slope is sited in the upper part of the massif, where regularly higher rainfall occurs (Cascini et al. 2005).

The slope has been modeled assuming plain strain conditions and is discretized with 480 8-node elements (Figure 8.36). The developed mesh is rather coarse, with very long

Table 8.3 Geotechnical properties for soil layers (Cascini et al. 2003).

	Lower ashy	Upper ashy	Pumice
Young mod. (Pa)	5E + 05	4E + 05	2E + 05
Poisson ratio	0.316	0.294	0.30
Solid grain density (kg/m^3)	2393	2169	2039
Porosity	0.664	0.584	0.69
Hydraulic conductivity (S_w = 1) (m/s)	1E−6	1E−5	1E−4
Gravity accel. (m/s^2)	−9.81	−9.81	−9.81
Apparent cohesion (S_w = 1) (Pa)	4.7E+03	4.7E+03	3.3E+02
Hardening modulus (Pa)	6.5E+04	8.0E+04	1.1E+04
Friction angle	32°	36°	37°
Dilatant angle	−4.75°	−5.5°	5°
Shear angle	25°	25°	25°

Source: Based on Cascini et al. (2005).

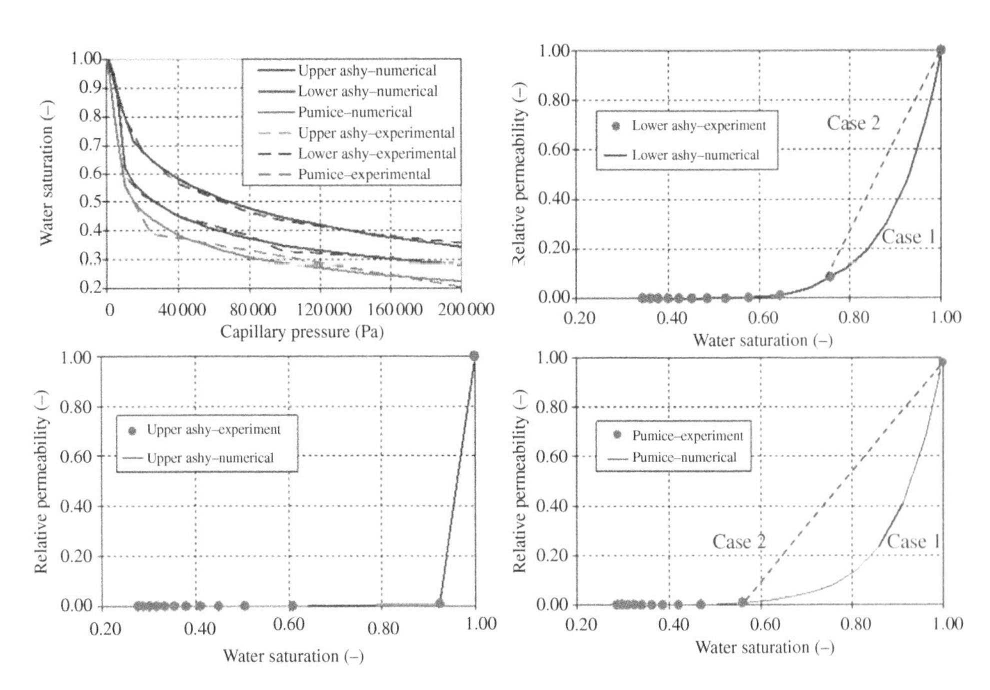

Figure 8.35 Capillary pressure–water saturation relationship and relative permeability of pyroclastic soils of Pizzo d'Alvano.

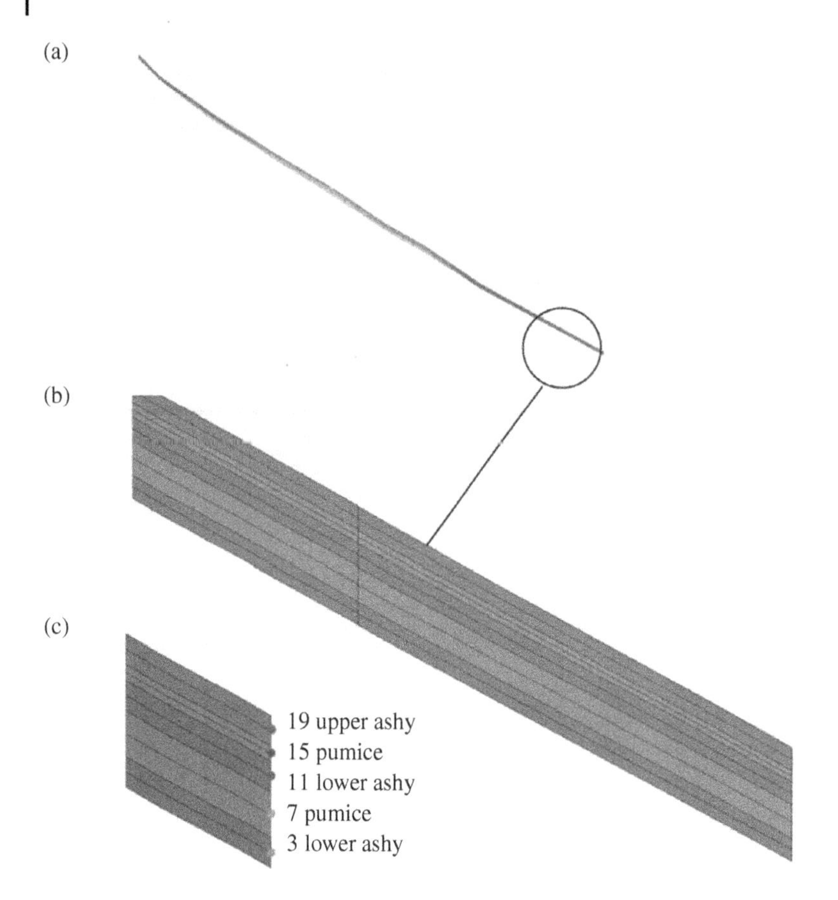

Figure 8.36 (a) F.E. mesh (2D – plain strain, 8 node element, 1565 nodes, 480 F.E.); (b) Stratigraphic section (basis of the slope); (c) Node location for plotting graphs (lower part of the slope).

elements, which have been adopted because the water flux occurs mainly along the longitudinal direction. This allowed minimizing the computational cost.

First, we have computed the stress state in equilibrium with an assumed initial uniform capillary pressure of 10 kPa and uniform gas pressure at atmospheric value due to the small thickness of the slope.

As expected, the results of the simulation when landslides occurred (6 May) show that the displacements are concentrated in the lower part of the slope, with a maximum value of 1.95 m (Figure 8.37). From the time history of the displacements norm of some nodes in the final part of the slope in Figure 8.38, one can observe that the onset of the landslide has been captured by the multiphase model because a very high displacement rate is obtained in the last two days of the period analyzed (5 and 6 May), in agreement with what observed in situ.

The numerical results reveal that collapse is due to the accumulation of the equivalent plastic strain in the lower part of the section (Figure 8.39) because the materials become fully saturated (Figure 8.40). In particular, plastic strain reaches higher values within the

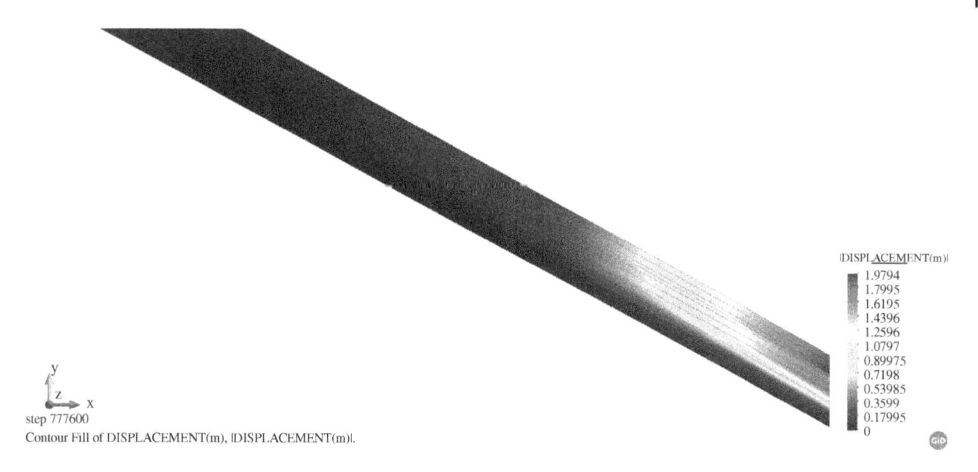

Figure 8.37 Displacement contours at the end of 6 May (lower part of the slope).

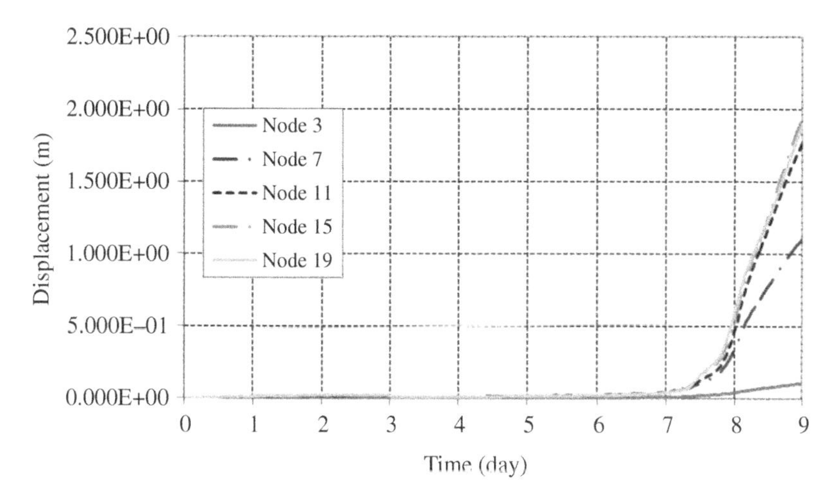

Figure 8.38 Displacement norm time history at the toe.

pumice layers, which are the less resistant and more permeable ones (Figure 8.40) and saturate even if dilatant (Table 8.3, Figure 8.41).

A closer look at the time histories of the displacement norms (Figure 8.38), equivalent plastic strain (Figure 8.42), water pressure (Figure 8.43), mean effective stress (Figure 8.44), and shear stress (Figure 8.45) in the lower part of the slope (Figure 8.36c) shows that the displacement rate increases sharply only when the plastic strains develop. This increase indicates the occurrence of landslide and is due to the increment of water pressure: a higher rate of increase in water pressure is, in fact, observed during the development of plasticity. At the same time, we also observe a decrease of the mean effective pressure up to quasi zero, which corresponds to a quasi-zero volumetric resistance of the material.

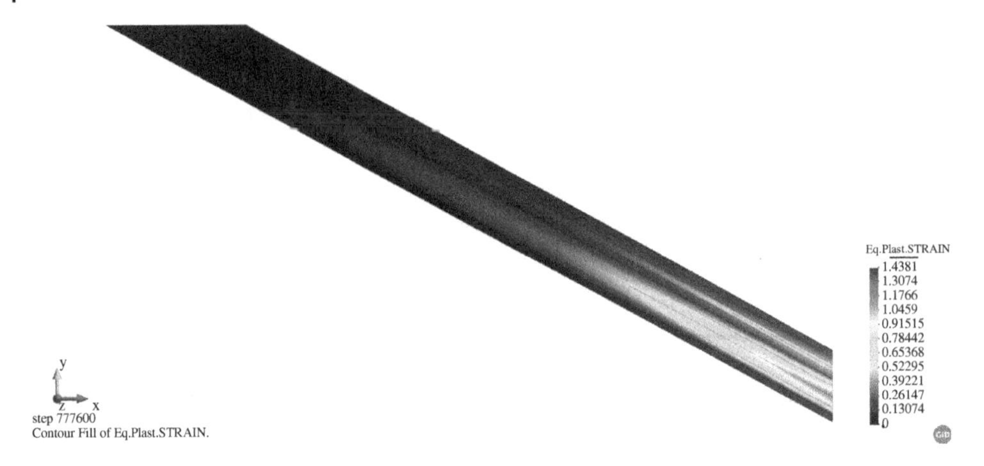

Figure 8.39 Equivalent plastic strain at the end of 6 May (lower part of the slope).

Figure 8.40 Saturation at the end of 6 May (lower part of the slope).

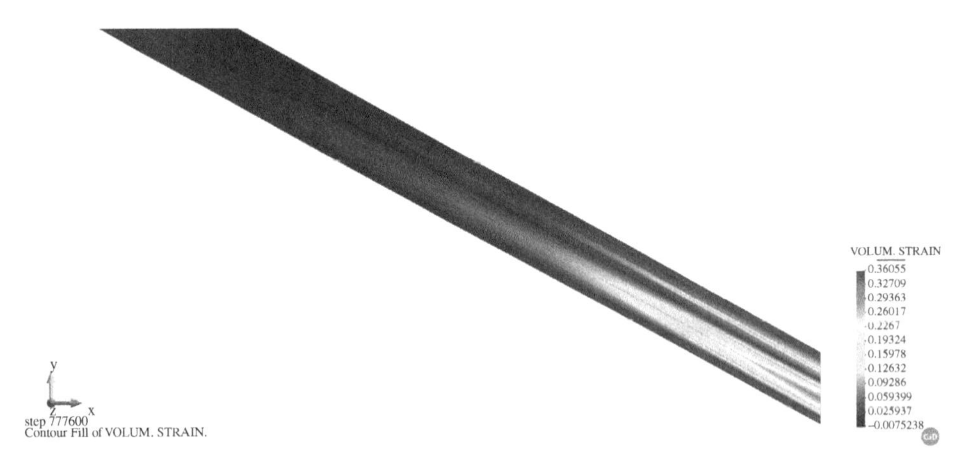

Figure 8.41 Equivalent volumetric strain at the end of 6 May (lower part of the slope).

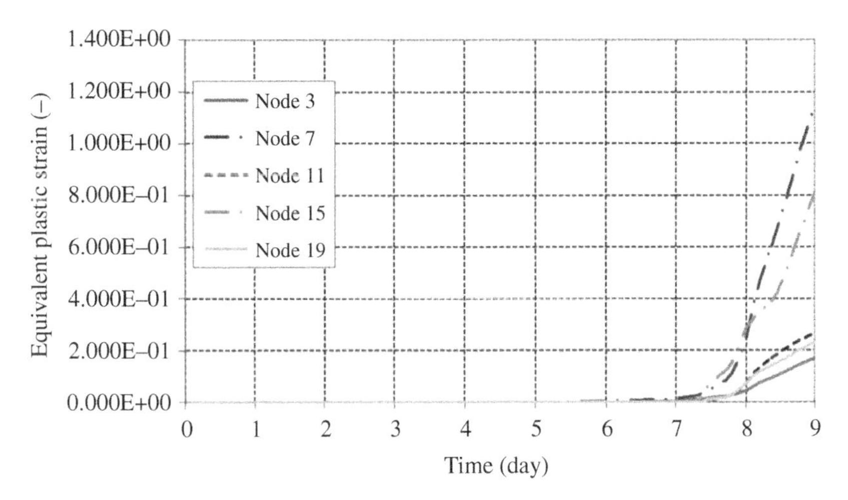

Figure 8.42 Equivalent plastic strain history of the toe.

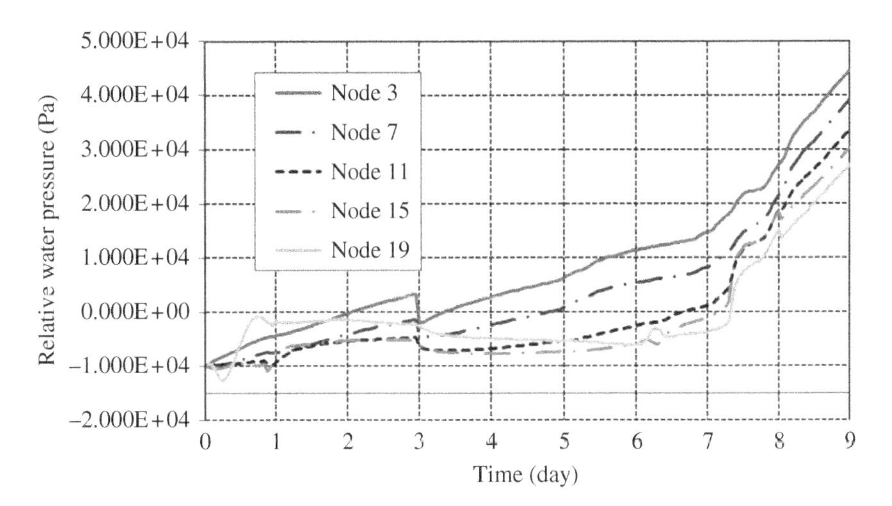

Figure 8.43 Water pressure time history at the toe.

A jump in the pressure as e.g. for nodes 3 and 19 in Figure 8.43 has also been observed experimentally by Klubertanz (1999). These simulations show that the first failure of the lower part of the slope is captured, as observed in situ, involving mainly isotropic plastic deformation in the first stage of the sliding followed by the development of deviatoric plastic strain caused by the continuous increase of the displacements. Once the landslide started, this model is no longer valid and a switch to the model of Chapter 10 is necessary.

The onset of the flow slide can also be described by the amplitude of the unstable zone at the material-point level by applying the second-order work criterion based on Hill's sufficient condition of stability (Hill 1958). This criterion consists in computing the sign of the second-order work for each material point of the domain, enabling to judge the potential

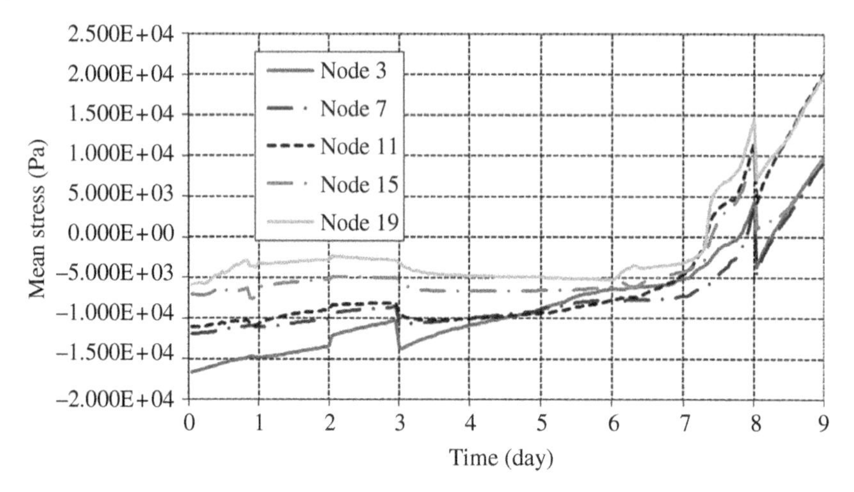

Figure 8.44 Mean effective stress at the toe.

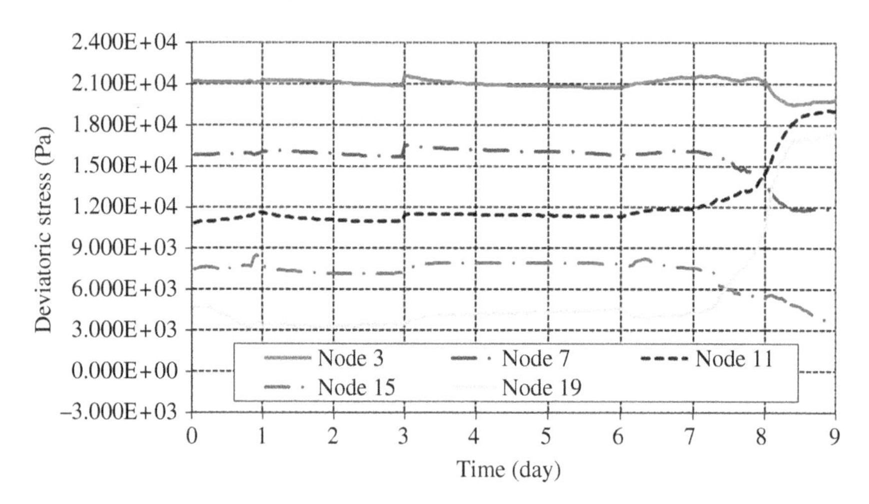

Figure 8.45 Mean shear stress at the toe.

instability of a spatial domain in which a set of local negative values occurs and giving information for the detection of the onset of the failure (e.g. Laouafa and Darve 2002; Nicot et al. 2007).

For the slope studied in this section, it is shown in Kakogiannou et al. (2016) that the set of the local negative values of the second-order work gives a good indication of the spatial extent of the potentially unstable domains (Figure 8.46, in which the contour of negative values of the second-order work in terms of effective stresses, $W_2 = d\sigma' : d\varepsilon$, is plotted at the beginning of 6 May) and is consistent with the extent of the plastic zones (Figure 8.39) and the zone with high displacements (Figure 8.37).

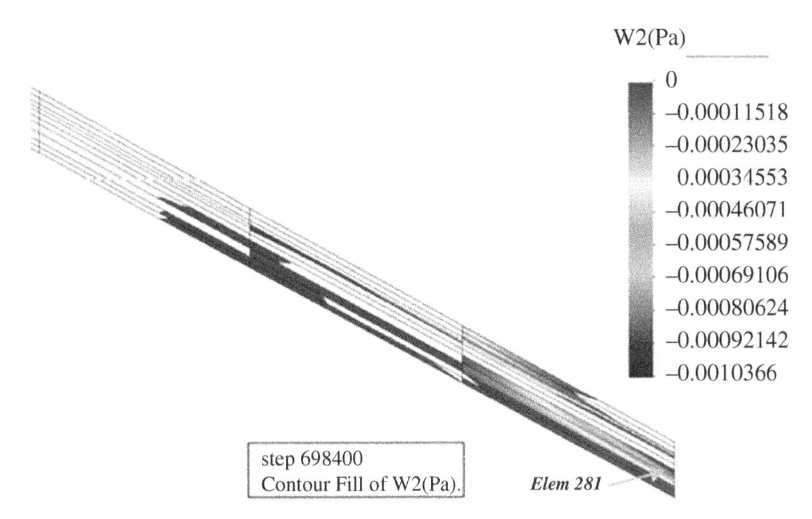

Figure 8.46 Negative values of the second-order work contours in terms of effective stresses at the beginning of 6 May (lower part of the slope). *Source:* Redrawn with permission from Kakogiannou et al. (2016) Copyright Springer.

8.8 Conclusion

The formulation developed in Chapter 2 and discretized in Chapter 3 is used to analyze various one-dimensional and two-dimensional problems in partially saturated conditions. The results obtained from these analyses are highly satisfactory and compared well with available numerical and experimental solutions.

References

Advani, S. H., Lee, T. S., Lee, J. H. W. and Kim, C. S. (1993). Hygrothermomechanical evaluation of porous media under finite deformation. Part I-Finite element formulations, *Int. J. Num. Meth. Eng.*, **36**, 147–160.

Bathe, K. J. and Ozdemir, H. (1976). Elastic-plastic large deformation static and dynamic analysis, *Comp. Struct.*, **6**, 81–92.

Bathe, K. J., Ramm, E. and Wilson, E. L. (1975). Finite element formulations for large-deformation dynamic analysis, *Int. J. Num. Meth. Eng.*, **9**, 353–386.

Booker, J. R. (1974). The consolidation of a finite layer subject to surface loading, *Int. J. Solids Struct.*, **10**, 1053–1065.

Brooks, R. N. and Corey, A. T. (1966). Properties of porous media affecting fluid flow, *ASCE IR*, **92**, IR2, 61–68.

Carter, J. P., Booker, J. R. and Small, J. C. (1979). The analysis of finite elasto-plastic consolidation, *Int. J. Num. Anal. Geomech.*, **3**, 107–129.

Cascini, L., Sorbino, G., and Cuomo, S. (2003). Modelling of flowslides triggering in pyroclastic soils, in *Proceedings of the International Conference on "Fast Slope Movements, Prediction and Prevention for Risk Mitigation"*, L. Picarelli (Ed), 11–14 May 2003 Napoli, Patron Editore, ISBN 88-555-2699-5, v. 1, pp. 93–100.

Cascini, L., Cuomo, S. and Sorbino, G. (2005). Flow-like mass movements in pyroclastic soils: remarks on the modelling of triggering mechanisms, *Ital. Geotech. J. (RIG)*, **4**, 11–31

Chen, W. F. and Mizuno, E. (1990). *Nonlinear Analysis in Soil Mechanics-Theory and Implementation*, Elsevier, Amsterdam.

Drucker, D. C. and Prager, W. (1952). Soil mechanics and plastic analysis or limit design, *Quart. Appl. Math.*, **10**, 157–165.

Eichenberger, J., Nuth, M., and Laloui, L. (2011). Modeling the onset of shallow landslides in partially saturated slopes subjected to rain infiltration. *Geo-Frontiers 2011* (13–16 March 2011), ASCE.

Fredlund, D. G., Morgenstern, N. R. and Widger, R. A. (1978) The shear strength of unsaturated soils, *Can. Geotech. J.*, **15**, 313–321.

Gambolati, G. Ricceri, G., Bertoni, W. Brighenti, G. and Vuillermin, E. (1991). Mathematical simulation of the subsidence of Ravenna, *Water Resour. Res.*, **27**, 2899–2918.

Gawin, D. and Schrefler, B. A. (1996). Thermo-hydro-mechanical analysis of partially saturated porous materials, *Eng. Comput.*, **13**, 7, 113–143.

Gawin, D., Baggio, P. and Schrefler, B. A. (1995). Coupled heat, water and gas flow in deformable porous media, *Int. J. Num. Meth. Fluids.*, **20**, 969–987.

Gawin, D., Simoni, L. and Schrefler, B. A. (1997). Numerical model for hydro-mechanical behaviour in deformable porous media: a benchmark problem, in *Computer Methods and Advance in Geomechanics*, M. W. Yuan (Ed), Balkema, Rotterdam, 1143–1148.

Heyliger, P. R. and Reddy, J. N. (1988). On a mixed finite element model for large deformation analysis of elastic solids, *Int. J. Non Linear Mech.*, **23**, 131–145.

Hill, R. (1958). A general theory of uniqueness and stability in elastic-plastic solids, *J. Mech. Phys. Solids*, **6**, 239–249.

Kakogiannou, E., Sanavia, L., Nicot, F., Darve, F. and Schrefler, B. A. (2016). A porous media finite element approach for soil instability including the second-order work criterion, *Acta Geotech.*, **11**, 805–825.

Kim, C. S., Lee, T. S., Advani, S. H. and Lee, J. H. W. (1993). Hygrothermomechanical evaluation of porous media under finite deformation: Part II – Model validations and field simulations, *Int. J. Num. Methods Eng.*, **36**, 161–179.

Klubertanz, G. (1999). Couplage hydro-mécanique dans les mileux triphasiques avec application au déclenchement des laves torrentielles. Ph.D. Thesis. No 2027, EPFL, Lausanne.

Klubertanz, G., Laloui, L. and Vulliet, L. (2009). Identification of mechanisms for landslide type initiation of debris flows. *Eng. Geol.*, **109**, 1-2, 114–123.

Laouafa, F. and Darve, F. (2002). Modelling of slope failure by a material instability mechanism. *Comput. Geotech.* **29**, 301–325.

Lewis, R. W. and Schrefler, B. A. (1982). A finite element simulation of the subsidence of gas reservoirs undergoing a waterdrive, in *Finite Element in Fluids*, **4**, R. H. Gallagher, D. H. Norrie, J. T. Oden and O. C. Zienkiewicz (Eds), Wiley, 179–199.

Lewis, R. W. and Schrefler, B. A (1998). *The Finite Element Method in the Static and Dynamic Deformation and Consolidation of Porous Media*, Wiley, Chichester.

Li, C., Borja, R. I., and Regueiro, R. A. (2004). Dynamics of porous media at finite strain, *Comput. Methods Appl. Mech. Eng.*, **193**, (36-3), 3837–3870.

Liakopoulos, A. C. (1965). Transient flow through unsaturated porous media, D. Eng. dissertation, University of California, Berkeley, USA.

Meijer, K. L. (1984). Comparison of finite and infinitesimal strain consolidation by numerical experiments, *Int. J. Num. Anal. Geomech.*, **8**, 531–548.

Meiri, D. and Karadi, G. M. (1982). Simulation of air storage aquifer by finite element model, *Int. J. Num. Anal. Geomech.*, **6**, 339–351.

Menin, A., Salomoni, V. A., Santagiuliana, R., Simoni, L., Gens, A. and Schrefler, B. A. (2008). A mechanism contributing to subsidence above gas reservoirs and an application to a case study, *Int. J. Comput. Methods Eng. Sci. Mech.*, **9**, 270–287.

Meroi, E. A., Schrefler, B. A. and Zienkiewicz, O. C. (1995). Large strain static and dynamic semisaturated soil behaviour, *Int. J. Num. Anal. Geomech.*, **19**, 81–106.

Monte, J. L. and Krizen, R. J. (1976) One-dimensional mathematical model for large-strain consolidation, *Géotechnique*, **26**, 495–510.

Narasimhan, T. N. and Witherspoon, P. A. (1978). Numerical model for saturated-unsaturated flow in deformable porous media 3. Applications, *Water Resour. Res.*, **14**, 1017–1034.

Nicot, F., Darve, F. and Khoa, H. D. V. (2007). Bifurcation and second-order work in geomaterials, *Int. J. Numer. Anal. Methods Geomech.*, **31**, 1007–1032.

Papamichos, E., Brignoli, M. and Schei, G. (1998). Compaction in soft weak rocks due to water flooding. *SPE/ISRM* 47389 (8 July 1998), Trondheim.

Pastor, M. and Zienkiewicz, O. C. (1986). A generalised plasticity hierarchical model for sand under monotonic and cyclic loading. *NUMOG II*, Ghent (April), 131–150.

Prévost, J. H. (1981). Consolidation of anelastic porous media, *ASCE EM*, **107**, 169–186.

Prévost, J. H. (1982). Nonlinear transient phenomena in saturated porous media, *Comp. Meth. Appl. Mech. Eng.*, **30**, 3–18.

Prévost, J. H. (1984). Non-linear transient phenomena in soil media, in *Mechanics of Engineering Materials*, C. S. Desai and R. H. Gallagher (Eds), Wiley, Chichester 515–533, Chapter 26.

Sanavia, L., Schrefler, B. A. and Steinmann, P. (2002). A formulation for an unsaturated porous medium undergoing large inelastic strains, *Comput. Mech.*, **28**, 2, 137–151.

Sanavia, L., Pesavento, F. and Schrefler, B. A. (2006). Finite element analysis of non-isothermal multiphase geomaterials with application to strain localisation simulation, *Comput. Mech.*, **37**, 331–348.

Santagiuliana, R. and Schrefler, B. A. (2006). Enhancing the Bolzon-Schrefler-Zienkiewicz constitutive model for partially saturated soil, *Transp. Porous Media*, **65**, 1–30.

Santagiuliana, R., Fabris, M. and Schrefler, B. A. (2015). Subsidence above depleted gas fields, *Eng. Comput.*, **32**, 3, 863–884, http://dx.doi.org/10.1108/EC-12-2013-0308.

Schrefler, B. and Delage, P. (Eds) (2010). *Environmental Geomechanics*, ISTE, Wiley.

Schrefler, B. A. and Scotta, R. (2001). A fully coupled dynamic model for two-phase fluid flow in deformable porous media, *Comput. Methods Appl. Mech. Eng.*, **190**, 3223–3246.

Schrefler, B. A. and Simoni, L. (1988). A Unified approach to the analysis of saturated – unsaturated elastoplastic porous media, in *Numerical Methods in Geomechanics*, G. Swoboda (Ed), A.A. Balkema, 205–212 Innsbruck, Rotterdam.

Schrefler, B. A. and Zhan, X. (1993). A fully coupled model for waterflow and airflow in deformable porous media, *Water Resour. Res.*, **29**, 1, 155–167.

Shantaram, D., Owen, D. R. J. and Zienkiewicz, O. C. (1976). Dynamic transient behaviour of two-and three-dimensional structures including plasticity, large deformation effects and fluid interaction, *Earthq. Eng. Struct. Dyn.*, **4**, 561–578.

Simoni, L. and Schrefler, B. A. (2001). Parameters identification for a suction dependent plasticity model, *Int. J. Num. Anal. Methods Geomech.*, **25**, 273–288.

Simoni, L., Salomoni, V. and Schrefler, B. A. (1999). Elastoplastic subsidence models with and without capillary effects, *Comp. Methods Appl. Mech. Eng.*, **171**, 491–502.

Terzaghi, K. (1943). *Theoretical Soil Mechanics*, Wiley, New York.

Uzuoka, R. and Borja, R. I. (2012). Dynamics of unsaturated poroelastic solids at finite strain. *Int. J. Numer. Anal. Methods Geomech.*, DOI: 10.1002/nag.1061

Xie, Y. M. (1990). Finite element solution and adaptive analysis for static and dynamic problems of saturated-unsaturated porous media. Ph.D. Dissertation, University College of Swansea, Wales.

Zienkiewicz, O. C., Xie, Y. M., Schrefler, B. A., Ledesma, A. and Bicanic, N. (1990). Static and dynamic behaviour of soils: a rational approach to quantitative solutions. Part II: Semi-saturated problems, *Proc. Roy. Soc. Lond.*, **A429**, 310–323.

9

Prediction Application and Back Analysis to Earthquake Engineering

Basic Concepts, Seismic Input, Frequency, and Time Domain Analysis

9.1 Introduction

This chapter presents some interesting case studies of real earthquake engineering problems, most of them in Japan. Here, the focus is on practical applications and special attention is paid to describing the seismic input. The applications involve both standard methods also showing their shortcomings and the method outlined in this book.

Earthquake response analyses are mainly performed for three practical purposes; i.e. prediction for plans of disaster prevention, seismic design, and prediction/back analysis. The choice of the method used largely depends on the purpose accordingly. A quick and simple calculation is necessary for disaster prevention purpose. In most cases, little information of the site condition and underlying materials is available. For example, thousands of analyses would be needed to be performed to produce the liquefaction map of Tokyo. A resolution of a 250 m grid on the ground surface may be needed. More detailed analyses are required for structural design purpose and engineering designers should be familiar with the basic principles and the limitations of the numerical method used. Engineers still tend to employ well-established methods such as the equivalent linear analysis as the ones used in the program SHAKE (Schnabel et al. 1972). However, the method outlined in Chapters 2 and 3 can give much more insight. The same is true for simulation and research purpose, where the method presented in this book gives much more accurate, detailed, and valuable information. Many examples of such analyses have been published by different research groups using the software derived from the method, e.g. Liu and Song (2005, 2006), Aydingun and Adalier (2003), Madabhushi and Zeng (1998a, 1998b, 2006, 2007), Liu and Won (2009), Ling et al. (2004, 2005, 2008), Liu and Ling (2007), Haigh et al. (2005), Ghosh et al. (2005), Madabhushi and Chandrasekaran (2006), Elia and Rouainia (2010), Elia et al. (2010, 2011), Ye et al. (2012, 2013, 2017), and Dunn et al. (2006).

In order to solve practical problems, it is important to obtain as much actual information about the soil layers, material properties, and structures as possible before the numerical modeling. The soil layers can be idealized as one-, two-, or three-dimensional model with half-space boundary condition as seen in the previous chapters. The idealization of material properties is one of the most controversial issues which we shall discuss more in detail.

Stiffness of soil shows nonlinear behavior at a very small strain and it is usually considered as nonlinear material in earthquake response analysis. However, as it was expensive to perform nonlinear analyses at the time, in the 1970s, the equivalent linear analysis was

Computational Geomechanics: Theory and Applications, Second Edition. Andrew H. C. Chan, Manuel Pastor, Bernhard A. Schrefler, Tadahiko Shiomi and O. C. Zienkiewicz.

proposed and the computer program SHAKE (Schnabel et al. 1972) was developed incorporating such an approach. This became the most commonly used tool in practical earthquake engineering in many countries and is still often used for structural design to obtain input motion from the bedrock to the structures as shown in Figure 9.1, soil behavior around piles (stiffness reduction around piles) under the earthquake motion, liquefaction risk analysis, etc. Analyses of SHAKE were well accepted for engineering design but it is well known that errors become larger for large earthquakes (Towhata and Routeix 1990) and nonlinear step-by-step integration analysis should be used for large-strain and failure-strain zone for cases as shown in Figure 9.2 (Ishihara 1996). This warning should be carefully heeded since large earthquakes did take place such as 6.786 m/s^2 (EW component at GL, 83 m Port Island) at Hyogoken-Nanbu 1995 Earthquake (Iwasaki and Tai 1996), 8.13 m/s^2 (synthesized component at NIG018 observation point of K-NET) for Niigata-ken Chuetsu-Oki 2007 Earthquake (NIED 2007) and 29.33 m/s^2 (synthesized component for MYG004 observation point of K-NET) for The 2011 off the Pacific coast of Tohoku Earthquake (NIED 2011). There are some concerns in the use of nonlinear step-by-step integration analysis like the one outlined in this book relating to the determination of soil parameters and confidence level for the constitutive models. Particular care has, therefore, to be taken in the choice of the constitutive models because some cannot properly simulate the liquefaction curve of laboratory tests and the combination of parameters is not unique for a given liquefaction curve (Uzuoka et al. 2003). However, the use of these methods is constantly growing and interesting examples are shown from Section 9.2 onwards.

This chapter begins with a critical assessment of the equivalent linear analysis and another method which uses conventional parameters for nonlinear analysis (Sections 9.2–9.4) while the remaining sections will deal with applications of the method outlined in Chapters 2 and 3 Hence, the analysis methods are going to be described in the following sequence:

- Equivalent linear analysis – example and error of the analysis
- An improvement of equivalent linear analysis – Half-wave cycle equivalent analysis
- A practical nonlinear analysis for 1D problem – Port Island (Hyogoken-Nanbu 1995 Earthquake)
- A practical nonlinear analysis for 3D problem – a large-scale liquefaction test in a coal mine
- Unsaturated effect on Lower San Fernando Dam during 1971 earthquake

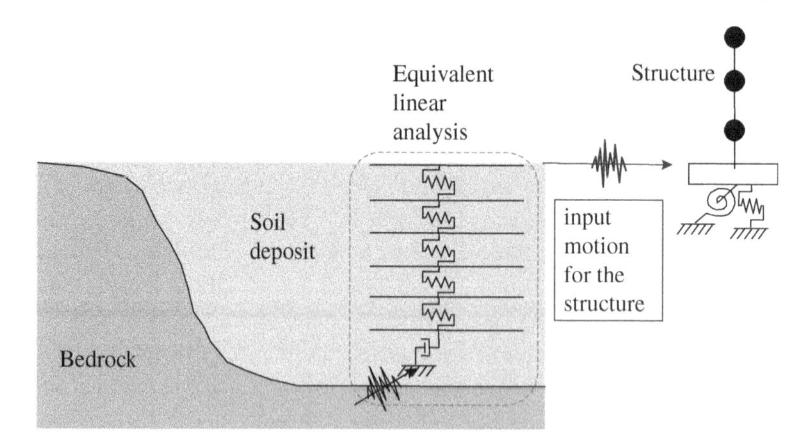

Figure 9.1 Obtaining input motion for the structure using the equivalent linear method.

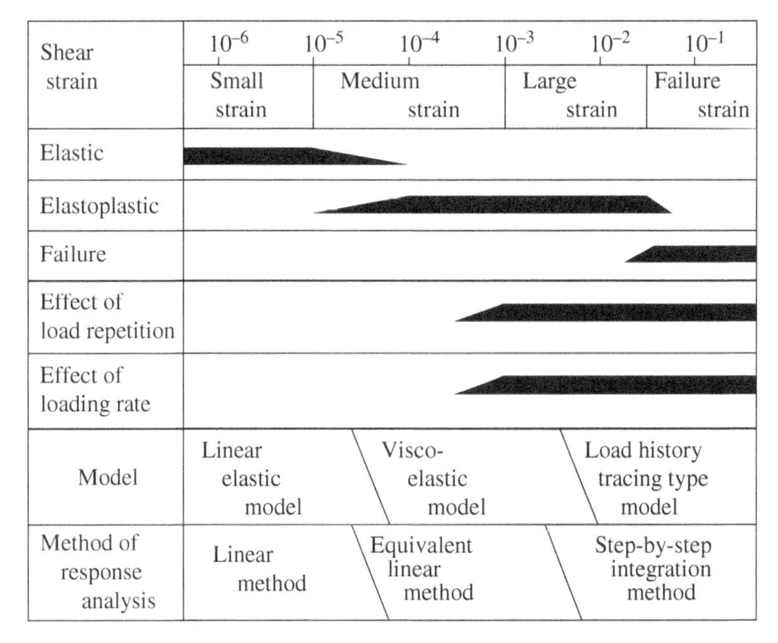

Figure 9.2 Modeling of soil behavior in compliance with strain-dependent deformation characteristics. *Source:* From Ishihara (1996).

9.2 Material Properties of Soil

In many instances, the engineer is interested in obtaining the first approximation to the seismic response of a geotechnical structure without resorting to the use of the more comprehensive models described in Chapter 4. We will illustrate here how the two key material parameters characterizing soil response can be obtained and employed in the equivalent linear analysis.

The key material properties of soil required for response analysis using the equivalent linear method are the secant stiffness and damping ratio h where both of them are strongly dependent on the magnitude of shear strain. Small strain shear modulus G_0 and Poisson's ratio ν can be determined from the velocities of P-wave (primary wave) and S-wave (shear wave) which can be measured using PS logging/Velocity survey.

$$G_0 = \rho V_s^2 \quad \text{and} \quad \nu = \frac{1}{2}\frac{1-2\left(\frac{V_s}{V_p}\right)^2}{1-\left(\frac{V_s}{V_p}\right)^2} \tag{9.1}$$

where ρ is bulk density, and V_p and V_s are velocity of the primary wave and shear wave.

Nonlinear behavior of soil can be broadly classified into consolidation, shear failure, and dilatancy. Consolidation occurs when clay foundation is settling and liquefaction happens during and after an earthquake.

Shear and damping behavior attracts the largest attention in most earthquake response analyses. The two behaviors are combined to govern the response magnitude. So, the dependency of secant shear modulus and damping ratio on shear strain are well investigated and used in the equivalent linear method (ELM), i.e. $\frac{G}{G_0} \sim \gamma$ and $h \sim \gamma$ relationship (Schnabel et al. 1972). Although the relationships obtained from the cyclic test defined in the standards of Japanese geotechnical society for laboratory shear test JGS 0541, 0542, and 0543 (JGS 2000) are directly used in the case of ELM, they are often expressed by mathematical function $f(\gamma)$. The function is defined as $\frac{\tau - \tau_a}{2} = f\left(\frac{\gamma - \gamma_a}{2}\right)$. The $f(\gamma)$ of Ramberg–Osgood model is used for pre-failure region of strain since it cannot express the failure phenomenon. Hyperbolic function can be used to express the shear failure but damping may be overestimated if the hysteresis curve is defined by the Masing rule (Masing 1926).

$$\text{Hyperbolic function} \quad \tau = \frac{G_0 \gamma}{1 + \frac{\gamma}{\gamma_r}} \tag{9.2}$$

(Hardin and Drnevich, 1972) where γ_r is shear strength/G_0

$$\text{Ramberg-Osgood model} \quad \tau = \frac{G_0 \gamma}{1 + \alpha \left| \frac{\tau}{G_0} \right|^{\beta - 1}} \tag{9.3}$$

(Ramberg and Osgood 1943), where α and β are material constants to define the curvature of the function.

There is no typical dilatancy parameter commonly used for the earthquake response analysis but a liquefaction strength curve is usually obtained to assess the risk of liquefaction disaster. The liquefaction strength curve is similar to the strength diagram (S–N curve), where S is the stress ratio $\left(\tau/\sigma'_m\right)$ and N is number of loading cycles (Figure 9.3).

9.3 Characteristics of Equivalent Linear Method

The Equivalent Linear Method (ELM) is the most commonly used method to calculate input motion from bedrock to the base of the structures and/or soil ground behavior under earthquake from given earthquake motion at bedrock considering material nonlinearity of soil (Figure 9.1). The ELM is a linear analysis with equivalent stiffness and damping of each layer taking into account the nonlinearity of the soil behavior in some heuristic manner. The equivalent values of each layer are determined from the maximum strain of the layer from the previous calculation starting from the shear modulus (G_0) and damping at a very small strain as shown in Figure 9.4. The calculation is repeated until the maximum strains converge. The calculation flow of the method is shown in Figure 9.5.

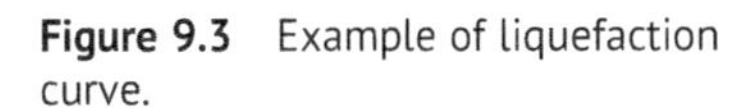

Figure 9.3 Example of liquefaction curve.

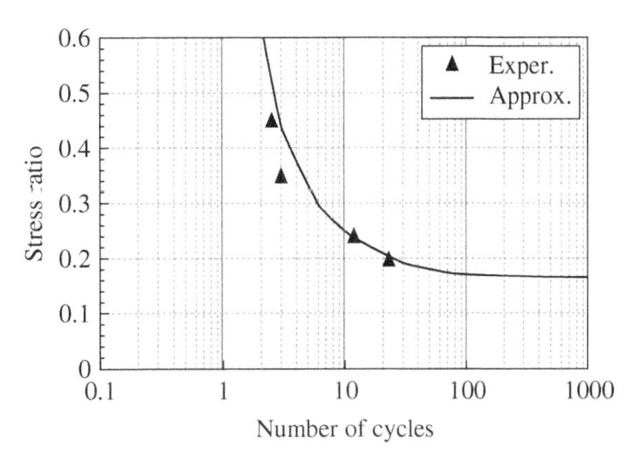

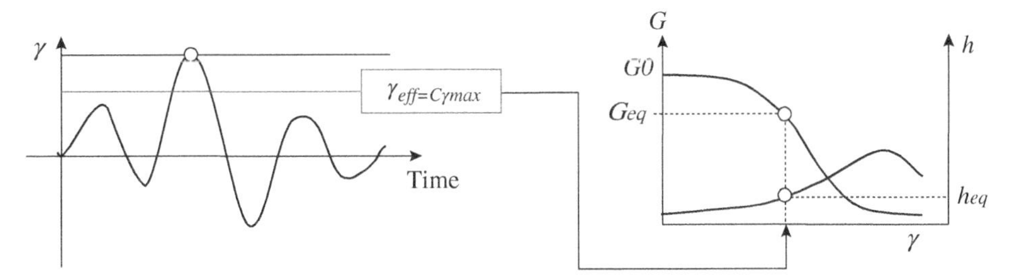

Figure 9.4 Equivalent shear stiffness G_{eq} and damping ratio h_{eq}.

Obviously, permanent deformation cannot be calculated using ELM because of its elastic nature. Furthermore, the precise time history of the response cannot be obtained too. This is because ELM attempts to solve nonlinear problems using constant material parameters and linear analysis for the whole duration of the earthquake. This will cause the high-frequency region of the response spectrum to be underestimated since equivalent parameters are calibrated for maximum strain and not for smaller strain periods of the calculation. Frequency-dependent damping was proposed (Sugito et al. 1994) to resolve this deficiency but this does not have a clear physical meaning (Shiomi and Fujiwara 2012).

We shall illustrate the shortcomings of ELM using a single degree of freedom problem, which is normally solved using ELM and nonlinear step-by-step analysis respectively using the same $\frac{G}{G_0} \sim \gamma$ and $h \sim \gamma$ curves. In the nonlinear analysis, the backbone ($\tau \sim \gamma$) curve is obtained as $\tau = \frac{G}{G_0} \times G_0 \gamma \sim \gamma$ and the hysteretic curve is calibrated to satisfy the $h \sim \gamma$ curve for the repeated cycle.

$$\tau = \frac{\hat{G}_0 \gamma}{1 + \gamma / \hat{\gamma}_r} \tag{9.4}$$

$$\hat{\gamma}_r = \frac{\gamma_R \tau_R}{\hat{G}_0 \gamma_R - \tau_R}, \tag{9.5}$$

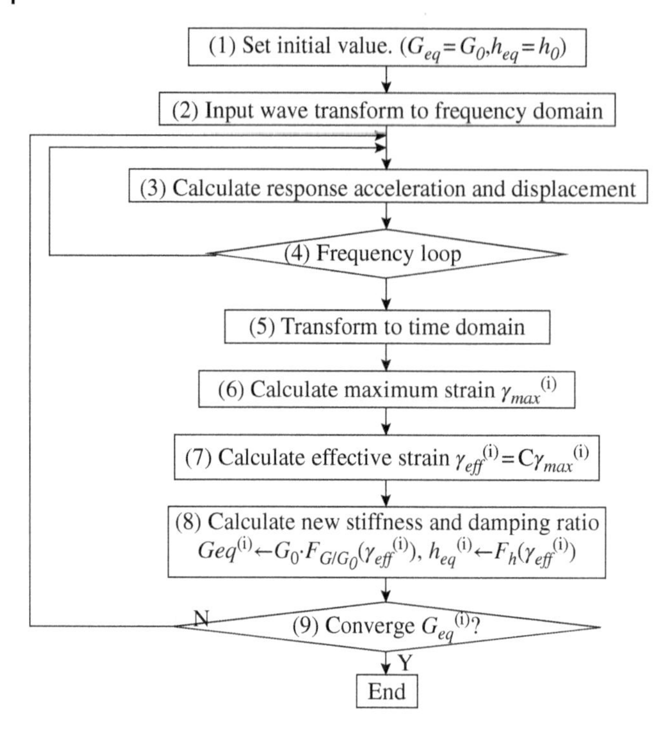

Figure 9.5 Flow diagram of the equivalent linear analysis. *Source:* Based on Schnabel et al. (1972).

$$\hat{G}_0 = \frac{\tau_R}{\gamma_R}\frac{1}{\hat{g}} \tag{9.6}$$

$$h = \frac{2}{\pi}\left\{\frac{1+\hat{g}}{1-\hat{g}} + \frac{2\hat{g}}{(1-\hat{g})^2}\ln\hat{g}\right\} \tag{9.7}$$

where τ_R and γ_R are the stress and strain at the stress reversal point. However, the value h is taken as 0.02 if $\gamma \leq 0.01\%$. $\hat{G}_0$ and $\hat{\gamma}_r$ are the stiffness and reference strain of repeated cycles calculated according to the damping ratio h corresponding to γ_R through the $h \sim \gamma$ curve in Figure 9.6.

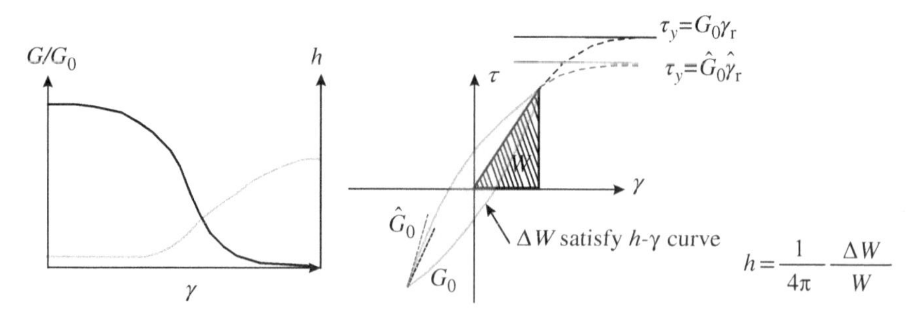

Figure 9.6 Evaluation of the additive hysteresis damping ratio.

Table 9.1 Material properties of used-for-single DOF problem.

Mass m (Mg)	**1**
Natural frequency f_0 (Hz)	1
Stiffness k_0 (kN/m)	39.48
Rayleigh damping	0
Reference strain (%)	0.1

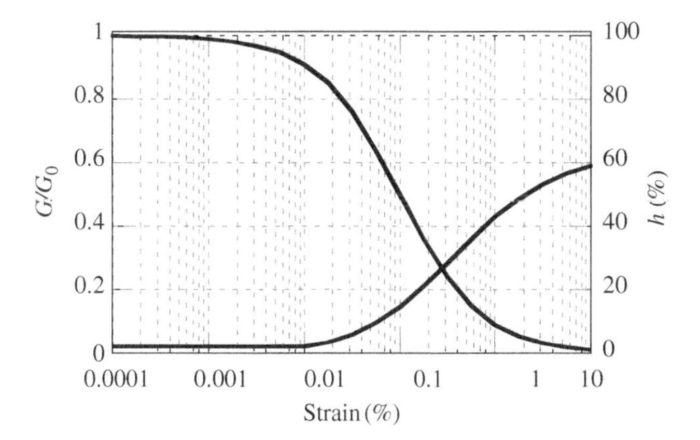

Figure 9.7 Dynamic characteristics of deformation

We demostrantes signle degree freedom analyses for ELM and non-linear method, where the soil parameters are shown in Table 9.1 and Figure 9.7.

Maximum response accelerations of both analyses for sinusoidal input motions where the frequencies vary from 0 to 2.5 Hz are shown in Figure 9.8. The responses are almost identical.

When input motion is a random wave, response of acceleration may become different from each other since it is not possible to synthesize a sinusoidal wave response. Results are shown for analysis with random input motion (Figure 9.9) with changing amplitude. When the amplitude of input motion is small, the responses from the two methods are close to each other since the stiffness is nearly constant (Figures 9.10 and 9.11a). But when the input motion becomes larger, the difference becomes larger as shown in Figure 9.11b and c. The acceleration obtained from ELM is more than 1.5 times that of the nonlinear analysis.

It is obvious that step-by-step nonlinear analysis should be employed for large strain and failure strain zone analysis as Ishihara (1996) has recognized and as shown by the above example. But for practical engineering applications, ELM is preferred as used in the seismic safety design of the Kyoto University Research Reactor (Kyoto University Research Institute 2000) where step-by-step nonlinear analyses were only used for cases with the maximum shear strain of a soil layer exceeding 1% and ELM was used for other cases. It should however be noted that the method outlined in the previous chapters can give much more insight as shown by examples from Section 9.5 onwards.

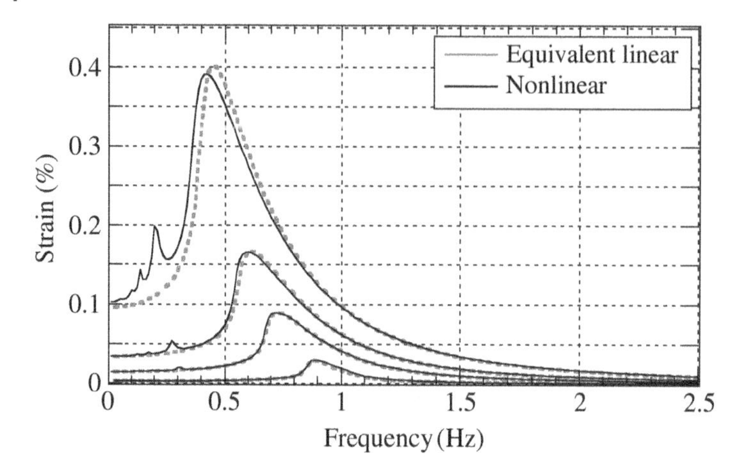

Figure 9.8 Comparison with resonance curve of the equivalent linear and nonlinear analysis. *Source:* From Fujiwara and Shiomi (2011).

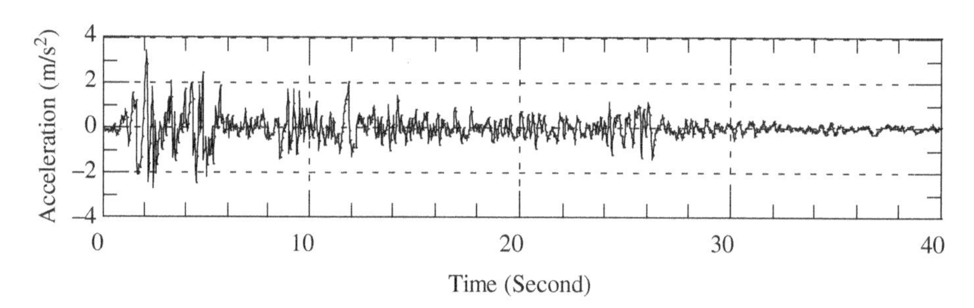

Figure 9.9 Time history of the input wave. Earthquake record El Centro 1940 NS (*Source:* e.g. http://www.vibrationdata.com/elcentro.htm http://www.vibrationdata.com/elcentro_NS.dat)

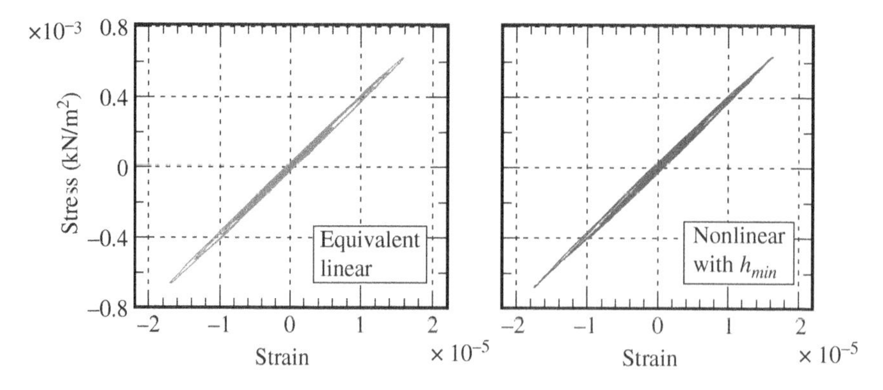

Figure 9.10 Comparison of stress–strain relationship of ELM and nonlinear analysis under small strain. *Source:* From Fujiwara and Shiomi (2011).

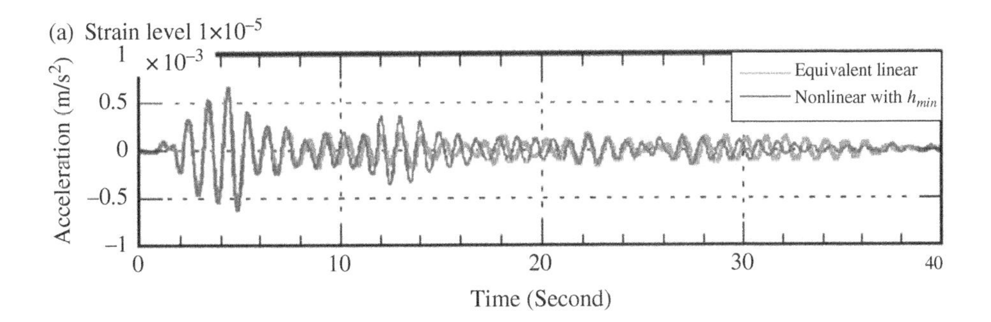

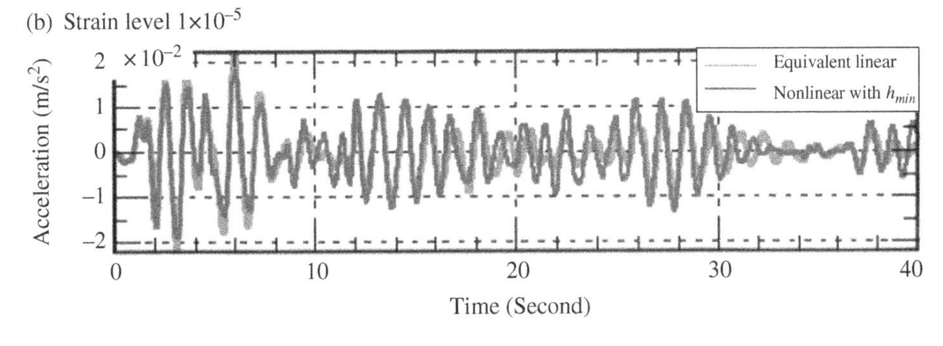

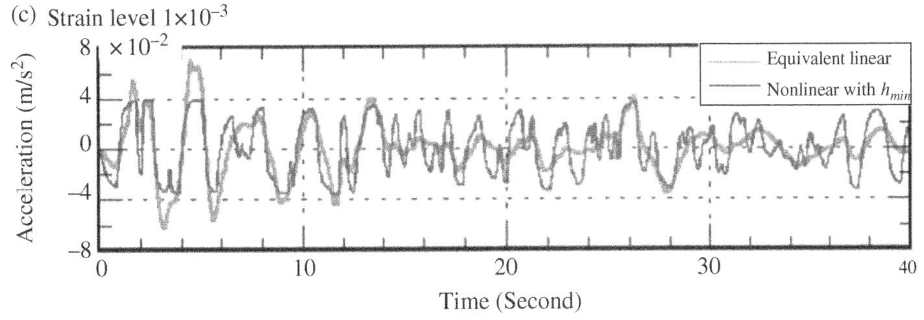

Figure 9.11 Comparison of acceleration of ELM and nonlinear analysis under the various levels of strain. (a) Strain level 1×10^{-5}. (b) Strain level 1×10^{-3}. (c) Strain level 1×10^{-2}. *Source:* From Fujiwara and Shiomi (2011)

9.4 Port Island Liquefaction Assessment Using the Cycle-Wise Equivalent Linear Method (Shiomi et al. 2008)

It is not possible for ELM to analyze the liquefaction problem since the change of the state parameters with time cannot be adequately represented using constant parameters over the earthquake duration.

In order to overcome this deficiency of ELM, "Cycle-Wise Equivalent Linear method" was introduced (Shiomi et al. 2008). The method modifies the soil parameters during a half-cycle of stress/strain history in the time domain to model progressive damage such as liquefaction problem.

9.4.1 Integration of Dynamic Equation by Half-Cycle of Wave

The basic idea of equivalent linear analysis is to solve a nonlinear problem using linear analysis with a reduced secant stiffness at the largest amplitude of a cycle to obtain an appropriate secant stiffness; the linear analysis is repeated until the stiffness at each layer converges satisfying the material relationship of G/G_0-γ and h-γ curve. Therefore, the equivalent linear analysis can well simulate the response for the cycle at which the maximum magnitude occurs. However, the response of the other cycles will not be correct since the stiffness used is not correct for the amplitude of the other cycle.

This problem can be improved by determining the secant shear modulus at each half cycle using the Cycle-Wise Equivalent Linear (CWEL) method as schematically shown in Figure 9.12, where the half cycle is defined as the period from a shear strain crossing of zero line to the next crossing.

Secant modulus can be changed with time for the CWEL method so it is possible for the method to model the liquefaction phenomena which changes material properties from time to time. The schematic figure of CWEL liquefaction (CWELL) method is shown in Figure 9.14. By solving a free field soil layer problem as shown at the top of the left-hand side corner, shear strain and stress are obtained. Secant stiffness and damping factor due to strain hardening are obtained using G/G_0-γ and h-γ curves from the shear stress (stress ratio) as shown at the top right-hand side.

$$\frac{G}{G_0} = F_G\left(\frac{\gamma}{\gamma_{50_{ref}}\left(\sigma'_{m_{ref}}\right)^n}\right), h = F_h\left(\frac{\gamma}{\gamma_{50_{ref}}\left(\sigma'_{m_{ref}}\right)^n}\right) \tag{9.8}$$

where these curves are obtained by an ordinary dynamic property test such as "Method for Cyclic Triaxial Test to Determine Deformation Properties of Geomaterials" of Japanese Geotechnical Society (JGS 2000). The excess pore pressure due to liquefaction is then

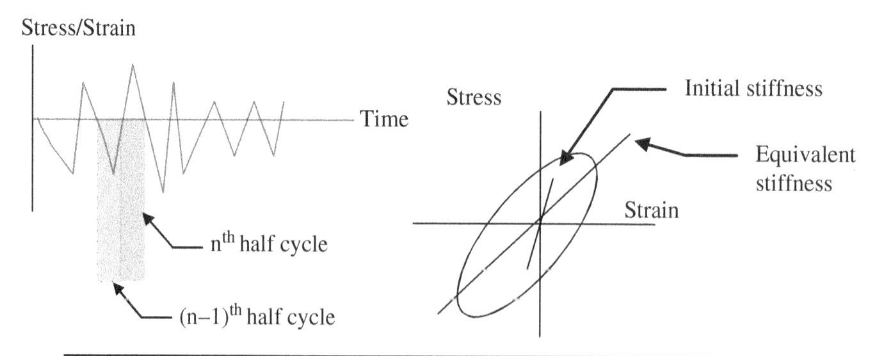

Figure 9.12 Concept of CWEL. *Source:* From Shiomi et al. (2008).

calculated from the accumulated damage parameter, which is, in turn, calculated from the shear strain. Here the dependency of initial shear stiffness on effective stress, i.e. pore pressure is defined by the empirical formula shown in Equation 9.9.

$$G_0 = G_{0\_ref}\left(\frac{\sigma'_{m0}}{\sigma'_{m\_ref}}\right)^n \tag{9.9}$$

where σ'_m is effective mean stress, G_{0_ref} is G_0 at reference effective mean stress σ'_{m_ref}, G_0 is elastic shear stiffness, γ is shear strain, γ_{50_ref} is γ_{50} at the reference effective mean stress, and γ_{50} is the strain where the shear stiffness is half of the initial value. The mean effective stress is obtained from the excess pore pressure ratio r_u

$$\sigma'_m = \sigma'_{mo}(1 - r_u) \tag{9.10}$$

The excess pore pressure ratio r_u is obtained by an empirical formula of the accumulated damage ratio D. This formula was originally proposed by Seed et al. (1976) in the form of Equation 9.11 and modified to Equations (9.12a) and (9.12b) for cases of "before cyclic mobility" and "after cyclic mobility" (Yoshida et al. 2008a)

$$r_u = \frac{1}{2} + \frac{1}{\pi}\sin^{-1}\left(2D^{\frac{1}{\alpha}} - 1\right) \tag{9.11}$$

$$r_u = f_{ru}(D, \alpha) = \frac{1}{2} + \frac{1}{\pi}\sin^{-1}\left(\left(2D^{\frac{1}{\alpha}} - 1\right)\left(\frac{((\cos(2\pi D) + 1)/2)^n + \beta}{1 + \beta}\right)\right) \tag{9.12a}$$

$$r_u = 1 - \left(1 - r_{u(D = D_{CM})}\right)\cdot\left(\frac{D - D_{CM}}{1 - D_{CM}} + 1\right)^4 \tag{9.12b}$$

where $\alpha = (0.0177D_r)^3$, β, n are constants 2.0 and 5.0, respectively. D_r is the relative density (%) of sand. $r_{u(D = DCM)}$ is the excess pore pressure ratio at $D = D_{CM}$ and D_{CM} is the accumulated damage parameter when cyclic mobility starts. Those equations are well adapted to sample sand as shown in Figure 9.13. The schematic flow of the CWEL Liquefaction method is shown in Figure 9.14. The secant stiffness G is degraded by the magnitude of the maximum strain of the half-cycle and the accumulated damage parameter D.

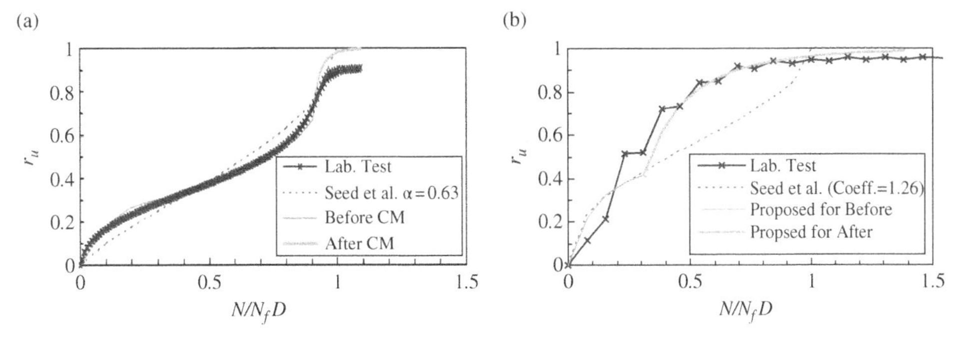

Figure 9.13 Excess pore pressure ratio r_u against damage parameter D (a) In case when excitation force is small, (b) In case when excitation force is large. *Source:* From Shiomi, et al. (2008).

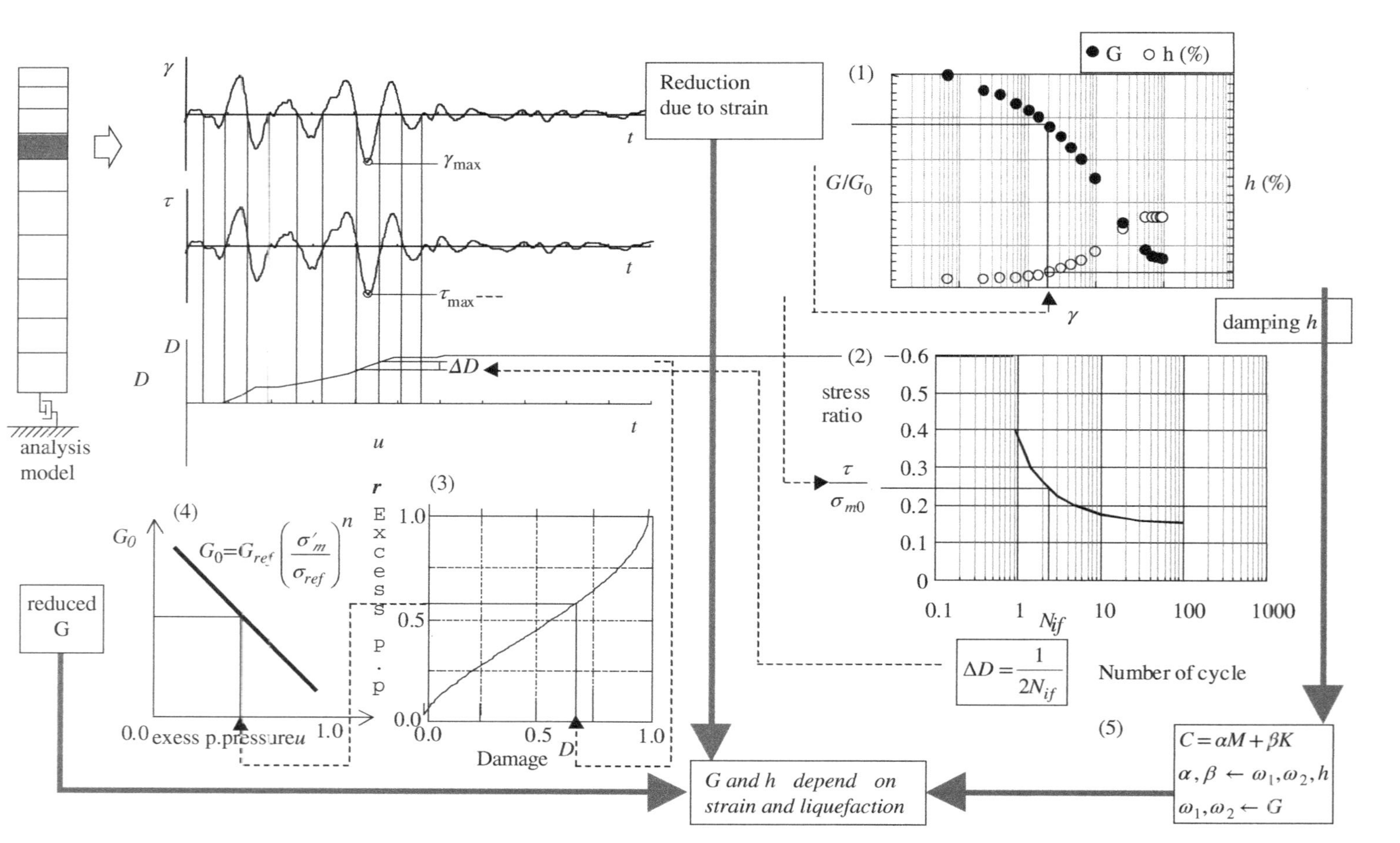

Figure 9.14 Schematic flow to calculate stiffness under liquefying process. *Source:* From Shiomi et al. (2008).

9.4.2 Example of Analysis

Simulation using the above-mentioned method is performed for a ground response of Kobe Port Island during the 1995 Hyogoken Nanbu earthquake (Iwasaki and Tai 1996). The soil layer data is taken from information obtained through frozen sampling by Hatanaka et al. (1997). The material properties of the soil layers are shown in Table 9.2. The water table is 2.3 m below the ground surface and the soil layers consist of the upper liquefiable layer (−10 to −18.6 m) and the middle layer comprising clay and dense sand and gravel. The layer from −67 to −83.8 m consists of alluvial clay. The liquefiable layer was investigated using the frozen sampling method, so the liquefaction strength curves KPU, KPM, and KPL as shown in Figure 9.15 were obtained using the most accurate acquisition method for the current state of the art. The shear wave velocity measured is 210 m/s.

The G-γ curve is also obtained with the frozen sampling. Figure 9.16 shows the shear modulus of the layers of sand and clay. Although shear modulus was directly obtained from the laboratory tests, only reduction ratio G/G_0 is used for the analysis. Strain when shear modulus becomes half of the initial value (γ_{50}) is about 0.05% for sand and about 0.2% for clay. The strain at damping coefficient of 10% is about 0.1% for sand and about 0.15% for clay.

Input motion is applied at the depth −83.3 m as the motion of the sum of incoming and outgoing waves. The waveform is shown in Figure 9.17. An impulsive wave takes place at about four seconds that may trigger the liquefaction phenomenon. Two sets of response accelerations on surface are shown in Figure 9.18. Case 1 is with no considerations of stress recovery due to cyclic mobility which is considered for Case 2. Both accelerations agree well with the recorded acceleration until the largest peak acceleration. After that, the acceleration of Case 2 shows closer wave period to the observed value than Case1. The reason for this is that the pore pressure ratio is assumed to be unity at full liquefaction in Case 1 but it recovers and becomes less than one due to cyclic mobility in Case 2.

Table 9.2 Material properties of soil layer (Shiomi et al. 2008).

Layer	Depth (m)	V_s (m/s)	γ (kN/m^3)	Shear modulus G_0 (kPa)	$G \sim \gamma$ curve	Liquefaction strength curve
Sand and gravel	-2.3	170	19.60	57 800	S-1	
	-10	210	19.60	88 200	S-1	KPU
	-14.7	210	19.60	88 200	S-2	KPM
	-18.6	210	19.60	88 200	S-2	KPL
Alluvial clay	-23	180	16.07	53 136	C-1	
	-27	180	16.17	53 460	C-2	
Sand	-32.8	245	17.15	105 044	S-2	
Sand and gravel	-50	305	17.93	170 236	S-2	
Sand	-61	350	18.23	227 850	S-2	
Diluvial clay	-67	303	16.66	156 075	C-5	
	-79	303	16.66	156 075	C-5	
Sand and gravel	-83.8	320	19.60	204 800		

Source: From Shiomi et al. (2008).

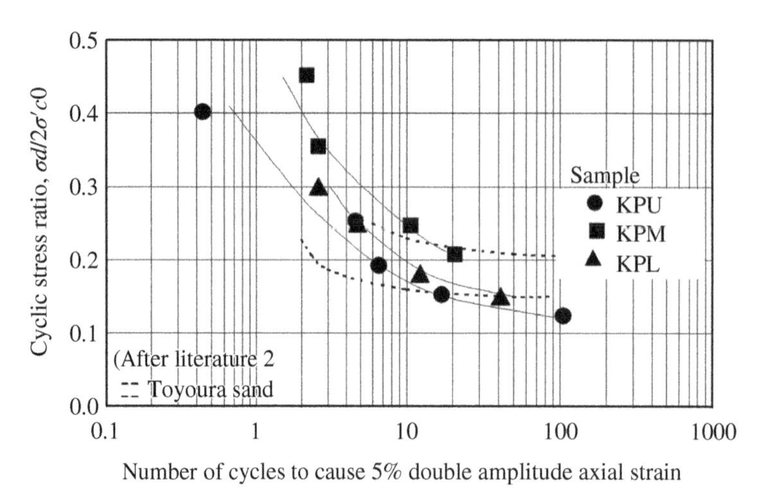

Figure 9.15 Liquefaction strength of sand. *Source:* From Shiomi et al. (2008).

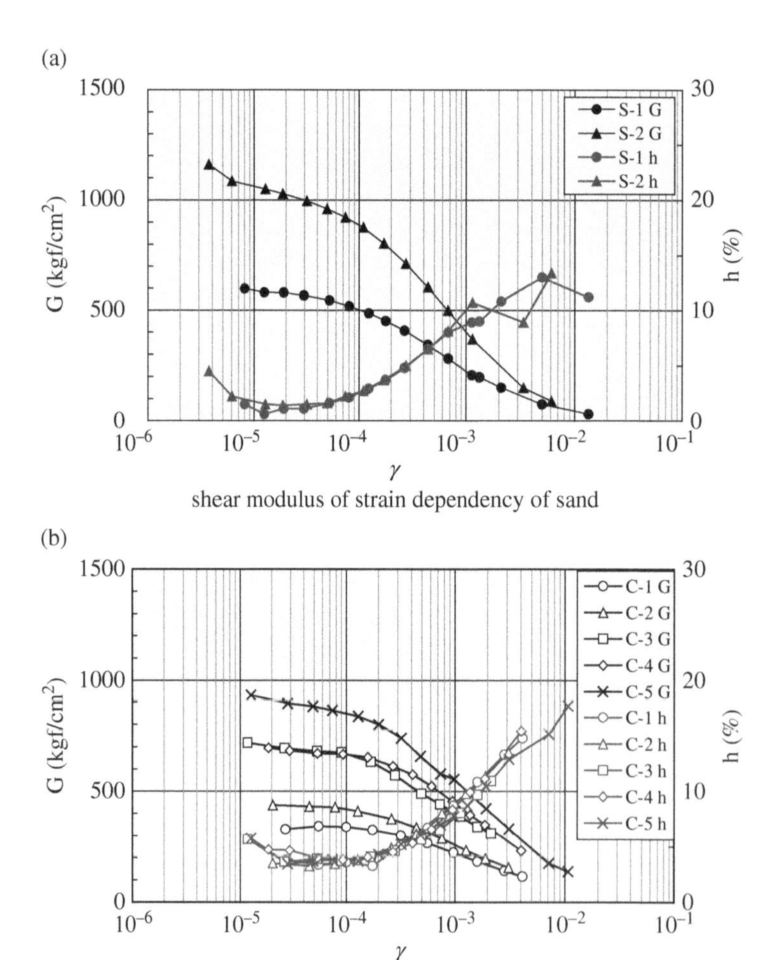

Figure 9.16 Secant shear modulus for soil layers. *Source:* From Shiomi et al. (2008). (a) Shear modulus of strain dependency of sand (b) Shear modulus of strain dependency of clay.

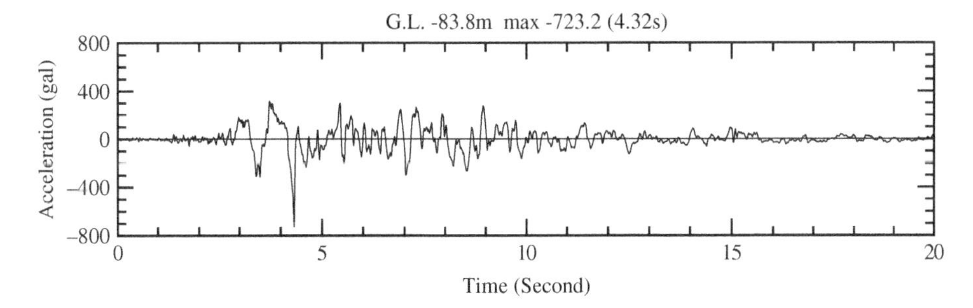

Figure 9.17 Recorded earthquake at Port Island −83.8 m. *Source:* From Shiomi et al. (2008).

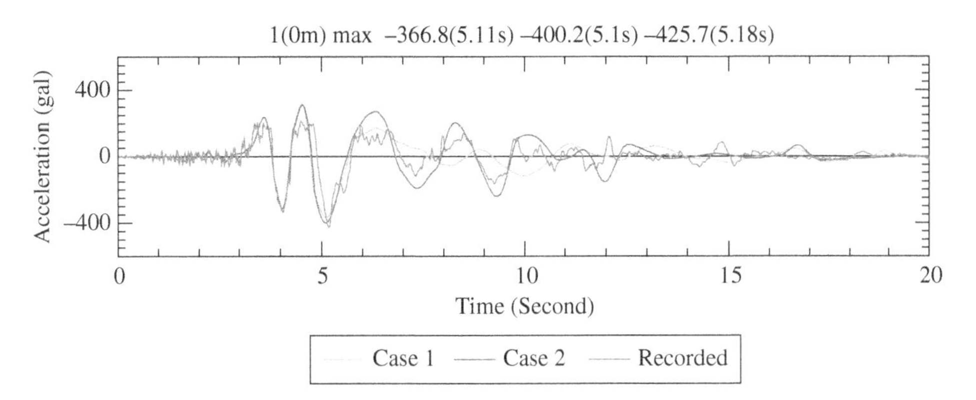

Figure 9.18 Acceleration at the ground surface. *Source:* From Shiomi et al. (2008).

9.5 Port Island Liquefaction Using One-Column Nonlinear Analysis in Multi-Direction

9.5.1 Introductory Remarks

Although a real earthquake inevitably consists of multidirectional components, most analyses are done using only one- or two-dimensional models. The aim of this section is to investigate the importance of multidirectional loading (MDL) in engineering practice. The liquefaction events that occurred during the Hyogoken-Nanbu Earthquake of 1995 are taken as case studies in this section. Two-phase dynamic equations used in the examples are those derived in Chapters 2 and 3. It was found that the unidirectional loading along the principal axis of earthquake orbit agreed well with results from the horizontally multidirectional loading for the maximum response acceleration, except for the process of pore water pressure buildup and other details of response. It was also found that the effect of vertical loading is not significant, howsoever the effect of initial (static) shear stress (ISS) was, and this will be discussed in Section 9.6.

MDL and initial stress obviously play important roles in the geotechnical numerical analysis if material nonlinearity occurs. In reality, an earthquake, inevitably, has multidirectional movement and initial static stresses with a shear component will always be found in such soil structures as dams, dykes, and in natural ground. Most studies of liquefaction analysis in the past have considered only one horizontal component of the earthquake and K_0 in-situ stress state, however, multidirectional earthquake loading and the initial stress condition are important for precise numerical prediction and studies of these are reported in the individual sections. Firstly, MDL will be discussed and followed by ISS in the next section.

The effect of MDL has been studied experimentally by several researchers. Settlement of a thin dry sand layer was studied by Pyke et al. (1975) for the case of a sand layer shaken on the shaking table in one or two horizontal directions. Circular and random motions were applied for the two-component tests. Their conclusion is that the settlements caused by combined horizontal motions are almost equivalent to the sum of settlements caused by components acting separately. While vertical accelerations less than 1 g cause no settlement if acting alone, vertical accelerations superimposed on the horizontal accelerations could cause a marked increase in the settlements. The effect of MDL on the liquefaction strength (stress ratio to induce a certain shear strain, e.g. 3%, under a given number of cyclic loading, e.g. 15 or 20) has been studied at the end of the 1970s (Seed et al. 1978; Ishihara and Yamazaki 1980). Seed et al. attributed the settlement on the shaking table test to the liquefaction strength. On the other hand, Ishihara and Yamazaki obtained directly the liquefaction strength with a two-directional simple shear test apparatus under undrained conditions. They found that the cyclic stress ratio dropped approximately 25–35%, causing 3% strain in a direction depending upon the pattern of the two components for loading. This liquefaction strength is also influenced by the loading irregularity (Nagase and Ishihara 1988). The volumetric strain due to consolidation following liquefaction also differed due to MDL (Ishihara and Nagase 1988).

From the numerical analysis point of view, there has to be an influence of MDL on the liquefaction induced by earthquakes since negative dilatancy, which causes liquefaction, depends on the accumulated shear strain. The accumulated shear strain is the combination, though not simply an arithmetic sum, of the six components of the strain (three deviatoric strains and three shear strains). Ghaboussi and Dikmen (1981) first studied the effect of MDL on a soil layer problem with their proposed numerical method using the fully coupled Biot's equation with u-U formulation (Zienkiewicz and Shiomi 1984) as the dynamic equation and a nonlinear material model. The material model was based on the hyperbolic stress–strain relation for shear (Konder and Zelasko 1963) and the effective stress path approach for dilatancy (Ishihara et al. 1975). In this model, the decrease of effective stress is a function of both components of horizontal shear strains. A hypothetical horizontally layered ground subjected to the El Centro Earthquake was solved as a case study. Two-dimensional analysis showed a marked difference from 1D analysis in the buildup behavior of pore water pressure and some differences in surface velocity spectra. Even when the amplitude of the acceleration is increased to the resultant peak accelerations from the two directions, the 1D results were still different from the results of the two-dimensional analysis. As an alternative, Fukutake et al. (1995) suggested the use of input motions 1.3 times larger than the earthquake of the stronger direction between north–south (NS) or east–west (EW) directions, although this conclusion is considered premature.

Therefore, there is a need to survey the three-dimensional behavior of the level ground subjected to MDL together with other variations such as soil properties and input motions.

This section investigates the effects of MDL on a real site. The site is a typical soil "column" on the Port Island in Kobe City, and earthquake motions for the studies were recorded during Hyogoken-Nanbu Earthquake on 17 January 1995, where the liquefaction phenomena has been observed throughout the island and settlement was about 20 cm (estimated from the relative gap between buildings supported by piles and ground surface after the earthquake). At first, back analysis of observed data is explained and then the parameter study for MDL reported.

9.5.2 Multidirectional Loading Observed and Its Numerical Modeling – Simulation of Liquefaction Phenomena Observed at Port Island

During the Hyogoken-Nanbu Earthquake, liquefaction took place along most of the coastal area of Kobe City. Sand boiling and flushing water due to liquefaction occurred at many locations including Port Island where an array of seismometers was installed at four different depths. Figure 9.19 shows the orbit of the records at Port Island during the

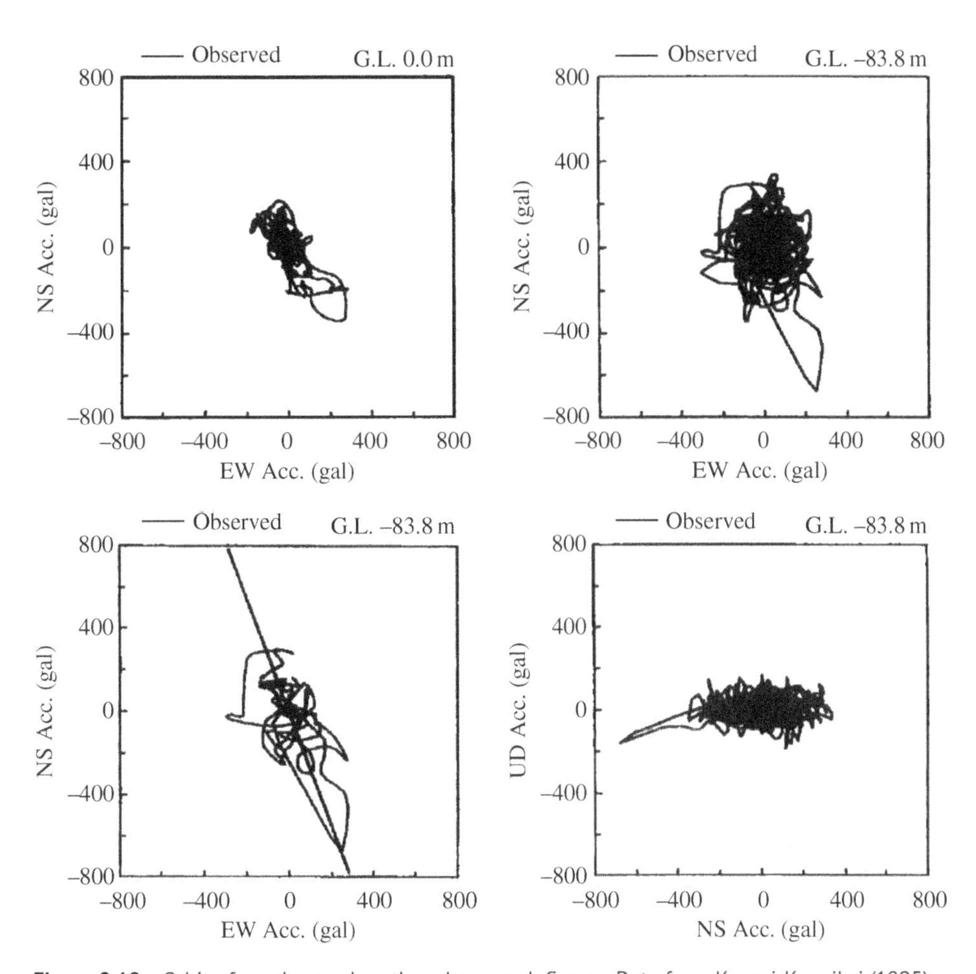

Figure 9.19 Orbit of an observed earthquake record. *Source:* Data from Kansai-Kyogikai (1995).

Hyogoken-Nanbu Earthquake in 1995. Two to four peaks with very large amplitude can be seen from the figures. The maximum acceleration at the surface (GL 0.0 m) was 314 gal (3.14 ms^2) for NS and 288 gal for EW direction. They are about half of the peak value record at GL-83.8 m. The diagram in the bottom left of Figure 9.19 shows the orbit at GL-83.8 m during zero to five seconds. Several peaks with large amplitude were clearly seen. The direction, when the maximum peak occurred, was about 20° from north to west. This direction is considered to be the principal axis of the earthquake components. The time histories of the NS, EW, and principal axis components were shown in Figure 9.20. The diagram in the bottom right of Figure 9.19 shows the orbit of NS (north-south)-UD (up-down) motion. At GL-83.8 m, the UD component was not considered significant.

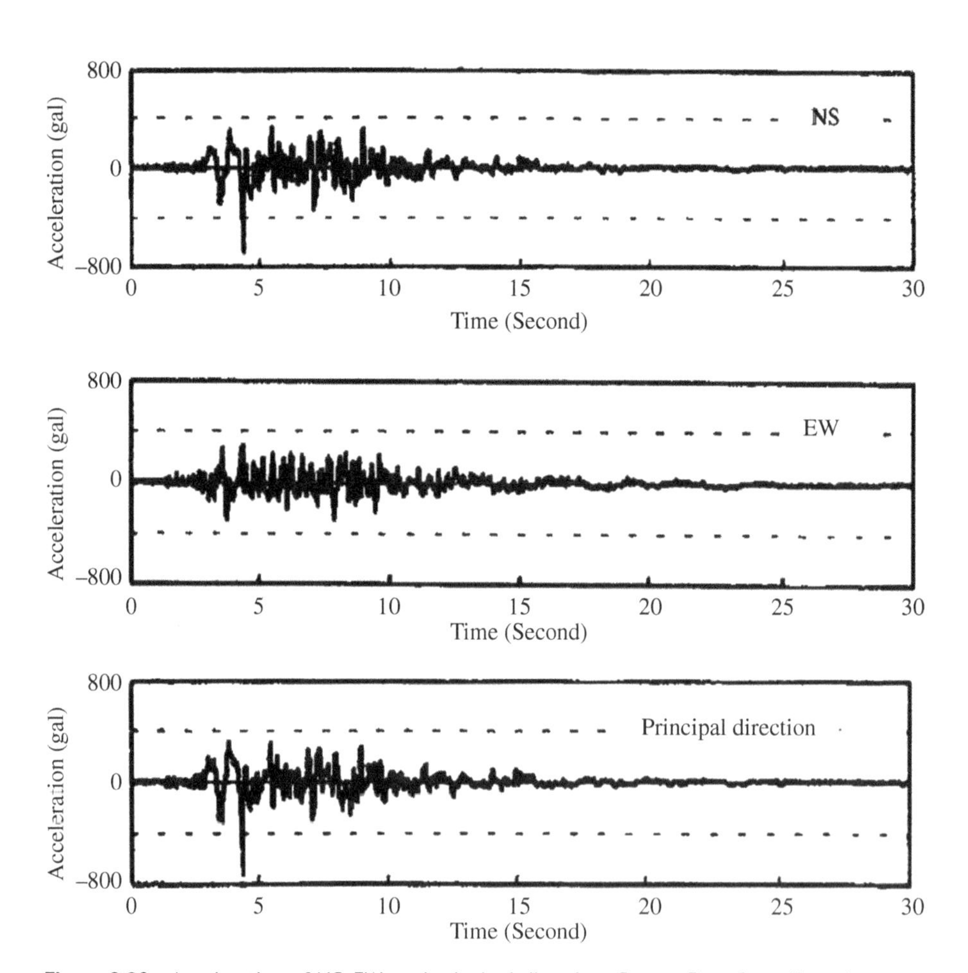

Figure 9.20 Acceleration of NS, EW, and principal direction. *Source:* Data from Kansai-Kyogikai (1995).

9.5.2.1 Conditions and Modeling

The effect of MDL was studied by simulating the above records using a column of finite elements. Case studies performed are shown in Table 9.3. The column of soil used in the numerical modeling is shown in Figure 9.21. The record at GL-83.8 m was introduced as the input motion. Four cases are studied. Case 1 simulates the observed record by incorporating all three components of the earthquake motion. Case 2 was studied to investigate the influence of vertical input motion on the liquefaction phenomena. Cases 3–5 are used to compare between two- and three-dimensional modeling.

Table 9.4 shows the material properties of the soil layer at Port Island. The other properties are derived from the data shown in the table or obtained from soil properties at a similar site. For example, friction angle and liquefaction strength were calculated through the N value of standard penetration tests (Tokimatsu and Yoshimi 1983).

9.5.2.2 Results of Simulation

Figure 9.22 shows the time history of response accelerations in NS direction overlaying the observed accelerations. These agreed well initially. However, some difference in the response is found at the surface after five seconds. The observed acceleration shows a very long period wave, while the calculated acceleration has a component of higher frequencies. Figure 7.5 shows the time history of the excess pore pressure in Cases 1 (NS+EW+UD) and 2 (NS+EW). The layer between GL-12.6 and -14.0 m was fully liquefied at about six seconds and kept liquefied. But the rate of buildup of the excess pore pressure above the layer slowed down after six seconds. This is because the input motion transferred from the bottom to the surface was significantly reduced by the sudden loss of the shear strength at the layer between GL-12.6 and -14.0 m.

On the other hand, at a deeper level of GL-15.4 m, pore pressure continued to build up and almost reached full liquefaction. Therefore, contrary to popular belief, a tendency of liquefaction at a deeper layer was found in this large impact-type earthquake. In order to investigate the effect of vertical input motion, Case 2 (NS+EW only) was then analyzed. The excess pore pressure response using the up-down (UD) component has a very high-frequency content as shown by the thin solid line in Figure 9.23. Nevertheless, excess pore pressure response without UD motion (bold solid line) passes through nearly the mean value of the excess pore pressure obtained from the analysis with UD motion. This implied that vertical input motion induced only compressive wave in the water but did not affect the

Table 9.3 Cases studied.

Case	Analysis type	Input motion
1	3D analysis	NS, EW, and UD
2	3D analysis	NS and EW only
3	2D analysis	NS and UD only
4	2D analysis	EW and UD only
5	2D analysis	Principal axis and UD only

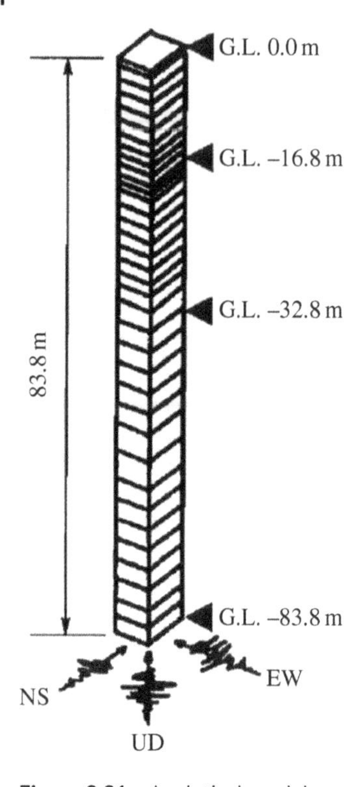

Figure 9.21 Analytical model.

Table 9.4 Soil layer and material properties.

Layer no.	Depth (m)	Soil type	V_p (m/s)	V_s (m/s)	Average N-value	Density (kN/m^3)
1	0.0 to −5.0	Man-made fill	330	170	5.2	17.64
2	−5.0 to −12.6		780	210	6.5	17.64
3	−12.6 to −16.8		1480	210	6.5	17.64
4	−16.8 to −19.0		1180	210	6.5	17.64
5	−19.0 to −27.0	Silty clay	1180	180	3.5	16.17
6	−27.0 to −32.8	Layers of gravelly Sand and silt	1330	245	13.5	17.15
7	−32.8 to −50.0		1530	305	36.5	18.13
8	−50.0 to −61.0		1610	350	61.9	18.13
9	−61.0 to −79.0	Silty clay	1610	303	11.7	16.66
10	−79.0 and deeper	Diluvium sand	2000	320	68	23.52

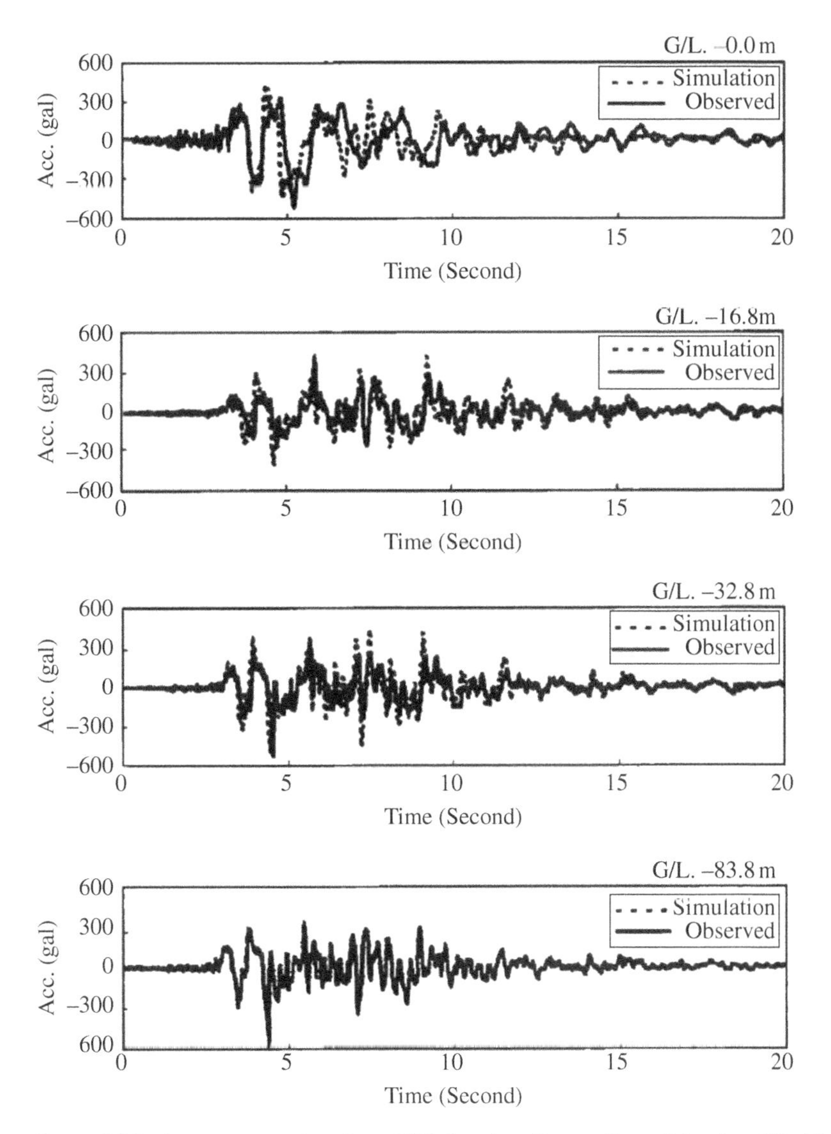

Figure 9.22 Response acceleration of NS direction. *Source:* From Shiomi and Yoshizawa (1996).

response of the soil skeleton. Therefore, the result of Case 1 is found to be very similar to Case 2 when the high-frequency content is filtered out.

Next, the results of unidirectional loading (Figure 9.23) and multidirectional loading (Figures 9.24 and 9.25) were compared. The response obtained with unidirectional EW direction loading showed a different response while the response obtained with unidirectional NS direction loading was close to the one obtained with multidirectional loading. This is because the NS direction was very close to the principal direction of the earthquake motion. The buildup of pore water pressure could have been dominated by the NS component of the earthquake. In order to find out the effect of the direction of input motion,

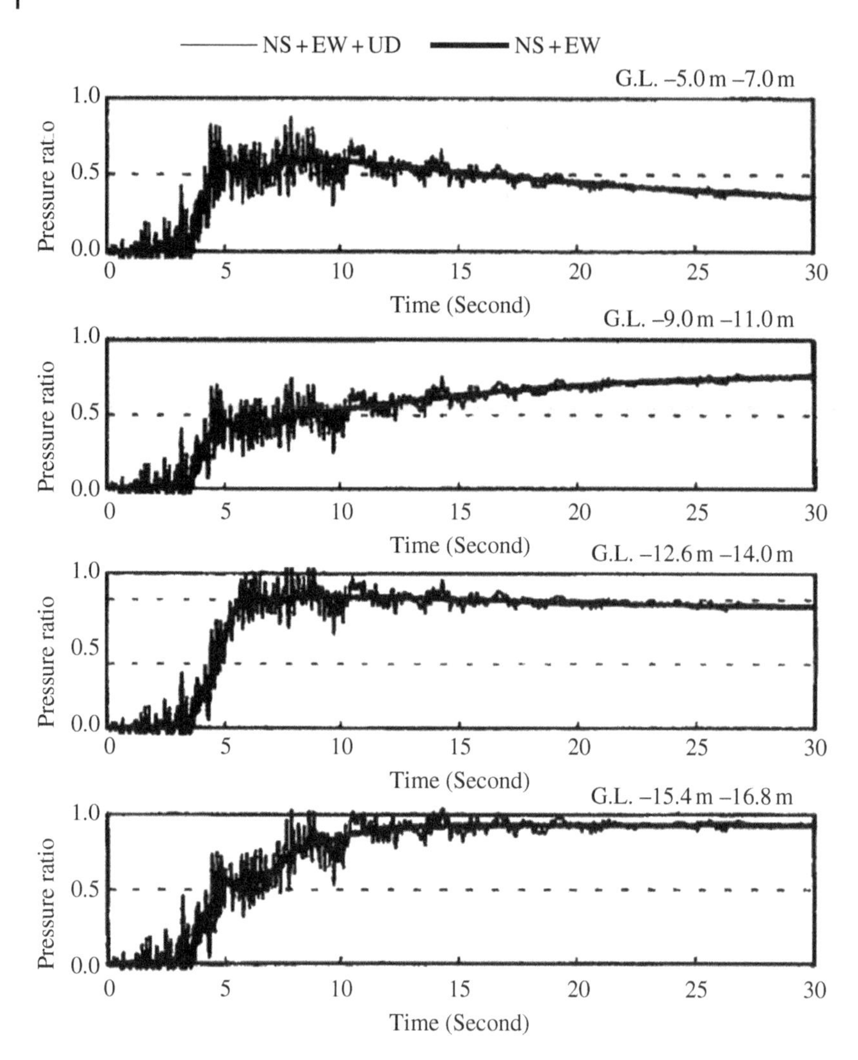

Figure 9.23 Excess pore pressure ratio (NS). *Source:* From Shiomi and Yoshizawa (1996).

response with unidirectional loading in the principal direction was then calculated and shown in Figure 9.26. In this case, the results were very similar to those obtained using the NS component alone.

9.5.2.3 Effects of Multidirectional Loading

Maximum accelerations for uni- and multidirectional loading are plotted in Figure 9.27. Results of the NS direction for Case 1 (NS + EW + UD) and Case 3 (NS + UD) showed similar behavior. Response of Case 5 with input motion in the principal axis was similar to the response of Case 1 (NS + EW) at the same (principal) direction. Maximum vertical acceleration was hardly influenced by the horizontal excitation. This tendency was found to be the same as the results of Ghaboussi and Dikmen (1981).

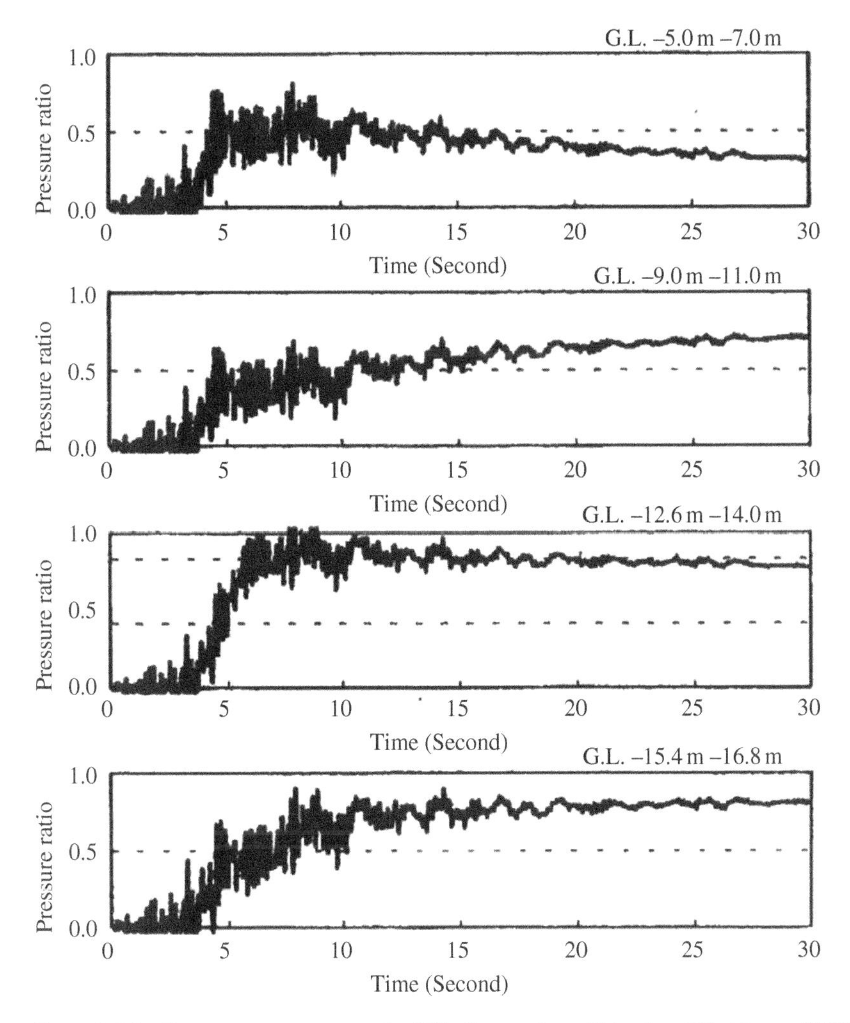

Figure 9.24 Excess pore pressure ratio (EW). *Source:* From Shiomi and Yoshizawa (1996).

9.6 Simulation of Liquefaction Behavior During Niigata Earthquake to Illustrate the Effect of Initial (Shear) Stress

In addition to MDL, Initial Shear Stress (ISS) also has significant effects on the results of the liquefaction analysis. ISS due to self-weight is usually assumed through K_0 (earth pressure coefficient at rest) or static analysis due to self-weight but the actual value could not be determined by the current measurement. Only a few studies have been done for in situ K_0 (Hatanaka and Uchida 1995). Consequently, there are very few reports of numerical analyses investigating the effect of ISS. Effects of ISS could be modeled implicitly by adjusting the shear strength of the material if only shear resistance is important and in static problems. But it is not straightforward since not only shear strength but also dilatancy characteristic is affected by the ISS.

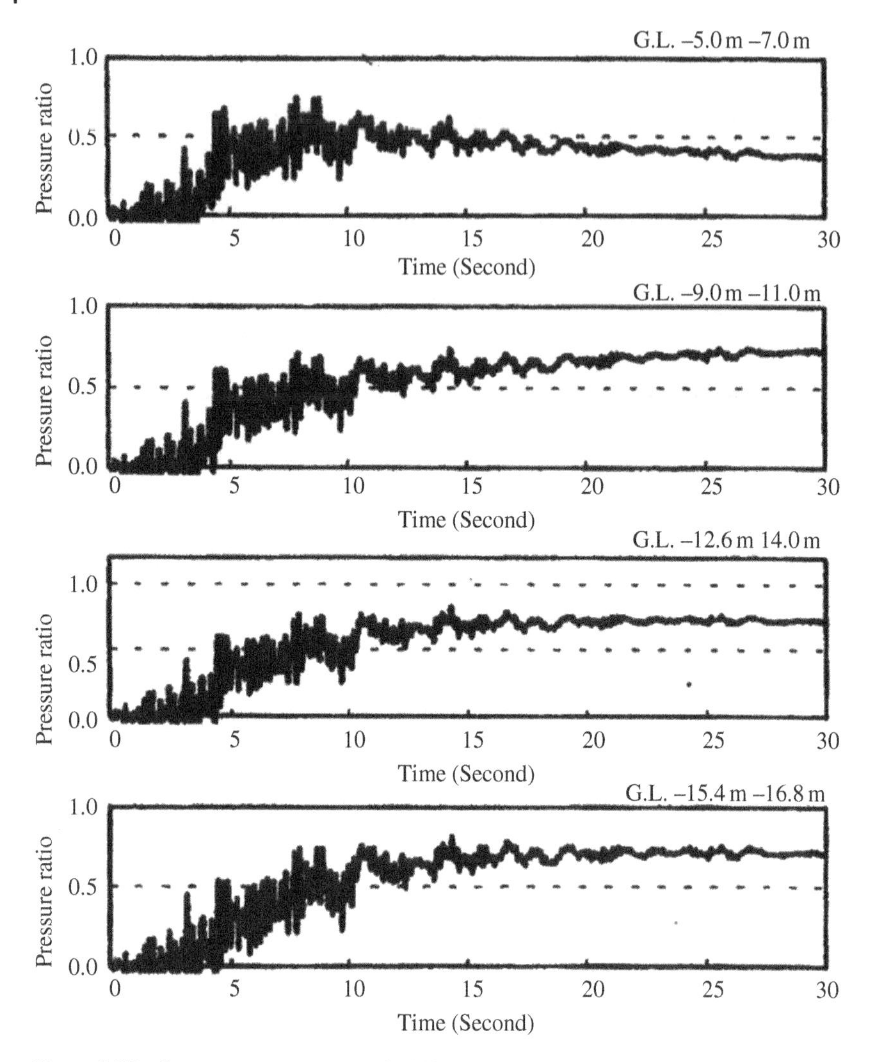

Figure 9.25 Excess pore pressure ratio (NS + EW + UD). *Source:* From Shiomi and Yoshizawa (1996).

In this section, a layered soil ground was analyzed with two different initial conditions, i.e. "with ISS" and "without ISS" conditions. The earth pressure at rest K_0 is assumed to be 1.0 for the case of condition "without ISS" and 0.5 for the case "with ISS." The mean stresses were kept the same at the same depth since liquefaction strength has long been assumed to be the same for the same mean stress in the soil column problem (Yoshimi 1991). This means that vertical stress has to be changed to maintain the mean stress. This assumption was made to simulate the liquefaction safety factor. The analysis of these examples was very significant. The same results were obtained for most of the constitutive models in which criteria depend on the maximum shear stress in three-dimensional space. However, there are some analysis codes that do not detect any significance for this problem. This difference in response is obviously caused by the different constitutive equation used in different codes.

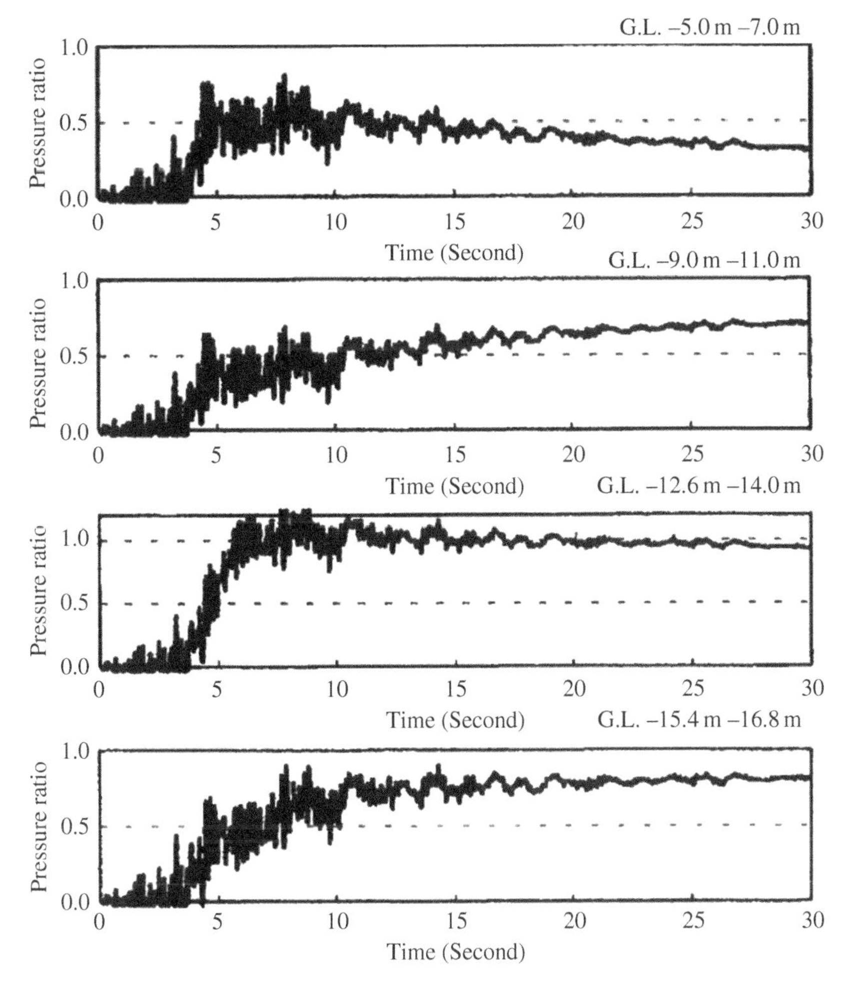

Figure 9.26 Excess pore pressure ratio (principal direction). *Source:* From Shiomi and Yoshizawa (1996).

9.6.1 Influence of Initial Shear Stress

Initial stress is calculated or modeled differently depending upon analysis codes and/or analysts (Shiomi and Shigeno 1993). However, it is very difficult to evaluate the initial stress since no one knows how the ground was formed geologically and no measurement is available even for the ratio of lateral earth pressure ratio to the vertical stress. Therefore, several calculations could be made to find appropriate initial stress. Then the choice would be made according to the experience of the engineer and this could result in completely different results. The differences in results are mainly caused by the initial shear stress (ISS) component of the initial stress. We are going to comment on this using an example of a one-dimensional layer problem and show how significant the difference can be.

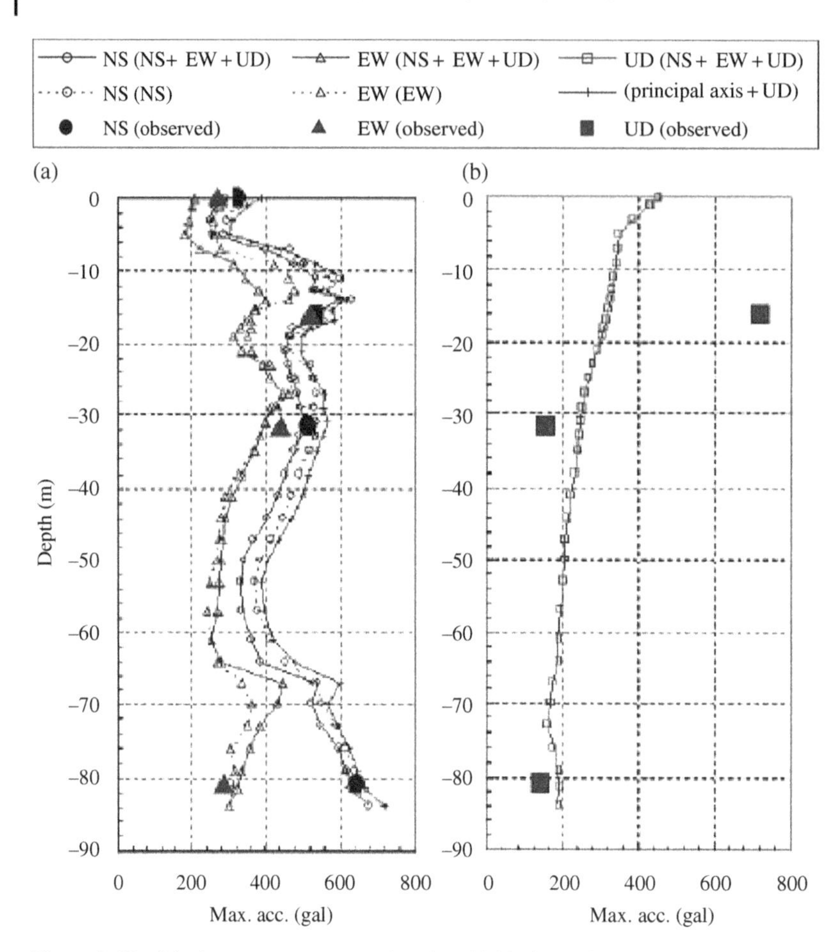

Figure 9.27 Maximum response acceleration (a) Horizontal component (b) Vertical component. *Source:* From Shiomi and Yoshizawa (1996).

9.6.1.1 Significance of ISS Component to the Responses

9.6.1.1.1 Response Acceleration

Figure 9.28 shows the maximum response acceleration and displacement. The response near the surface of "with ISS", i.e. $K_0 = 0.5$, is larger for acceleration and smaller for displacement than those of "without ISS". This means that the stiffness of the soil material for the case "with ISS" is reduced more than in the case "without ISS." The existence of ISS places the stress closer to the failure line so the material with ISS is then "weaker" and this is the main reason for smaller acceleration and larger displacement response for the case with ISS. The difference is about 30% for acceleration and 20% for displacement.

9.6.1.2 Excess Pore Water Pressure

Figure 9.29 shows the time history of the liquefaction ratio. The pore water pressure of the case "with ISS" is built up quickly and higher. Both cases reach full liquefaction but the

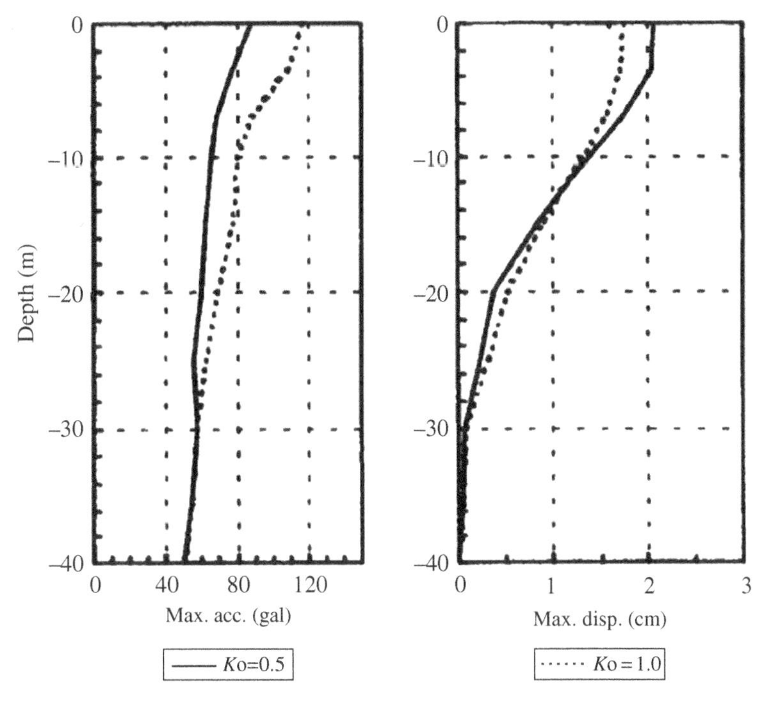

Figure 9.28 Profile of maximum response acceleration. *Source:* From Shiomi and Yoshizawa (1996).

excess pore water pressure without ISS is about 10% less than in the case of "without ISS." This tendency was almost the same as the laboratory element tests by Vaid and Finn (1979) and Hyodo et al. (1988) except for the value of the final pore water pressure. The liquefaction ratio is 75% for the "with ISS" and 40% for the "without ISS" at 3.5 seconds.

Figure 9.29 may give a clue as to why the liquefaction strength curve gives the same value for the same mean stress. The liquefaction strength curves were determined by the number of cycles for the samples to reach the final liquefaction. The criteria for the final liquefaction is either 5% shear strain in deformation or pore pressure reached 95% of the initial vertical stress (JGS 2000). As the pore water pressure ratios reached 1.0 at almost the same time at about eight seconds, so the liquefaction strength could be the same for the same mean stress despite the difference in initial shear stress.

9.6.1.2.1 Theoretical Consideration

ISS can, therefore, be classified into two types. Type I, ISS is seen in the case where an external force is applied in a perpendicular direction to ISS. In this case, the incremental shear stress has mainly the effect to rotate the principal stress with an increment of equivalent stress being small. For example, the soil beneath a structure has almost no shear stress in the horizontal direction; however, deviatoric shear stress (vertical stress minus horizontal stress) is relatively large. The maximum shear stress acts for the direction of $45 + \phi/2°$. External force due to earthquake acceleration produces a large horizontal shear stress and the principal stress direction is then rotated. Type II, ISS is the case where an external

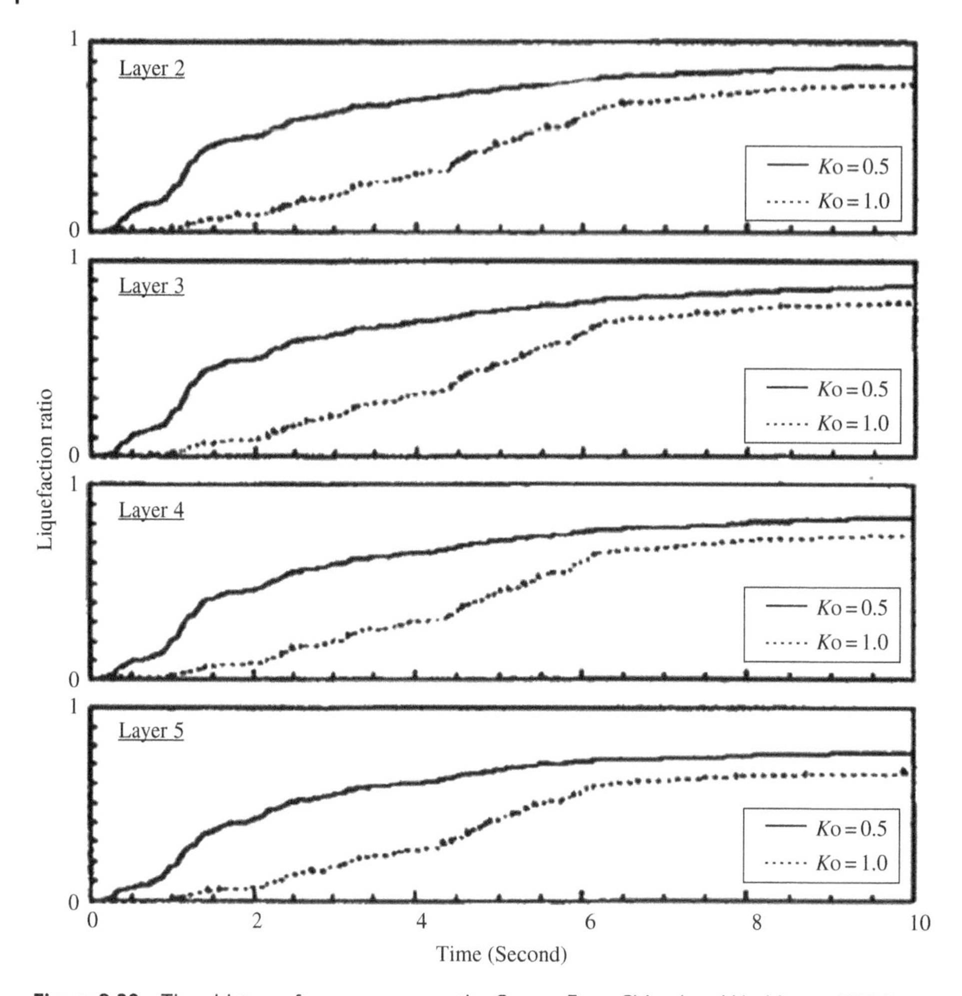

Figure 9.29 Time history of pore pressure ratio. *Source:* From Shiomi and Yoshizawa (1996).

force is applied to the direction parallel to the maximum shear direction of ISS. In this case, the incremental equivalent shear stress is equal to the external incremental shear stress. For example, a slope such as that of a dam has an ISS close to the horizontal direction. These two types of ISS might work differently. There is no substantial evidence, however, that ISS should affect liquefaction differently. A soil-column-type dynamic effective analysis can obviously be classified into Type I.

It was found that the existence of ISS results in a slower buildup of the pore water pressure as indicated by the numerical experiments. In the example problem, the upper soil layers with ISS were weakened more than in the case without ISS. It should be noticed that ground layer analyses frequently neglect ISS. Constitutive models, based on a typical elastoplastic theory that employs a plastic flow rule, produce different results for the conditions "with ISS" and "without ISS." The constitutive models developed as an extension of a one-dimensional shear soil column model often ignored ISS (deviatoric stress due to the difference of vertical stress and horizontal stress) since most of the models use hyperbolic stress–strain model for the shear behavior and the strain always starts from zero. That means that ISS is not involved in the formulation.

9.7 Large-Scale Liquefaction Experiment Using Three-Dimensional Nonlinear Analysis

In the above, we mentioned that nonlinear analysis is inevitable in order to solve large earthquake problems and the three-dimensional effect on liquefaction problems. At the same time, it is also important that the parameters of constitutive equation can be determined from standard soil tests, i.e. in situ tests and laboratory tests for ordinary engineering practice. Yoshida et al. (2008b) have reported an example of simulation analysis for a large-scale liquefaction experiment.

9.7.1 Analytical Model and Condition

The experiment took place at Black Thunder Coal Mine, Wyoming, USA. The experiment simulated is a two-story frame structure that was supported by piles at four corners. The piles stand on the excavated clay and mudstone and surrounded by sand, where liquefaction may take place. The frame structure and ground are shaken by an explosive blast of the mine at some distance (Figure 9.30). The structure and soil mass by 16 m times 16 m and 4 m depth are modeled as shown in Figure 9.31. The piles and columns are modeled by beam elements and floors and ground are modeled using eight noded-solid element.

9.7.1.1 Constitutive Model

The constitutive model used here is an accumulated damage model which includes dilatancy behavior by adding it to shear failure criteria as Zienkiewicz et al. (1978) has done in

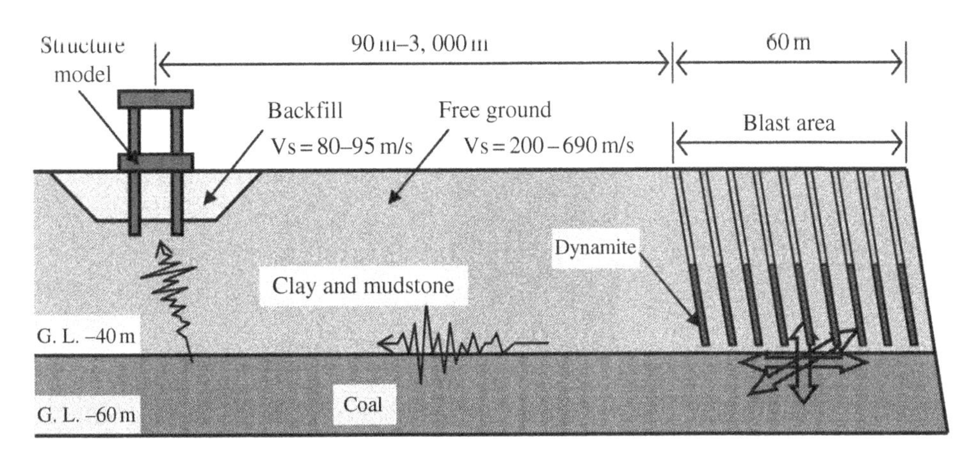

Figure 9.30 Liquefaction experiment by blast vibration in a coal mine. *Source:* From Yoshida et al. (2008b).

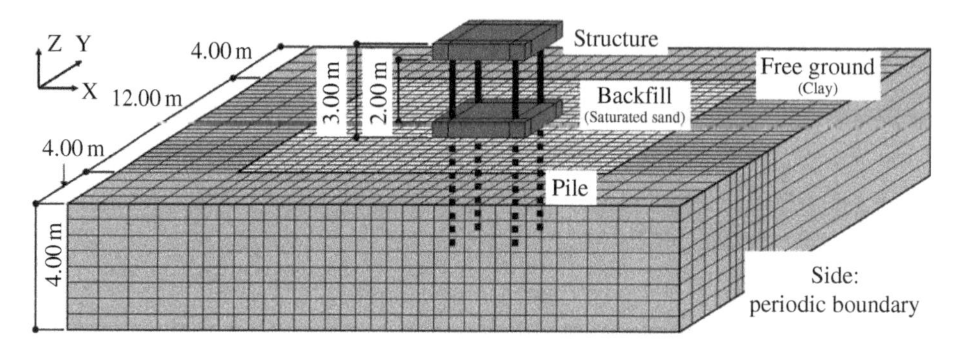

Figure 9.31 Three-dimensional FEA model. *Source:* From Yoshida et al. (2008b).

the densification model. The accumulated damage model here is an extension of the accumulated damage parameter model of the cyclic-wise equivalent method (Section 9.4.1) corresponding to time step instead of half-cycle of response that is defined only with the liquefaction strength curve. Therefore, users are able to determine soil parameters directly from field test and soil tests (Yoshida et al. 2008a).

The shear behavior of the constitutive model is defined in shear stress on π-plane with the Drucker–Prager yield surface (dependent on effective mean stress and is influenced by dilatancy) with a kind of sub-loading mechanism by Hashiguchi (2009) except that the hardening rule of the sub-loading is defined by a skeleton curve, which is calculated from G/G_0~γ and h~γ curve. The curve is determined using data obtained from an ordinary soil test. The hardening rule is explained as follows.

During virgin loading, isotropic loading takes place and the radius of sub-loading (i) is governed by a skeleton curve as is done in a one-dimensional problem. The failure surface corresponds to the shear strength. The shear strain of a one-dimensional problem corresponds to the effective strain in a three-dimensional problem. During hysteretic loading, the radius of the new sub-loading that shares a tangent line with outer sub-loading expands to the outer sub-loading following Masing's hysteretic rule (Figure 9.32).

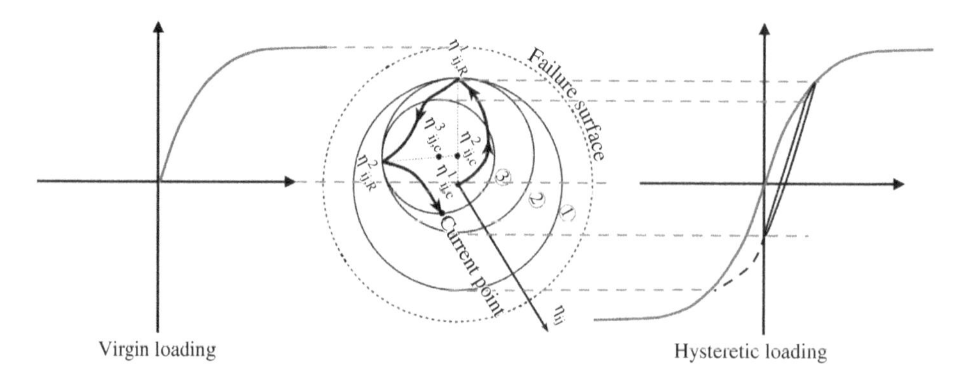

Figure 9.32 Stress path on Pi plane. *Source:* Modified from Yoshida et al. (2008a).

9.7.1.2 Dilatancy Modeling

The accumulated damage model is similar to the densification model, which was introduced in the first edition of this book (Zienkiewicz et al. 1999). It is calculated from the stress ratio and the liquefaction curve for the accumulated damage model.

$$D_t = \sum \Delta D_i + \Delta D\, \sigma'_m \tag{9.13}$$

where D_t is the accumulated damage parameter at time t and ΔD_i is the increment of the damage parameter at ith half cycle, where it is obtained as $\frac{1}{2N}$ of the corresponding maximum stress ratio ($R = \bar{\sigma}/\sigma_{m0}$) between zero crossings of shear strain in the time history. The relationship D and excess pore-pressure ratio r_u can be empirically modeled by Equation (9.14) before cyclic mobility takes place and Equation (9.15) for after cyclic mobility experienced once (Yoshida et al. 2008a) that are modification of Equation (9.12) considering cyclic mobility. The increment of plastic volumetric strain is calculated from the effective mean stress increment.

$$r_u = (1 - D) + \frac{2\sin^{-1}\left(2D^{1/0.8} - 1\right)}{\pi} \tag{9.14}$$

$$r_u = \left(1 - r_{u(D = D_{CM})}\right)\left(\frac{D - D_{CM}}{1 - D_{CM}}\right)^{D_{CM}} + r_{u(D = D_{CM})} \tag{9.15}$$

Shear modulus during cyclic mobility is also modeled empirically as follows:

$$\begin{aligned} G_T &= G_{CM} + f_b \cdot \frac{\sigma'_m - \sigma'_{mp}}{\sigma'_{m0}} \cdot G_{m0} \\ G_{CM} &= G_{m0} \cdot 10^{6\left(\sigma'_{mp}/\sigma'_{m0}\right)^{0.15} - 6.6} \\ f_b &= 5b \cdot \exp\left(-40 \cdot \gamma_{max}\right) + b \\ b &= 350 \cdot \left(G_{m0}/\sigma'_{m0}\right)^{-1.4} + 0.045 \end{aligned} \tag{9.16}$$

where G_T is shear modulus during loading state of the cyclic mobility (the stress ratio is over a phase transformation line), G_{CM} is shear modulus when the cyclic mobility starts and determined as a function of the effective mean stress σ'_{mp} when stress crosses the phase transformation line ϕ_p. σ'_{m0} and G_{m0} are the initial effective mean stress and shear modulus corresponding to it. γ_{max} is the maximum shear strain experienced until the current time. σ'_m is obtained by Equation (9.17) where ϕ_f is friction angle. These equations only require the angle of phase transformation line as soil parameter in addition to the ordinary soil parameters such as initial shear modulus, liquefaction strength curve, shear strength τ_{max}, and the friction angle ϕ.

$$\sigma'_m = \frac{|\tau/\tau_{max}| - \tan\phi_p/\tan\phi_f}{\tan\phi_f - \tan\phi_p} \tag{9.17}$$

9.7.1.3 Determination of the Material Parameters

Obtaining soil parameters for liquefaction analysis can be an annoying task for engineers. Many constitutive equations require several simulation analyses of laboratory tests (cyclic triaxial/hollow cylinder test). The constitutive law mentioned above only needs soil parameters which can be directly determined by site investigation and soil tests. The parameters are determined in order of elastic modulus, hardening parameters for strain, and liquefaction strength.

Elastic modulus (small strain shear stiffness G and Poisson's ratio ν) are mostly determined from PS Logging.

The hardening parameter $\left(\text{skeleton curve } {}^{\tau}/_{\sigma_m} - {}^{\gamma}/_{\gamma_{ref}}\right)$ is determined from the cyclic deformation test (Figure 9.33a), which measures the secant shear modulus ratio vs. shear strain $G/G_0 - \gamma$ relationship. This one-dimensional relationship extended to effective stresses ratio against effective strain ${}^{\overline{\sigma}}/_{\sigma_m} - {}^{\overline{\varepsilon}}/_{\overline{\varepsilon}_{ref}}$. The hysteresis curve is determined to satisfy $h - \gamma$

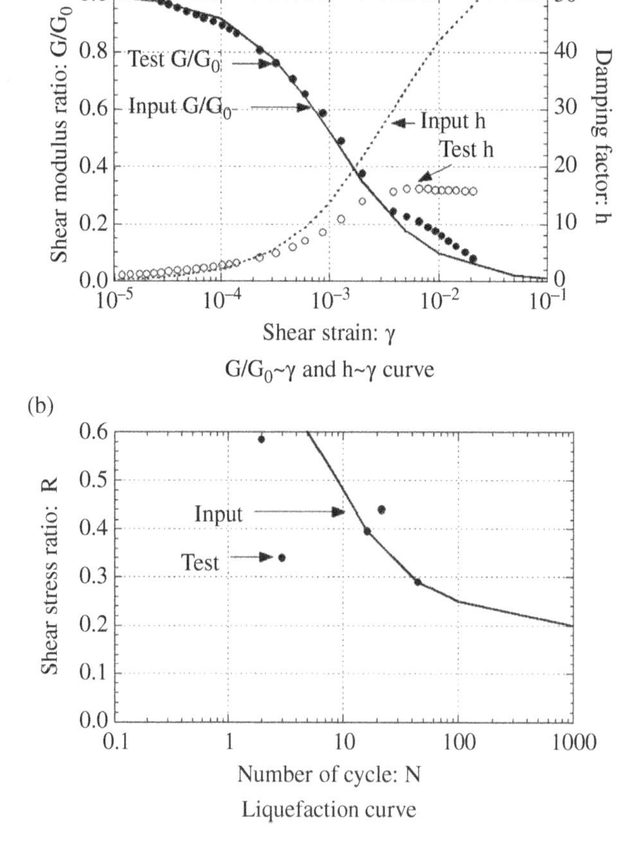

Figure 9.33 Diagram obtained by cyclic shear test and liquefaction test. *Source:* From Yoshida et al. (2008b). (a) G/G_0~γ and h~γ curve (b) Liquefaction curve.

relationship (Figure 9.33a). The shear strength (τ_{yield}) is obtained from $G/G_0 - \gamma$ relationship at the largest strain. This can be compared and modified if a simple loading shear test (such as CU test) exists. In this analysis, consolidation is not considered, so coefficients of consolidation are not required.

Dilatancy is defined by liquefaction strength curve (shear stress ratio τ/σ_{m0} vs. number of loading cycle N) which is obtained from the liquefaction strength test (Figure 9.33b). The phase transformation line angle which governs cyclic mobility behavior is then determined. The commonly used soil parameters are summarized in Table 9.5.

9.7.2 Input Motion

Input motion is obtained from the observed record at G.L. −4 m depth and 18 m away from the edge of the backfill (Figure 9.34).

9.7.3 Analysis Results

Figure 9.35 shows the excess pore water pressure-time history of the analysis at G.L.-1.4 m in the backfill with the observation. It reached the initial effective stress at 4.5 seconds. Both showed good agreement.

The acceleration time histories at the surface of the backfill (Figure 9.36) also show good agreement till the initial liquefaction at 4.5 seconds. Thereafter, the wave period of the observation becomes longer due to the initial liquefaction but the one from the analysis does

Table 9.5 Typical soil parameters (Yoshida et al. 2008b).

	S-wave velocity Vs (m/s)	P-wave velocity Vp (m/s)	Poisson ratio ν_d	Density of soil γ (t/m^3)	Density of water γ^f (kN/m^3)	Bulk modulus of soil Ks (kN/m^2)	Bulk modulus of water K^f (kN/m^2)
Free ground	215	480	0.37	1.67	—	—	—
Backfill	80	1530	0.33	1.89	9.81	1.0E+10	2.2E+6
	Shear modulus of soil G (kN/m^2)	Reference pressure σ_{ref} (kN/m^2)	Cohesion c (kN/m^2)	Internal friction angle ϕ_f (°)	Phase transformation Φ_p (°)	Porosity n	Permeability k (m/s)
Free ground	77 196	41.00	130	0	—	—	—
Backfill	12 096	13.80	0	17.44	17.27	0.44	1.00E−5

Source: From Yoshida et al. (2008b).

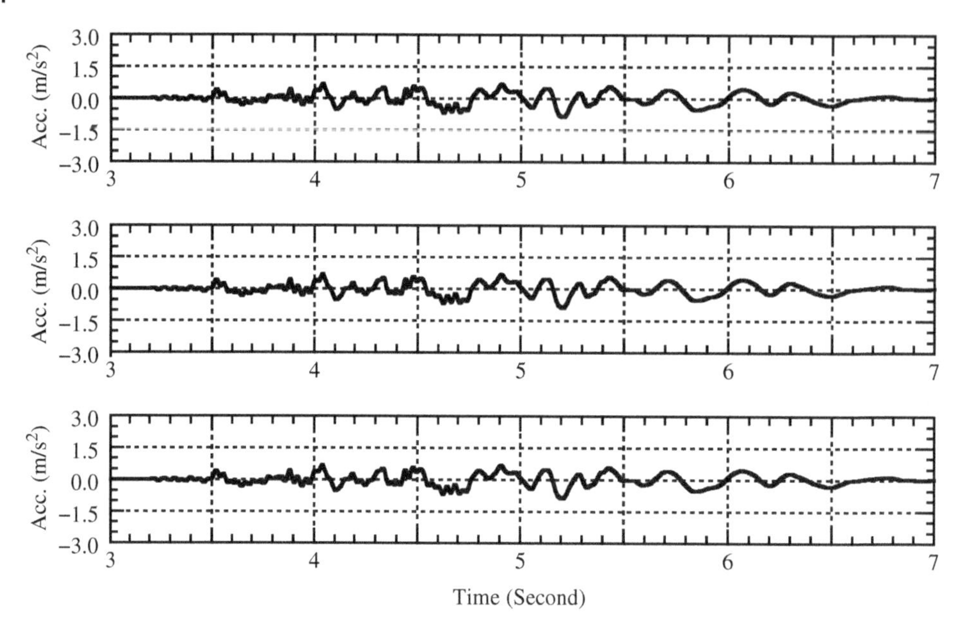

Figure 9.34 Input motion. *Source:* From Yoshida et al. (2008b).

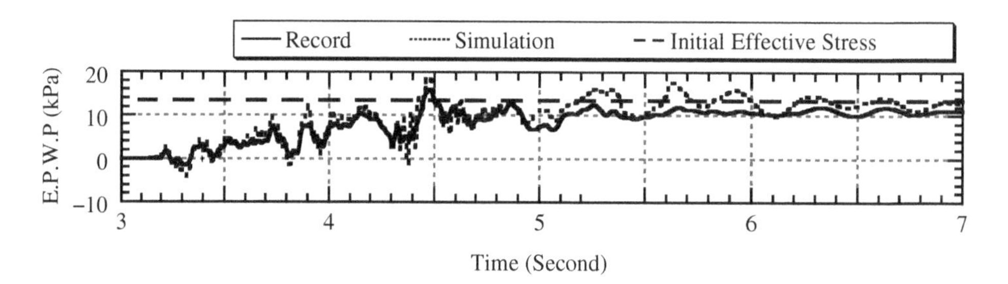

Figure 9.35 Excess pore water pressure at GL−1.4 m. *Source:* From Yoshida et al. (2008b).

not lengthen as much as the observation. The calculated shear stiffness at the liquefied soil is a little higher after 4.5 seconds in the analysis.

The computed acceleration time histories on the base slab show good agreement with the observed ones as shown in Figure 9.37. This may mean that the reduction of soil stiffness caused by the full liquefaction does not much affect the pile's behavior.

The time history of the computed curvature at the pile head after complete liquefaction simulates well the observed one (Figure 9.38). It is conceivable that the influence of the difference of the shear modulus of soil at liquefaction is small from the point of view of the response evaluation of the structure.

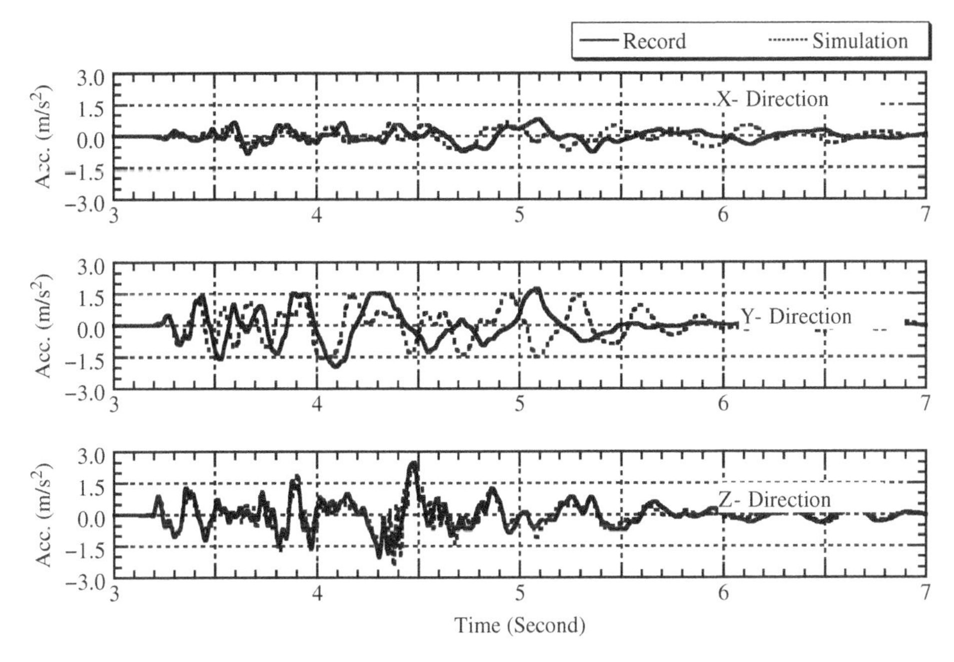

Figure 9.36 Acceleration on the surface of the backfill. *Source:* From Yoshida et al. (2008b).

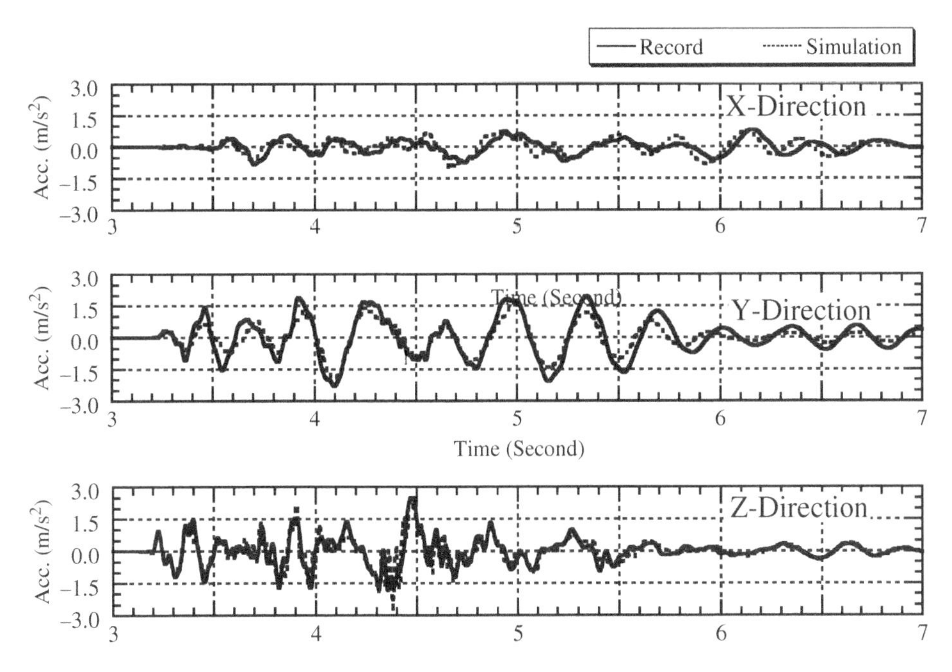

Figure 9.37 Acceleration at the Base slab. *Source:* From Yoshida et al. (2008b).

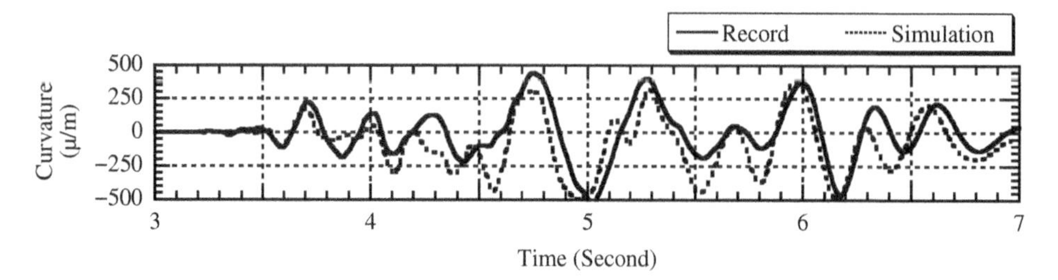

Figure 9.38 Curvature at the pile head. *Source:* From Yoshida et al. (2008b).

9.8 Lower San Fernando Dam Failure

The failure of the lower San Fernando earth dam in 1971 with nearly catastrophic consequences is typical of what can occur in a poorly consolidated soil structure affected by shaking, resulting from an earthquake. Zienkiewicz and Xie (1991) have reported some results of numerical simulation of the failure. Here, the details are presented. In particular, the effect of cohesion resulting from negative pore pressure and the influence of such parameters as permeability and relative density on the dynamic response of the dam are illustrated.

Although full comparative measurements are not available, the reconstruction of the event by Seed et al. (1975) and Seed (1979) is remarkable in attempting to explain why the failure occurred apparently some 60–90 seconds after the start of the earthquake. The actual collapsed dam and a "reconstructed" cross section are shown in Figure 9.39 following Seed (1979). The hypothesis made here was that the important pressure buildup occurring as a result of cyclic loading which manifested itself first in the central portions of the dam "migrates" in the post-earthquake period to regions closer to the "heel" of the dam where it triggers the failure.

We show in Figure 9.40 the material idealization, finite element meshes, and boundary conditions used in the present computations. First, an initial, elastic static analysis is carried out by the full program considering the semi-saturated condition and assuming the gravity and external water pressure to be applied without dynamic effects. Figure 9.41 shows such an initial steady-state solution for the saturation and the pore pressure distribution, indicating clearly the "phreatic" line and the suction pressures developing above.

Starting with the above-computed effective stress and pressure distribution, a full nonlinear dynamic computation is carried out for the period of the earthquake and continued for a further time of 200 seconds. The material properties assumed to describe the various zones of the dam using the constitutive model described in Chapter 4 and also found in Pastor and Zienkiewicz (1986) and Pastor et al. (1988,1990) are summarized in Table 9.6.

Figure 9.42 shows the displaced form of the dam at various times. The displacements at some characteristic points and the development and decay of excess pore pressures are shown in Figures 9.43 and 9.44. It is noted that deformations continue to increase for a considerable period after the end of the earthquake. This undoubtedly is aided by the

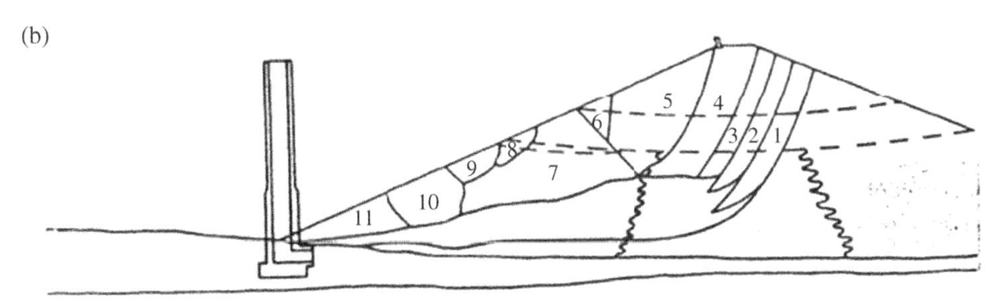

Figure 9.39 Failure and reconstruction of original conditions of the Lower San Fernando dam. *Source:* From Seed (1979).

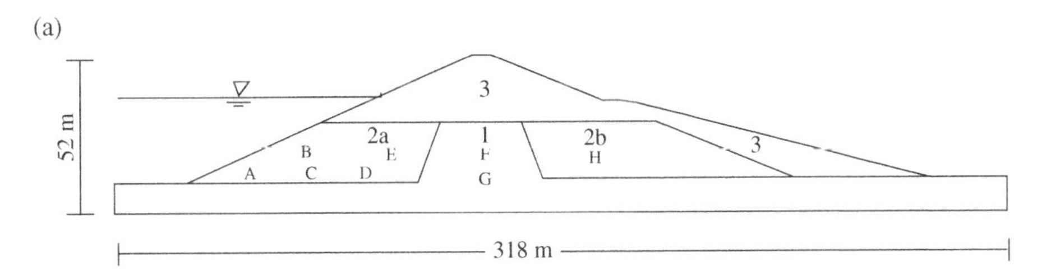

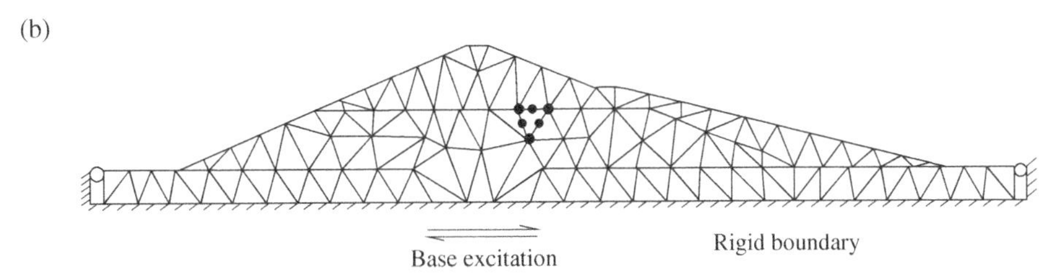

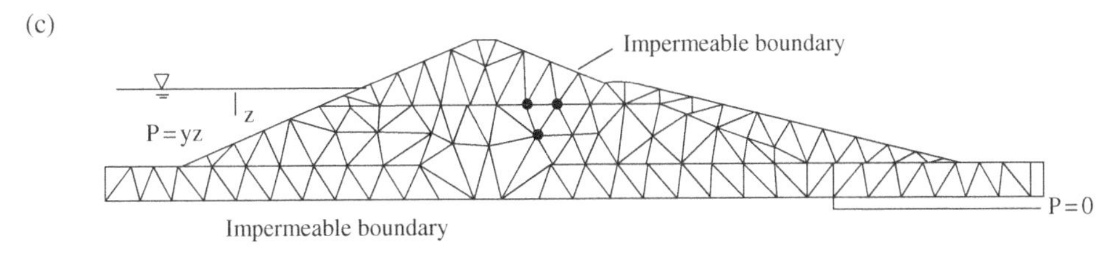

Figure 9.40 Idealization of San Fernando dam for analysis: (a) material zones (see Table 9.6); (b) displacement discretization and boundary conditions; and (c) pore pressure discretization and boundary conditions

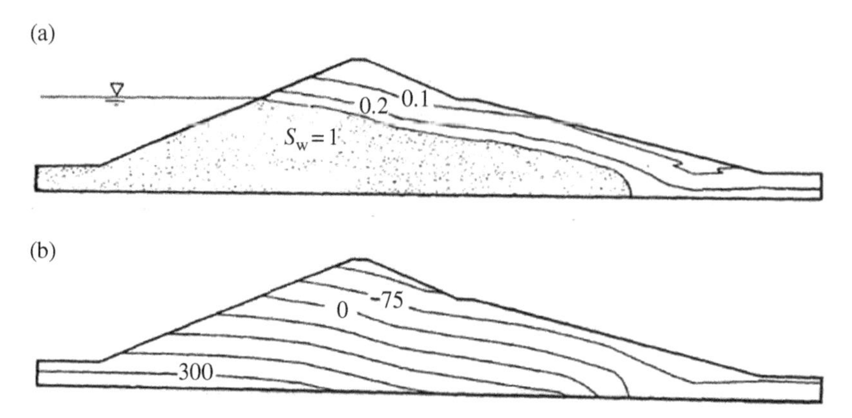

Figure 9.41 Initial steady-state solution: (a) pressure (kPa); and (b) saturation contours

Table 9.6 Material properties used in the Lower San Fernando dam analysis.

Material zone	ρ_s (kg/m^3)	ρ_f (kg/m^3)	K_s (Pa)	K_f (Pa)	ν	Porosity	k (m/s)
1	2090.0	980.0,	10^{12}	2×10^9	0.2857	0.375	10^{-3}
2a	2020.0	980.0	10^{12}	2×10^9	0.2857	0.375	10^{-2}
2b	2020.0	980.0	10^{12}	2×10^9	0.2857	0.375	10^{-2}
3	2020.0	980.0	10^{12}	2×10^9	0.2857	0.375	10^{-3}

Material zone	K_{evo}	K_{eso}	M_g	M_f	α_g	α_f	β_o	β_1	H_o	H_{uo} (Pa)	γ_u	γ_{DM}
1	120.0	180.0	1.55	1.400	0.45	0.45	4.2	0.2	700.0	6×10^7	2.0	2.0
2a	70.0	105.0	1.51	0.755	0.45	0.45	4.2	0.2	408.3	3.5×10^7	2.0	2.0
2b	75.0	112.6	1.51	0.906	0.45	0.45	4.2	0.2	437.5	3.75×10^7	2.0	2.0
3	80.0	120.0	1.51	1.133	0.45	0.45	4.2	0.2	467.0	4.00×10^7	2.0	2.0

redistribution of pore pressures. Near the upstream surface, the pore pressures continue to rise well after the passage of the earthquake. This indeed was conjectured by Seed (1979). It is also noted that the pattern of deformation is very similar to that which occurred in the actual case showing large movements near the upstream base and indicating the motion along the failure plane.

The suction pressures which developed above the phreatic line give a substantial cohesion there. Indeed, preliminary computation indicates that without such cohesion, an almost immediate local failure would develop in the dry material upon shaking.

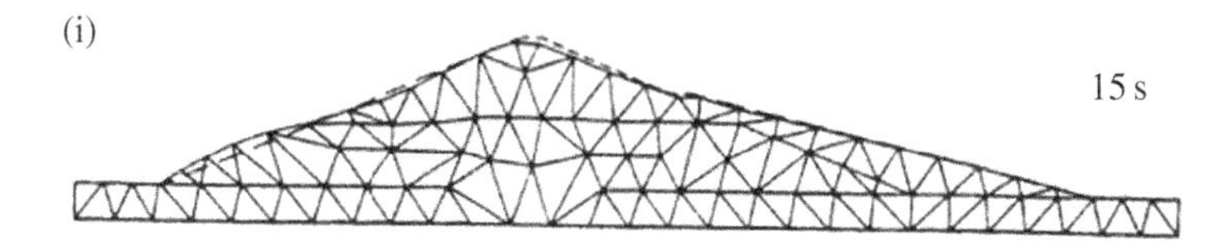

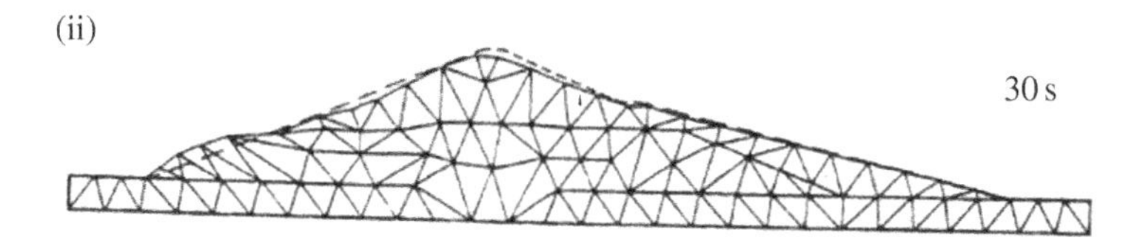

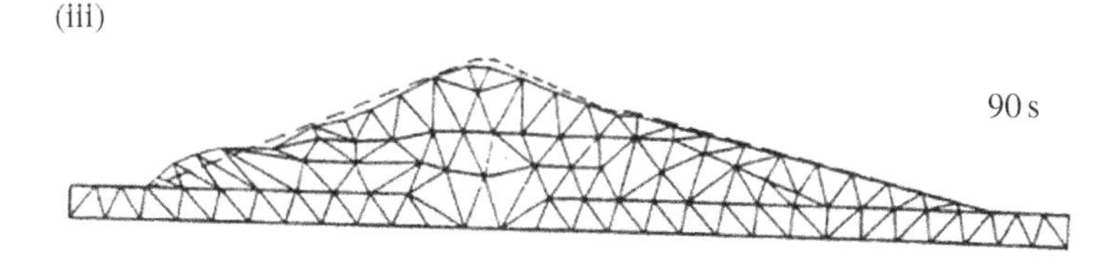

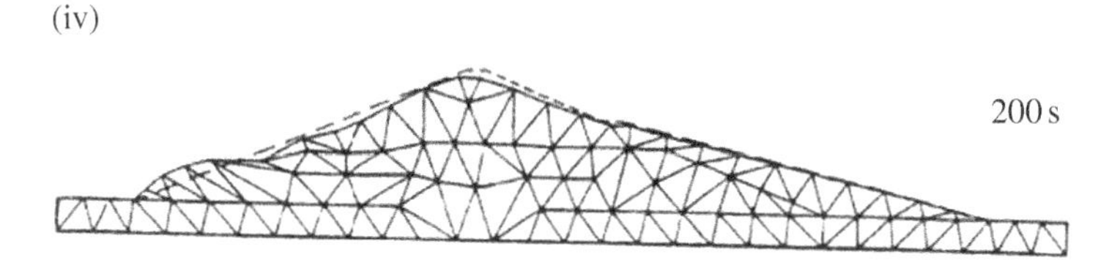

Figure 9.42 Deformed shapes of the dam at various times: (i) 15 s (end of earthquake); (ii) 30 s; (iii) 90 s; 120 s.

The amount of cohesion depends on the S_w–p_w (or S_w–h_w) curve in the following equation. Since p_a is assumed as zero, it is seen that the cohesion is of the value $S_w\ p_w$ in the otherwise cohesionless granular soil since the effective stress is defined as (see Equation (2.69))

$$\sigma'_{ij} = \sigma_{ij} + \delta_{ij}[S_w p_w + (1 - S_w)p_a] \tag{9.18}$$

when we reduce the parameter b of Van Genuchten's formula (Table 9.7) by a factor of 100, the pore pressure distribution of the static solution will be almost the same as in Figure 9.41b, but the saturation in the semi-saturated zones will be close to 1.

$$S_w = \delta + (1 - \delta)[1 + (\beta|h_w|)^{\gamma}]^{-1} \tag{9.19}$$

$$k_r = \left[1 + \left(a|h_w|^b\right)\right]^{-\alpha} \tag{9.20}$$

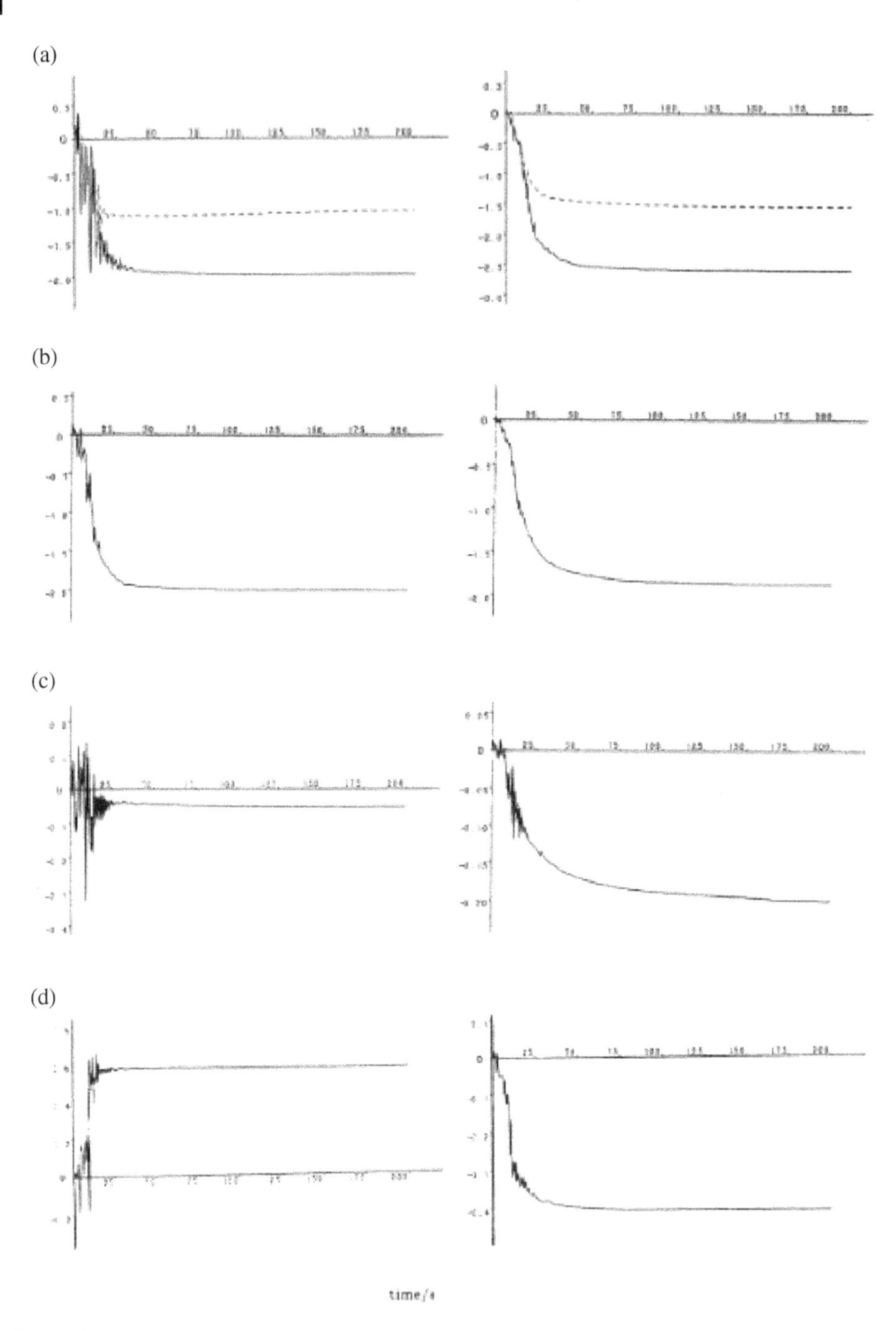

Figure 9.43 Horizontal (left) and vertical (right) displacements: (a) at the crest (dashed line represents the result of computation with increased "cohesion"); (b) at point E; (c) at point H; (d) at point I (see Figure 9.40a)

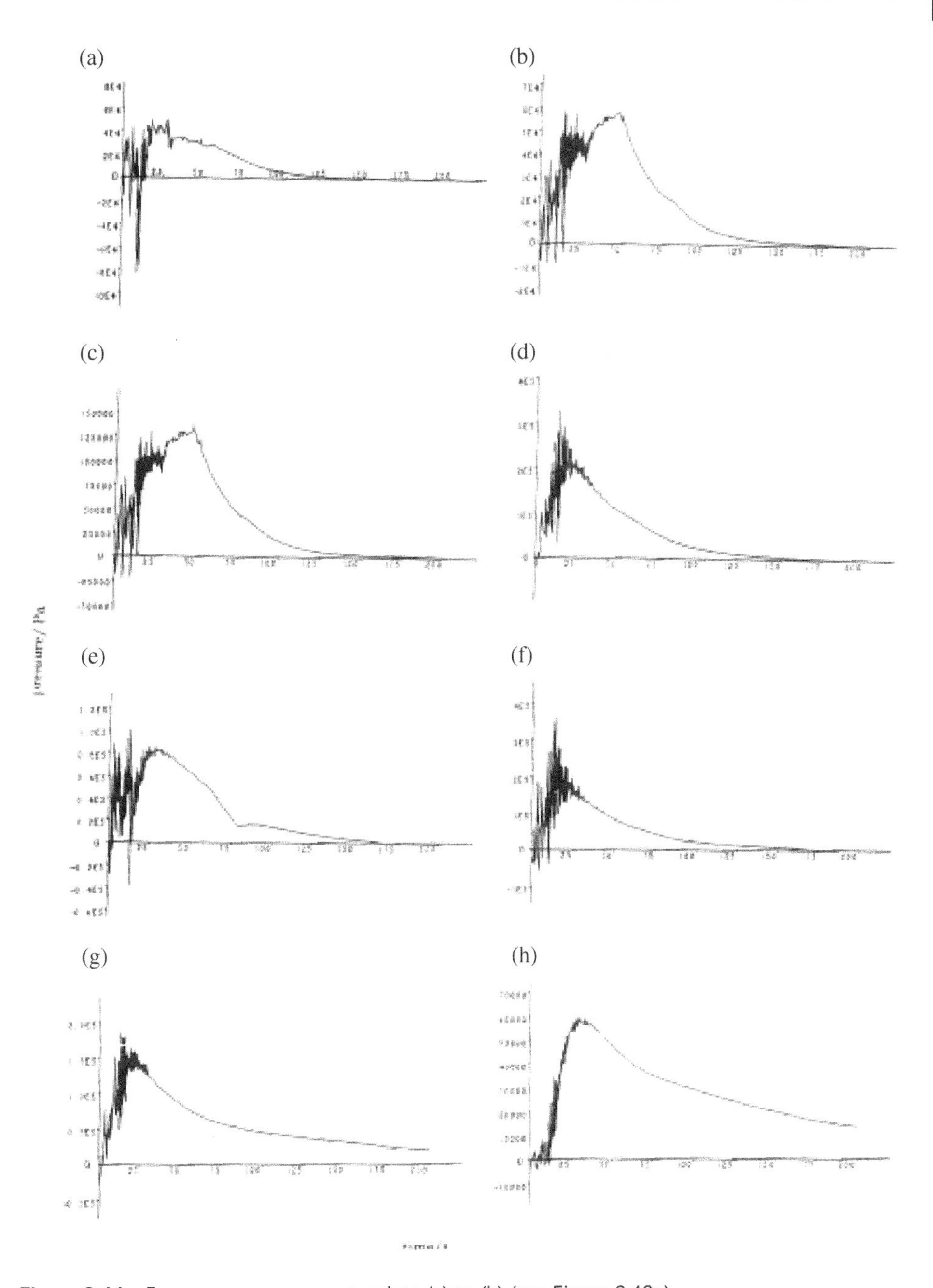

Figure 9.44 Excess pore pressure at points (a) to (h) (see Figure 9.40a)

The higher value of S_w results in stronger cohesion in the upper part of the dam. The dotted lines in Figure 9.43a are the results of computation occurred because of the now increased cohesion in the upper regions of the dam.

If the permeability of the dam material is sufficiently high, it may be impossible for an earthquake to cause any substantial buildup of pore pressures in the embankment since

Table 9.7 Coefficients of saturation function

	δ	β (cm^{-1})	γ	a (cm^{-1})	b	α
For sand[a]	0.068 9	0.017 40	2.5	0.066 7	5.0	1.00
For San Fernando	0.084 2	0.007 00	2.0	0.050 0	4.0	0.90

[a] After Van Genuchten et al. (1977)

the pore pressure can dissipate by drainage as rapidly as the earthquake can generate them by shaking. Figure 9.45 shows the results which indicate a rapid dissipation of pore pressures and much reduced permanent deformations.

In an additional analysis, the relative density of the dam material is assumed lower, which implies in the present constitutive model that the ratio M_f/M_g is considered equivalent to

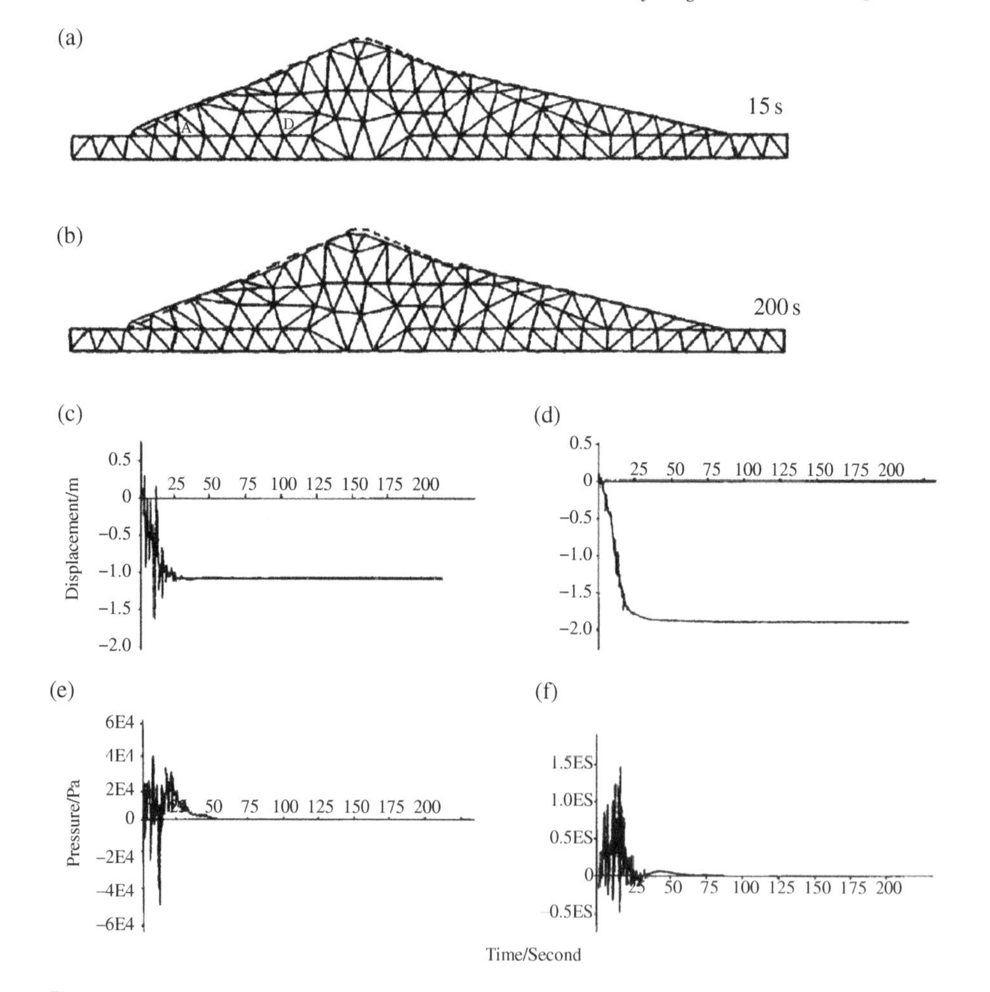

Figure 9.45 Results of analysis with increased permeabilities: (a) deformed shape of the dam after 15 seconds; (b) deformed shape of the dam at 200 seconds; (c) horizontal displacement on the crest; (d) vertical displacement on the crest; (e) excess pore pressure at point A; (f) excess pore pressure at point D (see Figure 9.40a)

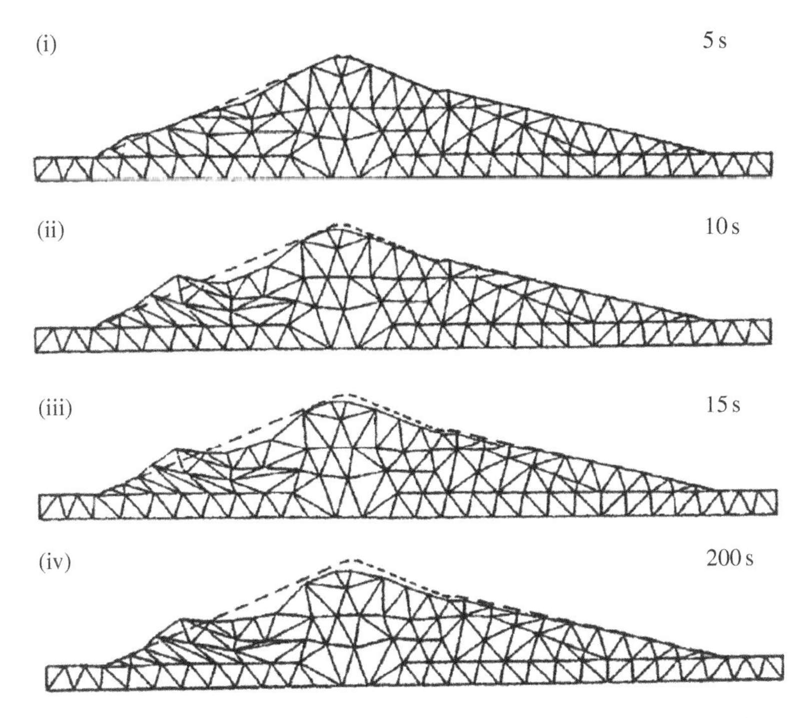

Figure 9.46 Results of analysis with softer materials, showing deformed shapes

the relative density (Pastor et al. 1985). With M_g values in Table 9.6 fixed, M_f values are now reduced to 1.24, 0.453, 0.604, and 0.906 for material zones 1, 2a, 2b, and 3, respectively. In this case, significantly larger displacements are recorded at the early stages of the earthquake shaking as shown in Figures 9.45 and 9.46.

References

Aydingun, O. and Adalier, K. (2003). Numerical analysis of seismically induced liquefaction in earth embankment foundations. Part I. Benchmark model. *Can. Geotech. J.* **40** (4): 753–765.

Dunn, S. L., Vun, P. L., Chan, A. H. C., and Damgaard, J.S. (2006). Numerical modeling of wave-induced liquefaction around buried pipelines. *J. Waterw. Port Coast. Ocean Eng.* **132** (4): 276–288. https://doi.org/10.1061/(ASCE)0733-950X(2006)132:4(276).

Elia, G. and Rouainia, M. (2010). 2D Finite element analysis of the seismic response of an earth embankment. *SECED Newsl.* **22** (2) May 2010: 1–5.

Elia, G., Amorosi, A., Chan, A. H. C., and Kavvadas, M. J. (2011a). Numerical prediction of the dynamic behaviour of two earth dams in Italy using a fully-coupled non-linear approach. *Int. J. Geosci.* **11**: 2010.

Elia, G., Amorosi, A., Chan, A. H. C., and Kavvadas, M. J. (2011b). Fully coupled dynamic analysis of an earth dam. *Géotechnique* **61** (7): 549–563. July 2011. (DOI:10.1680/geot.8.P.028.

Fujiwara, Y. and Shiomi, T. (2011). Practical hysteretic damping model of soil dynamic analysis. *The 61st National Congress of Theoretical and Applied Mechanics*, No.61-120-, 2012/03/28, Tokyo, Japan (in Japanese).

Fukutake, K. and Ohtsuki, A., and Fujikawa, S. (1995). Applicability of 2D analysis and merit of 3D analysis in liquefaction phenomena. *Proceedings of Symposium on Three Dimensional Evaluation of Ground Failure*, Tokyo, 229–236 (in Japanese).

Ghaboussi, J. and Dikmen, S. U. (1981). Liquefaction analysis for multidirectional shaking. *J. Geotech. Eng. Div.*, Proc. of ASCE, 197, GT5 **107**: 605–627.

Ghosh, B., Klar, A., and Madabhushi, S. P. G. (2005). Modification of site response in presence of localised soft layer. *J. Earthq. Eng.* **9** (6): 855–876. Imperial College Press, London.

Haigh, S. K., Ghosh, B., and Madabhushi, S.P.G. (2005). The effect of time step discretisation on dynamic finite element analysis. *Can. Geotech. J.* **42** (3): 957–963.

Hardin, B. O. and Drnevich, V. P. (1972). Shear modulus and damping in soils. *J. Soil Mech. Found. Div.* **98** (7): 667–692.

Hashiguchi, K. (2009). *Elastoplasticity Theory, Lecture Notes in Applied and Computational Mechanics*, vol. **42**. Springer.

Hatanaka, M. and Uchida, A. (1995). Earthquake geotechnical engineering: proceedings of IS-Tokyo '95, the first International Conference on Earthquake Geotechnical Engineering, Tokyo, 14 –16 November 1995/ed. by Kenji Ishihara.

Hatanaka, M., Uchida, A., and Ohara, J. (1997). Liquefaction characteristics of a gravelly fill liquefied during the 1995 Hyogo-Ken Nanbu earthquake. *Soils Found.* **37** (3): 107–115.

Hyodo, M., Yamamoto, Y., and Sugiyama, M. (1994). Undrained cyclic shear behaviour of normally consolidated clay subjected to initial static shear stress. *Soils and Foundations* **34** (4): 1–11.

Ishihara, K. (1989). Dynamic behaviour of ground and earth structure – numerical method and problem specification. *Proceedings of Symposium on Behaviour of Ground and Earth Structure During Earthquake held by Japanese Society of Soil and Foundation Engineering in Tokyo*, 50–63, (in Japanese).

Ishihara, K. (1996). *Soil Behaviour in Earthquake Geotechnics.* Oxford Science Publications.

Ishihara, K. and Nagase, H. (1988). Multi-directional irregular loading tests on sand. *Soil Dyn. Earthq. Eng.* **7**, 1988: 201–212.

Ishihara, K. and Yamazaki, F. (1980). Cyclic simple shear tests on saturated sand in multi-directional loading. *Soils Found.* **20** (1): 45–59.

Ishihara, K., Tatsuoka, F., and Yasuda, S. (1975). Undrained deformation and liquefaction of sand under cyclic stresses. *Soils Found.* **15** (1): 29–44.

Ishihara, K., Yoshida, N., and Tsujino, S. (1985). Modeling of stress-strain relationship of soils in cyclic loading. *Proceeding of the 5th Internaltional Conference for Numerical Method in Geomechanics*, Nagoya (1–5 April), Vol. **1**, pp.373–380.

Iwasaki, Y. and Tai, M. (1996). Strong motion records at Kobe Port Island. *Soils Found.* (Sepcial Issue): 29–40.

JGS (The Japanese Geotechnical Society) (2000). Standards of Japanese geotechnical society for laboratory shear test.

Kansai-Kyogikai (1995). CEORKA:Kansai Earthqukae Observation Research Kyogikai, http://www.ceorka.org/.

Konder, R. L. and Zelasko, J. S. (1963). A hyperbolic stress-strain formulation for sands. *Proceedings 2nd Pan American Conference on Soil Mechanics and Foundations Engineering Brazil, 1963* (Vol. 1, pp. 289-324), 289–324.

Kyoto University Research Reactor Institute (2000). Report of Seismic Safty of the Kyoto University Research Ractor Based on Revised Regulation of Seismic Design for Nuclear Power Plant (in Japanese).

Ling, H. I., Liu, H., Kaliakin, V. N., and Leshchinsky, D. (2004). Analyzing dynamic behavior of geosynthetic-reinforced soil retaining walls. *J. Eng. Mech. ASCE.* **130** (8): 911–920. doi:http://dx.doi.org/10.1061/(ASCE)0733-9399(2004)130:8(911)).

Ling, H. I., Liu, H., and Mohri, Y. (2005). Parametric studies on the behavior of reinforced soil retaining walls under earthquake loading. *J. Eng. Mech. ASCE.* **131** (10): 1056–1065. doi:http://dx.doi.org/10.1061/(ASCE)0733-9399(2005)131:10(1056).

Ling, H. I., Sun, L., Liu, H. Mori, Y., and Kawabata, T. (2008). Finite element analysis of pipe buried in saturated soil deposit subject to earthquake loading. *J. Earthq. Tsunami* **2** (1): 1–17.

Liu, H. and Ling, H.I. (2007). A unified elastoplastic-viscoplastic bounding surface model of geosynthetics and its applications to GRS-RW analysis. *J. Eng. Mech. ASCE.* **133** (7): 801–815.

Liu, H. and Song, E. (2005). Seismic response of large underground structures in liquefiable soils subjected to horizontal and vertical earthquake excitations. *Comput. Geotech.* **32** (4): 223–244. https://doi.org/10.1016/j.compgeo.2005.02.002.

Liu, H. and Song, E. (2006). Working mechanism of cutoff walls in reducing uplift of large underground structures induced by soil liquefaction. *Comput. Geotech.* **33** (4–5) June–July 2006: 209–221. https://doi.org/10.1016/j.compgeo.2006.07.002.

Liu, H. and Won, M.-S. (2009). Long-term reinforcement load in geosynthetic-reinforced soil walls. *J. Geotech. Geoenviron. Eng. ASCE* **135** (7) July 2009, https://doi.org/10.1061/(ASCE)GT.1943-5606.0000052.

Madabhushi, S. P. G. and Chandrasekaran, V.S. (2006). On modelling the behaviour of flexible sheet pile walls. *Indian Geotech. J.* **36** (2): 160–180.

Madabhushi, S. P. G. and Zeng, X. (1998a). Seismic response of gravity quay wall II: numerical modelling. *J Geotech. Geoenviron. Eng., ASCE* **124** (5): 418–427. http://dx.doi.org/10.1061/(ASCE)1090-0241(1998)124:5(418).

Madabhushi, S. P. G. and Zeng, X. (1998b). Behaviour of gravity quay walls subjected to earthquake loading. Part II: Numerical modelling. *J. Geotech. Eng. ASCE* **124** (5): 418–428. USA.

Madabhushi, S. P. G. and Zeng, X. (2006). Seismic response of flexible cantilever retaining walls with dry backfill. *Intt. J. Geomech. Geoeng.* **1** (4): 275–290.

Madabhushi, S. P. G. and Zeng, X. (2007). Simulating seismic response of cantilever retaining walls with saturated backfill. *ASCE J. Geotech. GeoEnv. Eng.* **133** (5): 539–549.

Masing, G. (1926). Eigenspannungen und verfestigung beim messing, *Proceedings of the Second International Congress of Applied Mechanics*, Zurich (12–17 Sep 1926), pp. 332–335

Nagase, H. and Ishihara, K. (1988). Liquefaction-induced compaction and settlement of sand during earthquakes. *Soils Found.* **28** (1): 65–76.

NIED (National Research Institute for Earth Science and Disaster Prevention) (since 1996). Strong motion seismograph networks (K-NET,KiK-net). http://www.kyoshin.bosai.go.jp/ (10 May 2021)

NIED (National Research Institute for Earth Science and Disaster Prevention) (2007). 2007/07/16-10:13, The Niigataken Chuetsu-oki Earthquake in 2007, Past Large Earthquake, Strong motion seismograph networks (K-NET,KiK-net). http://www.kyoshin.bosai.go.jp/

NIED (National Research Institute for Earth Science and Disaster Prevention) (2011). 2011/03/11-14:46, The 2011 off the Pacific coast of Tohoku Earthquake, Past Large Earthquake, Strong motion seismograph networks (K-NET,KiK-net). http://www.kyoshin.bosai.go.jp/

Pastor, M. and Zienkiewicz, O. C. (1986). A generalized plasticity, hierarchical model for sand under monotonic and cyclic loading. In *International symposium on numerical models in geomechanics*. 2 (NUMOG II), Ghent, 31 Marth–4 April 1986, Eds. G.N. Pand and W.F. Van Impe., M.Jackson and Son, England (pp. 131–150).

Pastor, M. Zienkiewicz, O. C. and Leung, K. H. (1985). Simple model for transient soil loading in earthquake analysis, II Non-associative models for sands. *Int. J. Numeri. Anal. Methods Geotech.* **9**: 477–498.

Pastor, M., Zienkiewicz, O. C. and Chan, A. H. C. (1988). Simple models for soil behaviour and applications to problems of soil liquefaction, in: *Numerical Methods in Geomechanics* (ed. G. Swaboda), 169–180. A.A. Balkhema Geomechanics.

Pastor, M., Zienkiewicz, O. C., and Chan, A. H. C. (1990). Generalised plasticity and the modelling of soil behaviour. *Int. J. Numer. Anal. Methods Geomech.* **14**: 151–190.

Pyke, R. M., Chan, C. K., and Seed, H. B. (1975). Settlement of sands under multidirectional shaking. *J. Geotech. Eng. Div.* **101** (4): 379–398.

Ramberg, W. and Osgood, W. R. (1943). *Description of Stress-Strain Curves by Three Parameters*, Technical Note No. 902. National Advisory Committee For Aeronautics. Washington DC.

Schnabel, P. B., Lysmer, J. and Seed, H. B. (1972). SHAKE: A Computer Program for Earthquake Response Analysis of Horizontally Layered Sites. Report No. EERC 72-12, College of Engineering, University of California.

Seed, H.B. (1979). Consideration in the earthquake resistant design of earth and rockfill dams. *Geotechnique* **29**: 215–263. Geotechnique.

Seed, H. B., Lee, K. L., Idriss, I. M., and Makdisi, F. I. (1975). Analysis of slides of the San Fernando dams during the earthquake of February 9, (1971). *J. Geotech. Eng. Div. ASCE* **101** (GT7): 651–688.

Seed, H. B., Martin, P. P., and Lysmer, J. (1976). Pore-water pressure changes during soil liquefaction. *J. Geotech. Geoenviron. Eng.* **102** (4): 323–346.

Seed, H. B., Pyke, R. M., and Martin, G. R. (1978). Effect of multidirectional shaking on pore water pressure development on sands. *J. Geotech. Eng. Div. Proc. ASCE* **104** (GT1): 27–44.

Shiomi, T. (1994). Practical consideration on initial shear stresses on liquefaction analysis, in: *Computer Methods and Advances in Geomechanics* H.J. Siriwardane and M.M. Zaman (Eds), 973–980.

Shiomi, T. (1995). *User Manual of MuDIAN*. Takenaka Corp.

Shiomi, T. and Fujiwara, Y. (2012). Accuracy of frequency dependent equivalent linear analysis, geotechnical engineering magazine. *Jpn. Geotech. Soc.* **60** (1): 34–35. (in Japanese).

Shiomi, T. and Shigeno, Y. (1993). Consideration of initial shear stress on ground liquefaction, in *Proceedings of the Asian-Pacific Conference*, 3–6 August 1993 S. Valliappan, V. A. Pulmano, E. Tin-Loi (Eds), Balkema, Sydney, 1077–1082.

Shiomi, T. and Yoshizawa, M. (1996) Effect of multi-directional loading and initial stress of liquefaction behavior. *Proceedings of the 11th World Conference on Earthquake Engineering*, Acapulco, Mexico, June 23-28, 1996, Pergamon. ISBN:0 08 042822 3

Shiomi, T., Shigeno, Y., Sugimoto, M. and Suzuki, Y. (1991). Influence of liquefaction to pile-soil-structure interaction. *International Conference on Recent Advances in Geotechnical Earthquake Engineering and Soil Dynamics*, St. Louis, 11–15 March 1991, 465–472

Shiomi, T., Shigeno, Y., and Zienkiewicz, O. C. (1993). Numerical prediction for model No.1. In: *Verification of Numerical Procedures for the Analysis of Soil Liquefaction Problems*, (eds. K. Arulanandan and R.F. Scott) (Eds), Vol. 1. Balkema, pp. 213–219.

Shiomi, T., Nukui, Y., and Yagishita, F. (2008) Practical effective stress analysis and determination of its dynamic property. *Proceedings of the 14th World Conference of Earthquake Engineering*, Bejing, China (12–17 October 2008).

Sugito, M., Goda, H., and Masuda, T. (1994). Frequency dependent equi-linearized technique for seismic response analysis of multi-layered ground. *Doboku Gakkai Ronbunshu* **1994** (493): 49–58.

Tokimatsu, K. and Yoshimi, Y. (1983). Empirical correlation of soil liquefaction based on SPT N-value and fines content. *Soil Found.* **23** (4): 56–74.

Towhata, N. and Routeix, S. (1990). Applicability of melti-reflection theory in earthquake response of layered ground. *Proceedings of the 35th Symposium on Soil Mechanics and Foundation Engineering* (1 November 1990), 1–8 (in Japanese).

Uzuoka, R., Tateishi, A., and Yashima, A. (2003). Liquefaction mechanism, method of prediction and structural design. *Geotech. Eng. Mag.* **51** (2): 49–54. (in Japanese).

Vaid, Y. P. and Finn, W. D. L. (1979). Effect of static shear on liquefaction potential. *J. Geotech. Engng. Div. ASCE* **105** (GT10): 1233–1246.

Van Genuchten, M. T., Pinder, G. F. and Saukin, W. P. (1977). Modeling of leachate and soil interactions in an aquifer. Management of gas and leachate in landfills. *Municipal Solid Waste Research Symposium*, Rep. EPA-600/9-77-026. USEPA(United State Environmental Protection Agency), Cincinnati, OH, 95–103.

Ye, J. H., Jeng, D. S., and Chan, A. (2012). Consolidation and dynamics of 3D unsaturated porous seabed under rigid caisson breakwater loaded by hydrostatic pressure and wave. *Sci. China Technol. Sci.*, August 2012 **55** (8): 2362–2376.

Ye, J., Jeng, D., Liu, P.L.-F., Chan, A. H. C., Wang, R. and Zhu, C. (2013). Breaking wave-induced response of composite breakwater and liquefaction in seabed foundation. *Coast. Eng.* **85** (72–86): 2013. http://dx.doi.org/10.1016/j.coastaleng.2013.08.003.

Ye, J., Jeng, D.-S., Chan, A. H. C., Wang, R. and Zhu, Q. C. (2017). 3D integrated numerical model for Fluid-Structures-Seabed Interaction (FSSI): Loosely deposited seabed foundation. *Soil Dyn. Earthq. Eng.* **92**, 1 January 2017: 239–252. https://doi.org/10.1016/j.soildyn.2016.10.026.

Yoshida, H., Hijikata, K., Sugiyama, T., Takahashi, S., Shiomi, T. and Tokimatsu, K. (2007). Study on influence of multi-directional input by three dimensional effective stress analysis. *Summaries of technical papers of annual meeting, Architectural Institute of Japan, Fukuoka.*

Yoshida, Y., Tokimatsu, K., Hijikata, K. Sugiyama, T. and Shiomi, T. (2008a). Practical effective stress-strain model of sand considering cyclic mobility behavior. *J. Struct. Constru. Eng. Archit. Inst. Jpn.* **630**: 1257–1264.

Yoshida, H., Tokimatsu, K., Sugiyama, T., and Shiomi, T. (2008b). Practical three-dimensional effective stress analysis considering cyclic mobility behavior. *Proceedings of 14th World Conference on Earthquake Engineering*, Paper_ID (04–01).

Yoshimi (1991). Liquefaction of sany soil deposit (2nd Ed), *Influence of confining Confining pressurePressure*, Liquefaction of sand Sand, ISBN: 978-4-7655-1511-5, Gihodo-Shuppan.

Zienkiewicz, O. C. and Shiomi, T. (1984). Dynamic behaviour of saturated porous media, the generalized Biot formula and its numerical solution. *Int. J. Num. And Anal. Meth. Geomech.* **8**: 71–96.

Zienkiewicz, O. C. and Xie, Y. M. (1991). Analysis of lower san fernando dam failure under earthquake. *Dam Eng.* **2** (4): 307–322.

Zienkiewicz, O. C., Chang, C. T., and Hinton, E. (1978). Nonlinear seismic response and liquefaction. *Int. J. Numer. Anal. Methods Geomech.* **2** (4): 381–404.

Zienkiewicz, O. C., Chan, A. H. C., Pastor, M., Schrefler, B. A. and Shiomi, T. (1999). *Computational Geomechanics with Special Reference to Earthquake Engineering*. Chichester: Wiley February 1999.

10

Beyond Failure: Modeling of Fluidized Geomaterials

Application to Fast Catastrophic Landslides

10.1 Introduction

Engineers analyze the behavior of geostructures under design loads and predict the conditions under which failure will take place in order to assess how far are design conditions from failure (which is quantified with a safety factor), the limit load and the failure mechanism. This will allow reinforcing the weakest zones. The analysis is not usually carried to post-failure behavior. However, there are cases where the analysis of post-failure phenomena is of paramount importance, such as fast catastrophic landslides traveling long distances at high speeds which present a high destruction power. There engineers have to provide an estimation of velocity, run-out distance and path, depth, etc.

Indeed, fast catastrophic landslides can cause a large number of fatalities and extensive economic damage. Among the most destructive events, we can mention those of Kansu (China 1920), with 180 000 victims, the Huascarán lahar in Perú (1970), with 67 000, the 1949 landslide at Khait (Tajikistan), with 12 000, the debris flows of Venezuela (1999) with 30 000 and that of Vajont in northern Italy with 2000. Here, the victims were caused by the wave originated by the landslide when entering the reservoir, which overflowed the dam and reached the village of Longarone (see plate 1 and Figure 1.1).

Landslides can be defined as failures of natural slopes and require the techniques used in geomechanics to analyze failure. The study of failure conditions can be tracked down to the early works of Coulomb (1773). Historically, researchers have followed three main directions: (i) The slip line method, (ii) Limit theorems for plastic collapse, and (iii) limit equilibrium methods. All of them provide information on failure loads and mechanisms, but not on post-failure phenomena. Because of both the type of material behavior assumed and the kinematics of failure, they are not applicable in all cases, as it will be explained below.

Limit theorems and limit equilibrium methods are based on two main assumptions: (i) Failure takes place at effective stress states located on failure surfaces, such as that of Mohr–Coulomb, and (ii) It exists in the soil mass a surface where failure takes place – the failure surface. These two conditions correspond to what is known as "localized failure," which is found in overconsolidated materials exhibiting softening.

Localized failure is characterized by a concentration of strain in very narrow and limited zones, where the phenomenon is idealized by assuming that there exists a discontinuity in

Computational Geomechanics: Theory and Applications, Second Edition. Andrew H. C. Chan, Manuel Pastor, Bernhard A. Schrefler, Tadahiko Shiomi and O. C. Zienkiewicz.

the strain or its rate (weak discontinuity). In the case of soils, the most frequent case is that of discontinuities in the shear strain, which these features are referred to as shear bands. Shear bands can evolve (see Chapter 5 where shear bands are treated extensively). Shear bands can evolve into discontinuities in the displacement and velocity fields (strong discon tinuity). The failure mechanism is interpreted as a relative sliding of two regions where deformations are small.

These conditions are not always satisfied, as in the case of loose, collapsible materials, where instability and failure are observed at effective stresses below Mohr–Coulomb strength envelope and where failure is observed to occur in a much larger volume of soil. This type of failure has been described as "diffuse."

Diffuse failure is related to instabilities of material behavior in loose soils. The study of these instabilities is still a young area of research. It is worth mentioning the work of Darve (1995) and Nova (1994) on constitutive instabilities and di Prisco et al. (1995) on the stability of shallow submerged slopes where liquefaction induced by sea waves can be triggered, and Darve and Laouafa (2000, 2001) and Fernández Merodo et al. (2004) on diffuse mechanisms of failure in catastrophic landslides. An updated account of work done in this area can be found in Nicot and Wan (2008). An example of diffuse failure is given in Section 8.7.

There exist several alternative classifications of landslides, such as that provided by Dikau et al. (1996), where the main types are: (a) fall, (b) topple, (c) slides, (d) lateral spreading, (e) flows, and (f) complex movements. Of course, this is a simplification of the complex pattern which is often observed in reality. For instance, we could mention the case of rock falls where the falling blocks can break into smaller ones due to fragmentation. What is interesting is that slides have a failure mechanism of localized type while flows are related to diffuse modes of failure. In the case of slides, a mass of soil moves along a surface below which material deforms much less. Depending on the shape of the sliding surface, slides are said to be translational or rotational.

There are several types of landslides where material can flow in a fluid-like manner once the failure has been triggered. We will present succinctly the most relevant cases, which are: (i) rock avalanches, (ii) mudflows, (iii) debris flows, and (iv) flow slides.

Rock avalanches involve a large amount of material, which can disaggregate during the process as a consequence of impacts between blocks or with the irregular basal surface, generating smaller blocks. We can mention the case of Mount Turtley in Canada (1903), where the rock mass consisted of 36 million cubic meters. The front of the avalanche was 700 m wide, reaching the city of Frank, where it caused 70 fatalities. The deposit formed, after the descent of 800 m, was 2000 m long, 1700 m wide and reached a maximum depth of 18 m. Rock avalanches can be considered as a drained process, as the pore fluid is air and the voids ratio can be high. Other important avalanches are those of Valtellina (Italy, 1987) and Randa (Switzerland, 1991).

Mudflows are mixtures of water and fine soils, which, depending on the slope, can reach several kilometers of propagation. Failure of tailing dams originates mudflows with much smaller run-out distances of the order of several hundred meters if both the slope and the amount of flowing material are not large.

Debris flows are a type of fast landslides where a mixture of soil and water propagates along narrow channels. The main characteristics are: (i) important relative displacements between the solid and fluid phases, and (ii) development of pore water pressures in excess to hydrostatic. They are associated with heavy rain events and shallow landslides, and can transport large blocks, which cause severe damage to building structures when they hit them. Debris flows and floods occur in many regions of the world. Recent cases of debris floods have been studied by Yumuang (2006), Chen et al. (2017), and Li et al. (2017) in Thailand, Taiwan, and China, respectively. The cases of Venezuela (1999) and those of Hong Kong in the past decades are examples worth mentioning. A special case is that of lahars occurring in volcanic soils. As examples, we can mention those of Popocatépl, Mount Saint Helen, and Pinatubo (1991). Water in lahars can be provided by melting snow and ice.

Flow slides are fast landslides where an important coupling exists between the sliding soil and the pore fluids. Flow slides are a particular case of fast propagating mass movements, characterized by an important coupling between the solid skeleton and the pore fluids resulting in high pore pressures on the basal surface. These pore pressures generated during the initiation phase dissipate during propagation, being the consolidation and the propagation times of the same order of magnitude. It is possible therefore to have flow slides in a dry collapsible material, as will be shown later. This is an important difference with other phenomena. In the case of dry granular avalanches, the permeability can be large enough to dissipate the generated pore pressures within the body of the avalanche. Falling blocks disaggregate and evolve into granular fluids. The process of grain breakage provides energy to the system, increasing its mobility, which results in apparent friction angles smaller than the basal friction angle. In the case of mudflows, mixtures of fine-grained soils, and water with high water content – such as clays and water – the material is often studied as a monophasic cohesive-frictional fluid.

When fast landslides arrive at water bodies such as lakes, reservoirs, fjords, etc., they can generate impulse waves, which can be considered as a case of tsunamis. Two representative cases are those of Vajont and Gilbert inlet. The Vaiont disaster in 1963 involved two mechanisms of destruction: the water wave generated by the approximately 270 million m^3 slide from Mount Toc first reached the villages of Erto and Casso, located about 100 m above the lake level in the opposite shore. The dam was overtopped by as much as 245 m. An estimated 30 million m^3 of water then descended the valley as a wave, initially more than 70 m in height, destroying the villages of Longarone, Pirago, Villanova, Rivalta, and Fae. About 2500 people were killed.

Modeling of these phenomena presents important difficulties from mathematical, constitutive, rheological, and numerical points of view.

Regarding mathematical modeling, the main difficulty is how to describe all different types and stages of flows. Until recently, modeling of landslides was not done in a consistent manner describing both initiation and propagation phases. The key for covering triggering and propagation, and the different types of flows is the use of a consistent mathematical model, describing in the most general situation flows of water and solid particles with large relative mobilities and pore water pressures. This general model can be applied to simpler conditions like those of flow slides, and to one-phase models such as rock avalanches and mudflows.

The second difficulty concerns constitutive and rheological modeling. Triggering is usually modeled using well-established constitutive models while the flow phase is described using rheological models. The latter are still based on simple shear or rheometrical flow experiments carried out in laboratories. Due to the limitation imposed by their size, it has not been possible so far to perform tests with real materials found in landslides. One important difficulty, therefore, is the lack of constitutive models able to describe the behavior of the material in the whole range, from solid to fluid. The transition from soil at Critical State conditions to a fluidized mixture is an area where much research is still needed.

Concerning multiphase numerical modeling, the situation has very much advanced in the past decade, with the development of particle-based continuum models, such as the material point model (MPM) and smoothed particle hydrodynamics (SPH) among others, improving Eulerian methods such as Volume of Fluid (VOF) or Level Set (Quecedo et al. 2004) allowing for tracking free surfaces with good accuracy.

Regarding the Material Point Method, it was introduced by Sulsky et al. (1994) and further developed and applied to geotechnical problems by Wieckowski (1999), Coetzee et al. (2007), Andersen and Andersen (2009), Zabala and Alonso (2011), and Yerro et al. (2015). Two interesting books have been written in the last years, which is an indication of both interest in this approach and the degree of maturity reached (Zhang et al. 2016; Fern et al. 2019).

The Smoothed Particle Hydrodynamics (SPH) method was introduced independently by Lucy (1977) and Gingold and Monaghan (1977) for astronomical modeling. Since then, it has been applied to model a large variety of problems in hydrodynamics, flow through porous media, shallow water flows, geotechnical (Bui et al. 2007; Bui and Nguyen 2017) and, finally, avalanche propagation (McDougall and Hungr 2004; Rodriguez-Paz and Bonet 2005; Pastor et al. 2009a), just to mention a few representative cases. It can be said that SPH presents here the advantage of a much simpler determination of free surfaces and interfaces, avoiding special techniques such as the Level Set.

10.2 Mathematical Model: A Hierarchical Set of Models for the Coupled Behavior of Fluidized Geomaterials

The materials found in fast landslides are mixtures including various combinations of soil, rocks, water, and air. The behavior of the mixture can be described using alternative approximations, which range in a hierarchical manner, from the more complex based on mixture theory, to the simpler depth-integrated models. Here, we will describe both the models used in our computations, a coupled depth-integrated model incorporating pore pressures, and the framework within which it has been derived.

The first mathematical model describing the coupling between solid and fluid phases was proposed by Biot (1941, 1955) for linear elastic materials. This work was followed by further development at Swansea University, where Zienkiewicz and coworkers (1980, 1984, 1990a, 1990b, 1999) extended the theory to nonlinear materials and large deformation problems. It is also worth mentioning the work of Lewis and Schrefler (1998), Coussy (1995) and de Boer (2000). It can be concluded that the geotechnical community has incorporated coupled formulations to describe the behavior of foundations and geostructures. Indeed, analyses of earth dams, slope failures, and landslide-triggering mechanisms have been carried out using such techniques during the last decades.

This theoretical framework has not been applied to model the propagation of landslides until recently. We can mention here the work of Hutchinson (1986) who proposed a sliding consolidation model to predict run-out of landslides, Iverson and LaHusen (1989), Iverson (1993), Iverson and Denlinger (2001), Pastor et al. (2002), and Quecedo et al. (2004).

It is important to notice that in the area of granular media, Anderson and Jackson (1967) proposed a general coupled model which has been applied to industrial problems such as fluidized beds. Pitman and Le (2005) and Pudasaini (2012) have proposed depth-integrated two-phase models for debris flows. It is worth mentioning the work of Córdoba et al. (2015) who have applied Pitman and Le model to debris flow propagation and Bui and coworkers who have recently proposed a 3D SPH model for geomechanical problems incorporating two separate phases (Bui and Nguyen 2017).

10.2.1 General 3D Model

The general model is based on the assumption that the mixture is composed of a solid phase and a fluid phase. The equations are: (i) Balance of mass, (ii) Balance of linear momentum for the constituents and the mixture, (iii) Constitutive or rheological laws describing the material behavior of all constituents, and (iv) Kinematics relations linking velocities to rate of deformation tensors. The main problem with this approach for modeling fast catastrophic landslides is the computational cost because of the high number of unknowns and the difficulty in tracking all the interfaces. The main advantage is its general character as it can describe phenomena involving large relative displacements between solid and fluid phases.

We will recall here equations and magnitudes defined in preceding chapters devoted to 3D modeling of multiphase problems.

Regarding the 3D-coupled model, the basic magnitudes and concepts are:

i) the porosity n
ii) the phase densities for solid and fluid, which depend on the porosity and on the densities of soil grains ρ_s and pore fluid ρ_w as:

$$\begin{aligned} \rho^{(s)} &= (1-n)\rho_s \\ \rho^{(w)} &= n\rho_w \end{aligned} \tag{10.1}$$

iii) material derivatives for soil and water phases are used

$$\begin{aligned} \frac{\mathrm{d}^{(s)}}{\mathrm{d}t} &= \frac{\partial}{\partial t} + \boldsymbol{v}_s{}^T.\text{grad} \\ \frac{\mathrm{d}^{(w)}}{\mathrm{d}t} &= \frac{\partial}{\partial t} + \boldsymbol{v}_w{}^T.\text{grad} \end{aligned} \tag{10.2}$$

where $\boldsymbol{v}_s$ and $\boldsymbol{v}_w$ are the velocities of solid and fluid particles. They are related by

$$\frac{\mathrm{d}^{(w)}}{\mathrm{d}t} = \frac{\mathrm{d}^{(s)}}{\mathrm{d}t} + (\boldsymbol{v}_w - \boldsymbol{v}_s)^T.\text{grad} \tag{10.3}$$

iv) Regarding stresses, the total stress acting on the mixture is decomposed as

$$\boldsymbol{\sigma} = (1-n)\boldsymbol{\sigma}_s + n\boldsymbol{\sigma}_w = \boldsymbol{\sigma}^{(s)} + \boldsymbol{\sigma}^{(w)} \tag{10.4}$$

where the partial stresses $\boldsymbol{\sigma}^{(s)}$ and $\boldsymbol{\sigma}^{(w)}$ have been introduced as:

$$\boldsymbol{\sigma}^{(s)} = (1-n)\boldsymbol{\sigma}_s \; \boldsymbol{\sigma}^{(w)} = n\boldsymbol{\sigma}_w \tag{10.5}$$

If the stress in the fluid can be decomposed as:

$$\boldsymbol{\sigma}^{(w)} = -np_w\boldsymbol{I} + n\boldsymbol{\tau}_w \tag{10.6}$$

where p_w is the total pressure in the fluid (hydrostatic plus excess pore water pressure) and τ_w characterizes viscous behavior. We will neglect it in what follows.

Effective stress can be written as

$$\boldsymbol{\sigma}' = \boldsymbol{\sigma} + p_w\boldsymbol{I} = (1-n)(\boldsymbol{\sigma}_s + p_w\boldsymbol{I}) \tag{10.7}$$

From here, partial stresses result in:

$$\boldsymbol{\sigma}^{(s)} = \boldsymbol{\sigma}' - (1-n)p_w\boldsymbol{I} \quad \boldsymbol{\sigma}^{(w)} = -np_w\boldsymbol{I} \tag{10.8}$$

The balance equations are:

(v.a) Balance of mass equations for fluid and solid phases can be cast as:

$$\begin{aligned} &-\frac{\mathrm{d}^{(s)}n}{\mathrm{d}t} + (1-n)\operatorname{div}\boldsymbol{v}_s = 0 \\ &\frac{1}{Q}\frac{\mathrm{d}^{(w)}p_w}{\mathrm{d}t} + \frac{\mathrm{d}^{(w)}n}{\mathrm{d}t} + n\operatorname{div}\boldsymbol{v}_w = 0 \end{aligned} \tag{10.9}$$

where the mixed volumetric stiffness Q depends on volumetric stiffnesses of solid grains and water as:

$$\frac{1}{Q} = \frac{(1-n)}{K_s} + \frac{n}{K_w} \approx \frac{n}{K_w} \tag{10.10}$$

(v.b) Balance of momentum for solid and fluid phases

$$\begin{aligned} &n\rho_w\frac{\mathrm{d}^{(w)}\boldsymbol{v}_w}{\mathrm{d}t} = \{-n\,\mathrm{grad}\,p_w\} + n\rho_w\boldsymbol{b} - \boldsymbol{R} \\ &(1-n)\rho_s\frac{\mathrm{d}^{(s)}\boldsymbol{v}_s}{\mathrm{d}t} = \operatorname{div}\boldsymbol{\sigma}' - (1-n)\mathrm{grad}\,p_w + (1-n)\rho_s\boldsymbol{b} + \boldsymbol{R} \end{aligned} \tag{10.11}$$

where $\boldsymbol{b}$ is the gravity acceleration, $\boldsymbol{R}$ is the interaction solid-fluid force,

$$\boldsymbol{R} = -n\boldsymbol{R}_w = (1-n)\boldsymbol{R}_s \tag{10.12}$$

and $\boldsymbol{R}_{(\alpha)}$ that acting on phase (α).

Regarding interaction laws, for a Darcy flow, $\boldsymbol{R}$ is given by:

$$\boldsymbol{R} = n^2\boldsymbol{k}_w{}^{-1}(\boldsymbol{v}_w - \boldsymbol{v}_s) = nk_w{}^{-1}\boldsymbol{w} \tag{10.13}$$

where $\boldsymbol{k}_w$ is the permeability tensor. Other alternatives, such as that used by Pitman and Le (2005) (see Anderson and Jackson 1967), can be used for a wider range of porosities, and when the relative velocity is larger:

$$R = \frac{n(1-n)}{V_T n^m}(\rho_s - \rho_w)g(\boldsymbol{v}_w - \boldsymbol{v}_s) \tag{10.14}$$

where

V_T is the terminal velocity of solid particles falling in the fluid
g the acceleration of gravity
m a constant
It is convenient to express the interaction term as

$$\boldsymbol{R} = C_d(\boldsymbol{v}_w - \boldsymbol{v}_s) \tag{10.15}$$

where C_d is

$$C_d = n^2 k_w^{-1} \text{ (Darcy)} \tag{10.16}$$

or

$$C_d = \frac{n(1-n)}{V_T n^m}(\rho_s - \rho_w) g \text{ (Anderson)} \tag{10.17}$$

From here, the first simplification consists of assuming that the velocity of fluid phases relative to the solid skeleton is small. In this case, the equations can be cast in terms of the displacements or velocities of the solid skeleton, the velocities of the pore water relative to the skeleton and the averaged pore pressure of the interstitial fluids. This model is referred to as $u - p_w - w$. It was proposed by Zienkiewicz and Shiomi (1984) for the case of saturated soils. The case of unsaturated soils with air at atmospheric pressure was proposed by Zienkiewicz et al. (1990b). This model consists of the three equations of balance given above, plus a suitable constitutive or rheological law providing the stress tensor and a kinematical relation relating displacements or velocities to strain or rate of deformation tensors. The main variables are: (i) the velocity of the solid skeleton v_s, (ii) the Darcy velocity of the pore water, w and the pore pressure p.

Under certain assumptions, which were used for the analysis of soil mechanics problems by Zienkiewicz et al. (1980), it is possible to eliminate the Darcy velocity from the model. This is the most celebrated $u - p$ model of Zienkiewicz, which has been widely used in geomechanics being the basis of many computer codes and is dealt with extensively in Chapter 2. The resulting model consists of the equations of (i) balance of mass and momentum of pore water, which is obtained eliminating Darcy's velocity of the pore water, and (ii) the balance of moment of the mixture, which are completed here by suitable rheological and kinematical relations relating (iii) the stress tensor to the rate of deformation tensor D (see Section 10.3), and (iv) the rate of deformation tensor to the velocity field.

10.2.2 A Two-Phase Depth-Integrated Model

Depth-integrated models are a convenient simplification of 3D models, providing an acceptable compromise between computational cost and accuracy. They have been extensively used in the fields of coastal, harbor, oceanographic, and hydraulics engineering since the work of Barré de Saint Venant in 1871. In the case of avalanche dynamics, Savage and Hutter (1989, 1991) proposed their much-celebrated 1D Lagrangian model, where a simple Mohr–Coulomb model allowed a description of the granular material behavior. This work was extended to 2D and more complex terrains in Hutter et al. (1993) and Gray et al. (1999). It has been applied by Laigle and Coussot (1997), Mc Dougall and Hungr (2004), Pastor et al. (2002, 2009a), and Quecedo et al. (2004). Concerning the limitations of the model, Hutter et al. (2005) provide a detailed discussion, being worth mentioning the textbook by Pudasaini and Hutter (2007).

One key aspect is a proper description of pore water evolution within the sliding mass. As solid particles and pore fluid interact, the coupling results in pore pressure change, affecting effective stresses. In consequence, basal friction and mobility of the soil mass will be much affected.

The two-phase, depth-integrated model which we describe in this chapter allows (i) considering two-phase debris flows, and (ii) pore water pressure generation and dissipation along with propagation.

In our opinion, 3D models are to be applied when the flow is 3D, as during initiation or when the flow arrives at obstacles, while depth-integrated models can describe well the propagation. As the accuracy will be that of the weaker link of the chain, depth-integrated models must have the highest degree of accuracy possible.

In some cases, where mobility of the pore fluid is important, it can abandon the solid skeleton. Two-phase models based on the evolution of porosity cannot be used after the skeleton void ratio or porosity has reached a lower limit. Once reached this point, porosity will not decrease more and the water will be abandoning the material, causing its desaturation.

We will use the reference system with axes $\{x_1, x_2, x_3\}$ sketched in Figure 10.1. Z will denote the basal surface elevation, and h the depth of the flowing mass. Velocities will be denoted as $\{v_1, v_2, v_3\}$, and subindices s and w will refer to solid and fluid phases.

An over bar over a magnitude indicates it is a depth-averaged value. For instance:

$$\overline{\theta} = \frac{1}{h} \int_{Z}^{Z+h} \theta(x_1, x_2, x_3)\, \mathrm{d}x_3$$

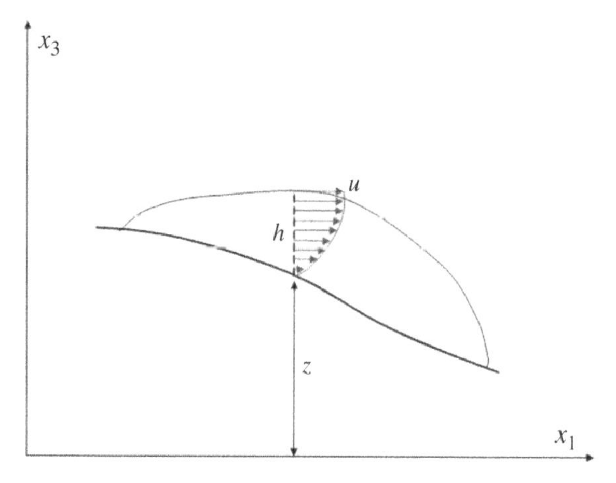

Figure 10.1 Sketch of the reference axes and main magnitudes

We will define the mixture-averaged velocity $\bar{\mathbf{v}}$

$$\bar{\mathbf{v}} = (1-\bar{n})\bar{\mathbf{v}}_s + \bar{n}\,\bar{\mathbf{v}}_w \tag{10.18}$$

and the "quasi-material derivative" as:

$$\frac{\bar{\mathrm{d}}}{\mathrm{d}t} = \frac{\partial}{\partial t} + \bar{v}_j \frac{\partial}{\partial x_j} \quad j = 1,2 \tag{10.19}$$

Depth integration is performed taking into account Leibnitz's rule

$$\int_a^b \frac{\partial}{\partial s} F(r,s)\,\mathrm{d}r = \frac{\partial}{\partial s}\int_a^b F(r,s)\,\mathrm{d}r - F(b,s)\frac{\partial b}{\partial s} + F(a,s)\frac{\partial a}{\partial s} \tag{10.20}$$

We will introduce the next two auxiliary variables h_s and h_w which characterize the solid and fluid contents in a column of water of height h (see Figure 10.2)

$$\begin{aligned} h &= h_s + h_w \\ h_s &= (1-n)h \quad h_w = nh \end{aligned} \tag{10.21}$$

After applying Leibnitz's rule to balance of mass equations, we obtain the balance of mass equations for both phases – solid (s) and fluid (w) as:

$$a\frac{\bar{\mathrm{d}}^{(\alpha)}}{\mathrm{d}t}(h_\alpha) + h_\alpha \operatorname{div}\bar{\mathbf{v}}_\alpha = \bar{n}_\alpha e_R \tag{10.22}$$

where

$$\alpha = \{s, w\} \quad h_\alpha = \bar{n}_\alpha h$$

and

$$\bar{n}_s = (1-\bar{n}) \quad \bar{n}_w = \bar{n}$$

In the above equations, α refers to the phase, $\frac{\bar{d}^{(\alpha)}}{dt}$ is the derivative following phase α, $\bar{n}_\alpha$ the volume fraction, h the depth of the flow, $\bar{\mathbf{v}}_\alpha$ the depth-averaged velocity and e_R the erosion rate, defined as the increment of the height of the moving soil per unit time. There are laws such as those proposed by Hungr et al. (2005), which relate it to the depth-averaged velocity of the flowing material.

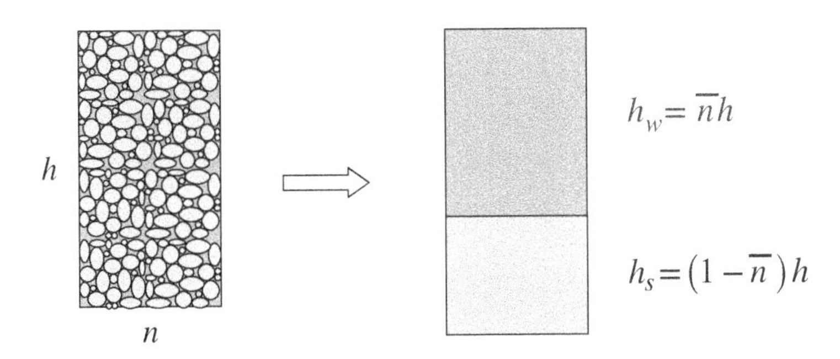

Figure 10.2 Definition of auxiliary variables h_s and h_w

Regarding the **balance of momentum equations** for both phases, after integrating along the depth (local x_3 axis), we arrive at:

$$\begin{aligned}
\rho_s h_s \frac{\overline{\mathrm{d}}^{(s)} \overline{\boldsymbol{v}}_s}{\mathrm{d}t} &= \operatorname{div}\left(h\overline{\boldsymbol{\sigma}}^{(s)}\right) - h\overline{p}_w \operatorname{grad}\overline{n} \\
&\quad - \boldsymbol{\tau}_b^{(s)} + \rho_s h_s \boldsymbol{b} + h_s \overline{\boldsymbol{R}}_s - (1-\overline{n})\rho_s\left(\overline{\boldsymbol{v}}_s - \overline{\boldsymbol{v}}_s{}^{(b)}\right) e_R \\
\rho_w h_w \frac{\overline{\mathrm{d}}^{(w)} \overline{\boldsymbol{v}}_w}{\mathrm{d}t} &= -\operatorname{grad}\left(h\overline{p}_w\right) + h\overline{p}_w \operatorname{grad}\overline{n} \\
&\quad - \boldsymbol{\tau}_b^{(w)} + \rho_w h_w \boldsymbol{b} + h_w \overline{\boldsymbol{R}}_w - \overline{n}\rho_w\left(\overline{\boldsymbol{v}}_w - \overline{\boldsymbol{v}}_w{}^{(b)}\right) e_R
\end{aligned} \tag{10.23}$$

where we have introduced the shear basal stresses of the solid and fluid phases as:

$$\tau_{bi}^{(s)} = -\sigma_{i3}^{(s)}\Big|_Z \quad \tau_{bi}^{(w)} = -\sigma_{i3}^{(w)}\Big|_Z \tag{10.24}$$

The terms $\overline{\boldsymbol{v}}_s{}^{(b)}$ and $\overline{\boldsymbol{v}}_w{}^{(b)}$ denote the basal slip velocities of solid and water phases.

The depth-averaged pore pressure $\overline{p}_w$ will be decomposed as described when presenting the 3D mathematical model into a hydrostatic part and an excess pore pressure as:

$$\overline{p}_w = \overline{p}_{w,hydr} + \Delta\overline{p}_w \tag{10.25}$$

One special case of interest is that where the stresses in the solid phase can be considered as hydrostatic, the pore fluid being in viscid.

$$\begin{aligned}
\sigma_{ii} &= ((1-n)\rho_s + n\rho_w)b_3(h-x_3) \quad i = 1..3 \\
\sigma_{ii}^{(w)} &= n\rho_w b_3(h-x_3) - n\,\Delta p_w
\end{aligned} \tag{10.26}$$

from where we obtain:

$$\begin{aligned}
\sigma_{ii}^{(s)} &= (1-n)\rho_s b_3(h-x_3) + n\Delta p_{w\ i=1..3} \\
\sigma_{ii}' &= (1-n)(\rho_s - \rho_w)b_3(h-x_3) + \Delta p_{w\ i=1..3}
\end{aligned} \tag{10.27}$$

The depth-integrated equations are then:

$$\begin{aligned}
\rho_s h_s \frac{\overline{\mathrm{d}}^{(s)} \overline{\boldsymbol{v}}_s}{\mathrm{d}t} &= \operatorname{grad}\left\{\frac{1}{2}(1-\overline{n})\rho_s h^2 b_3\right\} + \operatorname{grad}(\overline{n}\,h\,\Delta\overline{p}_w) \\
&\quad + \frac{1}{2}\rho_w h^2 b_3 \operatorname{grad}\overline{n} - h\,\Delta\overline{p}_w \operatorname{grad}\overline{n} \\
&\quad - \boldsymbol{\tau}_b^{(s)} + \rho_s \boldsymbol{b} h_s + h_s \overline{\boldsymbol{R}}_s - (1-\overline{n})\rho_s\left(\overline{\boldsymbol{v}}_s - \overline{\boldsymbol{v}}_s{}^{(b)}\right) e_R
\end{aligned} \tag{10.28}$$

for thc solid phase, and

$$\begin{aligned}
\rho_w h_w \frac{\mathrm{d}^{(w)} \boldsymbol{v}_w}{\mathrm{d}t} &= \operatorname{grad}\left\{\frac{1}{2}\overline{n}\rho_w h^2 b_3\right\} - \operatorname{grad}(\overline{n}\,h\,\Delta\overline{p}_w) \\
&\quad - \frac{1}{2}\rho_w h^2 b_3 \operatorname{grad}\overline{n} + h\,\Delta\overline{p}_w \operatorname{grad}\overline{n} \\
&\quad - \boldsymbol{\tau}_b^{(w)} + \rho_w \boldsymbol{b} h_w + h_w \overline{\boldsymbol{R}}_w - \overline{n}\rho_w\left(\overline{\boldsymbol{v}}_w - \overline{\boldsymbol{v}}_w{}^{(b)}\right) e_R
\end{aligned} \tag{10.29}$$

for the fluid.

The above equations can be written in a more compact manner by introducing

i) the pressure terms P_s and P_w defined as:

$$
\begin{aligned}
P_s &= \left\{ -\frac{1}{2}(1-\overline{n})h^2 b_3 - \frac{1}{\rho_s}\overline{n}\,h\,\Delta\overline{p}_w \right\} \\
P_w &= \left\{ -\frac{1}{2}\overline{n}\,h^2 b_3 + \frac{1}{\rho_w}\overline{n}\,h\,\Delta\overline{p}_w \right\}
\end{aligned}
\tag{10.30}
$$

ii) F_s and F_w:

$$
\begin{aligned}
F_s &= \left\{ \frac{1}{2}\frac{\rho_w}{\rho_s} h^2 b_3 - h\frac{1}{\rho_s}\Delta\overline{p}_w \right\} \\
F_w &= \left\{ \frac{1}{2} h^2 b_3 + h\frac{1}{\rho_w}\Delta\overline{p}_w \right\}
\end{aligned}
\tag{10.31}
$$

iii) and the source terms

$$
\begin{aligned}
S_s &= \frac{1}{\rho_s h_s}\left\{ \boldsymbol{\tau}_b^{(s)} + \rho_s \mathbf{b} h_s + h_s \overline{R}_s - (1-\overline{n})\rho_s\left(\overline{\boldsymbol{v}}_s - \mathbf{v}_s^b\right) e_R \right\} \\
S_w &= \frac{1}{\rho_w h_w}\left\{ \boldsymbol{\tau}_b^{(w)} + \rho_w \mathbf{b} h_w + h_w \overline{R}_w - \overline{n}\rho_w\left(\overline{\boldsymbol{v}}_w - \mathbf{v}_w^b\right) e_R \right\}
\end{aligned}
\tag{10.32}
$$

The balance of momentum equations are now written as:

$$
\begin{aligned}
\frac{\overline{\mathrm{d}}^{(s)}\overline{\boldsymbol{v}}_s}{dt} &= \frac{1}{h_s}\operatorname{grad} P_s + \frac{1}{h_s} F_s \operatorname{grad}\overline{n} + S_s \\
\frac{\overline{\mathrm{d}}^{(w)}\overline{\boldsymbol{v}}_w}{dt} &= \frac{1}{h_w}\operatorname{grad} P_w + \frac{1}{h_w} F_w \operatorname{grad}\overline{n} + S_w
\end{aligned}
\tag{10.33}
$$

From the above equations, we can obtain those proposed by Pitman and Le (2005) just by assuming that excess pore pressures $\Delta\overline{p}_w$ are zero.

For convenience, from now on, we will drop the over-bar, all magnitudes being depth-integrated unless otherwise stated.

Regarding the excess pore pressure evolution, it is given by:

$$
\begin{aligned}
\frac{\mathrm{d}^{(s)}\Delta p_w}{\mathrm{d}t} &= -\rho' b_3 \frac{\mathrm{d}^{(s)}h}{\mathrm{d}t}\left(1 - \frac{x_3}{h}\right) \\
&\quad + \frac{K_v}{\alpha}\frac{\partial}{\partial x_3}\left(\frac{n}{\overline{C}_\mathrm{d}}\frac{\partial \Delta p_w}{\partial x_3}\right) - \frac{K_v}{\alpha}\frac{1}{1-\overline{n}}\frac{d^{(s)}\overline{n}}{dt}
\end{aligned}
\tag{10.34}
$$

where $K_v = \dfrac{E}{3(1-2\nu)}$ is the volumetric stiffness of soil skeleton, E Young's modulus, ν Poisson's ratio, α is a constitutive parameter and $\overline{C}_d = \dfrac{C_d}{n}$.

If the state of stress is purely hydrostatic, $\alpha = 1$, while under the state of stress, $(k_0\sigma_1, k_0\,\sigma_1, \sigma_3)$ $\alpha = k_0$.

Equation (10.47) consists of three terms:

i) the increment of excess pore pressure caused by an increase in the debris flow height.
ii) the consolidation along x_3.
iii) the changes of averaged porosity obtained in the depth-integrated equations.

An alternative description of pore pressure evolution is based on assuming an approximation of the pore pressure as

$$p_w(x_1, x_2, x_3, t) = \sum_{k=1}^{Npw} P_k(x_1, x_2, t) N_k(x_3)$$

Taking

$$N_k(x_3) = \cos \frac{(2k-1)}{2h} \pi(x_3 - Z) \quad k = 1, Npw$$

and keeping only the first term, we have:

$$p_w(x_1, x_2, x_3, t) = P_1(x_1, x_2, t) \cos \frac{\pi}{2h}(x_3 - Z)$$

from where we obtain the evolution of basal pore pressure as:

$$\frac{\bar{d}P_1}{dt} = \frac{\pi^2}{4h^2} c_v P_1$$

We have not included, for simplicity, variations caused by changes of porosity or height of the soil mass.

10.2.3 A Note on Reference Systems

In the depth-integrated model, the flow velocities are parallel to the x_1, x_2 plane. However, fast flows propagate downhill on curved beds with big gradients; flow velocities are parallel to the bed of the slide. Changes in slope gradient and curvature produce a pore pressure distribution different from the hydrostatic one; therefore these effects must be incorporated into the general equations.

Savage and Hutter (1991) proposed a one-dimensional Lagrangian model that takes into account the slope and curvature effect. They showed that the curvature effect leads to a vertical stress increment, above all in granular materials.

The movement equations were also solved by Hungr (1995) using a Lagrangian finite differences scheme. The mass was divided into a number of blocks in contact with each other, which could deform during the propagation. The curvature effect was incorporated into the bottom friction force adding to the normal component of the contribution of the centrifuge force.

If a frictional material is considered, the curvature effect can be taken into account through the two methods based on the models of Savage and Hutter (1991) and Hungr (1995). The first method uses a curvilinear reference system oriented in the normal and tangent directions in relation to the bed line at each point. Quecedo and Pastor (2003) have used this reference system in order to model two-dimensional fast flows using an Eulerian formulation. The second one incorporates the curvature effect in the friction law, adding the centrifuge force to the normal force.

In the last method, the formulation used for the friction law is:

$$|t_B| = \left(g + \frac{v^2}{R}\right) h \cdot tg\, \phi$$

In order to calculate the curvature radius R in the point of the surface which is considered (Figure 10.3), the approximation is:

$$K = \frac{1}{R} = \left.\frac{d^2\overline{r}}{ds^2}\right|_M = \frac{L\, \cos^2 \alpha + M\, \cos\alpha\, \mathrm{sen}\, \alpha + N\, \mathrm{sen}^2\, \alpha}{E\, \cos^2 \alpha + F\, \cos\alpha\, \mathrm{sen}\, \alpha + G\, \mathrm{sen}^2\, \alpha} = \frac{L + M\lambda + N\lambda^2}{E + F\lambda + G\lambda^2}$$

where:

$$\lambda = tg\, \alpha, L = \overline{n}X_{uu}, M = \overline{n}X_{uv}, N = \overline{n}X_{vv},$$
$$E = X_u X_u, F = X_u X_v, G = X_v X_v$$

And if: $u \equiv x\ y\ v \equiv y$, we obtain:

$$\overline{n} = \frac{(-Z_x, -Z_y, 1)}{\sqrt{1 + Z_x^2 + Z_y^2}}$$
$$X_u = (1, 0, Z_x) \qquad X_v = (0, 1, Z_y)$$
$$X_{uu} = (0, 0, Z_{xx}) \quad X_{uv} = (0, 0, Z_{xy}) \quad X_{vv} = (0, 0, Z_{yy})$$

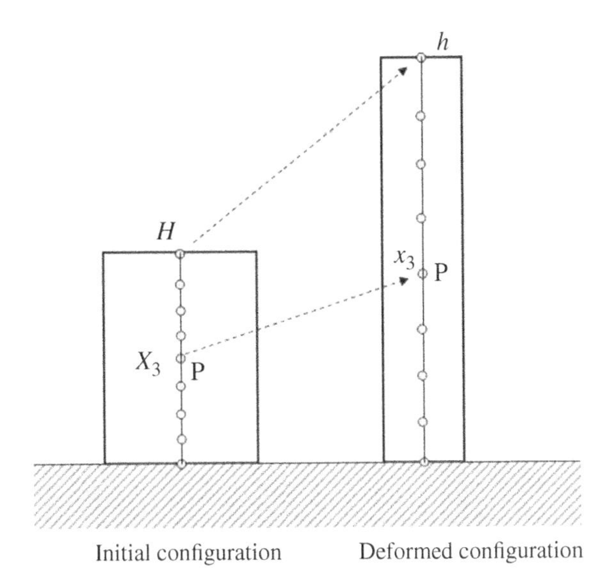

Figure 10.3 Curvature approximation and the values of E, F, G, L, M, and N are:

$$E = \left(1 + Z_x^2\right) \qquad F = \left(Z_x Z_y\right) \qquad G = \left(1 + Z_y^2\right)$$
$$L = \frac{Z_{xx}}{\sqrt{1 + Z_x^2 + Z_y^2}} \qquad M = \frac{Z_{xy}}{\sqrt{1 + Z_x^2 + Z_y^2}} \qquad N = \frac{Z_{yy}}{\sqrt{1 + Z_x^2 + Z_y^2}} \tag{10.35}$$

10.3 Behavior of Fluidized Soils: Rheological Modeling Alternatives

When obtaining the depth-integrated equations described in the preceding section, we have lost the flow structure along the vertical direction, which is needed to obtain both the basal friction and the depth-integrated stress tensor. A possible solution that is widely used consists of assuming that the flow at a given point and time, with known depth and depth-averaged velocities, has the same vertical structure as a uniform, steady-state flow. In the case of flow-like landslides, this model is often referred to as the infinite landslide, as it is assumed to have constant depth and move at constant velocity along a constant slope. This infinite landslide model is used to obtain necessary items in our depth-integrated model. It is a consistent model in the sense that the basal friction is exact for landslides moving at a constant velocity, and having a large length. The models used to obtain the friction are often referred to as "rheological models." Rheology describes the relations between stress and rate of deformation in fluids, while constitutive relations provide suitable relations between stress, rate of stress, and rate of strain. Landslide triggering is usually modeled with constitutive equations, while the propagation of the fluidized material is described by rheological laws. One possible solution has been explored by the authors in Pastor et al. (2013), and applied to both flow slides and rock avalanches. We will present next some models frequently found in landslide propagation modeling.

10.3.1 Bingham Fluid

In the case of Bingham fluids, there is an additional difficulty because it is not possible to obtain directly the shear stress on the bottom as a function of the averaged velocity. The expression relating the averaged velocity to the basal friction for the infinite landslide problem is given as

$$\overline{v} = \frac{\tau_B h}{6\mu}\left(1 - \frac{\tau_Y}{\tau_B}\right)^2\left(2 + \frac{\tau_Y}{\tau_B}\right) \tag{10.36}$$

where μ is the viscosity, τ_Y the yield stress, and τ_B the shear stress on the bottom. This expression can be transformed into

$$P_3(\eta) = \eta^3 - (3 + a)\eta + 2 = 0 \tag{10.37}$$

where we have introduced $\eta = h_P/h$ which is the ratio between the height of the constant velocity region – often referred to as plug – to the total height of the flow, and the nondimensional number a defined as

$$a = \frac{6\mu\bar{v}}{h\tau_Y}$$

It is first necessary to obtain the root of a third order polynomial. In order to reduce the computational load, several simplified formulae have been proposed in the past. Pastor et al. (2004) introduced a simple method based on obtaining the solution of a second-order polynomial which is the best approximation in the uniform distance sense of the third-order polynomial which is given by

$$P_2(\eta) = \frac{3}{2}\eta^2 - \left(\frac{57}{16} + a\right)\eta + \frac{65}{32} \tag{10.38}$$

Given the nondimensional number a, the root can be obtained immediately.

10.3.2 Frictional Fluid

One simple yet effective model is the frictional fluid, especially in the case where it is used within the framework of coupled behavior between soil skeleton and pore fluid. Without further additional data, it is not possible to obtain the velocity distribution. This is why, information concerning depth-integrated stresses $\bar{\sigma}$ cannot be obtained using depth-integrated models employing pure frictional models. Concerning the basal friction, it is usually approximated as

$$\tau_b = -\sigma_v \tan\phi \frac{\bar{v}_i}{|\bar{v}|} \tag{10.39}$$

where σ_v is the normal stress acting on the bottom. Sometimes, when there is high mobility of granular particles and drag forces due to the contact with the air are important, it is convenient to introduce the extra term proposed by Voellmy (1955), which includes the correction term $\frac{\rho g v^2}{\xi}$ where ξ is the Voellmy turbulence parameter.

In some cases, the fluidized soil flows over a basal surface made of a different material. If the friction angle between both materials δ is smaller than the friction angle of the fluidized soil, the basal shear stress is given by

$$\tau_b = -\rho'_d g h \tan\phi_b \frac{\bar{v}_i}{|\bar{v}|} \tag{10.40}$$

where the basal friction ϕ_b is

$$\phi_b = \min(\delta, \phi) \tag{10.41}$$

This simplified model can implement the effect of pore pressure at the basal surface. In this case, the basal shear stress will be:

$$\tau_b = -\left(\sigma'_v \tan\phi_b - p_w^b\right) \frac{\bar{v}_i}{|\bar{v}|} \tag{10.42}$$

We can see that the effect of the pore pressure is similar to that of decreasing friction angle.

10.3.3 Cohesive-Frictional Fluids

The cohesive-frictional model proposed in 3D by Pastor et al. (2009b) focuses on the case of a simple shear flow which is

$$\begin{aligned}\sigma_{11} &= \sigma_{22} = \sigma_{33} = -p \\ \sigma_{13} &= \sigma_{31} = s + \mu_{CF}\left(\frac{\partial v_1}{\partial x_3}\right)^m\end{aligned} \tag{10.43}$$

where

$$s = c\cos\varphi + p'\sin\varphi$$

Two particular cases of interest are the Bingham fluid ($\varphi = 0 \quad c = \tau_y \quad m = 1$) and the Herschel–Bulkley fluid. For flows of granular materials, we shall use $c = 0 \quad m = 2$. The basal friction term in Pastor et al. (2009b) is given as

$$\tau_b = \rho_d'^{*} gh\cos\theta\tan\varphi + \frac{25}{4}\mu_{CF}\frac{\overline{v}^2}{h^2} \tag{10.44}$$

which is a law with a similar structure as Voellmy's

$$\tau_b = \left\{\rho_d'^{*} gh\cos\theta\tan\varphi + \rho g\frac{\overline{v}^2}{\varsigma}\right\} \tag{10.45}$$

where ς is a material parameter. If we compare both expressions, we can see that both incorporate a quadratic term dependent on the averaged velocity.

In the above, we have defined for convenience $\rho_d'^{*} = \rho_d' - \beta_w \rho_w$, where $\rho_d' = (1-n)\rho_s'$, and the pore pressure in excess of hydrostatic is written as $\Delta p_w = \beta_w p_w$.

10.3.4 Erosion

One important aspect in the behavior of catastrophic landslides and related phenomena is erosion. This complex phenomenon requires a rheological or constitutive behavior of the interface and depends on variables such as the flow structure, density, size of particles, and on how close are the effective stresses at the surface of the terrain to failure. We have used here the simple yet effective law proposed by Hungr (1995) which gives the erosion rate as $E_t = E_s h\overline{v}$ where E_s can be obtained directly from the initial and final volumes of the material and the distance traveled as $E_s \approx \frac{\ln\left(V_{final}/V_0\right)}{\text{distance}}$. Units of erosion coefficient are L^{-1}.

It is also worth mentioning that there are other erosion laws proposed by Blanc (2008), Issler and Jóhannesson (2010), and Issler and Pastor (2011).

10.4 Numerical Modeling: 2-Phase Depth-Integrated Coupled Models

To analyze the propagation of a fast landslide over a terrain, there are two main alternatives. The first is Eulerian which can be based either on a structured (finite differences) or an unstructured (finite elements and volumes) grid within which the material flows. The main problem here is the need for a very fine computational mesh for both the terrain information and for the fluidized soil. Classical mesh-based Lagrangian methods present problems because as soon as the mesh deforms, mesh refinement is necessary. As an alternative, meshless methods avoid distortion problems in an elegant way. It is important to notice that mesh-based methods present advantages in some cases, such as those where the avalanche input is given as a hydrograph or those where material boundaries such as walls are to be included in the model.

Regarding Eulerian methods, the author has used in the past successfully (Quecedo et al. 2004) finite elements and the Taylor–Galerkin method which was introduced independently by Donea (1984) and Löhner et al. (1984). This method can be considered as the FEM counterpart of the Lax–Wendroff procedure in the FDM. It consists in performing a second-order expansion of the time derivative, followed by the spatial discretization of the resulting equation using the conventional Galerkin method.

We will use here the SPH to discretize the depth-integrated equations described in Section 2, introducing nodes (particles) that can move in the domain. Pore water pressure evolution will be analyzed, when necessary, using finite difference meshes associated with each solid node. The terrain is described using a structured quadrilateral mesh in which nodes we store the elevation, slopes, and second-order derivatives. The latter are necessary when the curvature of particle paths is required to compute centripetal accelerations.

10.4.1 SPH Fundamentals

SPH method is one of the meshless methods which have been developed in the last decades. These techniques have been applied since Nayrolles, Touzot, and Villon (1992) introduced the diffuse element method. Since then, new meshless methods have been proposed. It is worth mentioning the element-free Galerkin Method of Belytschko et al. (1994), the hp-cloud method of Duarte and Oden (1996), the partition of unity method of Melenk and Babuška (1996), the finite point method introduced by Oñate and Idelsohn (1998), material point model (Więckowski 2004, Coetzee et al. 2007; Andersen 2009; Zabala and Alonso (2011), Jassim et al. (2013) and, finally, the smoothed particle hydrodynamics method (SPH), which is the technique which will be described here.

10.4.2 An SPH Lagrangian Model for Depth-Integrated Equations

10.4.2.1 Introduction and Fundamentals of SPH

A meshless method referred to as the smoothed particle hydrodynamic or SPH is to be described in this section. As in any meshless method, information is linked to moving nodes. We shall describe next the method in a very succinct way. Smoothed particle hydrodynamics (SPH) is a meshless method introduced independently by Lucy (1977) and Gingold and Monaghan (1977) and firstly applied to astrophysical modeling, a domain where SPH presents important advantages over other methods. SPH is well suited for hydrodynamics, and researchers have applied it to a variety of problems, like those described in Gingold and Monaghan (1982), Monaghan et al. (1999), Bonet and Kulasegaram (2000) and Monaghan et al. (2003), just to mention a few. SPH has been also applied to model the propagation of catastrophic landslides (Bonet and Rodíguez-Paz, 2005; McDougall, 2006; McDougall and Hungr, 2004). However, in both cases, the analysis did not incorporate hydromechanical coupling between the solid skeleton and the pore fluid, which has been proposed by Pastor et al. (2018).

The approximation of a given function $\phi(x)$ is written as

$$\langle \phi(x) \rangle = \int_{\Omega} \phi(x') W(x' - x, h) \mathrm{d}x' \tag{10.46}$$

where $W(x' - x, h)$ is referred to as the kernel of the linear functional, h being a parameter describing its decay. A special class of kernels is that of functions having radial symmetry, i.e. depending only on r:

$$\begin{aligned} \frac{\mathrm{d}^{(s)} \Delta p_w}{\mathrm{d}t} &= -\rho' b_3 \frac{\mathrm{d}^{(s)} h}{\mathrm{d}t} \left(1 - \frac{x_3}{h}\right) \\ &+ \frac{K_v}{\alpha} \frac{\partial}{\partial x_3} \left(\frac{n}{\overline{C}_{\mathrm{d}}} \frac{\partial \Delta p_w}{\partial x_3} \right) - \frac{K_v}{\alpha} \frac{1}{1 - \overline{n}} \frac{\mathrm{d}^{(s)} \overline{n}}{\mathrm{d}t} \end{aligned} \tag{10.47}$$

It is convenient to introduce the notation:

$$\xi = \frac{|x' - x|}{h} = \frac{r}{h} \tag{10.48}$$

The functions $W(x,h)$ used as kernels in SPH approximations are required to fulfill the following conditions:

i)

$$\lim_{h \to 0} W(x' - x, h) = \delta(x) \tag{10.49}$$

ii)

$$\int_{\Omega} W(x' - x, h) \mathrm{d}x' = 1 \tag{10.50}$$

Condition (10.50), which also follows from (10.49), can be interpreted as well as the ability of the approximation to reproduce a constant or polynomial of degree zero (zero-order consistency).

iii) The kernel $W(x' - x, h)$ is positive and has compact support

$$W(x' - x, h) = 0 \quad if \quad |x' - x| \geq kh \tag{10.51}$$

where k is a positive integer which is usually taken as 2.

iv) The kernel $W(x' - x, h)$ is a monotonically decreasing function of ξ.

v) The kernel is an even function with respect to ξ.

It is possible to show that under the conditions specified above, the approximation is second-order accurate, i.e.

$$\langle \phi(x) \rangle = \phi(x) + O(h^2) \tag{10.52}$$

It is interesting to note that the integral approximation 10.46 is nothing but a linear function, which is usually denoted as

$$T_h[\phi] = \int_{\Omega} W(x', h)\phi(x')\mathrm{d}x' \tag{10.53}$$

Moreover, it can be seen that the limit of $T_h[\phi]$ as h tends toward zero

$$\lim_{h \to 0} T_h[\phi] = \delta[\phi] \tag{10.54}$$

is the Dirac delta, which is a singular transformation. It is possible to write:

$$\phi(x) = \int_{\Omega} \phi(x')\,\delta(x' - x)\mathrm{d}x' \tag{10.55}$$

In the framework of SPH formulations, several kernels have been proposed in the past. Among, them, it is worth mentioning: (i) The Gaussian kernel McDougall (2006) and (ii) the cubic spline (Pastor et al. 2015b; Pastor et al. 2018). Concerning the integral representation of the derivatives in SPH, we have:

$$\langle \phi'(x) \rangle = \int_{\Omega} \phi'(x') W(x' - x, h)\mathrm{d}x' \tag{10.56}$$

After integration by parts – in a one-dimensional problem, and taking into account that the kernel has compact support, 10.56 reads

$$\langle \phi'(x) \rangle = -\int_{\Omega} \phi(x') W'(x' - x, h)\mathrm{d}x' \tag{10.57}$$

Classical differential operators of continuum mechanics can be approximated in the same way. We list below the gradient of a scalar function, the divergence of a vector function, and the divergence of a tensor function:

$$\langle \mathrm{grad}\phi(x) \rangle = -\int_{\Omega} \phi(x')\,\frac{1}{h} W' \frac{x' - x}{r} \mathrm{d}\Omega \text{ with } r = |x' - x| \tag{10.58}$$

$$\langle \mathrm{div}\, \boldsymbol{u}(x) \rangle = -\int_{\Omega} \boldsymbol{u}(x')\,\mathrm{grad}\; W \mathrm{d}\Omega = -\int_{\Omega} \frac{1}{h} W' \frac{\boldsymbol{u}(x').(x' - x)}{r} \mathrm{d}\Omega \tag{10.59}$$

$$\langle \mathrm{div}\, \boldsymbol{\sigma}(x) \rangle = -\int_{\Omega} \boldsymbol{\sigma}.\mathrm{grad}\; W \mathrm{d}\Omega = -\int_{\Omega} \frac{1}{h} W' \frac{\boldsymbol{\sigma}.(x' - x)}{r} \mathrm{d}\Omega \tag{10.60}$$

These approximations of functions and derivatives are valid at the continuum level. If the information is stored in a discrete manner, for instance, in a series of points or nodes, it is necessary to construct discrete approximations. The SPH method introduces the concept of "particles," to which information concerning field variables and their derivatives is linked. But indeed, they are nodes, much in the same way than found in finite elements or finite differences. All operations are to be referred to as nodes. We will therefore introduce the set of particles or nodes with $K = 1...N$. Of course, the level of approximation will depend on how the nodes are spaced and on their location. The classical finite element strategy of having more nodes in those zones where larger gradients are expected is of application here.

As per the last paragraph, evaluation of an integral approximation like 10.46 can be performed using a numerical integration technique of the type:

$$\langle \phi(x_I) \rangle_h = \sum_{J=1}^{N} \phi(x_J) W(x_J - x_I, h)\, \omega_J \tag{10.61}$$

where the information of the given function $\phi(x)$ is only available at a set of N nodes within the domain Ω. In 10.61, the subindex "h" has been used to denote the discrete approximation, with ω_J denoting the weights of the integration formula – which can be shown to be $\omega_J = \Omega_J = m_J/\rho_J$, with Ω_J, m_J and ρ_J being the volume, mass, and densities associated to node J. In order to simplify the representation, the following notation is introduced:

$$\phi_I = \langle \phi(x_I) \rangle_h = \sum_{J=1}^{N} \phi(x_J) W(x_J - x_I, h)\, \Omega_J \tag{10.62}$$

Taking into account 10.51, that is to say, the kernel function has local support, expression 10.62 reads

$$\phi_I = \langle \phi(x_I) \rangle_h = \sum_{J=1}^{Nh} \phi(x_J) W(x_J - x_I, h)\, \Omega_J \tag{10.63}$$

where x_J with $J = 1...Nh$ is the set of nodes fulfilling the relation $|x_J - x_I| < kh$. In case the function ϕ represents the density, 10.63 becomes

$$\begin{aligned} \rho_I &= \sum_{J=1}^{Nh} \rho_J W_{IJ} \frac{m_J}{\rho_J} \\ \rho_I &= \sum_{J=1}^{n} W_{IJ}\, m_J \end{aligned} \tag{10.64}$$

with

$$W_{IJ} = W(x_J - x_I, h)$$

One interesting aspect of SPH is the existence of several alternative discretized forms for the differential operators. For instance, the gradient of a scalar function can be approximated as (basic form):

$$\text{grad}\phi_I = \sum_{J=1}^{Nh} \frac{m_J}{\rho_J} \phi_J \text{grad}\, W_{IJ} \tag{10.65}$$

and also by the following symmetrized forms

$$\text{grad}\phi_I = \frac{1}{\rho_I} \sum_J m_J (\phi_J - \phi_I) \text{grad} W_{IJ} \tag{10.66}$$

$$\text{grad}\phi_I = \rho_I \sum_J m_J \left\{ \frac{\phi_J}{\rho_J^2} + \frac{\phi_I}{\rho_I^2} \right\} \text{grad} W_{IJ} \tag{10.67}$$

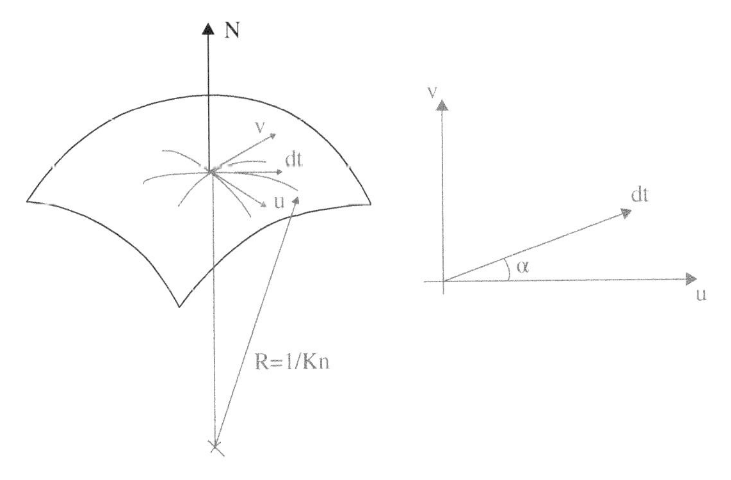

Figure 10.4 Nodes and numerical integration in an SPH mesh

Sometimes it is preferred to use a variant of this form, which is

$$\text{grad}\phi_I = \sum_{J=1}^{Nh} \frac{m_J}{\rho_J}(\phi_I + \phi_J)\text{grad}W_{IJ}$$

Figure 10.4 illustrates the numerical integration procedure performed.

10.4.2.2 SPH Discretization

We will introduce the notation:

$$\begin{aligned} x_{IJ} &= x_I - x_J \\ r_{IJ} &= |x_I - x_J| \\ \text{grad}\, W_{IJ} &= \frac{W'}{h}\frac{x_{IJ}}{r_{IJ}} \end{aligned} \tag{10.68}$$

The weight ω_J can be shown to be the volume Ω_J or area associated with the node. In the context of continuum mechanics (solids and fluids), it is convenient to introduce the density ρ_J associated with node J as

$$\rho_J = m_J/\Omega_J \tag{10.69}$$

where m_j is the mass associated with node j, which is nothing else than the mass of the volume associated with the considered node. The nodal variable ϕ_I is then:

$$\phi_I = \sum_{J=1}^{Nh} \phi(x_J) W_{IJ} \frac{m_J}{\rho_J} \tag{10.70}$$

which is a form commonly used in SPH. In case we choose the function $\phi(x)$ to represent the density, we will obtain after substituting in (10.71)

$$\rho_I = \sum_{J=1}^{Nh} \rho_J W_{IJ} \frac{m_J}{\rho_J}$$
$$\rho_I = \sum_{J=1}^{n} W_{IJ} m_J \tag{10.71}$$

10.4.2.3 SPH Modeling of Two-Phase Depth-Integrated Equations

The proposed method combines two sets of SPH nodes – for solid and water particles – with the finite difference (FD) meshes associated with SPH nodes. The former sets describe the behavior of depth-integrated columns of soil and water, while the latter allows the analysis along the depth of pore pressures. It is an improvement over models implementing pore pressures using simple approximations depending only on the value of the basal pore pressure.

We will introduce:

i) two sets of nodes $\{x_{\alpha K}\}$ with $K = 1. N_\alpha$ where N_s and N_w are the number of SPH nodes in the solid and fluid phases,
ii) The nodal variables:

$h_{\alpha I}$ heights of phases at node I
$\bar{\boldsymbol{v}}_{\alpha I}$ depth averaged, 2D velocities

and the pore water pressure at the nodes $p_{wp\,I}$ which is a vector including the finite difference nodes along the height.

In Figure 10.5, we sketch the SPH soil and water nodes together with the finite difference meshes associated with each solid point to describe pore pressure evolution.

If the 2D area associated with a general fluid or solid node I is Ω_I, we will introduce for convenience a fictitious volume m_I with dimensions L^3 moving with this node:

$$m_I = \Omega_I h_I \tag{10.72}$$

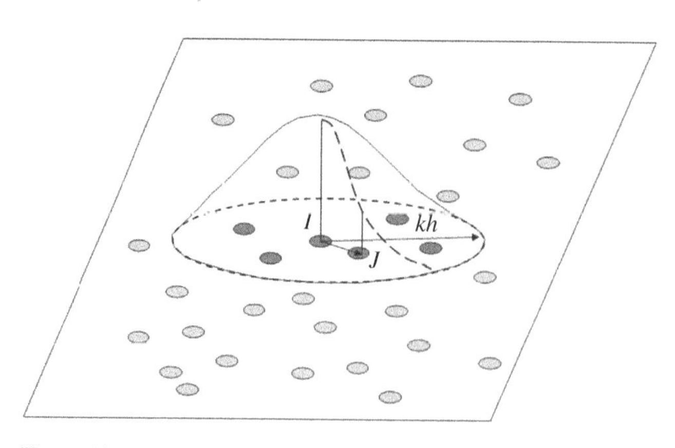

Figure 10.5 SPH nodes with FD meshes at solid nodes.

It is important to note that m_I has no physical meaning, as when node I moves, the material contained in a column of the base Ω_I does not move as if it were a solid column. Due to the shear velocity, there are parts of the column where the velocity of the fluid is larger than that of the column, and the relative movement will be such that these fluid particles will flow out of the column. The opposite case will be that of fluid particles with a velocity smaller than that of the column.

Regarding the balance of mass equations, we have used a simple alternative, computing the heights from the position of the neighboring particles as:

$$h_I = \langle h(x_I) \rangle = \sum_J h_J \Omega_J W_{IJ} = \sum_J m_J W_{IJ} \tag{10.73}$$

The height can be normalized, which allows improving the approximation when SPH nodes are close to the boundaries

$$h_I = \frac{\sum_J m_J W_{IJ}}{\sum_J \left(\frac{m_J}{h_J}\right) W_{IJ}} \tag{10.74}$$

We will recall here for convenience the momentum equations, writing them in a more compact form as:

$$\frac{\overline{\mathrm{d}}^{(\alpha)} \overline{\boldsymbol{v}}_\alpha}{\mathrm{d}t} = \frac{1}{h_\alpha} \operatorname{grad} P_\alpha + \frac{1}{h_\alpha} F_\alpha \operatorname{grad} \overline{n} + S_\alpha \quad \alpha = \{s, w\} \tag{10.75}$$

where the pressure terms are

$$\begin{aligned} P_s &= \left\{ -\frac{1}{2}(1-\overline{n})h^2 b_3 - \frac{1}{\rho_s} \overline{n}\, h\, \Delta \overline{p}_w \right\} \\ P_w &= \left\{ -\frac{1}{2} \overline{n} h^2 b_3 + \frac{1}{\rho_w} \overline{n}\, h\, \Delta \overline{p}_w \right\} \end{aligned} \tag{10.76}$$

and

$$\begin{aligned} F_s &= \left\{ \frac{1}{2}\frac{\rho_w}{\rho_s} h^2 b_3 - h \frac{1}{\rho_s} \Delta \overline{p}_w \right\} \\ F_w &= \left\{ \frac{1}{2} h^2 b_3 + h \frac{1}{\rho_w} \Delta \overline{p}_w \right\} \end{aligned} \tag{10.77}$$

Finally, the source terms are:

$$\begin{aligned} S_s &= \frac{1}{\rho_s h_s} \left\{ \boldsymbol{\tau}_b^{(s)} + \rho_s \mathbf{b} h_s + h_s \overline{R}_s - (1-\overline{n}) \rho_s \left(\overline{\boldsymbol{v}}_s - \mathbf{v}_s^b \right) e_R \right\} \\ S_w &= \frac{1}{\rho_w h_w} \left\{ \boldsymbol{\tau}_b^{(w)} + \rho_w \mathbf{b} h_w + h_w \overline{R}_w - \overline{n} \rho_w \left(\overline{\boldsymbol{v}}_w - \mathbf{v}_w^b \right) e_R \right\} \end{aligned} \tag{10.78}$$

We will consider next how to discretize each of the three terms $\operatorname{grad} P_\alpha$, $\frac{1}{h_\alpha} F_\alpha \operatorname{grad} \overline{n}$ and S_α.

Regarding the gradient terms, we will write only one of the symmetrized forms (see Monaghan 1982; Monaghan and Lattanzio 1985; Monaghan et al.1999):

$$\frac{1}{h_{\alpha I}} \operatorname{grad} P_{\alpha I} = -\sum_{1}^{N_{ah}} m_J \left(\frac{P_{\alpha I}}{h_{\alpha I}^2} + \frac{P_{\alpha J}}{h_{\alpha J}^2} \right)$$
$$\frac{1}{h_{\alpha I}} \operatorname{grad} \overline{n}_{\alpha I} = -\sum_{1}^{N_{ah}} m_J \left(\frac{\overline{n}_{\alpha I}}{h_{\alpha I}^2} + \frac{\overline{n}_{\alpha J}}{h_{\alpha J}^2} \right) \tag{10.79}$$

which results in:

$$\frac{\overline{\mathrm{d}}^{(\alpha)}}{\mathrm{d}t} \overline{\mathbf{v}}_{\alpha I} = \sum_J m_J \left(\frac{P_{\alpha I}}{h_I^2} + \frac{P_{\alpha J}}{h_J^2} \right) \operatorname{grad} W_{IJ} + F_{\alpha I} \sum_J m_J \left(\frac{n_{\alpha I}}{h_I^2} + \frac{n_{\alpha J}}{h_J^2} \right) \operatorname{grad} W_{IJ} + \mathbf{S}_{\alpha I} \tag{10.80}$$

10.4.2.4 Boundary Conditions in Two-Phase Depth-Integrated Equations

One of the main issues in SPH is the formulation and application of boundary conditions.

In depth-integrated models applied to fast landslide propagation and similar phenomena, the more frequent situations are:

i) Walls that cannot be described by digital terrain models, including houses which can divert the flow
ii) Injection of nodes into the domain using hydrograms
iii) Absorbing wave boundary conditions in lakes and reservoirs where a landslide has generated impulsive waves

Regarding the walls, this type of condition has been studied by SPH researchers since 1982, when Monaghan (1982) introduced a type of virtual particles, often referred to as "type 1 virtual particles," which applies a repulsive force on particles approaching the boundary, which prevents boundary penetration by them.

Later, Libersky et al. (1993) and Randles and Libersky (1996) introduced the second type of virtual particles, which are located symmetrically with respect to the real particle approaching the boundary.

In our case, the condition consists in making zero the normal velocity only, leaving free the tangential velocity to the wall, being equivalent to a free slip condition. The wall boundary conditions are introduced using a set of wall nodes that define the wall. Once an interaction between a wall node and a fluid node is detected, the normal velocity to the wall is made zero.

Regarding the injection and absorption nodes, conditions have been applied to depth-integrated problems using finite elements by Peraire et al. (1986), Quecedo and Pastor (2004) or SPH, by Lastiwka et al. (2005), Vacondio et al. (2011), and Lin et al. (2019).

The boundary conditions to be imposed depend on the Froude number defined as

$$F = \frac{|\overline{v}|}{\sqrt{gh}}$$

In case Froude number is higher than 1, we will have supercritical conditions and we have to prescribe both the height and the averaged velocity of the incoming particles.

In case the flow at the boundary is subcritical, only one variable, velocity or height is prescribed, the remaining variable being obtained by using the Riemann invariants

$$R^{(1)} = 2\sqrt{gh} + \bar{v}$$
$$R^{(2)} = 2\sqrt{gh} - \bar{v}$$

which characterize the waves coming into the domain or leaving it respectively.

At time *n+1*, we will obtain the values of heights and velocities in the domain without applying this boundary condition. They will be called h^* and $\bar{v}^*$. From here, we obtain the outgoing Riemann invariant $R^{(2)*}$ as

$$R^{(2)*} = 2\sqrt{gh^*} - \bar{v}^*$$

If we know the hydrograph $\bar{h}(t)$, assuming that the outgoing Riemann invariant will not change, we will have

$$R^{(2)\,n+1} = 2\sqrt{g\bar{h}^{n+1}} - \bar{v}^{n+1} = R^{(2)*} = 2\sqrt{gh^*} - \bar{v}^*$$

from where we will obtain the depth-averaged velocity at time *(n+1)* as

$$\bar{v}^{n+1} = 2\sqrt{g\bar{h}^{n+1}} - 2\sqrt{gh^*} + \bar{v}^*$$

In case we wish to prescribe the velocity, we will obtain the flow depth by

$$2\sqrt{g\bar{h}^{n+1}} = \bar{v}^{n+1} - \bar{v}^* + 2\sqrt{gh^*}$$

In conclusion, Riemann invariants characterize magnitudes exiting and entering the domain. If we consider the normal to the domain, the second invariant corresponds to the outgoing wave. This is the reason why the boundary condition is imposed on the second invariant only. The time-step algorithm is implicit and we compute heights and velocities at time $(n+1)$ at the boundary without imposing any boundary condition. From these values, we obtain the information of the outgoing wave, i.e. the second invariant. Knowing it together with the normal velocity at time $(n+1)$, it is possible to obtain the velocity.

Alternatively, if flow properties at the places where the flow enters the domain are known, it is possible to prescribe both height and velocity at all injected particles.

This method is implemented by defining (i) gates, through which particles are injected, and (ii) pools, which are fictitious reservoirs of particles used in the analysis to apply the boundary condition at the gate. Indeed, the values of h^* and $\bar{v}^*$ are assigned to these pool particles.

Particles are injected into the domain when the distance between the gate and the last series of injected particles is equal to or larger than the average particle spacing (Figure 10.6).

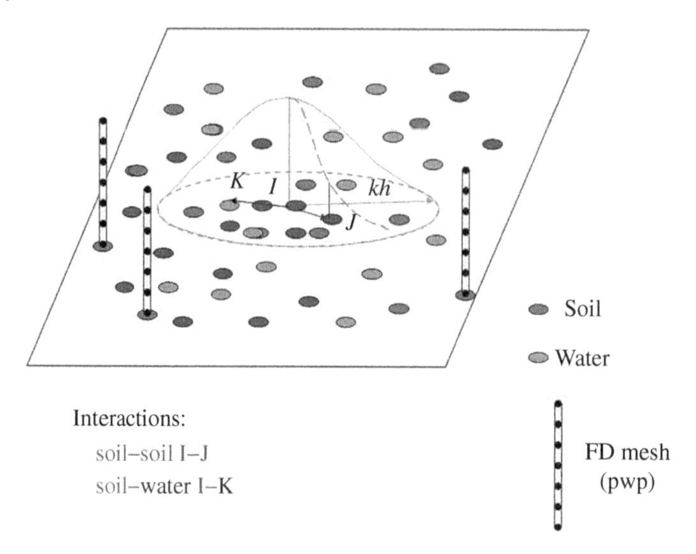

Figure 10.6 Injection strategy.

10.4.2.5 Excess Pore Water Pressure Modeling in Two-Phase Depth-Integrated Equations

In the above equations, there is a term describing basal excess pore pressure at node I Δp_{wbI} which has to be obtained at each node and time step. One alternative is to use simple shape functions fulfilling boundary conditions at the surface and the basal surface. This has been used by Iverson and Denlinger (2001), Pastor et al. (2002, 2004, 2009a, 2015b), and Quecedo et al. (2004). This approach presents the limitation of not being able to model changes of boundary conditions at the bottom. For instance, when a landslide runs over a very permeable basal layer – or a rack – pore pressure becomes zero there, while in the body of the landslide, it is not zero. If a single shape function is used, once the basal value is set to zero, the pressure becomes zero in the whole depth.

To overcome this limitation, we have proposed to introduce FD meshes associated with each SPH solid node (Pastor et al. 2015a, b). Figure 10.7 provides a sketch of the SPH nodes' and FD meshes' layout.

The analysis of excess pore pressure evolution is based on Equation 10.34, which is a classical parabolic partial differential equation that includes two source terms related to variations of height and porosity.

Initial conditions describe the excess pore pressure distribution in all FD meshes. Here we have assumed simple linear laws with values of zero at the top and $\Delta p_{wp,0}^{(b)}$ at the basal surface. The latter has to be estimated either from field data or from the results of a model describing the triggering of the landslide. This initial condition plays a fundamental role in debris flow propagation characteristics. When no data are available, it has to be estimated.

Concerning boundary conditions at the top and the base, we have assumed a zero value at the top, while at the base, it is usual to assume an impermeable boundary (zero flow).

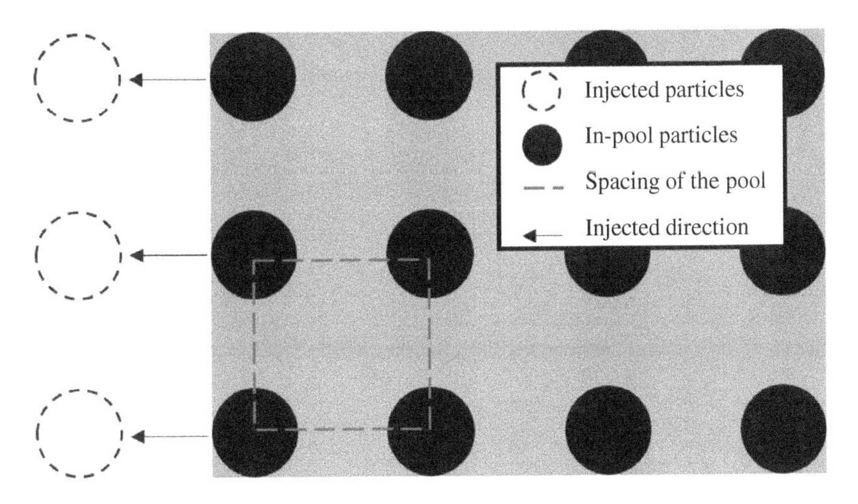

Figure 10.7 Deformation of a soil column.

However, there are situations where the debris flow arrives at mitigation structures such as basal grids, where total pore pressure will be equal to the atmospheric pressure. This makes the flow slow down or even stop as basal friction will increase. Once the flow exits the grid, the flux is made zero again. Here, we have boundary conditions that depend on the position of the nodes on the terrain.

10.5 Examples and Applications

We will devote this final section to present a series of examples where we assess the properties of the proposed models.[1] When assessing the quality of a model, it is advisable to compare its predictions with: (i) problems for which there exists an analytical solution, (ii) laboratory experiments, and (iii) real events for which there exists information. Here, we will concentrate on the latter type. We have selected a series of cases that correspond to three representative types of fast landslides: (i) a rock avalanche, where pore fluid is air and we neglect pore pressures, (ii) a lahar, and (iii) debris flow with high relative mobility of the phases and pore pressure coupling.

10.5.1 The Thurwieser Rock Avalanche

This case is a rock avalanche that occurred in the Central Italian Alps on 18 September 2004. The location was the south slope of Punta Thurwieser and it propagated through Zebrú valley. Its propagation path extended from 2300 m to 3500 m of altitude and traveled a distance of 2.9 km. The rock avalanche involved 2.2 million cubic meters. Sossio and Crosta (2008) have provided information concerning this avalanche, including a detailed digital terrain model. Figure 10.8 from Sossio and Crosta 2008 provides a general view of the avalanche and its location.

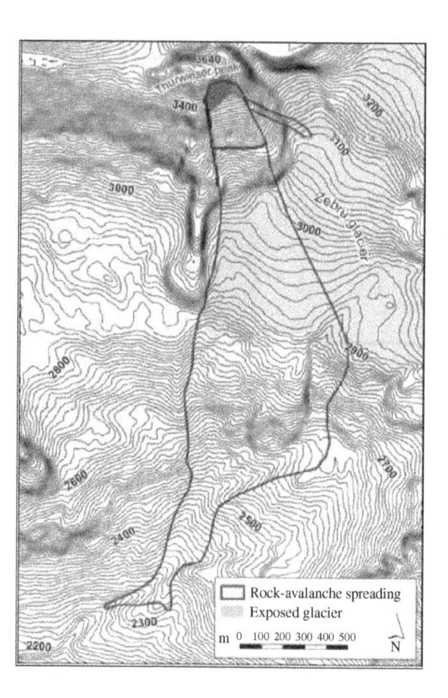

Figure 10.8 General view of Thurwieser rock avalanche. *Source:* From Sossio and Crosta (2008)

This avalanche presents several modeling difficulties, such as the crossing of terrains of different materials, as is the case with the Zebrú glacier. There the basal friction is very small, and erosion of ice and snow is possible. This entrained material can melt due to the heat generated by basal friction, providing extra water, and probably originating basal pore pressures. We have used here a simple frictional model including Voellmy turbulence. Concerning erosion, we have used the law proposed by Hungr 1995. The rheological parameters chosen are: $\tan\phi = 0.39$, Voellmy coefficient 1000 m/s^2, and erosion coefficient 0.00025 m^{-1}.

The results are given in Figure 10.9 where we have plotted the avalanche evolution along time and the computed final extension together with the one observed in the field.

10.5.2 A Lahar in Popocatépetl Volcano

Lahars are a special case of debris flows of volcanic origin that can be triggered by eruptions or/and heavy rainfall. We will study here the case of a lahar that occurred at Popocatépetl volcano in 2001 (Haddad et al. 2010). This case was proposed for the benchmarking activity of the 2nd JTC1 Workshop (Ho et al., 2018) on Triggering and Propagation of rapid flow-like landslides which took place in Hong Kong in 2018.

Popocatépetl is a stratovolcano (19°1′N, 98°37′W; 5450 m) located in the Trans-Mexican Volcanic Belt in central Mexico, with a glacier of 0.23 million cubic meters at the summit. Lahars are not uncommon and they travel downhill along gorges and channels which can

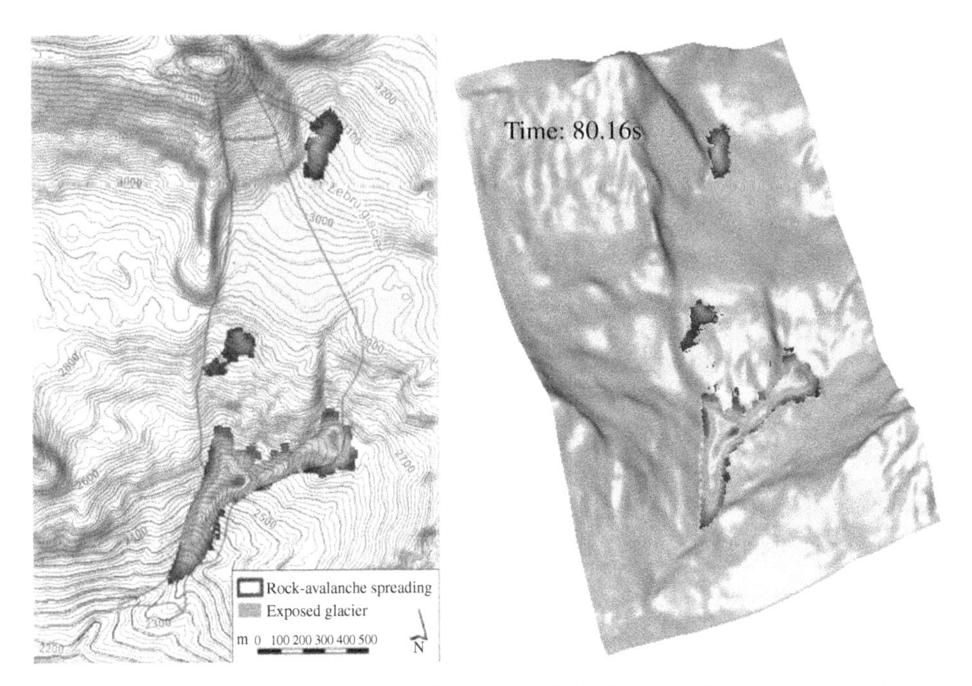

Figure 10.9 Thurwieser avalanche after 80 seconds with friction angle 26 : computed results versus field measurements *Source:* From Geotechnical Engineering Office, Civil Engineering and Development Department of the Government of the Hong Kong SAR

reach villages (Huilouac, Tenenepanco). Indeed, Huilouac gorge crosses the village of Santiago Xalitzintla, which is 17 km away from Popocatépetl crater. (Figure 10.10). Mexico city (population 24 million) and Puebla (population 4 million) are located at 70 km and 45 km, respectively (Figure 10.10).

In the data package provided for the benchmarking activity, it is reported that " ... *On 22 January 2001, the destruction of a dome inside the crater triggered a pyroclastic flow that crossed the glacier and flowed down through the drainage gorges. The flow abraded the glacier and triggered the melting and release of 717 000 m*^{3} *of water, based on calculations from photogrammetry records of the reduction of the mass of the glacier (Andrés et al. 2007). Four hours later, a lahar was initiated in the valley head areas; this carried the pyroclastic flow materials down Huiloac Gorge for approximately 14 km to the immediate surroundings of Santiago Xalitzintla. The lahar transported 160 000 m*^{3} *of material, including 68 000 m*^{3} *of water (Muñoz-Salinas et al. 2009), and reached a maximum velocity of 13.8 m/s that decreased to 1.4 m/s at 9.5 km from its starting point (Muñoz-Salinas et al. 2007). This lahar behaved as a debris flow throughout its course (Capra et al. 2004). Initial lahar volume was estimated by different methods (Sheridan et al. 2001 and Muñoz-Salinas et al. 2009) leading to oscillating values from 1.85 to 3.30 10*^{5} *m*^{3} *for 1997 lahar and 1.57 to 2.44 10*^{5} *m*^{3} *for 2001 lahar...*"

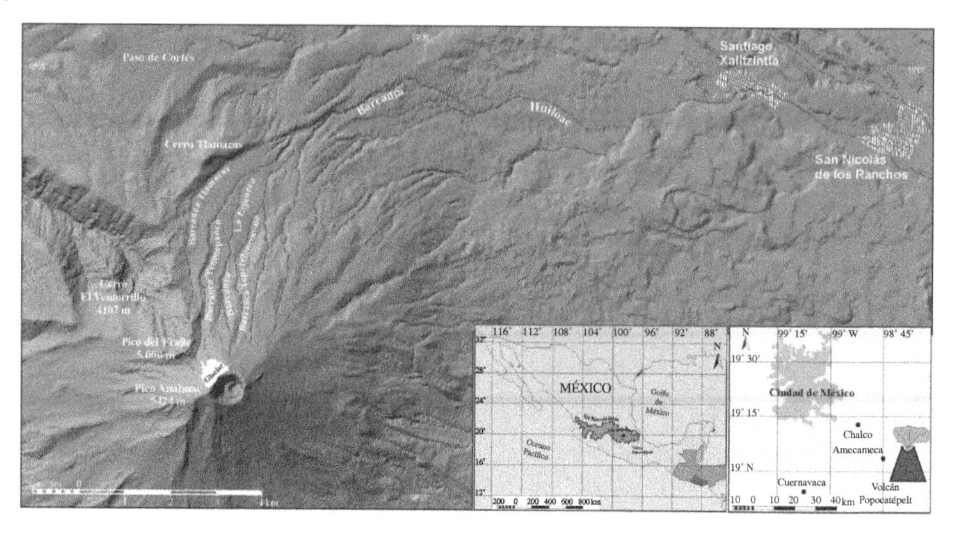

Figure 10.10 Shaded relief map of Popocatépetl volcano and surrounding areas with the main drainage systems *Source:* From Engineering Office, Civil Engineering and Development Department of the Government of the Hong Kong SAR

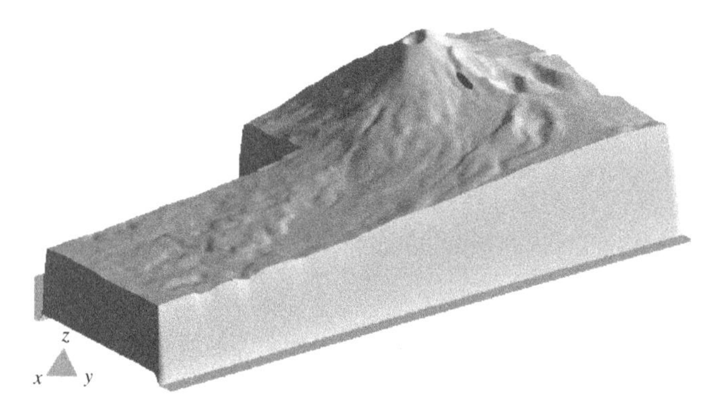

Figure 10.11 Initial conditions *Source:* From Geotechnical Engineering Office, Civil Engineering and Development Department of the Government of the Hong Kong SAR

The 2001 lahar has been simulated by (Haddad et al. 2010) using the SPH code described above. The initial mass considered is shown in Figure 10.11 and the observed propagation path is given in Figure 10.12.

The material has been assumed to be of Bingham type, its properties being selected as: $\tau_y = 60$ Pa, $\mu = 45$ Pa.s. The results are shown in Figure 10.13. Time of propagation from triggering to stop is close to 8000 s (2 h approximately), providing time to evacuate the small villages in the lahar path.

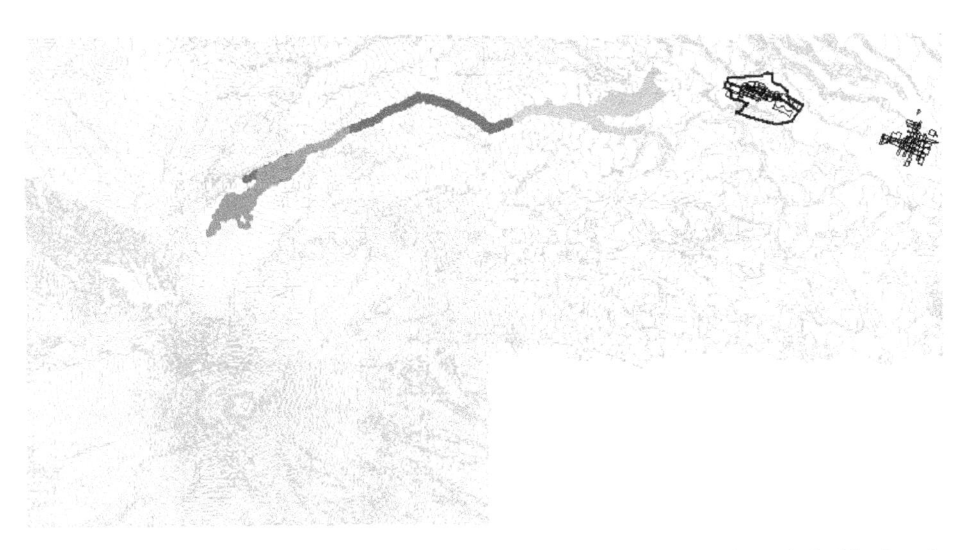

Figure 10.12 Propagation of the lahar along Huilouac gorge *Source:* From Geotechnical Engineering Office, Civil Engineering and Development Department of the Government of the Hong Kong SAR

10.5.3 Modeling of Yu Tung Road Debris Flow

This case was proposed for the benchmarking activity of the 2nd JTC1 Workshop on Triggering and Propagation of Rapid Flow-like Landslides which took place in Hong Kong in 2018. On 7 June 2008, an intense rainstorm triggered a debris flow on the natural hillside in Lantau Island, Hong Kong (Figure 6.9). Over $3000m^3$ of debris ran down the hillside and traveled $510m$ at an apparent travel angle of 17° to the Yu Tung Road which resulted in significant entrainment and deposition along the debris path and serious road blockage.

For the analysis, we have used the debris flow two-phase model including pore water pressure evolution. We provide in Figure 10.14 a view of the results of our simulations, depicting the landslide path and positions of the sliding mass produced by GeoFlow_SPH model from initial to final deposition. The rheological parameter of basal friction angle of 8° with a Voellmy coefficient equal to $500\frac{m}{s^2}$ has been considered.

Debris flow velocities were estimated at various locations along the flow path (see the dots in Figure 6.11) based on the super-elevation data and, the video footage which was captured by a member of the public.

The velocity distribution at different times is shown in Figure 10.16. The flow velocity was computed was computed to be about 12 m/s at $t = 4$s with a runout distance of 100 m. Then, the debris flow travelled at a higher speed with an average velocity of about 17 m/s reaching after 13s to the distance of 350 m. Finally, it slowed down to 11 m/s at $t = 23$s.

Field mapping revealed that significant entrainment of loose materials and erosion of the side slopes had occurred. It is estimated that the active volume increased to about 340 m^3. Therefore, in this case, bed entrainment was a key aspect in effecting the dynamics of the moving mass.

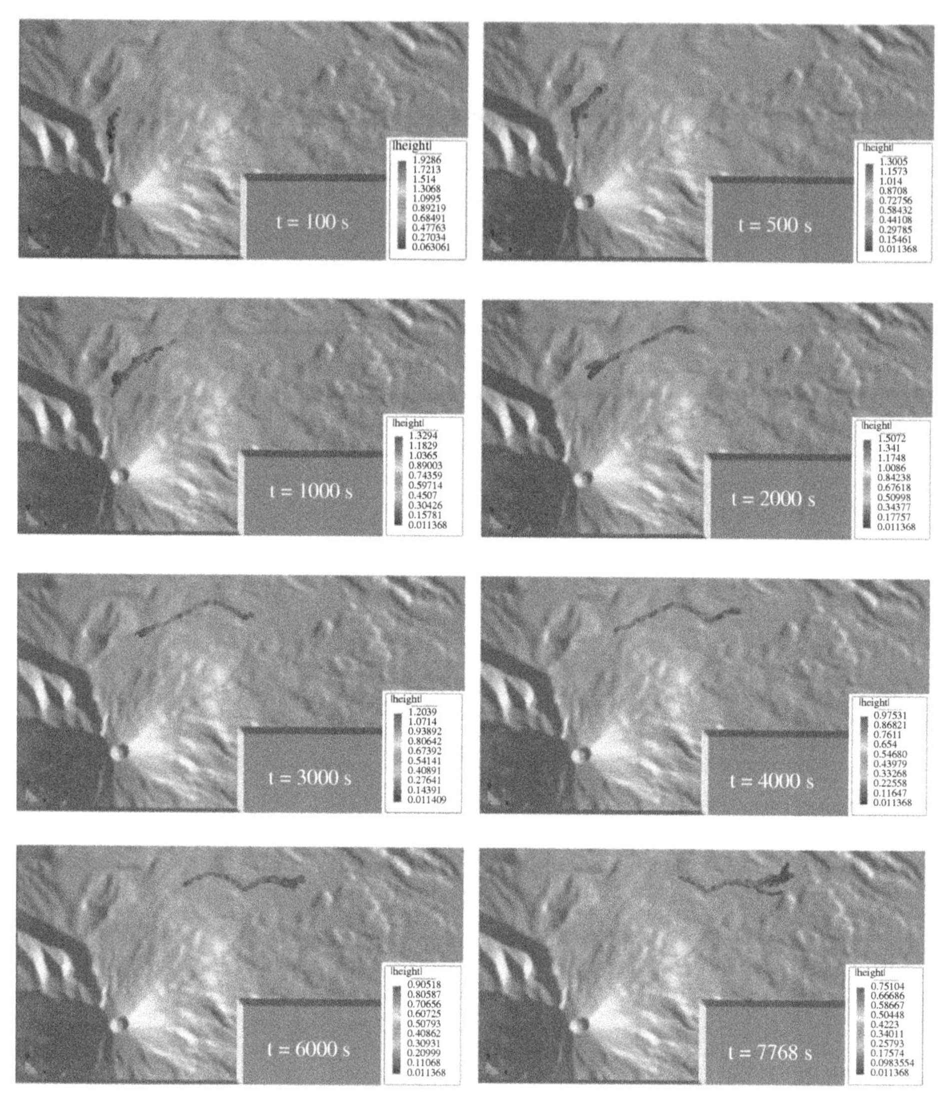

Figure 10.13 Propagation of the 2001 lahar *Source:* From Geotechnical Engineering Office, Civil Engineering and Development Department of the Government of the Hong Kong SAR

The model takes into account bed entrainment along the landslide path and decreasing the ground surface elevation consistently over time. Figure 10.16 shows the amount of erosion. We have used Hungr's erosion law (1995) with an entrainment coefficient of 0.0011 ms^{-1}.

Finally, Figure 10.18 compares the computed frontal velocity with these field data. Debris flow velocities were estimated at various locations along the flow path (see the dots in Figure 10.18) based on the super-elevation data and the video footage that a public member captured. The proposed model is capable of obtaining reasonable results and properly reproduce the propagation velocities of the debris flow at different time steps.

Figure 10.14 The aerial view of the debris flow event after the landslide incident (with permission of Geotechnical Engineering Office, Civil Engineering and Development Department of the Government of the Hong Kong SAR).

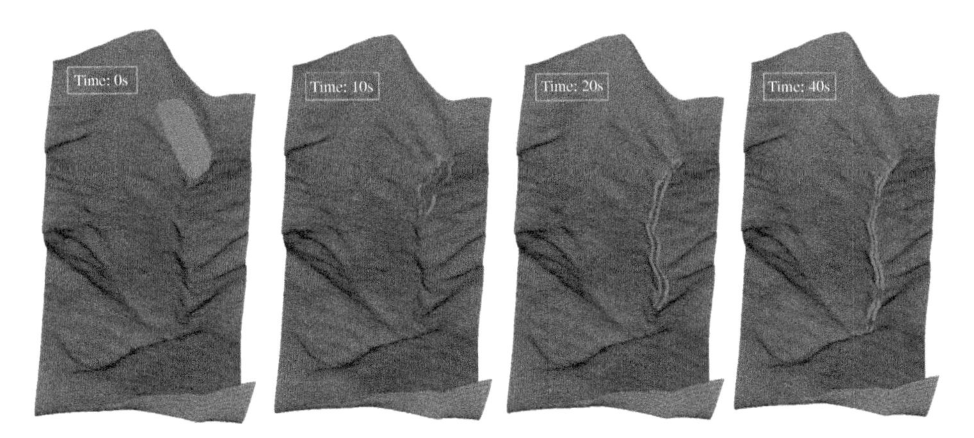

Figure 10.15 Results sequence of the debris flow simulation at different positions *Source:* From Geotechnical Engineering Office, Civil Engineering and Development Department of the Government of the Hong Kong SAR

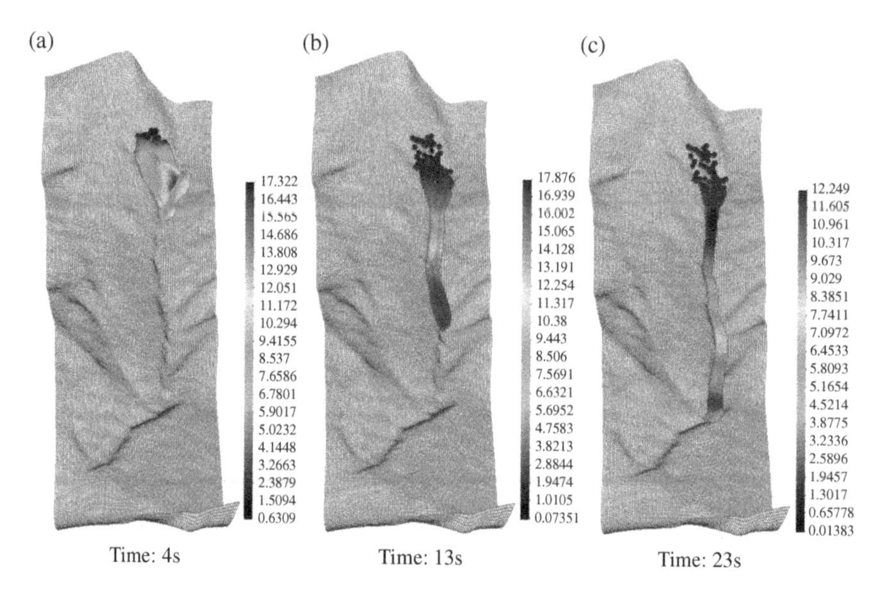

Figure 10.16 Computed velocities at times (a) 4s, (b) 13s and (c) 23s. (with permission of Geotechnical Engineering Office, Civil Engineering and Development Department of the Government of the Hong Kong SAR).

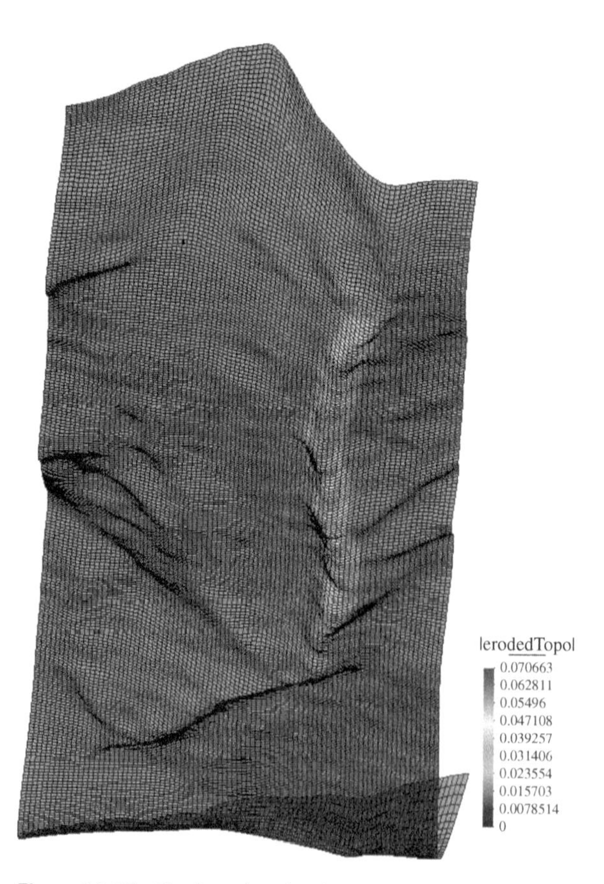

Figure 10.17 Final erosion depths at time 40second *Source:* From Geotechnical Engineering Office, Civil Engineering and Development Department of the Government of the Hong Kong SAR

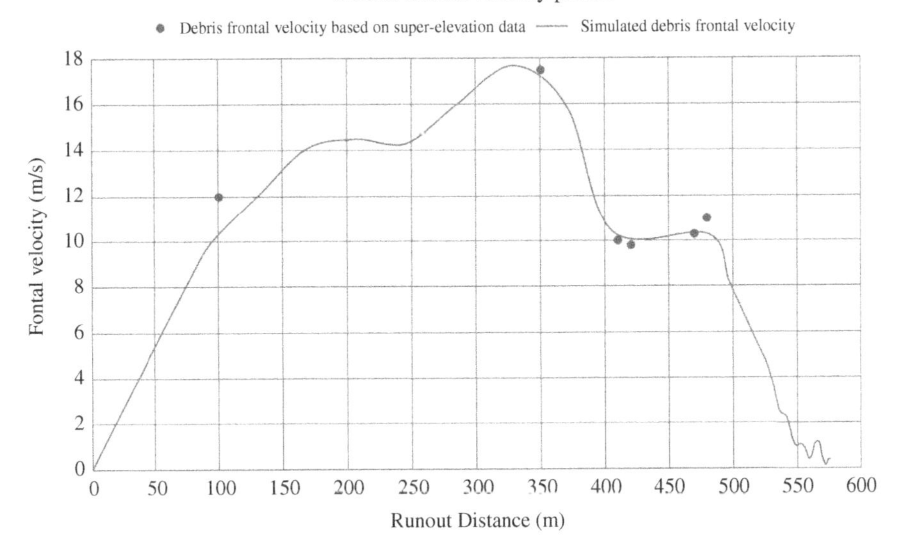

Figure 10.18 Comparison between observed and computed frontal velocities. *Source:* From Geotechnical Engineering Office, Civil Engineering and Development Department of the Government of the Hong Kong SAR

10.6 Conclusion

Rock avalanches, flow slides, debris flows, lahars, and other similar events are the most complex phenomena involving complex physical mechanisms such as segregation, comminution, basal erosion, coupling with pore water, the evolution of fluid properties, and thermal effects, just to mention some of them. Complete 3D models based on mixture theory and incorporating submodels for the above-mentioned phenomena are still very expensive from a computational point of view. Depth-integrated models provide a good combination of simplification and accuracy and can provide useful results for scientists and engineers. There are suitable discretization techniques for depth-integrated models, such as finite differences, finite elements, finite volumes, or the more recent meshless methods such as the SPH model used in this chapter. All of them provide accurate numerical approximations of the depth-integrated equations. In the author's experience, the SPH model allows to separate the computational mesh consisting of moving nodes or particles, from the topographical mesh which can have a structured nature, simplifying many computations. The computational time can be reduced up to 30 times as compared with unstructured finite element meshes.

Note

1 The authors gratefully acknowledge the support of the Geotechnical Engineering Office, Civil Engineering and Development Department of the Government of the Hong Kong SAR in the provision of the digital terrain models for the Hong Kong landslide cases.

References

Andersen, S. and Andersen, L. (2009). Modelling of landslides with the material-point method, *Comput. Geosci.*, **14**, 137–147.

Anderson, T. B. and Jackson, R. (1967). Fluid mechanical description of fluidized beds. Equations of motion, *Ind. Eng. Chem. Fundam.*, **6**, 527–539.

Andrés, N., Zamorano, J. J., Sanjosé, J. J., Atkinson, A. and Palacios, D. (2007). Glacier retreat during the recent eruptive period of Popocatépetl volcano (Mexico), *Ann. Glaciol.*, **45**, 73–82.

Belytschko, T., Lu, Y. Y. and Gu, L. (1994). Element-free Galerkin methods, *Int. J. Numer. Methods Eng.*, **37**, 229–256.

Biot, M. A. (1941). General theory of three-dimensional consolidation, *J. Appl. Phys.*, **12**, 155–164.

Biot, M. A. (1955). Theory of elasticity and consolidation for a porous anisotropic solid, *J. Appl. Phys.*, **26**, 182–185.

Blanc, T. (2008). Numerical simulation of debris flows with the 2D SPH depth-integrated model. Master's thesis. Institute for Mountain Risk Engineering, University of Natural Resources and Ap-plied Life Sciences, Vienna, Austria.

de Boer, R. (2000). *Theory of Porous Media*, Springer-Verlag, Berlin.

Bonet, J. and Kulasegaram, S. (2000). Correction and stabilization of smooth particle hydrodynamics methods with applications in metal forming simulations, *Int. J. Numer. Meth. Engng.*, **47**, 1189–1214.

Bonet, J. and Rodíguez Paz, M. X. (2005). A corrected smooth particle hydrodynamics formulation of the shallow-water equations, *Comput. Struct.*, **83**, 1396–1410.

Bui, H. H. and Nguyen, G. D. 2017. A coupled fluid-solid SPH approach to modelling flow through deformable porous media, *Int. J. Solids Struct.*, **125**, 244–264. https://doi.org/10.1016/j.ijsolstr.2017.06.022

Bui, H. H., Sako, K. and Fukagawa, R. (2007). Numerical simulation of soil–water interaction using smoothed particle hydrodynamics (SPH) method, *J. Terramechanics.*, **44**, 339–346.

Capra, L., Poblete, M. A. and Alvarado, R. 2004. The 1997 and 2001 lahars of Popocatépetl volcano (Central Mexico): textural and sedimentological constraints on their origin and hazards. *J. Volcanol. Geotherm. Res.*, **131**, 351–369.

Chen, C. H., Lin, Y. T., Chung, H. R., Hsieh, T. Y., Yang, J. C. and Lu, J. Y. (2017). Modelling of hyperconcentrated flow in steep-sloped channels, *J.Hydraul. Res.*, **56**, 1–19.

Coetzee, C. J., Basson, A. H. and Vermeer, P.A. (2007). Discrete and continuum modelling of excavator bucket filling *J. Terramechanics.*, **44**, 177–186.

Córdoba, G. Sheridan, M. F. and Pitman, E. B. (2015). TITAN2F: a pseudo-3-D model of 2-phase debris flows, *Nat. Hazards Earth Syst. Sci. Discuss.*, **3**, 3789–3822, www.nat-hazards-earth-syst-sci-discuss.net/3/3789/2015/ doi: 10.5194/nhessd-3-3789-2015.

Coulomb, C. A. (1773). Essai sur une application des régles de maximis et minimis à quelques problémes de statique relatifs a l'architecture". Mémoires de Mathématique et de Physique, présentés à l'Académie Royale des Sciences par Divers Savants et lus dans ses Assemblées, 7, Paris, France, 343–382.

Coussy, O. (1995). *Mechanics of Porous Media*, John Wiley and Sons, Chichester.

Darve, F. (1995). Liquefaction phenomenon of granular materials and constitutive instability, *Int. J. Eng. Comp.*, **7**, 5–28.

Darve, F. and Laouafa, F. (2000). Instabilities in granular materials and application to landslides, *Mech. Coh.-Frict. Mater.*, **5**, (8), 627–652

Darve, F. and Laouafa, F. (2001). Modelling of slope failure by a material instability mechanism, *Comput. Geotech.*, **29**, 301–325.

Di Prisco, C., Matiotti, R. and Nova, R. 1995. Theoretical investigation of undrained stability of shallow submerged slopes, *Geotechnique*, **45**, 479–496.

Dikau, R., Brundsen, D., Schrott, L. and Ibsen, M. L. (1996). *Landslide Recognition*, John Wiley and Sons.

Donea, J. (1984). A Taylor-Galerkin method for convective transport problems, *Int. J. Numer. Methods Eng.*, **20**,1001–1119.

Duarte, C. A. and Oden, J. T. (1996). An hp adaptive method using clouds, *Comput. Methods Appl. Mech. Eng.*, **139**, 237–262.

Fern, J., Rohe, A., Soga, K. and Alonso, E. (2019). (Eds) *The Material Point Method for Geotechnical Engineering, A Practical Guide*, CRC Press.

Fernández Merodo, J. A., Pastor, M., Mira, P., Tonni, L., Herreros, I., González, E. and Tamagnini, R. (2004). Modelling of diffuse failure mechanisms of catastrophic landslides, *Comp.Methods Appl. Mech.Engrg.*, **193**, 2911–2939

Gingold, R. A. and Monaghan, J. J. (1977). Smoothed particle hydrodynamics – theory and application to non-spherical stars, *MNRAS*, **181**, 375–389.

Gingold, R. A. and Monaghan, J. J (1982). Kernel estimates as a basis for general particle methods in hydrodynamics, *J. Comput. Phys.*, **46**, (1982), 429.

Gray, J. M. N. T., Wieland, M. and Hutter, K. (1999). Gravity-driven free surface flow of granular avalanches over complex basal topography, *Proc. R. Soc. A Math. Phys. Eng. Sci.*, **455**, 1841–1874.

Guinot, V. (2003). *Goudunov-Type Schemes: An Introduction for Engineers*, Elsevier, Amsterdam.

Haddad, B., Pastor, M., Palacios, D. and Muñoz-Salinas, E. (2010). A SPH depth integrated model for Popocatépetl 2001 lahar (Mexico): sensitivity analysis and runout simulation, *Eng. Geol.*, **114**, 312–329.

Ho, K., Leung, A., Kwan, J., Koo, R. and Law, R. (2018). (Eds) Second JTC1 Workshop Triggering and Propagation of Rapid Flow-like Landslides Proceedings of the Second JTC1 Workshop Triggering and Propagation of Rapid Flow-like Landslides, Hong Kong (3–5 December 2018).

Hungr, O. (1995). A model for the runout analysis of rapid flow slides, debris flows, and avalanches, *Can. Geotech. J.*, **32**, 610–623.

Hutchinson, J. N. (1986). A sliding-consolidation model for flow slides. *Can. Geotech. J.*, **23**, 115–126.

Hutter, K., Siegel, M., Savage, S. B. and Nohguchi, Y. (1993). Two-dimensional spreading of a granular avalanche down an inclined plane Part I. Theory, *Acta Mech.*, **100**, 37–68.

Issler, D. and Jóhannesson, T. (2010). Dynamically consistent entrainment and deposition rates in depth-averaged gravity mass flow models. (submitted for publication)

Issler, D. and Pastor, M. (2011). Interplay of entrainment and rheology in snow avalanches – a numerical study, *Ann. Glaciol.*, **52**, (58), 143–147.

Iverson, R. M. (1993). Differential equations governing slip-induced pore pressure fluctuations in a water-saturated granular medium, *Math. Geol.*, **25**, (8), 1027–1048.

Iverson, R. I. and Denlinger, R. P. (2001). Flow of variably fluidized granular masses across three dimensional terrain. 1. Coulomb mixture theory, *J. Geophys. Res.*, **106**, B1, 537–552.

Iverson, R. M. and LaHusen, R. G. (1989). Dynamic pore-pressure fluctuations in rapidly shearing granular materials, *Science*, **246**, 796–799.

Jassim, I., Stolle, D. and Vermeer, P. (2013). Two-phase dynamic analysis by material point method, *Int. J. Numer. Anal. Methods Geomech.*, **37**, 2502–2522.

Kamphuis, J. W. and Bowering, R. J. (1970). Impulse waves generated by landslides, *Coast Eng.*, **1970**, 575–588.

King, J. P. (2001). The 2000 Tsing Shan Debris Flow. Report LSR 3/2001. Planning Division, Geotechnical Engineering Office, Civil Engineering and Development Department, The Government of the Hong Kong Special Administrative

Laigle, D. and Coussot, P. (1997). Numerical modelling of mudflows, *J. Hydraul. Eng., ASCE*, **123**, (7), 617–623.

Lastiwka, M., Quinlan, N. and Basa, M. (2005). Adaptive particle distribution for smoothed particle hydrodynamics, *Int. J. Numer. Methods Fluids*, **47**, (10–11), 1403–1409.

Lewis, R. L. and Schrefler, B. A. (1998). *The Finite Element Method in the Static and Dynamic Deformation and Consolidation of Porous Media*, John Wiley and Sons.

Li, W., Su, Z., van Maren, D. S., Wang, Z. and de Vriend, H. J. (2017). Mechanisms of hyperconcentrated flood propagation in a dynamic channel-floodplain system, *Adv. Water Resour.*, **107**

Libersky, L. D., Petschek, A. G., Carney, T. C., Hipp, J. R. and Allahdadi, F. A. (1993). High strain lagrangian hydrodynamics: a three-dimensional SPH code for dynamic material response, *J. Comput. Phys.*, **109**, (1), 67–75.

Lin, C., Pastor, M., Yagüe, A., Moussavi, S., Martín Stickle, M., Manzanal, D., Li, T. and Liu, X. D. (2019). A depth-integrated SPH model for debris floods: application to Lo Wai (Hong Kong) debris flood of August 2005, *Géotechnique*, **69**, 12, 1035–1055, https://doi.org/10.1680/jgeot.17.P.267

Löhner, R., Morgan, K. and Zienkiewicz, O. C. (1984). The solution of non-linear hyperbolic equation systems by the finite element method, *Int. J. Numer. Methods Fluids*, **4**, 1043–1063.

Lucy, L. B. (1977). A numerical approach to the testing of fusion process, *Astronomical J.*, **82**, 1013–1024.

McDougall, S. (2006). A new continuum dynamic model for the analysis of extremely rapid landslide motion across complex 3D terrain. Ph.D Thesis. University of British Columbia.

McDougall, S. and Hungr, O. (2004). A model for the analysis of rapid landslide motion across three-dimensional terrain, *Can. Geotech. J.*, **41**, 12, 1084–1097.

Melenk, J. M. and Babuška, I. (1996). The partition of unity finite element method: Basic theory and applications, *Comput. Methods Appl. Mech. Eng.*, **139**, 289–314.

Monaghan, J. J. (1982). Why particle methods work, *SIAM J. Sci. Statistical Comput.*, **3**, (4), 422–433.

Monaghan J. J. and Lattanzio J. C. (1985). A refined particle method for astrophysical problems, *Astron. Astrophys.*, **149**, 135–143.

Monaghan, J. J, Cas, R. F., Kos, A. and Hallworth, M. (1999). Gravity currents descending a ramp in a stratified tank, *J. Fluid Mech.*, **379**, 36–39.

Monaghan, J. J.; Kos, A. and Issa, N. (2003). Fluid motion generated by impact, *J. Waterw. Port, Coast. Ocean Eng. ASCE*, **129**, 250–259.

Muñoz-Salinas, E.,Manea, V. C., Palacios, D. and Castillo-Rodriguez, M. 2007. Estimation of lahar flow velocity on popocatepetl volcano (Mexico), *Geomorphology*, **92**, (1–2), 91–99.

Muñoz-Salinas, E., Renchler, C. and Palacios, D. 2009. A GIS-based method to determine the volume of lahars: Popocatépetl volcano, Mexico, *Geomorphology*, **111**, 61–69.

Nayroles, B., Touzot, G. and Villon, P. (1992). Generalizing the finite element method: diffuse approximation and diffuse elements, *Comput. Mech.*, **10**, 307–318.

Nicot, F. and Wan, R. (2008). (Eds) *Micromécanique De La Rupture Dans Les Milieux Granulaires*, Hermes Science Lavoisier.

Nova, R. (1994). Controllability of the incremental response of soil specimens subjected to arbitrary loading programmes, *J. Mech. Behav. Mater.*, **5**, 193–201.

Oñate, E. and Idelsohn, S. (1998). A mesh-free finite point method for advective-diffusive transport and fluid flow problems, *Comput. Mech.*, **21**, 283–292.

Pastor, M., Quecedo, M., Fernández Merodo, J. A., Herreros, M. I., González, E. and Mira, P. (2002). Modelling tailing dams and mine waste dumps failures, *Geotechnique*, **LII**, 8, 579–592.

Pastor, M., Fernández Merodo, J. A., González, E., Mira, P., Li, T. and Liu, X. (2004). Modelling of landslides I. Failure mechanisms, in *Degradation and Instabilities in Geomaterials*, CISM Courses and Lectures n. 461, F. Darve and I. Vardoulakis (Eds), pp. 287–318, Springer Verlag, Wien New York.

Pastor, M., Quecedo, M., González, E., Herreros, I., Fernández Merodo, J.A. and Mira, P. (2004). A simple approximation to bottom friction for bingham fluid depth integrated models, *J. Hydraul. Eng. ASCE*, **130**, 2, 149–155.

Pastor, M., Haddad, B., Sorbino, G. and Cuomo, S. (2009a). A depth integrated coupled SPH model for flow-like landslides and related phenomena, *Int. J. Num. Anal. Meth. Geomech.*, **33**, 143–172.

Pastor, M., Blanc, T. and Pastor, M. J (2009b). A depth-integrated viscoplastic model for dilatant saturated cohesive-frictional fluidized mixtures: application to fast catastrophic landslides, *J. Non-Newtonian Fluid Mech.*, **158**, (2009), 142–153.

Pastor, M., Chan, A. H. C., Mira, P., Manzanal D., Fernández Merodo, J. A. and Blanc, T. (2011). Computational geomechanics: the heritage of Olek Zienkiewicz, *Int. J. Numer. Methods Eng.*, **87**, 1–5, 457–489. doi: 10.1002/nme.3192

Pastor, M., Martin Stickle, M., Dutto, P., Mira, P., Fernández Merodo, J. A., Blanc, T., Sancho, S. and Benítez, A. S. 2013. A viscoplastic approach to the behaviour of fluidized geomaterials with application to fast landslides, *Contin. Mech. Thermodyn.*, **27**, 21 47.

Pastor, M., Blanc, T., Haddad, B., Petrone, S., Sanchez Morles, M., Drempetic, V., Issler, D., Crosta, G. B., Cascini, L., Sorbino, G. and Cuomo, S. (2014). Application of a SPH depth-integrated model to landslide run-out analysis, *Landslides*, October 2014, **11**, 793–812 doi: 10.1007/s10346-014-0484-y

Pastor, M., Blanc, T., Haddad, B., Drempetic, V., Morles, M. S., Dutto, P., Stickle, M. M., Mira, P. and Merodo, J. A. F. (2015a). Depth averaged models for fast landslide propagation: mathematical, rheological and numerical aspects. *Arch. Comput. Methods Eng.*, **22**, 1, 67–104. doi: 10.1007/s11831-014-9110-3

Pastor, M., Blanc, T., Haddad, B., Drempetic, V., Morles, M. S., Dutto, P. and Merodo, J. F. (2015b). Depth averaged models for fast landslide propagation: mathematical, rheological and numerical aspects, *Arch. Comput. Methods Eng.*, **22**, (1), 67–104.

Pastor, M., Yague, A., Martin Stickle, M., Manzanal, D. and Mira, P. (2018). A two-phase SPH model for debris flow propagation, *Int. J. Numer. Anal. Methods Geomech.*, doi: 10.1002/nag.2748.

Peraire, J., Zienkiewicz, O. C. and Morgan, K. (1986). Shallow water problems: a general explicit formulation, *Int. J. Numer. Methods Eng.*, **22**, (3), 547–574.

Pitman, E. B. and Le, L. (2005). A two-fluid model for avalanche and debris flows, *Philos. Trans. A. Math. Phys. Eng. Sci.*, **363**, 1573–601.

Pudasaini, S. P. (2012). A general two-phase debris flow model, *J. Geophys. Res. Earth Surface (2003–2012)*, **117**, F3, F03010.

Pudasaini, S. P. and Hutter, K. (2007). *Avalanche Dynamics: Dynamics of Rapid Flows of Dense Granular Avalanches*, Philosophical Transactions of the Royal Society A.

Quecedo, M. and Pastor, M. (2003). Finite element modelling of free surface flows on inclined and curved beds, *J. Comput. Phys.*, **189**, 45–62.

Quecedo, M., Pastor, M., Herreros, M. I. and Fernández. Merodo, J. A. (2004). Numerical modelling of the propagation of fast landslides using the finite element method, *Int. J. Numer Methods Eng.*, **59**, 6, 755–794.

Randles, P. W. and Libersky, L. D. (1996). Smoothed particle hydrodynamics: some recent improvements and applications, *Comput. Methods Appl. Mech. Eng.*, **139**, (1–4), 375–408.

Rodriguez-Paz, M. and Bonet, J. (2005). A corrected smooth particle hydrodynamics formulation of the shallow-water equations, *Comput. Struct.*, **83**, 1396–1410.

Saint-Venant, A. D. (1871). Theorie du mouvement non permanent des eaux, avec application aux crues des rivieres et a l'introduction de marees dans leurs lits. *C. R. Seances Acad. Sci.*, **36**, 174–154.

Savage, S. B. and Hutter, K. (1989). The motion of a finite mass of granular material down a rough incline, *J. Fluid Mech.*, **199**, 177.

Savage, S. B. and Hutter, K. (1991). The dynamics of avalanches of granular materials from initiation to runout. Part I: analysis, *Acta Mech.*, **86**, 201–223.

Sheridan, M. F., Stinton, A. J., Patra, A. K., Bauer, A. C., Nichita, C. C. and Pitman, E. B. 2005. Evaluating TITAN2D mass-flow model using the 1963 little tahoma peak avalanches, Mount Rainier, Washington, *J. Volcanol. Geotherm. Res.*, **139**, 89–102.

Sosio, R., Crosta, G. B. and Hungr, O. (2008). Complete dynamic modeling calibration for the Thurwieser rock avalanche (Italian Central Alps), *Eng. Geol.*, **100**, (1–2), 11–26, doi: 10.1016/j.enggeo.2008.02.012.

Sulsky, D., Chen, Z. and Schreyer, H. (1994). A particle method for history-dependent materials, *Comput. Methods Appl. Mech. Eng.*, **118**, 179–196.

Vacondio, R., Rogers, B. D., Stansby, P. K. and Mignosa, P. (2011). SPH modeling of shallow flow with open boundaries for practical flood simulation, *J. Hydraul. Eng.*, **138**, (6), 530–541.

Voellmy, A. (1955). Über die Zerstörungskraft von Lawinen, *Schweizerische Bauzeitung*, **73**, 212–285.

Wieckowski, Z., Youn, S.-K. and Yeon, J.-H. (1999). A particle-in-cell solution to the silo discharging problem, *Int. J. Numer. Methods Eng.*, **45**, 1203–1225.

Yerro, A., Alonso, E. E., Pinyol, N. M. (2015). The material point method for unsaturated soils, *Géotechnique*, **65**, 3, March 2015 201–217.

Yumuang, S. (2006). 2001 debris flow and debris flood in Nam Ko area, Phetchabun province, central Thailand, *Environ. Geol.*, **51**, (4), 545–564.

Zabala, F. and Alonso, E. E. (2011). Progressive failure of Aznalcóllar dam using the material point method, *Géotechnique*, **61**, 9, 795–808.

Zhang, X., Chen, Z. and Liu, Y. (2016). *The Material Point Method. A Continuum-Based Particle Method for Extreme Loading Cases*, Academic Press.

Zienkiewicz, O. C. and Shiomi, T. (1984). Dynamic behaviour of saturated porous media: the generalised Biot formulation and its numerical solution, *Int. J. Numer. Anal. Methods Geomech.*, **8**, 71–96.

Zienkiewicz, O. C., Chang, C. T. and Bettess, P. (1980). Drained, undrained, consolidating dynamic behaviour assumptions in soils, *Geotechnique*, **30**, 385–395.

Zienkiewicz, O. C., Chan, A. H. C., Pastor, M., Paul, D. K and Shiomi, T. (1990a). Static and dynamic behaviour of soils: a rational approach to quantitative solutions. I. Fully saturated problems, *Proc. R. Soc. Lond. A Math. Phys. Sci.*, **429**, 285–309.

Zienkiewicz, O. C, Xie, Y. M., Schrefler, B. A., Ledesma, A. and Bicanic, N. (1990b). Static and dynamic behaviour of soils: a rational approach to quantitative solutions. II. Semi-saturated problems, *Proc. R. Soc. Lond. A Math. Phys. Sci.*, **429**, 311–321.

Zienkiewicz, O. C., Chan, A. H. C., Pastor, M., Shrefler, B. A. and Shiomi, T. (1999). *Computational Geomechanics*, John Wiley and Sons.

Index

Computational Geomechanics: Theory and Applications, Second Edition. Andrew H. C. Chan, Manuel Pastor, Bernhard A. Schrefler, Tadahiko Shiomi and O. C. Zienkiewicz.

e

t